Semiconductor Memories

Semiconductor Memories
Technology, Testing, and Reliability

Ashok K. Sharma

IEEE Solid-State Circuits Council, *Sponsor*

The Institute of Electrical and Electronics Engineers, Inc., New York

A JOHN WILEY & SONS, INC., PUBLICATION

Published by John Wiley & Sons, Inc., Hoboken, New Jersey.

Library of Congress Cataloging-in-Publication Data

Sharma, Ashok K.
    Semiconductor memories : technology, testing, and reliability /
Ashok K. Sharma.
        p.   cm.
    Includes index.
    ISBN 0-7803-1000-4
    1. Semiconductor storage devices. I. Title
TK7895.M4S49 1996
621.39′732—dc20   96-6824
                                                    CIP

Contents

Preface xi

Chapter 1   Introduction 1

Chapter 2   Random Access Memory Technologies 10
    2.1   Introduction 10
    2.2   Static Random Access Memories (SRAMs) 12
        2.2.1   SRAM (NMOS and CMOS) Cell Structures 12
        2.2.2   MOS SRAM Architectures 14
        2.2.3   MOS SRAM Cell and Peripheral Circuit Operation 15
        2.2.4   Bipolar SRAM Technologies 17
            2.2.4.1   Direct-Coupled Transistor Logic (DCTL) Technology 18
            2.2.4.2   Emitter-Coupled Logic (ECL) Technology 19
            2.2.4.3   BiCMOS Technology 20
        2.2.5   Silicon-on-Insulator (SOI) Technology 24
        2.2.6   Advanced SRAM Architectures and Technologies 28
            2.2.6.1   1–4 Mb SRAM Designs 28
            2.2.6.2   16–64 Mb SRAM Development 32
            2.2.6.3   Gallium Arsenide (GaAs) SRAMs 34
        2.2.7   Application-Specific SRAMs 35
            2.2.7.1   Serially Accessed Memory (Line Buffers) 35
            2.2.7.2   Dual-Port RAMs 36
            2.2.7.3   Nonvolatile SRAMs 38
            2.2.7.4   Content-Addressable Memories (CAMs) 38
    2.3   Dynamic Random Access Memories (DRAMs) 40
        2.3.1   DRAM Technology Development 40
        2.3.2   CMOS DRAMs 45
            2.3.2.1   1 Mb DRAM (Example) 47
        2.3.3   DRAM Cell Theory and Advanced Cell Structures 50
            2.3.3.1   Trench Capacitor Cells 52
            2.3.3.2   Stacked Capacitor Cells (STC) 55
        2.3.4   BiCMOS DRAMs 58
        2.3.5   Soft-Error Failures in DRAMs 60
        2.3.6   Advanced DRAM Designs and Architectures 62

		2.3.6.1	A 16 Mb DRAM (Example)	63
		2.3.6.2	ULSI DRAM Developments	64
	2.3.7	Application-Specific DRAMs		69
		2.3.7.1	Pseudostatic DRAMs (PSRAMs)	69
		2.3.7.2	Silicon File	69
		2.3.7.3	Video DRAMs (VRAMs)	70
		2.3.7.4	High-Speed DRAMs	71
		2.3.7.5	Application-Specific RAM Glossary and Summary of Important Characteristics	75

Chapter 3   Nonvolatile Memories   81

3.1	Introduction		81
3.2	Masked Read-Only Memories (ROMs)		83
	3.2.1	Technology Development and Cell Programming	83
	3.2.2	ROM Cell Structures	85
	3.2.3	High-Density (Multimegabit) ROMs	87
3.3	Programmable Read-Only Memories (PROMs)		87
	3.3.1	Bipolar PROMs	87
	3.3.2	CMOS PROMs	91
3.4	Erasable (UV)-Programmable Read-Only Memories (EPROMs)		93
	3.4.1	Floating-Gate EPROM Cell	93
	3.4.2	EPROM Technology Developments	96
		3.4.2.1 1 Mb EPROM (Example)	97
	3.4.3	Advanced EPROM Architectures	98
	3.4.4	One-Time Programmable (OTP) EPROMs	103
3.5	Electrically Erasable PROMs (EEPROMs)		104
	3.5.1	EEPROM Technologies	105
		3.5.1.1 Metal–Nitride–Oxide Silicon (MNOS) Memories	105
		3.5.1.2 Silicon–Oxide Nitride–Oxide Semiconductor (SONOS) Memories	109
		3.5.1.3 Floating-Gate Tunneling Oxide (FLOTOX) Technology	110
		3.5.1.4 Textured-Polysilicon Technology	115
	3.5.2	EEPROM Architectures	116
	3.5.3	Nonvolatile SRAM (or Shadow RAM)	120
3.6	Flash Memories (EPROMs or EEPROMs)		122
	3.6.1	Flash Memory Cells and Technology Developments	123
	3.6.2	Advanced Flash Memory Architectures	128

Chapter 4   Memory Fault Modeling and Testing   140

4.1	Introduction		140
4.2	RAM Fault Modeling		142
	4.2.1	Stuck-At Fault Model	142
	4.2.2	Bridging Faults	145
	4.2.3	Coupling Faults	147
	4.2.4	Pattern-Sensitive Faults	151
	4.2.5	Miscellaneous Faults	155
	4.2.6	GaAs SRAM Fault Modeling and Testing	155
	4.2.7	Embedded DRAM Fault Modeling and Testing	156
4.3	RAM Electrical Testing		158
	4.3.1	DC and AC Parametric Testing	158
	4.3.2	Functional Testing and some Commonly Used Algorithms	158
	4.3.3	Functional Test Pattern Selection	174
4.4	RAM Pseudorandom Testing		176

4.5   Megabit DRAM Testing 178
4.6   Nonvolatile Memory Modeling and Testing 180
    4.6.1   DC Electrical Measurements 181
    4.6.2   AC (Dynamic) and Functional Measurements 182
        4.6.2.1   256K UVEPROM 182
        4.6.2.2   64K EEPROM 183
4.7   IDDQ Fault Modeling and Testing 185
4.8   Application Specific Memory Testing 189
    4.8.1   General Testing Requirements 189
    4.8.2   Double-Buffered Memory (DBM) Testing 191

Chapter 5   **Memory Design for Testability and Fault Tolerance** **195**
5.1   General Design for Testability Techniques 195
    5.1.1   Ad Hoc Design Techniques 196
        5.1.1.1   Logic Partitioning.. 196
        5.1.1.2   Input/Output Test Points 196
    5.1.2   Structured Design Techniques 197
        5.1.2.1   Level-Sensitive Scan Design................................... 197
        5.1.2.2   Scan Path.. 199
        5.1.2.3   Scan/Set Logic.. 199
        5.1.2.4   Random Access Scan... 200
        5.1.2.5   Boundary Scan Testing.. 202
5.2   RAM Built-In Self-Test (BIST) 203
    5.2.1   BIST Using Algorithmic Test Sequence 205
    5.2.2   BIST Using 13N March Algorithm 207
    5.2.3   BIST for Pattern-Sensitive Faults 209
    5.2.4   BIST Using Built-In Logic Block Observation (BILBO) 210
5.3   Embedded Memory DFT and BIST Techniques 211
5.4   Advanced BIST and Built-In Self-Repair Architectures 216
    5.4.1   Multibit and Line Mode Tests 216
    5.4.2   Column Address-Maskable Parallel Test (CMT) Architecture . . 219
    5.4.3   BIST Scheme Using Microprogram ROM 220
    5.4.4   BIST and Built-In Self-Repair (BISR) Techniques 222
5.5   DFT and BIST for ROMs 228
5.6   Memory Error-Detection and Correction Techniques 230
5.7   Memory Fault-Tolerance Designs 241

Chapter 6   **Semiconductor Memory Reliability** **249**
6.1   General Reliability Issues 249
    6.1.1   Semiconductor Bulk Failures 252
    6.1.2   Dielectric-Related Failures 252
    6.1.3   Semiconductor–Dielectric Interface Failures 253
    6.1.4   Conductor and Metallization Failures 255
    6.1.5   Metallization Corrosion-Related Failures 256
    6.1.6   Assembly- and Packaging-Related Failures 257
6.2   RAM Failure Modes and Mechanisms 258
    6.2.1   RAM Gate Oxide Reliability 258
    6.2.2   RAM Hot-Carrier Degradation 260
    6.2.3   DRAM Capacitor Reliability 262
        6.2.3.1   Trench Capacitors 262
        6.2.3.2   Stacked Capacitors 264
    6.2.4   DRAM Soft-Error Failures 264

       6.2.5   DRAM Data-Retention Properties 267
6.3   Nonvolatile Memory Reliability 268
       6.3.1   Programmable Read-Only Memory (PROM) Fusible Links ... 268
       6.3.2   EPROM Data Retention and Charge Loss 270
       6.3.3   Electrically Erasable Programmable
                Read-Only Memories (EEPROMs) 275
       6.3.4   Flash Memories 280
       6.3.5   Ferroelectric Memories 283
6.4   Reliability Modeling and Failure Rate Prediction 287
       6.4.1   Reliability Definitions and Statistical Distributions 287
             6.4.1.1   Binomial Distribution.................................. 289
             6.4.1.2   Poisson Distribution.................................. 289
             6.4.1.3   Normal (or Gaussian) Distribution...................... 289
             6.4.1.4   Exponential Distribution............................... 291
             6.4.1.5   Gamma Distribution.................................. 291
             6.4.1.6   Weibull Distribution.................................. 291
             6.4.1.7   Lognormal Distribution 292
       6.4.2   Reliability Modeling and Failure Rate Prediction 292
6.5   Design for Reliability 296
6.6   Reliability Test Structures 300
6.7   Reliability Screening and Qualification 304
       6.7.1   Reliability Testing 304
       6.7.2   Screening, Qualification, and Quality Conformance Inspections
             (QCI) 310

Chapter 7   **Semiconductor Memory Radiation Effects 320**
7.1   Introduction 320
7.2   Radiation Effects 322
       7.2.1   Space Radiation Environments 322
       7.2.2   Total Dose Effects 325
       7.2.3   Single-Event Phenomenon (SEP) 330
             7.2.3.1   DRAM and SRAM Upsets............................. 333
             7.2.3.2   SEU Modeling and Error Rate Prediction.............. 335
             7.2.3.3   SEP In-Orbit Flight Data............................. 337
       7.2.4   Nonvolatile Memory Radiation Characteristics 344
7.3   Radiation-Hardening Techniques 347
       7.3.1   Radiation-Hardening Process Issues 347
             7.3.1.1   Substrate Effects.................................... 347
             7.3.1.2   Gate Oxide (Dielectric) Effects........................ 347
             7.3.1.3   Gate Electrode Effects................................ 348
             7.3.1.4   Postgate-Electrode Deposition Processing 349
             7.3.1.5   Field Oxide Hardening................................ 349
             7.3.1.6   Bulk CMOS Latchup Considerations.................... 350
             7.3.1.7   CMOS SOS/SOI Processes........................... 351
             7.3.1.8   Bipolar Process Radiation Characteristics 352
       7.3.2   Radiation-Hardening Design Issues 352
             7.3.2.1   Total Dose Radiation Hardness........................ 353
             7.3.2.2   Single-Event Upset (SEU) Hardening.................. 358
       7.3.3   Radiation-Hardened Memory Characteristics (Example) 363
7.4   Radiation Hardness Assurance and Testing 367
       7.4.1   Radiation Hardness Assurance 367
       7.4.2   Radiation Testing 369
             7.4.2.1   Total Dose Testing 370
             7.4.2.2   Single-Event Phenomenon (SEP) Testing.............. 372

                7.4.2.3   Dose Rate Transient Effects 376

                7.4.2.4   Neutron Irradiation... 377

        7.4.3   Radiation Dosimetry 377

        7.4.4   Wafer Level Radiation Testing and Test Structures 378

                7.4.4.1   Wafer Level Radiation Testing 378

                7.4.4.2   Radiation Test Structures 379

Chapter 8   **Advanced Memory Technologies 387**

    8.1   Introduction 387

    8.2   Ferroelectric Random Access Memories (FRAMs) 389

        8.2.1   Basic Theory 389

        8.2.2   FRAM Cell and Memory Operation 390

        8.2.3   FRAM Technology Developments 393

        8.2.4   FRAM Reliability Issues 393

        8.2.5   FRAM Radiation Effects 395

        8.2.6   FRAMs Versus EEPROMs 397

    8.3   Gallium Arsenide (GaAs) FRAMs 397

    8.4   Analog Memories 398

    8.5   Magnetoresistive Random Access Memories (MRAMs) 401

    8.6   Experimental Memory Devices 407

        8.6.1   Quantum–Mechanical Switch Memories 407

        8.6.2   A GaAs n-p-n-p Thyristor/JFET Memory Cell 408

        8.6.3   Single-Electron Memory 409

        8.6.4   Neuron-MOS Multiple-Valued (MV) Memory Technology 409

Chapter 9   **High-Density Memory Packaging Technologies 412**

    9.1   Introduction 412

    9.2   Memory Hybrids and MCMs (2-D) 417

        9.2.1   Memory Modules (Commercial) 417

        9.2.2   Memory MCMs (Honeywell ASCM) 421

        9.2.3   VLSI Chip-on-Silicon (VCOS) Technology 421

    9.3   Memory Stacks and MCMs (3-D) 424

        9.3.1   3-D Memory Stacks (Irvine Sensors Corporation) 427

                9.3.1.1   4 Mb SRAM Short Stack™ (Example)..................................... 429

        9.3.2   3-D Memory Cube Technology (Thomson CSF) 430

        9.3.3   3-D Memory MCMs (GE-HDI/TI) 431

                9.3.3.1   3-D HDI Solid-State Recorder (Example)............................... 432

        9.3.4   3-D Memory Stacks (n CHIPS) 432

    9.4   Memory MCM Testing and Reliability Issues 435

        9.4.1   VCOS DFT Methodology and Screening Flow (Example) 437

    9.5   Memory Cards 440

        9.5.1   CMO SRAM Card (Example) 442

        9.5.2   Flash Memory Cards 442

    9.6   High-Density Memory Packaging Future Directions 446

    Index 451

Preface

Semiconductor memories are usually considered to be the most vital microelectronics component of the digital level designs for main frame computers and PCs, telecommunications, automotive and consumer electronics, and commercial and military avionics systems. Semiconductor memory devices are characterized as volatile random access memories (RAMs), or nonvolatile memory devices. RAMS can either be static mode (SRAMs) where the logic information is stored by setting up the logic state of a bistable flip-flop, or through the charging of a capacitor as in dynamic random access memories (DRAMs). The nonvolatile memory data storage mode may be permanent or reprogrammable, depending upon the fabrication technology used.

In the last decade or so, semiconductor memories have advanced both in density and performance because of phenomenal developments in submicron technologies. DRAMs are considered key technology drivers and predictors of semiconductor manufacturing process scaling and performance trends. According to the Semiconductor Industries Association's recent technology road map, the DRAMs are expected to increase in density from 64 Mb in 1995 to 64 Gb in 2010. The growing demands for PCs and workstations using high-performance microprocessors are the key targets of DRAM designers and manufacturers. The technical advances in multimegabit DRAMs have resulted in a greater demand for application-specific products such as the pseudostatic DRAM (PSRAM), video DRAM (VDRAM), and high speed DRAM (HSDRAM), e.g., synchronous, cached, and Rambus (configurations using innovative architectures). Each of these specialty memory devices has its associated testing and reliability issues that have to be taken into consideration for board and system level designs.

Nonvolatile memories such as read-only memories (ROMs), programmable read-only memories (PROMs), and erasable and programmable read-only memories (EPROMs) in both ultraviolet erasable (UVPROM) and electrically erasable (EEPROM) versions have also made significant improvements in both density and performance. Flash memories are being fabricated in 8 and 16 Mb density devices for use in high-density nonvolatile storage applications such as memory modules and memory cards.

The continuously evolving complexity of memory devices makes memory fault modeling, test algorithms, design for testability (DFT), built-in self-test (BIST), and fault tolerance areas of significant concern. The general reliability issues pertaining to semiconductor devices in bipolar and MOS technologies are applicable to memories also. In addition, there are some special RAM failure modes and mechanisms, nonvolatile memory reliability issues, and reliability modeling and failure prediction techniques that are reviewed in this book.

Radiation effects on semiconductor memories is an area of growing concern. In general, the space radiation environment poses a certain radiation risk to electronic components on earth orbiting satellites and planetary mission spacecrafts, and the cumulative effect of damage from charged particles such as electrons and protons on semiconductor memories can be significant. The memory scaling for higher densities has made them more susceptible to logic upsets and failures. This book provides a detailed coverage of those radiation effects, including single-event phenomenon (SEP) modeling and error rate prediction, process and circuit design radiation hardening techniques, radiation testing procedures, and radiation test structures.

Several advanced memory technologies that are in the development stages are reviewed, such as ferroelectric random access memories (FRAMs), GaAs FRAMs, analog memories, magnetoresistive random access memories (MRAMs), and quantum–mechanical switch memories. Another area of current interest covered in the book is the section on high-density memory packaging technologies, which includes memory hybrids and multichip module (MCM) technologies, memory stacks and 3-D MCMs, memory MCM testing and reliability issues, and memory cards. In high-density memory development, the future direction is to produce mass memory configurations of very high bit densities ranging from tens of megabytes to several hundred gigabytes.

There are very few books on the market that deal exclusively with semiconductor memories, and the information available in them is fragmented—some of it is outdated. This book is an attempt to provide a comprehensive and integrated coverage in three key areas of semiconductor memories: technology, testing and reliability. It includes detailed chapters on the following: Introduction; Random Access Memory technology, both SRAMs, DRAMs, and their application-specific architectures; Nonvolatile Memories such as ROMs, PROMs, UVPROMs, EEPROMs, and flash memories; Memory Fault Modeling and Testing; Memory Design for Testability and Fault Tolerance; Semiconductor Memory Reliability; Semiconductor Memory Radiation Effects; Advanced Memory Technologies; and High-Density Memory Packaging Technologies.

This book should be of interest and be useful to a broad spectrum of people in the semiconductor manufacturing and electronics industries, including engineers, system level designers, and managers in the computer, telecommunications, automotive, commercial satellite, and military avionics areas. It can be used both as an introductory treatise as well as an advanced reference book.

I am thankful to all of the technical reviewers who diligently reviewed the manuscript and provided valuable comments and suggestions that I tried to incorporate into the final version. I would like to thank several people at Goddard Space Flight Center, especially George Kramer, Ann Garrison, Ronald Chinnapongse, David Cleveland, Wentworth Denoon, and Charles Vanek. Special thanks and acknowledgments are also due to various semiconductor memory manufacturers and suppliers for providing product information specification sheets and permission to reprint their material. And the author and IEEE Press acknowledge the efforts of Rochit Rajsuman as a technical reviewer and consultant.

Finally, the acknowledgments would be incomplete without thanks to Dudley R. Kay, Director of Book Publishing, John Griffin, and Orlando Vélez, along with the other staff members at IEEE Press, including the Editorial Board for their invaluable support.

Ashok K. Sharma
Silver Spring, Maryland

Semiconductor Memories

1

Introduction

Semiconductor memories are usually considered to be the most vital microelectronic component of digital logic system design, such as computers and microprocessor-based applications ranging from satellites to consumer electronics. Therefore, advances in the fabrication of semiconductor memories including process enhancements and technology developments through the scaling for higher densities and faster speeds help establish performance standards for other digital logic families. Semiconductor memory devices are characterized as volatile random access memories (RAMs), or nonvolatile memory devices. In RAMs, the logic information is stored either by setting up the logic state of a bistable flip-flop such as in a static random access memory (SRAM), or through the charging of a capacitor as in a dynamic random access memory (DRAM). In either case, the data are stored and can be read out as long as the power is applied, and are lost when the power is turned off; hence, they are called volatile memories.

Nonvolatile memories are capable of storing the data, even with the power turned off. The nonvolatile memory data storage mode may be permanent or reprogrammable, depending upon the fabrication technology used. Nonvolatile memories are used for program and microcode storage in a wide variety of applications in the computer, avionics, telecommunications, and consumer electronics industries. A combination of single-chip volatile as well as nonvolatile memory storage modes is also available in devices such as nonvolatile SRAM (nvRAM) for use in systems that require fast, reprogrammable nonvolatile memory. In addition, dozens of special memory architectures have evolved which contain some additional logic circuitry to optimize their performance for application-specific tasks.

This book on semiconductor memories covers random access memory technologies (SRAMs and DRAMs) and their application-specific architectures; nonvolatile memory technologies such as read-only memories (ROMs), programmable read-only memories (PROMs), and erasable and programmable read-only memories (EPROMs) in both ultraviolet erasable (UVPROM) and electrically erasable (EEPROM) versions; memory fault modeling and testing; memory design for testability and fault tolerance; semiconductor memory reliability; semiconductor memory radiation effects; advanced memory technologies; and high-density memory packaging technologies.

Chapter 2 on "Random Access Memory Technologies," reviews the static and dynamic RAM technologies as well as their application-specific architectures. In the last two decades of semiconductor memory growth, the DRAMs have been the largest volume volatile memory produced for use as main computer memories because of their high density and low cost per bit advantage. SRAM densities have generally lagged a generation behind the DRAMs, i.e., the SRAMs have about one-fourth the capacity of DRAMs, and therefore tend to cost about four times per bit as the DRAMs. However, the SRAMs offer low-power consumption and high-performance features which make them practical alternatives to the DRAMs. Nowadays, a vast majority of SRAMs are being fabricated in the NMOS and CMOS technologies, and a combination of two technologies (also referred to as the mixed-MOS) for commodity SRAMs.

Bipolar memories using emitter-coupled logic (ECL) provide very fast access times, but consume two–three times more power than MOS RAMs. Therefore, in high-density and high-speed applications, various combinations of bipolar and MOS technologies are being used. In addition to the MOS and bipolar memories referred to as the "bulk silicon" technologies, silicon-on-insulator (SOI) isolation technology such as silicon-on-sapphire (SOS) SRAMs have been developed for improved radiation hardness. The SRAM density and performance are usually enhanced by scaling down the device geometries. Advanced SRAM designs and architectures for 4 and 16 Mb density chips with submicron feature sizes (e.g., 0.2–0.6 μm) are reviewed. Application-specific memory designs include first-in first-out (FIFO), which is an example of shift register memory architecture through which the data are transferred in and out serially. The dual-port RAMs allow two independent devices to have simultaneous read and write access to the same memory. Special nonvolatile, byte-wide RAM configurations require not only very low operating power, but have battery back-up data-retention capabilities. The content-addressable memories (CAMs) are designed and used both as the embedded modules on larger VLSI chips and as stand-alone memory for specific system applications.

The DRAMs store binary data in cells on capacitors in the form of charge which has to be periodically refreshed in order to prevent it from leaking away. A significant improvement in DRAM evolution has been the switch from three-transistor (3-T) designs to one-transistor (1-T) cell design that has enabled production of 4 and 16 Mb density chips that use advanced, 3-D trench capacitor and stacked capacitor cell structures. DRAMs of 64 Mb density have been sampled, and prototypes for 256 Mb are in development. The DRAMs are susceptible to soft errors (or cell logic upset) occurring from alpha particles produced by trace radioactive packaging material. In NMOS/CMOS DRAM designs, various techniques are used to reduce their susceptibility to the soft-error phenomenon. The BiCMOS DRAMs have certain advantages over the pure CMOS designs, particularly in access time. The technical advances in multimegabit DRAMs have resulted in greater demand for application-specific products such as the pseudostatic DRAM (PSRAM) which uses dynamic storage cells but contains all refresh logic on-chip that enables it to function similarly to a SRAM. Video DRAMs (VDRAMs) have been produced for use as the multiport graphic buffers. A high-speed DRAM (HSDRAM) has been developed with random access time approaching that of SRAMs while retaining the density advantage of 1-T DRAM design. Some other examples of high-speed DRAM innovative architectures are synchronous, cached, and Rambus™ DRAMs.

Chapter 3 reviews various nonvolatile memory (NVM) technologies. A category of NVM is read-only memories (ROMs) in which the data are written permanently during manufacturing, or the user-programmable PROMs in which the data can be written only once. The PROMs are available in both bipolar and CMOS technologies. In 1970, a floating polysilicon-gate-based erasable programmable read-only memory was developed in which hot electrons are injected into the floating gate and removed either by ultraviolet internal photoemission or Fowler–Nordheim tunneling. The EPROMs (also known as UVEPROMs) are erased by removing them from the target system and exposing them to ultraviolet light. Since an EPROM consists of single-transistor cells, they can be made in densities comparable to the DRAMs. Floating-gate avalanche-injection MOS transistors (FAMOS) theory and charge-transfer mechanisms are discussed. Several technology advances in cell structures, scaling, and process enhancements have made possible the fabrication of 4–16 Mb density EPROMs. A cost-effective alternative has been the one-time programmable (OTP) EPROM introduced by the manufacturers for the high-volume applications ROM market.

An alternative to EPROM (or UVEPROM) has been the development of electrically erasable PROMs (EEPROMs) which offer in-circuit programming flexibility. The several variations of this technology include metal–nitride–oxide–semiconductor (MNOS), silicon–oxide–nitride–oxide–semiconductor (SONOS), floating-gate tunneling oxide (FLOTOX), and textured polysilicon. Since the FLOTOX is the most commonly used EEPROM technology, the Fowler–Nordheim tunneling theory for a FLOTOX transistor operation is reviewed. The conventional, full functional EEPROMs have several advantages, including the byte-erase, byte-program, and random access read capabilities. The conventional EEPROMs used NOR-gate cells, but the modified versions include the NAND-structured cells that have been used to build 5 V-only 4 Mb EEPROMs. An interesting NVM architecture is the nonvolatile SRAM, a combination of EEPROM and SRAM in which each SRAM cell has a corresponding "shadow" EEPROM cell. The EPROMs, including UVEPROMs and EEPROMs, are inherently radiation-susceptible. The SONOS technology EEPROMs have been developed for military and space applications that require radiation-hardened devices. Flash memories based on EPROM or EEPROM technologies are devices for which the contents of all memory array cells can be erased simultaneously through the use of an electrical erase signal. The flash memories, because of their bulk erase characteristics, are unlike the floating-gate EEPROMs which have select transistors incorporated in each cell to allow for the individual byte erasure. Therefore, the flash memories can be made roughly two or three times smaller than the floating-gate EEPROM cells. The improvements in flash EEPROM cell structures have resulted in the fabrication of 8 and 16 Mb density devices for use in high-density nonvolatile storage applications such as memory modules and memory cards.

Chapter 4 on "Memory Fault Modeling and Testing," reviews memory failure modes and mechanisms, fault modeling, and electrical testing. The memory device failures are usually represented by a bathtub curve and are typically grouped into three categories, depending upon the product's operating life cycle stage where the failures occur. Memory fault models have been developed for the stuck-at faults (SAFs), transition faults (TFs), address faults (AFs), bridging faults (BFs), coupling faults (CFs), pattern-sensitive faults (PSFs), and the dynamic (or delay) faults. A most commonly used model is the single-stuck-at fault (SSF) which is also referred to as the classical or standard fault model. However, the major shortcoming of the stuck-at fault model is that the simulation using this model is no longer an accurate quality indicator for the ICs like memory chips. A large percentage of physical faults occurring in the ICs can be considered as the bridging faults (BFs) consisting of shorts between two or more cells or lines. Another important category of faults that can cause the semiconductor RAM cell to function erroneously is the coupling or PSFs. March tests in various forms have been found to be quite effective for detecting the SAFs, TFs, and CFs. Many algorithms have been proposed for the neighborhood pattern-sensitive faults (NPSFs) based on an assumption that the memory array's physical and logical neighborhoods are identical. This may not be a valid assumption in state-of-the-art memory chips which are being designed with the spare rows and columns to increase yield and memory array reconfiguration, if needed. The embedded RAMs which are being used frequently are somewhat harder to test because of their limited observability and controllability. A defect-oriented (inductive fault) analysis has been shown to be quite useful in finding various defect mechanisms for a given layout and technology.

In general, the memory electrical testing consists of the dc and ac parametric tests and functional tests. For RAMs, various functional test algorithms have been developed for which the test time is a function of the number of memory bits (n) and range in complexity from $O(n)$ to $O(n^2)$. The selection of a particular set of test patterns for a given RAM is influenced by the type of failure modes to be detected, memory bit density which influences the test time, and the memory ATE availability. These are the deterministic techniques which require well-defined algorithms and memory test input patterns with corresponding measurements of expected outputs. An alternate test approach often used for the memories is random (or pseudorandom) testing which consists of applying a string of random patterns simultaneously to a device under test and to a reference memory, and comparing the outputs. Advanced megabit memory architectures are being designed with special features to reduce test time by the use of the multibit test (MBT), line mode test (LMT), and built-in self-test (BIST). The functional models for nonvolatile memories are basically derived from the RAM chip functional model, and major failure modes in the EPROMs such as the SAFs, AFs, and BFs can be detected through functional test algorithms. Recent studies have shown that monitoring of the elevated quiescent supply currents (IDDQ) appears to be a good technique for detecting the bridging failures. The IDDQ fault models are being developed with a goal to achieve 100% physical defect coverage. Application-specific memories such as the FIFOs, video RAMs, synchronous static and dynamic RAMs, and double-buffered memories (DBMs) have complex timing requirements and multiple setup modes which require a suitable mix of sophisticated test hardware, the DFT and BIST approach.

Chapter 5 reviews the memory design for testability (DFT) techniques, RAM and ROM BIST architectures, memory error-detection and correction (EDAC) techniques, and the memory fault-tolerance designs. In general, the memory testability is a function of variables such as circuit complexity and design methodology. The general guidelines for a logic design based on practical experience are called the ad hoc design techniques such as logic partitioning, and addition of some I/O test points for the embedded RAMs to increase controllability and observability. Structured design techniques are based upon a concept of providing uniform design for the latches to enhance logic controllability, and commonly used methodologies include the level-sensitive scan design (LSSD), scan path, scan/set logic, random access scan, and the boundary scan testing (BST). The RAM BIST techniques can be classified into two categories as the "on-line BIST" and "off-line BIST." The BIST is usually performed by applying certain test patterns, measuring the output response using linear feedback shift registers (LFSRs), and compressing it. The various methodologies for the BIST include exhaustive testing, pseudorandom testing, and the pseudoexhaustive testing. For the RAMs, two BIST approaches have been proposed that utilize either the random logic or a microcoded ROM. The major advantages associated with a microcoded ROM over the use of random logic are a shorter design cycle, the ability to implement alternative test algorithms with minimal changes, and ease in testability of the microcode. The RAM BIST implementation strategies include the use of the algorithmic test sequence (ATS), the 13N March algorithm with a data-retention test, a fault-syndrome-based strategy for detecting the PSFs, and the built-in logic block observation (BILBO) technique. For the embedded memories, various DFT and BIST techniques have been developed such as the scan-path-based flag-scan register (FLSR) and the random-pattern-based circular self-test path (CSTP).

Advanced BIST architectures have been implemented to allow parallel testing with on-chip test circuits that utilize multibit test (MBT) and line mode test (LMT). An example is the column address-maskable parallel test (CMT) architecture which is suitable for the ultrahigh-density DRAMs. The current generation megabit memory chips include spare rows and columns (redundancies) in the memory array to compensate for the faulty cells. In addition, to improve the memory chip yield, techniques such as the built-in self-diagnosis (BISD) and built-in self-repair (BISR) have been investigated. BIST schemes for ROMs have been developed that are based on exhaustive testing and test response compaction. The conventional exhaustive test schemes for the ROMs use compaction techniques which are parity-based, count-based, or polynomial-division-based (signature analysis).

The errors in semiconductor memories can be broadly categorized into hard failures caused by permanent physical damage to the devices, and soft errors caused by alpha particles or the ionizing dose radiation environments. The most commonly used error-correcting codes (ECC) which are used to correct hard and soft errors are the single-error-correction and double-error-detection (SEC–DED) codes, also referred to as the Hamming codes. However, these codes are inadequate for correcting double-bit/word-line soft errors. Advanced 16 Mb DRAM chips have been developed that use redundant word and bit lines in conjunction with the ECC to produce an optimized fault tolerance effect. In a new self-checking RAM architecture, on-line testing is performed during normal operations without destroying the stored data. A fault tolerance synergism for memory chips can be obtained by a combined use of redundancy and ECC. A RAM fault tolerance approach with dynamic redundancy can use either the standby reconfiguration method, or memory reconfiguration by the graceful degradation scheme. To recover from soft errors (transient effects), memory scrubbing techniques are often used which are based upon the probabilistic or deterministic models. These techniques can be used to calculate the reliability rate $R(t)$ and MTTF of the memory systems.

Chapter 6 reviews general reliability issues for semiconductor devices such as the memories, RAM failure modes and mechanisms, nonvolatile memories reliability, reliability modeling and failure rate prediction, design for reliability, and reliability test structures. The reliability of a semiconductor device such as a memory is the possibility that the device will

perform satisfactorily for a given time at a desired confidence level under specified operating and environmental conditions. The memory device failures are a function of the circuit design techniques, materials, and processes used in fabrication, beginning from the wafer level probing to assembly, packaging, and testing. The general reliability issues pertaining to semiconductor devices in bipolar and MOS technologies are applicable to the memories also, such as the dielectric-related failures from gate–oxide breakdown, time-dependent dielectric breakdown (TDDB), and ESD failures; the dielectric-interface failures such as those caused by ionic contamination and hot carrier effects; the conductor and metallization failures, e.g., electromigration and corrosion effects; the assembly and packaging-related failures. However, there are special reliability issues and failure modes which are of special concern for the RAMs. These issues include gate oxide reliability defects, hot-carrier degradation, the DRAM capacitor charge-storage and data-retention properties, and DRAM soft-error failures. The memory gate dielectric integrity and reliability are affected by all processes involved in the gate oxide growth.

The high-density DRAMs use 3-D storage cell structures such as trench capacitors, stacked capacitor cells (STCs), and buried storage electrodes (BSEs). The reliability of these cell structures depends upon the quality and growth of silicon dioxide, thin oxide/nitride (ON), and oxide/nitride/oxide (ONO) composite films. The reduced MOS transistor geometries from scaling of the memory devices has made them more susceptible to hot carrier degradation effects. In DRAMs, the alpha-particle-induced soft-error rate (SER) can be improved by using special design techniques. Nonvolatile memories, just like volatile memories, are also susceptible to some specific failure mechanisms. For PROMs with fusible links, the physical integrity and reliability of fusible links are a major concern. In floating-gate technologies such as EPROMs and EEPROMs, data retention characteristics and the number of write/erase cycles without degradation (endurance) are the most critical reliability concerns. The ferroelectric memory reliability concerns include the aging effects of temperature, electric field, and the number of polarization reversal cycles on ferroelectric films used (e.g., PZT).

Reliability failure modeling is a key to the failure rate predictions, and there are many statistical distributions such as the Poisson, Normal (or Gaussian), Exponential, Weibull, and Lognormal that are used to model various reliability parameters. There are several reliability prediction procedures for predicting electronic component reliability such as *MIL-HDBK-217* and *Bellcore Handbook*. However, the failure rate calculation results for semiconductor memories may vary widely from one model to another. Design for reliability (DFR), which includes failure mechanisms modeling and simulation, is an important concept that should be integrated with the overall routine process of design for performance. The method of accelerated stress aging for semiconductor devices such as memories is commonly used to ensure long-term reliability. For nonvolatile memories, endurance modeling is necessary in the DFR methodology. An approach commonly employed by the memory manufacturers in conjunction with the end-of-line product testing has been the use of reliability test structures and process (or yield) monitors incorporated at the wafer level in kerf test sites and "drop-in" test sites on the chip. The purpose of reliability testing is to quantify the expected failure rate of a device at various points in its life cycle. The memory failure modes which can be accelerated by a combined elevated temperature and high-voltage stress are the threshold voltage shifts, TDDB leading to oxide shorts, and data-retention degradation for the nonvolatile memories. *MIL-STD-883, Method 5004 Screening Procedure* (or equivalent) are commonly used by the memory manufacturers to detect and eliminate the infant mortality failures. *MIL-STD-883, Method 5005 Qualification and QCI Procedures* (or equivalent) are used for high-reliability military and space environments.

Chapter 7, entitled "Semiconductor Memory Radiation Effects," reviews the radiation-hardening techniques, radiation-hardening design issues, radiation testing, radiation dosimetry, wafer level radiation testing, and test structures. The space radiation environment poses a certain radiation risk to all electronic components on earth orbiting satellites and the planetary mission spacecrafts. Although the natural space environment does not contain the high dose rate pulse characteristics of a weapon environment (often referred to as the "gamma dot"), the cumulative effect of ionization damage from charged particles such as electrons and protons on semiconductor memories can be significant. In general, the bipolar technology memories (e.g., RAMs, PROMs) are more tolerant to total dose radiation effects than the nonhardened, bulk MOS memories. For the MOS devices, the ionization traps positive charge in the gate oxide called the oxide traps, and produces interface states at the $Si–SiO_2$ interface. The magnitude of these changes depends upon a number of factors such as total radiation dose and its energy; dose rate; applied bias and temperature during irradiation; and postirradiation annealing conditions. Ionization radiation damage causes changes in the memory circuit parameters such as standby power supply currents, I/O voltage threshold levels and leakage currents, critical path delays, and timing specification degradations.

The single-event phenomenon (SEP) in the memories is caused by high-energy particles such as those present in the cosmic rays passing through the device to cause single-event upsets (SEUs) or soft errors, and single-event latchup (SEL) which may result in hard errors. The impact of SEU on the memories, because of their shrinking dimensions and increasing densities, has become a significant reliability concern. The number of SEUs experienced by a memory device in a given radiation environment depends primarily on its threshold for upsets, usually expressed by its critical charge Q_c or the critical LET and the total device volume sensitive to ionic interaction, i.e., creation of electron–hole (e–h) pairs. The Q_c is primarily correlated to circuit design characteristics. Critical LET for a memory is found experimentally by bombarding the device with various ion species (e.g., in a cyclotron). For the memory devices flown in space, radiation tolerance is assessed with respect to the projected total dose accumulated, which is the sum of absorbed dose contributions from all ionizing particles, and is calculated (in the form of dose-depth curves) by sophisticated environmental modeling based upon the orbital parameters, mission duration, and thickness of spacecraft shielding. It is important to verify ground test results obtained by observation of actual device behavior in orbit. Several in-orbit satellite experiments have been designed to study the effect of the radiation particle environment on semiconductor devices such as memories.

The nonvolatile MOS memories are also subject to radiation degradation effects. The radiation hardness of memories is influenced by a number of factors, both process- and design-related. The process-related factors which affect radiation response are the substrate effects, gate oxidation and gate electrode effects, post-polysilicon processing, and field oxide hardening. The CMOS SOI/SOS technologies which utilize insulator isolation as opposed to the junction isolation for bulk technologies offer a substantial advantage in the latchup, transient upset, and SEU characteristics. The memory circuits can be designed for total dose radiation hardness by using optimized processes (e.g., hardened gate oxides and field oxides) and good design practices. The bulk CMOS memories have been hardened to SEU by using an appropriate combination of processes (e.g., thin gate oxides, twin-tub process with thin epitaxial layers) and design techniques such as utilizing polysilicon decoupling resistors in the cross-coupling segment of each cell.

Radiation sensitivity of unhardened memory devices can vary from lot to lot, and for space applications, radiation testing is required to characterize the lot radiation tolerance. The ground-based radiation testing is based upon a simulation of space environment by using radiation sources such as the Cobalt-60, X-ray tubes, particle accelerators, etc. For example, total dose radiation testing on the memories is performed per *MIL-STD-883, Method 1019,* which

defines the test apparatus, procedures, and other requirements for effects from the Co-60 gamma ray source. Radiation testing requires calibration of the radiation source and proper dosimetry. Sometimes, radiation test structures are used at the wafer (or chip level) as process monitors for radiation hardness assurance.

Chapter 8 reviews several advanced memory technologies such as the ferroelectric random access memories (FRAMs), GaAs FRAMs, analog memories, magnetoresistive random access memories (MRAMs), and quantum-mechanical switch memories. In the last few years, an area of interest in advanced non-volatile memories has been the development of thin-film ferroelectric (FE) technology that uses magnetic polarization (or hysteresis) properties to build the FRAMs. The high-dielectric-constant materials such as lead zirconate titanate (PZT) thin film can be used as a capacitive, nonvolatile storage element similar to trench capacitors in the DRAMs. This FE film technology can be easily integrated with standard semiconductor processing techniques to fabricate the FRAMs which offer considerable size and density advantage. A FRAM uses one transistor and one capacitor cell. Although the FRAMs have demonstrated very high write endurance cycle times, the FE capacitors depolarize over time from read/write cycling. Therefore, thermal stability, fatigue from polarization reversal cycling, and aging of the FRAMs are key reliability concerns. In general, the FE capacitors and memories made from thin-film PZT have shown high-radiation-tolerance characteristics suitable for space and military applications. The FE element processing has also been combined with GaAs technology to produce ferroelectric nonvolatile (or FERRAMs) prototypes of 2K/4K bit density levels.

The memory storage volatile (or nonvolatile) usually refers to the storage of digital bits of information ("0"s and "1"s). However, recently, analog nonvolatile data storage has also been investigated using the EEPROMs and FRAMs in applications such as audio recording of speech and analog synaptic weight storage for neural networks. This nonvolatile analog storage is accomplished by using the EEPROMs which are inherently analog memories on a cell-by-cell basis because each floating gate can store a variable voltage. The sensed value of a cell's conductivity corresponds to the value of the analog level stored. This technology has been used in audio applications such as single-chip voice messaging systems. Another technology development for nonvolatile storage is the magnetoresistive memory (MRAM) which uses a magnetic thin-film sandwich configured in two-dimensional arrays. These MRAMs are based upon the principle that a material's magnetoresistance will change due to the presence of a magnetic field. The magnetoresistive technology has characteristics such as a nondestructive readout (NDRO), very high radiation tolerance, higher write/erase endurance compared to the FRAMs, and virtually unlimited power-off storage capability. Another variation on this technology is the design and conceptual development of micromagnet-Hall effect random access memory (MHRAM) where information is stored in small magnetic elements. The latest research in advanced memory technologies and designs includes the solid-state devices that use quantum–mechanical effects such as resonant-tunneling diodes (RTDs) and resonant-tunneling hot-electron transistors (RHETs) for possible development of gigabit memory densities. These devices are based upon the negative resistance (or negative differential conductance) property which causes a decrease in current for an increase in voltage. This effect has been used in the development of a SRAM cell that uses two RTDs and one ordinary tunnel diode (TD) for the complete cell.

Chapter 9, "High-Density Memory Packaging Technologies," reviews commonly used memory packages, memory hybrids and 2-D multichip modules (MCMs), memory stacks and 3-D MCMs, memory MCM testing and reliability issues, memory cards, and high-density memory packaging future directions. The most common high-volume usage semiconductor RAMs and nonvolatile memories use "through-the-hole" (or insertion mount) and the surface mount technology (SMT) packages. For high-

reliability military and space applications, hermetically sealed ceramic packages are usually preferred. For high-density memory layouts on the PC boards, various types of packaging configurations are used to reduce the board level memory package "footprint." However, increasing requirements for denser memories have led to further compaction of packaging technologies through the conventional hybrid manufacturing techniques and MCMs. For the assembly of MCMs, various interconnection technologies have been developed such as the wire bonding, tape automated bonding (TAB), flip-chip bonding, and high-density interconnect (HDI).

A commonly used multichip module configuration for the DRAMs is the single-in-line memory module (SIMM). Several variations on 3-D MCM technology have evolved for the memories, with a goal of improving storage densities while lowering the cost per bit. The density of chip packaging expressed as the "silicon efficiency" is determined by the ratio of silicon die area to the printed circuit board (or substrate) area. An example is the memory MCMs fabricated by Honeywell, Inc. for the Advanced Spaceborne Computer Module (ASCM) in the following two technologies: MCM-D using thin-film multilayer copper/polyimide interconnects on an alumina ceramic substrate mounted in a perimeter-leaded cofired ceramic flatpack, and MCM-C which used a multilayer cofired alumina package. In the chip-on-board (COB) packaging, the bare memory chip (or die) is directly attached to a substrate, or even PC board (such as FR4 glass epoxy). IBM (now LORAL Federal Systems) has developed VLSI chip-on-silicon (VCOS) MCMs which combine HDI technology with a flip-chip, C4 (controlled-collapse chip connect) attach process.

An extension of 2-D planar technology has been the 3-D concept in which the memory chips are mounted vertically prior to the attachment of a suitable interconnect. The 3-D approach can provide higher packaging densities because of reduction in the substrate size, module weight, and volume; lower line capacitance and drive requirements; and reduced signal propagation delay times. Four generic types of

3-D packaging technologies are currently being used by several manufacturers: layered die, die stacked on edge, die stacked in layers, and vertically stacked modules. An example is Texas Instruments (TI) 3-D HDI MCM packaging technology that has been used for the development of a solid-state recorder (SSR) for DARPA with initial storage capacity of 1.3 Gb expandable to 10.4 Gb.

However, MCM defects and failures can occur due to the materials, including the substrate, dice, chip interconnections, and manufacturing process variations; lack of proper statistical process control (SPC) during fabrication and assembly; inadequate screening and qualification procedures; and a lack of proper design for testability (DFT) techniques. Availability of "known-good-die" (KGD) and "known-good-substrate" (KGS) are important prerequisites for high-yield MCMs and minimizing the need for expensive module rework/repair. The MCM DFT techniques such as the boundary scan and level-sensitive scan design (LSSD) are often used in conjunction with the design for the reliability approach. Another application for high-density memory bare chip assembly has been the development of memory cards that are lightweight plastic and metal cards containing the memory chips and associated circuitry. They offer significant advantages in size, weight, speed, and power consumption. These cards integrate multiple volatile/nonvolatile memory technologies, and are intended to serve as alternatives for traditional hard disks and floppy drives in notebook computers and mobile communication equipment.

In high-density memory development, the future direction is to produce mass memory configurations of very high bit densities ranging from tens of megabytes to several hundred gigabytes by integrating 3-D technology into the MCMs.

2

Random Access Memory Technologies

2.1 INTRODUCTION

Semiconductor memory devices are generally categorized as volatile or nonvolatile random access memories (RAMs). In RAMs, the information is stored either by setting the state of a bistable flip-flop circuit or through the charging of a capacitor. In either of these methods, the information stored is destroyed if the power is interrupted. Such memories are therefore referred to as volatile memories. If the data are stored (i.e., written into the memory) by setting the state of a flip-flop, they will be retained as long as the power is applied and no other write signals are received. The RAMs fabricated with such cells are known as static RAMs, or SRAMs. When a capacitor is used to store data in a semiconductor RAM, the charge needs to be periodically refreshed to prevent it from being drained through the leakage currents. Hence, the volatile memories based on this capacitor-based storage mechanism are known as the dynamic RAMs, or DRAMs.

SRAM densities have generally lagged behind those for the DRAMs (1:4), mainly because of the greater number of transistors in a static RAM cell. For example, a 256 kb SRAM has about the same number of transistors as a 1 Mb DRAM. However, static RAMs are being widely used in systems today because of their low-power dissipation and fast data access time.

Early static RAMs were developed in three separate technologies: bipolar, NMOS, and CMOS. By the middle of the 1980s, the vast majority of SRAMs were made in CMOS technology. SRAMs are currently available in many varieties, ranging from the conventional MOS (NMOS or CMOS) to high-speed bipolar and GaAs SRAM designs. Full bipolar SRAMs, although less than 1% of the total SRAM market, are still available in lower densities for very high-speed applications. The lower end commodity SRAMs are represented by the "mixed-MOS" technology which is a combination of CMOS and NMOS for high-density applications, and by the full CMOS technology for a combination of high-density and low-power requirements. High-speed and high-density SRAMs are fabricated in both CMOS and mixed-MOS technologies, as well as combinations of bipolars and CMOS, called BiCMOS.

SRAM speed has been usually enhanced by scaling of the MOS process since shorter gate channel length, $L(\text{eff})$ translates quite linearly into faster access time. Figure 2-1(a) shows the plot of SRAM wafer average access time versus $L(\text{eff})$. This scaling of the process from first-generation to second-generation SRAMs has resulted in support of higher density products, as shown in the $L(\text{eff})$ versus density logarithmic plot of Figure 2-1(b). Section 2.2 of this chapter discusses in detail various

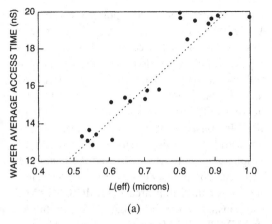

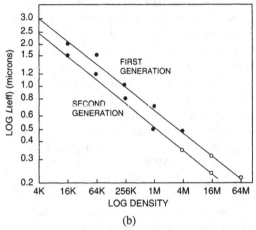

Figure 2-1. (a) SRAM wafer average access time versus L(eff). (b) L(eff) versus SRAM density logarithmic plot. (From [1], with permission of IEEE.)

SRAM (NMOS and CMOS) cell structures; MOS SRAM architectures, cell, and peripheral circuits operation; bipolar SRAM (ECL and BiCMOS) technologies; silicon-on-insulator (SOI) technologies; advanced SRAM architectures and technologies; and some application-specific SRAMs.

In the last decade of semiconductor memories growth, dynamic random access memories (DRAMs) have been produced in the largest quantities because of their high density and low cost per bit advantage. Main computer memories are mostly implemented using the DRAMs, even though they lag behind the SRAMs in speed.

Low-power dissipation and faster access times have been obtained, even with an increase in chip size for every successive generation of DRAMs, as shown in Figure 2-2 [2]. The DRAM memory array and peripheral circuitry such as decoders, selectors, sense amplifiers, and output drivers are fabricated using combinations of n-channel and p-channel MOS transistors. As a result, 1 Mb DRAMs have reached maturity in production, and nowadays, 4 and 16 Mb DRAMs are being offered by several manufacturers. While the density of DRAMs has been approximately quadrupling every three years, neither their access time nor cycle time has improved as rapidly. To increase the DRAM throughput, special techniques such as page mode, static column mode, or nibble mode have been developed. The faster DRAMs are fabricated with specially optimized processes and innovative circuit design techniques. In some new DRAMs being offered with wide on-chip buses, 1024–4096 memory locations can be accessed in parallel, e.g., cache DRAMs, enhanced DRAMs, synchronous DRAMs, and Rambus DRAMs. Section 2.3 of this chapter discusses the DRAM technology development; CMOS DRAMs; DRAM cell theory and advanced cell structures; BiCMOS DRAMs; soft-error failures

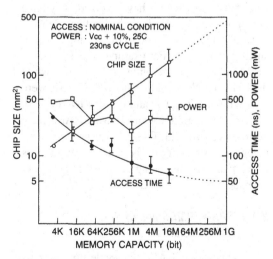

Figure 2-2. Trends in standard DRAM development. (From [2], with permission of IEEE.)

in DRAMs; advanced DRAM designs and architectures; and a few application-specific DRAMs.

2.2 STATIC RANDOM ACCESS MEMORIES (SRAMs)

SRAM is classified as a volatile memory because it depends upon the application of continuous power to maintain the stored data. If the power is interrupted, the memory contents are destroyed unless a back-up battery storage system is maintained. SRAM output width ranges from 1 to 32 b wide. Standard inputs and outputs include interfacing with CMOS, TTL, and ECL circuits. Power supply range includes standard 5 V and new 3.3 V standard for battery-powered applications. A SRAM is a matrix of static, volatile memory cells, and address decoding functions integrated on-chip to allow access to each cell for read/write functions. The semiconductor memory cells use active element feedback in the form of cross-coupled inverters to store a bit of information as a logic "one" or a "zero" state. The active elements in a memory cell need a constant source of dc (or static) power to remain latched in the desired state. The memory cells are arranged in parallel so that all the data can be received or retrieved simultaneously. An address multiplexing scheme is used to reduce the number of input and output pins. As SRAMs have evolved, they have undergone a dramatic increase in their density. Most of this has been due to the scaling down to smaller geometries. For example, the 4 kb MOS SRAM used 5 μm minimum feature size, while the 16 kb, 64 kb, 256 kb, and 1 Mb SRAMs were built with 3.0, 2.0, 1.2, and 0.8 μm feature size, respectively. Now, the 4 and 16 Mb SRAMs have been developed with 0.6–0.4 μm process technology. Also, scaling of the feature size reduces the chip area, allowing higher density SRAMs to be made more cost effectively.

2.2.1 SRAM (NMOS and CMOS) Cell Structures

The basic SRAM cell made up of cross-coupled inverters has several variations. The early NMOS static RAM cells consisted of six-transistor designs with four enhancement mode

and two depletion mode pull-up transistors. A significant improvement in cost and power dissipation was achieved by substituting ion-implanted polysilicon load resistors for the two pull-up transistors. These were called R-load NMOS, and a successor to these is called the "mixed-MOS" or "resistor-load CMOS." These SRAMs consisted of an NMOS transistor matrix with high ohmic resistor loads and CMOS peripheral circuits which allowed the benefit of lower standby power consumption while retaining the smaller chip area of NMOS SRAMs. The transition from the R-load NMOS to the R-load CMOS SRAMs occurred at the 4 kb density level. At the same time, the concept of power-down mode controlled by the chip enable pin ($\overline{CE}$) also appeared. In this low-voltage standby mode, the standby current for the mixed-MOS parts is typically in the microamp (μA) range, and for the full CMOS, in the nanoamp (nA) range. This low-power dissipation in the standby mode opened the potential for high-density battery back-up applications.

The mixed-MOS and the full NMOS designs have higher standby currents than the CMOS SRAMs. However, the mixed-MOS technique provides better scaling advantages and relatively lower power dissipation. The early CMOS RAMs with metal gate technology were mostly used in the aerospace and other high-reliability applications for their wider noise margins, wider supply voltage tolerance, higher operating temperature range, and lower power consumption. MOS SRAMs became more popular with the development of silicon gate technology in which the two levels of interconnections through polysilicon and aluminum allowed reduction in cell area, enabling the fabrication of larger and denser memory arrays. In the mid-1980s, a vast majority of all SRAMs were being made in several variations, such as the CMOS or mixed-MOS formation in n-well, p-well, or the twin-tub technologies.

Figure 2-3 shows an MOS SRAM cell with load devices which may either be the enhancement or depletion mode transistors as in an NMOS cell, PMOS transistors in a CMOS cell, or load resistors in a mixed-MOS or an R-load cell. The access and storage transistors are

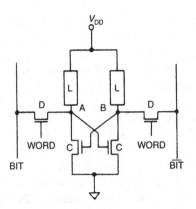

Figure 2-3. A general schematic of a SRAM memory cell.

enhancement mode NMOS. The purpose of load devices (L) is to offset the charge leakage at the drains of the storage and select transistors. When the load transistor is PMOS, the resulting CMOS cell has essentially no current flow through the cell, except during switching. The depletion load and resistive load have a low level of current flowing through them, and hence the standby power dissipation is always higher than that of the CMOS cell.

Figure 2-4(a) shows the basic CMOS SRAM cell consisting of two transistors and two load elements in a cross-coupled inverter configuration, with two select transistors added to make up a six-transistor cell. Figure 2-4(b) shows the use of polysilicon load resistors instead of PMOS transistors in the CMOS cell, which allows up to a 30% reduction in the cell size in double-polysilicon technology, using buried contacts.

There can be several application-specific variations in the basic SRAM cell. Figure 2-4(c) shows an eight-transistor, double-ended, dual-port static cell. This is useful in cache architectures, particularly as an embedded memory in a microprocessor chip. This chip can be simultaneously accessed through both of the

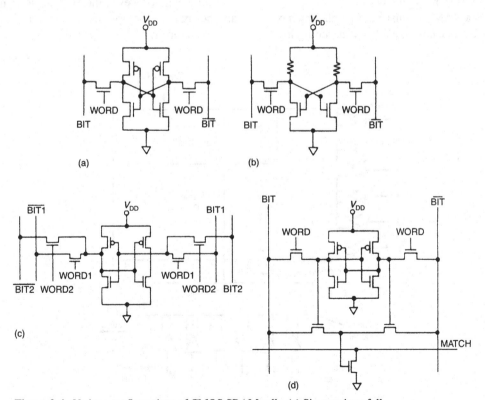

Figure 2-4. Various configurations of CMOS SRAM cells. (a) Six-transistor full CMOS. (b) Four transistors with *R*-load NMOS. (c) Dual-port with double-ended access. (d) Content-addressable memory (CAM).

14   Chap. 2   ■   Random Access Memory Technologies

ports. Figure 2-4(d) shows a nine-transistor content-addressable memory (CAM) cell. This is used in applications where knowledge of the contents of the cell, as well as location of the cell, are required.

2.2.2 MOS SRAM Architectures

Figure 2-5 shows (a) a typical SRAM basic organization schematic, and (b) the storage cell array details. Each memory cell shares electrical connections with all the other cells in its row and column. The horizontal lines connected to all the cells in a row are called the "word lines," and the vertical lines along which the data flow in and out of the cells are called the "bit lines." Each cell can be uniquely addressed, as required, through the selection of an appropriate word and a bit line. Some memories are designed so that a group of four or eight cells can be addressed; the data bus for such memories is called nibble or one byte wide, respectively.

In a RAM, the matrix of parallel memory cells is encircled by the address decoding logic and interface circuitry to external signals. The memory array nominally uses a square or a rectangular organization to minimize the overall chip area and for ease in implementation. The rationale for the square design can be seen by considering a memory device that contains 16K 1-bit storage cells. A memory array with 16K locations requires 14 address lines to allow selection of each bit ($2^{14} = 16,384$). If the array were organized as a single row of 16 Kb, a 14-to-16K line decoder would be required to allow individual selection of the bits. However, if the memory is organized as a 128-row $\times$ 128-column square, one 7-to-128 line decoder to select a row and another 7-to-128 line decoder to select a column are required. Each of these decoders can be placed on the sides of the square array. This 128-row $\times$ 128-column matrix contains 16,384 cross-points which allow access to all individual memory bits. Thus, the square memory array organization results in significantly less area for the entire chip. However, the 16K memory may also be organized as a 64-row $\times$ 256-column (or 256 $\times$ 64) array.

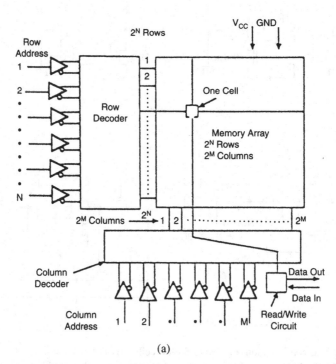

(a)

Figure 2-5. (a) A typical SRAM basic organization schematic.

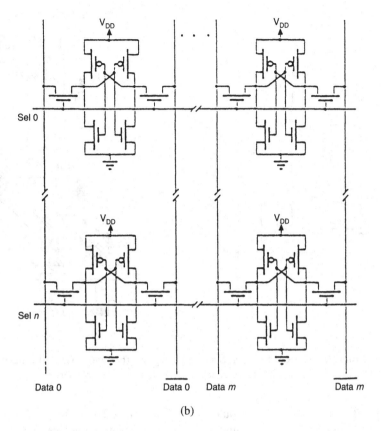

(b)

Figure 2-5 (cont.). (b) The storage cell array details.

Most RAMs operate such that the row address enables all cells along the selected row. The contents of these cells become available along the column lines. The column address is used to select the particular column containing the desired data bit which is read by the sense amplifier and routed to the data-output pin of the memory chip. When a RAM is organized to access n bits simultaneously, the data from n columns are selected and gated to n data-output pins simultaneously. Additional circuitry, including sense amplifiers, control logic, and tri-state input/output buffers, are normally placed along the sides of a cell array.

The two important time-dependent performance parameters of a memory are the "read-access time" and the "cycle time." The first timing parameter (access time) represents the propagation delay from the time when the address is presented at the memory chip until the data are available at the memory output. The cycle time is the minimum time that must be allowed after the initiation of the read operation (or a write operation in a RAM) before another read operation can be initiated.

2.2.3 MOS SRAM Cell and Peripheral Circuit Operation

A basic six-transistor CMOS SRAM cell and its layout are shown in Figure 2-6. The bit information is stored in the form of voltage levels in the cross-coupled inverters. This circuit has two stable states, designated as "1" and "0." If, in the logic state "1," the point C_5 is high and point C_6 is low, then T_1 is off and T_2 is on; also,

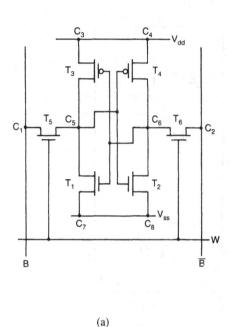

(a)

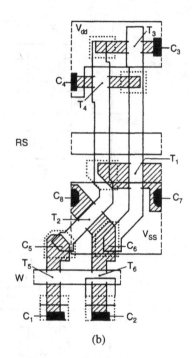

(b)

Figure 2-6. Six-transistor CMOS SRAM cell. (a) Schematic diagram. (b) Layout. (From Bastiaens and Gubbels [3], 1988, with permission of Philips Electronics, Eindhoven, The Netherlands.)

T_3 is on and T_4 is off. The logic "0" state would be the opposite, with point C_5 low and C_6 high. During the read/write operations, the row address of the desired cell is routed to the row address decoder which translates it and makes the correct word line of the addressed row high. This makes transistors T_5 and T_6 in all cells of the row switch "on." The column address decoder translates the column address, and makes connection to the bit line B and the inverse bit line $\overline{B}$ of all cells in the column addressed.

A READ operation is performed by starting with both the bit and $\overline{bit}$ lines high and selecting the desired word line. At this time, data in the cell will pull one of the bit lines low. The differential signal is detected on the bit and $\overline{bit}$ lines, amplified, and read out through the output buffer. In reference to Figure 2-6, reading from the cell would occur if B and $\overline{B}$ of the appropriate bit lines are high. If the cell is in state "1," then T_1 is off and T_2 is on. When the word line of the addressed column becomes high, a current starts to flow from $\overline{B}$ through T_6 and T_2 to ground. As a result, the level of $\overline{B}$ becomes

lower than B. This differential signal is detected by a differential amplifier connected to the bit and $\overline{bit}$ lines, amplified, and fed to the output buffer. The process for reading a "0" stored in the cell is opposite, so that the current flows through T_5 and T_1 to ground, and the bit line B has a lower potential than $\overline{B}$. The READ operation is nondestructive, and after reading, the logic state of the cell remains unchanged.

For a WRITE operation into the cell, data are placed on the bit line and $\overline{data}$ are placed on the $\overline{bit}$ line. Then the word line is activated. This forces the cell into the state represented by the bit lines, so that the new data are stored in the cross-coupled inverters. In Figure 2-6, if the information is to be written into a cell, then B becomes high and $\overline{B}$ becomes low for the logic state "1." For the logic state "0," B becomes low and $\overline{B}$ high. The word line is then raised, causing the cell to flip into the configuration of the desired state.

SRAM memory cell array's periphery contains the circuitry for address decoding and the READ/WRITE sense operations. Typical write circuitry consists of inverters on the input buffers and a pass transistor with a write control input signal to the bit and $\overline{\text{bit}}$ lines. Read circuitry generally involves the use of single-ended differential sense amplifiers to read the low-level signals from the cells. The data path is an important consideration with the SRAMs, since the power delay product is largely determined by the load impedance along the data path. The read data path circuitry can be static or dynamic. SRAMs have been designed which turn on or turn off the various sections of data path as needed to reduce the operating power of the device. The internal signals in an outwardly "static" RAM are often generated by a technique called "Address Transition Detection" (ATD) in which the transition of an address line is detected to generate the various clock signals. The input circuitry for SRAMs consists of the address decoders, word line drivers, and decoder controls. Figure 2-7 shows the various SRAM circuit elements.

RAMs are considered clocked or not clocked, based on the external circuitry. Asynchronous (nonclocked) SRAMs do not require external clocks, and therefore have a very simple system interface, although they have some internal timing delays. Synchronous (clocked) SRAMs do require system clocks, but they are faster since all the inputs are clocked into the memory on the edge of the system clock.

2.2.4 Bipolar SRAM Technologies

The earliest semiconductor memories were built in bipolar technology. Nowadays, bipolar memories are primarily used in high-speed applications. Bipolar RAMs are often "word-oriented" and require two-step decoding. For example, in a 1-kb $\times$ 1-b memory organized as a 32-row $\times$ 32-column array, the row decoder selects one of the 32 rows, and all of the 32 b (the "word") are read out and placed in a register. A second 5 b code is used to access the

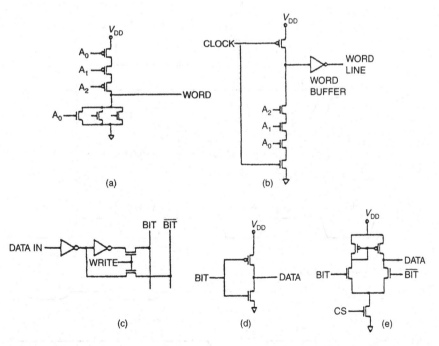

Figure 2-7. Various SRAM circuit elements. (a) Static row decoder. (b) Dynamic row decoder. (c) Simple write circuitry. (d) Inverter amplifier. (e) Differential sense amplifier. (From [4], with permission of Wiley, New York.)

register and select the desired bit. Similarly, the data are stored by writing an entire word simultaneously. The broad category of bipolar memories include the SRAMs fabricated in direct-coupled transistor logic (DCTL), emitter-coupled logic (ECL), and mixed BiCMOS technologies.

2.2.4.1 Direct-Coupled Transistor Logic (DCTL) Technology.

The architectures for bipolar SRAMs are basically similar to those of MOS SRAMs. Historically, transistor–transistor logic (TTL) has been the most commonly used bipolar technology. A simple bipolar DCTL RAM memory cell consists of two bipolar tran-

sistors, two resistors, and a power source. In this configuration, one of the transistors is always conducting, holding the other transistor OFF. When an external voltage forces the OFF transistor into conduction, the initially ON transistor turns off and remains in this condition until another external voltage resets it. Since only one of the cross-coupled transistors conducts at any given time, the circuit has only two stable states which can be latched to store information in the form of logic 1s and 0s. The cell state is stable until forced to change by an applied voltage.

The circuit in Figure 2-8 is an expanded version of the DCTL memory cell [5], [37]. In this figure, the data lines are connected to Q_1

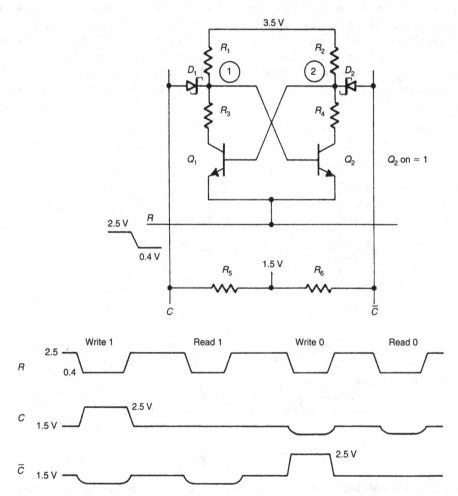

Figure 2-8. Direct-coupled memory cell (DCTL) with Schottky diodes. (From D. A. Hodges and H. Jackson [37], with permission of McGraw-Hill Inc., New York.)

and Q_2 through the Schottky diodes D_1 and D_2. To explain the operation of this cell, assume that a stored 1 corresponds to the state with Q_2 on. The row selection requires that the row voltage should be pulled low. To write a "1," the voltage on line C is raised, forward-biasing diode D_1. This forces sufficient current through R_1 so that the voltage at node 1 increases to turn on Q_2. The current gain of Q_1 is sufficiently high, and it remains in saturation so that most of the voltage drop appears across R_3. When Q_2 turns on, its collector voltage drops rapidly, turning off Q_1. The currents in R_1 and R_2 are always much smaller than the current used for writing, so that the voltage drops across R_3 and R_4 are much smaller than $V_{be(on)}$. In the standby condition, D_1 and D_2 are reverse-biased. To read a stored

"1," the row is pulled low, and current flows through C through D_2, R_4, and Q_2 to R. The resulting drop in voltage on C indicates the presence of a stored "1."

2.2.4.2 Emmiter-Coupled Logic (ECL) Technology.

Another popular bipolar technology is the emitter-coupled logic (ECL). The ECL memories provide very small access times with typical propagation delays of less than 1 ns and clock rates approaching 1 GHz. This high performance is achieved by preventing the cross-coupled transistors from entering into the saturation region. Figure 2-9 shows an ECL memory cell [37]. In this configuration, the data lines are connected to the emitters of the two transistors. Although both transistors have two

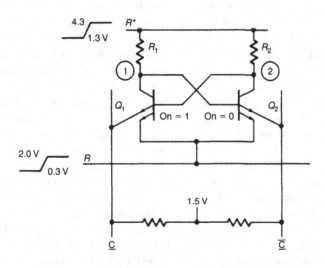

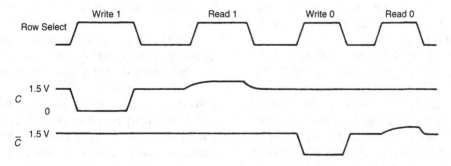

Figure 2-9. Emitter-coupled logic (ECL) memory cell. (From D. A. Hodges and H. Jackson [37], with permission of McGraw-Hill Inc., New York.)

emitters each, this cell is not a TTL circuit because they operate in their normal mode (as opposed to the inverted) modes. The operation of the cell is based on using the multiple-emitter transistors as the current switches. The voltage levels are selected such that these transistors never conduct simultaneously. The read and write operation is controlled by switching the current in the conducting transistor from the row line to the appropriate data line.

The basic operation of an ECL cell is explained with Figure 2-9. Assume that a logic "1" is stored with Q_1 on. The row selection requires that both R and R^* go to the positive levels as shown. To write a "1," the column line C must be held low, which forward-biases the emitter of Q_1, regardless of the previous state of the cell. As a result, the collector–emitter voltage of Q_1 drops quickly, removing the base drive from Q_2. When the row voltage returns to the standby levels, Q_1 remains "on," with its base current coming from R_2. Cell current flows through Q_1 and returns to ground through the line R. The emitters connected to C and $\overline{C}$ are reverse-biased in the standby condition. To read a stored "1," the cell is selected in the same way as for writing. The emitters connected to R become reverse-biased, and the current flowing in Q_1 transfers to the emitter connected to C. The resulting rise in voltage on C indicates the presence of a stored "1." The writing and reading operation for a "0" are complementary to those described for a "1."

A major drawback in the ECL technology is that a very low value of resistors is required, which results in a constant current in the storage cell. This constant current drain causes relatively higher power dissipation compared to the TTL or MOS memories.

Another bipolar logic developed in the 1970s was the integrated injection logic (I^2L). This evolved from the bipolar memory cell design in which the old direct-coupled transistor logic (DCTL) was shrunk to the single complementary transistor equivalent. In memory cells, lateral p-n-p transistors were used as the current source and multicollector n-p-n transistors as the inverters. The major attractive feature of this

technology was the high packaging density since the cells were quite compact, in contrast to the conventional bipolar memory cells using resistors for the load impedance of the flip-flop transistors. However, I^2L technology for memories failed to be commercially successful because of the process- and structure-related dependencies of its speed–power performance characteristics. The BiCMOS technology was preferred for high-speed and high-performance applications.

2.2.4.3 BiCMOS Technology.

In high density and high-speed applications, various combinations of bipolar and MOS technologies have been investigated. The BiCMOS process is more complex because of the additional steps required. Bipolar ECL input and output buffers have been used with the CMOS memories, both to interface to a bipolar circuit as well as to increase the performance. The CMOS memory cells have lower power consumption, better stability, and a smaller area compared to an equivalent bipolar cell. Therefore, the BiCMOS designs offer optimization of these parameters, since the various circuit elements are selectively chosen to maximize the performance [6]. For example, since the bipolar n-p-n output transistors provide large output drive, BiCMOS gates are effective with high capacitive nodes, such as those in the decoders, word line drivers, and output buffers. The control logic with small fan-out can still use CMOS gates. The sense amplifiers which require high gain and high input sensitivity for fast sensing of small differential bit line swings are built in bipolar logic. Figure 2-10 shows the BiCMOS circuit elements with a typical mix of various technologies [7]. There can be some variations on this technology mix, depending upon the manufacturer's process.

In a typical double-polysilicon technology CMOS SRAM process, the first poly level forms the MOS transistor gate; the second poly level forms highly resistive load resistors in the four-transistor CMOS memory cell, and makes contact to the silicon substrate. Since a poly-to-substrate contact already exists in such a process, a bipolar transistor with a polysilicon

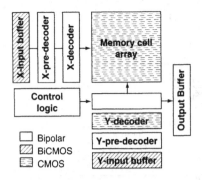

Figure 2-10. BiCMOS circuit elements.

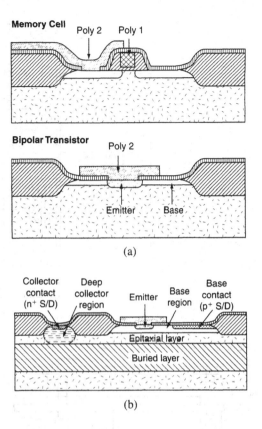

(a)

(b)

Figure 2-11. Schematic cross-sections. (a) A double-polysilicon SRAM with second poly as bipolar emitter. (b) Typical process enhancements for creating BiCMOS [8].

emitter can be produced with minimum complexity. The emitter is created by a dopant diffusion from a doped poly into the silicon substrate where an implanted base is present. Figure 2-11(a) shows the schematic cross-section of a double-polysilicon SRAM with second poly as a bipolar emitter [8].

A high-performance CMOS process can be obtained from a core CMOS process by the addition of a few steps such as: (1) use of a buried layer and an epitaxial layer, (2) a base region for bipolar transistors usually by ion implantation, and (3) an additional collector region (collector sink or collector plug) to reduce collector resistance. Typically, the added steps may increase process complexity by 10–20%. Figure 2-11(b) shows the typical enhancements to a core double-polysilicon process for creating high-performance BiCMOS memories.

The higher density commodity SRAMs tend to exhibit a balance of speed and power characteristics. The various density levels of very high-speed CMOS and BiCMOS SRAMs are specifically designed to optimize performance, often at the expense of chip area and power consumption. The new generation BiCMOS includes many improvements, such as the use of fast data bus in the sense circuit; parallel test circuits using the ECL sense amplifiers; optimized division of memory matrix to reduce parasitic capacitance on the signal lines (bit, word, and decoder); and column redundant circuits suitable for divided word line architectures. The use of low-resistance metal interconnects al-

lows more cells per bit-line pair without significantly increasing the delay.

A 4 Mb BiCMOS SRAM was developed by Fujitsu in 1991. It uses a wired-OR data bus in the sense circuit, a column redundant circuit, and a 16 b parallel test circuit [9]. Figure 2-12(a) shows a block diagram of this SRAM. It has a typical address access time of 7 ns, a write pulse width of 4 ns, and active current consumption of about 120 mA for a 4M × 1-b configuration. A symmetrical chip layout divides the bit lines, word lines, and decoder lines to reduce the ground noise from high-speed operation. The memory matrix is divided into four planes, each consisting of 16 blocks. The column redundant circuit is shown in Figure 2-12(b). The

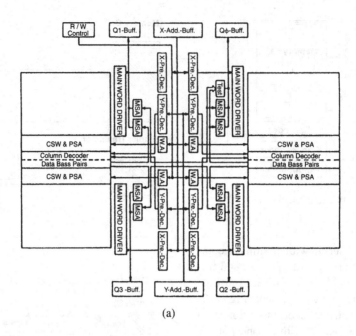

(a)

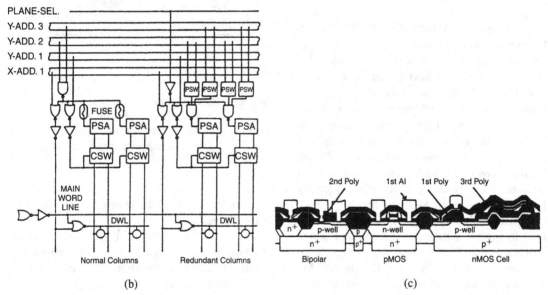

Normal Columns   Redundant Columns   Bipolar   pMOS   nMOS Cell

(b)   (c)

Figure 2-12. Fujitsu 4 Mb SRAM. (a) Block diagram. (b) Column redundant circuit. (c) Cross-section. (From [9], with permission of IEEE.)

redundancy is provided by two rows and eight columns, which provide more flexibility than the conventional architecture. The column arrays are distributed throughout each block and can replace cells within the block. In this 4 Mb

SRAM, only four redundant column arrays are placed in the 1 Mb plane to form a redundant section, and they are used for any column array in the plane. The process technology is 0.6 μm triple-polysilicon and double-metal BiCMOS.

The first polysilicon layer forms the gate electrodes of the MOS transistors, and the second layer forms the emitter electrodes of the bipolar transistors. For memory cells, the V_{EE} lines are formed with the second polysilicon layer, and the 50 GΩ loads are formed with the third polysilicon layer. The minimum gate lengths are 0.6 and 0.8 μm for the NMOS and PMOS transistors, respectively. The thickness of the gate oxide is 15 nm. Figure 2-12(c) shows the cross-section of this BiCMOS device.

A 0.5 μm BiCMOS technology has been developed for fast 4 Mb SRAMs in which bipolar transistors are added to an existing 0.5 μm CMOS process [10]. This process requires the growth of a thin epitaxial layer, as well as the addition of three masking steps as follows: self-aligned buried layer, deep collector, and active base. The original CMOS process featured self-aligned twin-well formation, framed mask poly-buffered LOCOS (FMPBL) isolation, a 150 Å thick gate oxide, surface-channel NMOS and buried-channel PMOS transistors, disposable polysilicon spacer module, three levels of polysilicon, and two layers of metallization. Figure 2-13 shows the schematic cross-section of this technology [10]. Three levels of polysilicon are used in this process. The first layer is the gate electrode for the CMOS transistors. The second layer is the tungsten–polycide/polysilicon stack

that performs the following three functions: creating self-aligned contact landing pads in the SRAM bit cell, forming an emitter of the bipolar n-p-n transistor, and providing a global interconnect. The third polysilicon layer forms the teraohm resistor load for the bit cell. This process provides a peak cutoff frequency (f_T) of 14 GHz with a collector–emitter breakdown voltage (BV_{CEO}) of 6.5 V, and ECL minimum gate delays of 105·ps at a gate current of 350 μA/μm² have been achieved.

The mainframe computers and supercomputers require high-speed ECL I/O SRAMs, whereas the high-performance workstations usually require TTL I/O SRAMs operating with microprocessors at low supply voltages. In order to meet both of these requirements, an approach has been developed that uses the metal mask operation technique to fabricate a 4 Mb BiCMOS SRAM that can have either a 6 ns access time at 750 mW and 50 MHz, or an 8 ns access time at 230 mW and 50 MHz using an ECL 100K or 3.3 V TTL interface, respectively [11]. A 3.3 V supply voltage is desirable for MOSFETs with a 0.55 μm gate length to ensure higher reliability. For an ECL 100K interface, ECL I/O buffers are supplied at −4.5 V V_{EE}, while the 4 Mb SRAM core circuit is supplied at −3.3 V $V_{EE'}$, which is generated by an internal voltage converter. The 4 Mb SRAM core consists of BiCMOS address decoders, MOS memory cells, and bipolar sense amplifiers.

In order to achieve high-speed address access times, a combination of the following technologies were used:

- A Bi-NMOS converter that directly converts the ECL level to a CMOS level
- A high-speed Bi-NMOS circuit with low-threshold-voltage NMOSFETs
- A design methodology for optimizing both the number of decoder gate stages and the size of gate transistors
- High-speed bipolar sensing circuits for 3.3 V supply voltage.

This process was used to fabricate a 4 Mb SRAM in 0.55 μm BiCMOS technology with

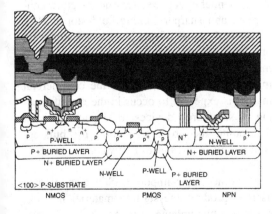

Figure 2-13. Schematic cross-section of the 0.5 μm BiCMOS technology. (From [10], with permission of IEEE.)

quadruple-level polysilicon, double-level metal, p-substrate, and triple-well. The first-level polysilicon layer was used for the MOSFET gate electrodes, the second poly for $V_{EE'}$ (-3.3 V), the third poly for high resistive loads in memory cells, and the fourth level poly for bipolar emitter electrodes. In memory cell arrays, double-level metal layers were used for bit lines (first level) and word lines (second level). The memory cell size is 5.8×3.2 μm^2 and consists of two high resistive loads and four NMOSFETs.

An experimental 16 Mb BiCMOS SRAM has been developed in 0.4 μm process to achieve 220 MHz fully random read–write operations [111]. The memory circuit design techniques include the following: (1) zigzag double word line, (2) centralized bit-line load layout, and (3) phase-locked loop (PLL) with a multistage-tapped (MST) voltage-controlled oscillator (VCO). This VCO consists of a voltage–current converter and a nine-stage current-limited CMOS inverter ring oscillator. The internal SRAM timing pulses are generated from the output signals at each stage and are proportional to the clock cycle. Since all the timing margins can be expanded by making the clock cycle time longer, this technique allows more flexibility for high-frequency timing designs. The PLL is used as an internal clock generator to achieve a clock frequency above 200 MHz.

In this 16 Mb BiCMOS memory design, the internal SRAM macro is made fully static (asynchronous) by using the bipolar sensing circuits which operate at high speed with only small bit-line voltage swing and do not require clocked equalization that is used in the CMOS sensing operations. It uses the wave pipeline technique, which implies that at a certain time period, two signals (waves) exist simultaneously on the data path. The bit-line delay is minimized by optimizing the location of bit-line load (BL load) PMOSFETs. To reduce the time for write to the same as that for the read operation, postwrite equalizer (PW Eq) PMOSFETs are distributed to three positions on a bit-line pair. A write cycle time of 4.5 ns is achieved using this layout.

2.2.5 Silicon-on-Insulator (SOI) Technology

The MOS and bipolar technologies discussed so far are referred to as the bulk silicon technologies. Traditionally, dielectrically isolated (DI) processing has been used to reduce the charge generation in the substrate and to obtain radiation hardness for military and space applications. This fabrication technique is particularly suited to bipolar devices which need a thick active layer. Since the 1960s, a significant amount of work has been done using sapphire for both substrate and insulator; this technology was called silicon-on-sapphire (SOS), and is a restricted version of the SOI technology. The enhanced device performance (increase in speed) of SOI device structures results from the reduced parasitic substrate capacitance, while the improved radiation hardness results from the absence of charge generation in the silicon substrate.

In SOI technology, small islands of silicon are formed on an insulator film. Using SOI fabrication processes coupled with lateral isolation techniques, completely latchup free circuits can be fabricated because there are no parasitic p-n-p-n paths. Also, the SOI films can be made extremely thin, resulting in fully depleted MOSFETs which have been shown to operate at very high speeds (2 GHz) [12]. Thin-film SOI/CMOS devices and circuits have reduced short channel effects, reduced hot electron generation, and sharper subthreshold slopes [13]. However, because of relatively low wafer yield problems, the major market for SOI chips is in military and aerospace applications. The real impact of SOI technology on the commercial market is expected to occur in the manufacturing of submicron devices (0.25 μm) since, at that point, the production costs for bulk/epi wafers will be greater or equivalent to the SOI wafers.

Some of the important SOI technologies used to produce the starting material include: separation by implanted oxygen (SIMOX), zone melt recrystallization (ZMR), full isolation by porous oxidized silicon (FIPOS), epitaxial lateral overgrowth (ELO), wafer bonding and etchback (a modified DI process), and silicon-on-sapphire (SOS) process.

SIMOX is the most commonly used technique, and requires a high dose of oxygen (O^-) ions, which provides the minimum concentration necessary to form a continuous layer of SiO_2. The energy of the implant must be high enough so that the peak of the implant is sufficiently deep within the silicon (0.3–0.5 μm). The wafer is normally heated to about 400°C during the implantation process to ensure that the surface maintains its crystallinity during the high implantation dose. A postimplant anneal is performed in neutral ambient (N_2) or He for a sufficient time (3–5 h) at a high temperature (1100–1175°C) to form a buried layer of SiO_2. This anneal step also allows excess oxygen to out-diffuse, thereby increasing the dielectric strength of the buried-oxide layer. After the annealing step, the crystalline silicon surface layer is typically only 100–300 nm thick [14]. Therefore, an additional layer of epitaxial silicon is deposited so that single-crystal regions (0.5 μm or greater) are available for fabrication.

A simplified cross-section of a thin-film SOI CMOS inverter is shown in Figure 2-14(a) [15]. A comparison was made between the device and the circuit performance of 0.5 μm CMOS on an ultra-thin SOI ring oscillator with 0.4 μm bulk CMOS devices. Figure 2-14(b) shows smaller

delays (or higher operating speeds) per stage for the SOI devices relative to bulk CMOS [16].

SOI 64 kb SRAM fully functional prototypes have been demonstrated by some companies. The SOI SRAMs with 256 kb and higher densities are under development. SIMOX is attractive for military and space electronic systems requiring high-speed, low-power, radiation-hardened devices that can operate reliably over a wide temperature range. It is a good candidate for power IC and telecommunications applications which must integrate low-voltage control circuits with the high-voltage devices [14]. This technology is well suited for integrating the bipolar and CMOS devices on the same chip. The use of wafers on which no epitaxial layer is deposited following oxygen implantation and annealing (referred to as thin-film SIMOX) has also been studied. Thin-film SIMOX processing is simpler than a bulk process because there is no need to form doped wells, and shallow junctions are automatically formed when the doping impurity reaches the underlying buried oxide. However, the SIMOX process does have a number of disadvantages, including the following:

- The process requires the availability of special oxygen implanters.
- The thickness of the buried oxide layer is limited to about 0.5 μm, whereas for

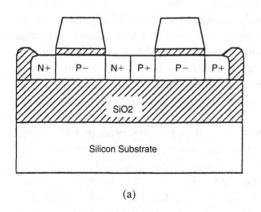

(a)

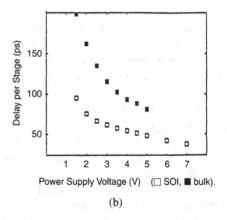

(b)

Figure 2-14. (a) Cross-section of a thin-film SOI CMOS inverter [15]. (b) Delay for bulk CMOS and SIMOX for a ring oscillator [16]. (From [15] and [16], with permission of IEEE.)

some high-voltage applications, a greater thickness may be required.

• The SOI surface silicon film and the deposited epitaxial layer may contain some residual damage from the implant process. The implantation parameters and annealing schedules must also be appropriately chosen to provide optimum IC performance because the microstructure of the surface silicon film is very sensitive to the oxygen dose and the postoxygen-implant annealing temperature.

Due to the floating substrate, SOI MOSFETs show a dip in their I–V characteristics, generally known as "kink effect." However, in the SOI RAMs, this kink effect has been eliminated by electrically shorting the source and substrate. In the zone-melting recrystallization (ZMR) process, which is currently in evolution, the basic steps consist of oxidation and poly-Si deposition, followed by a zone-melting technique to recrystallize the polysilicon. The melting of the polysilicon layer can be accomplished by lasers, electron beams, arc lamps, or strip heaters. The ZMR fabrication equipment is relatively inexpensive compared to that for the SIMOX process. The ZMR defect densities are being improved through seeding, the use of encapsulation layers, and other process-enhancement techniques [13].

The silicon-on-sapphire (SOS) technology has been extensively used for military and space applications. In SOS, the active device is built upon a silicon island on a sapphire substrate. Thus, all n-channel transistors are completely isolated from the p-channel transistors by the sapphire substrate. There are no parasitic transistors to latch up in the SOS, and the capacitance associated with the reverse-biased substrate junctions is eliminated. This gives the SOS process an important advantage over the bulk CMOS in terms of switching speed [17]. The sapphire substrate provides inherent radiation hardness against the alpha-particle-induced soft errors; ionizing total-dose, transient, and neutron radiation; and cosmic ray upsets. These

radiation effects and circuit-hardening techniques are discussed in more detail in Chapter 7.

The early CMOS/SOS logic devices such as the gates, flip-flops, buffers, and small counters were fabricated with an aluminum-gate process on n-type silicon islands, where the NMOS and PMOS transistors were formed and interconnected. The gate aligned over the channel after formation of the channel region overlapped both the source and drain regions to assure a total channel coverage. However, this overlap created gate-to-source and gate-to-drain capacitance, which increased the switching delays. This Al-gate process was subsequently replaced by the silicon-gate technology, which offered improved performance. In the silicon-gate process, the channel oxide and the polysilicon gate structure are defined prior to the source and drain. The polysilicon gate then serves as a mask to define both the source and the drain; thus, the gate is perfectly aligned with no overlap in this "self-aligned silicon-gate" technique. In the following steps, an oxide is deposited (or grown) over the gate, contact windows are defined in the oxide, and aluminum metallization is applied for the interconnections. Figure 2-15 shows the various processing steps for RCA (now Harris Corporation) CMOS/SOS technology [17].

In summary, the key features of a SIMOX-based process are the extended temperature range compared to bulk silicon, lower parasitic capacitance from reduced junction areas, and dielectric isolation which provides latchup immunity. SIMOX is the most commonly used SOI technique suitable for VLSI fabrication, although it requires expensive ion implanters and long implant times. Thomson-CSF Semiconductors uses an SOI-HD process with side and bottom oxides which provide total isolation of the device. Figure 2-16 shows the cross-section of this thin CMOS SOI-HD structure which uses a special gate oxide process to achieve high resistance against various types of radiation, including the neutrons and heavy ions [112]. This technology, available in two versions as 1.2 μm (HS013-HD) and 0.8 μm (HS014-HD) design rules with two and three metal levels, is used in fabricating 16, 64, and 256K SRAMs.

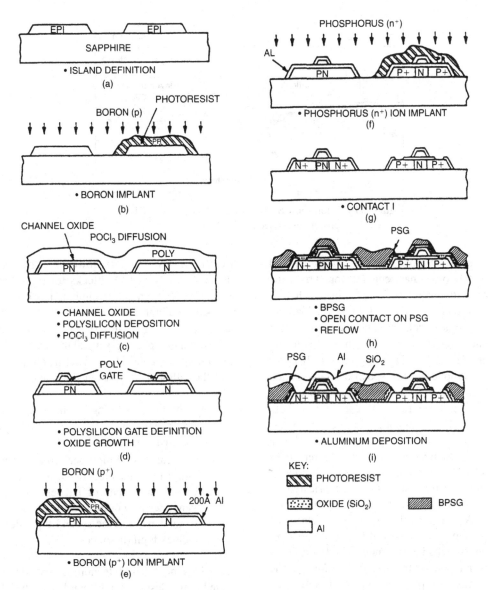

Figure 2-15. Typical RCA CMOS/SOS processing steps. (From *RCA Databook* [17], with permission of Harris Corporation.)

As bulk CMOS technology advances toward sub-0.25 μm channel length, the SOI technologies appear promising because of their reduced parasitic capacitance, lower threshold voltage and body-charging effects. In general, substantial advantages are being projected for the SOI memory technology, especially in the military and space environment, because of a substantial reduction in the soft-error rate and latchup elimination. The SOI MOSFETs can be made fully depleted, allowing only unidirectional current flow in the circuit. This virtually eliminates the kink effect normally associated with the SOS/SOI MOSFETs because of the floating substrate. The hot electron effect which causes threshold instabilities in the MOS devices through charge injection in the gate oxide is also reduced in the fully depleted SOI.

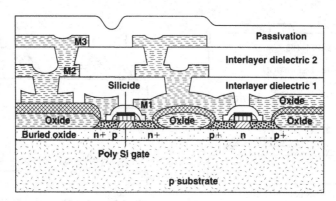

Figure 2-16. Thomson-CSF Semiconductors CMOS SOI-HD
process cross-section [112].

A new low-power and high-stability SOI SRAM technology has been developed by using laser recrystallization techniques in which self-aligned, p-channel SOI MOSFETs for load devices in memory cells are stacked over bottom n-channel bulk MOSFETs for both driver and transfer gates in a 3-D configuration [18]. For process simplicity, no specific planarization process is utilized for an intermediate insulating layer between the top SOI and bottom bulk MOSFETs. This technique increases the on/off ratio of SOI MOSFETs by a factor of 10^4, and reduces the source–drain leakage current by a factor of 10–100 compared with that for polysilicon thin-film transistors (TFTs) fabricated by using low-temperature regrowth of amorphous silicon. A test 256 kb SRAM was fabricated using this process with only memory cell regions in a stacked 3-D structure and peripheral circuits in conventional CMOS technology. The total number of process steps is roughly the same as that for polysilicon TFT load SRAMs. The process steps can be reduced to 83% of those for the TFT load SRAMs if both the peripheral circuit and memory cells are made with p-channel SOI and n-channel bulk MOSFETs.

In recent years, several circuits with operating frequencies above 400 MHz have been designed in the industry standard CMOS processes. Typically, these designs are highly pipelined, i.e., they use a clocking strategy with a single clock net. Also, these designs use careful device sizing and their logical blocks have low fan-in, low fan-out. However, it has been difficult to design RAM blocks for these high-speed CMOS circuits. A technique has been developed for a high-speed CMOS RAM structure based on a six-transistor (6-T) static memory cell, which is capable of several hundred million accesses per second [113]. This technique utilizes the following key design approaches:

- A proper pipelining of the address decoder with an option to include pre-charged logic in the latches.
- Integration of a high-performance sense amplifier in the clocking scheme that includes two types of latches—one that generates new output during a clock low (p-latches), and the second type, during a clock high (n-latches).

This new RAM technique was demonstrated in the design of a 32-word × 64-b (2048 b), three-ported (two read and one write) register file running at 420 MHz, using a standard 1.0 µm process.

2.2.6 Advanced SRAM Architectures and Technologies

2.2.6.1 1–4 Mb SRAM Designs. SRAM speed has usually been enhanced by scaling down the device geometries. The shorter gate channel length, $L(\text{eff})$ translates quite linearly into faster access times. The scaling of the process also results in higher density memories. Since 50–60% area of a SRAM is the memory array, the problem of reducing the chip size directly relates to reducing the cell area. Figure 2-17(a) shows the cell size as a function of SRAM density [4]. Figure 2-17(b) shows the evolution of MOS SRAMs: the shrinking line widths and a variety of process enhancements which have been primarily responsible for the density and performance improvements [1].

The polysilicon load resistors have been commonly used up to 1 Mb SRAM designs; however, their use at 4 Mb and higher densities to achieve both low power and high speed becomes quite challenging. The requirement for a small chip size demands a very small memory cell, which can only be obtained by the scaled polyresistor cell. The standby current requirement sets a minimum value to the load resistance. On the other hand, the resistance must be small enough to provide sufficient current to the storage node so that the charge does not leak away. Another factor in the design of an R-load cell is that the power supply of the memory array in the 0.5–0.6 μm technologies has to be less than 5 V, even though the external chip supply voltage is maintained at 5 V.

Several companies continued to optimize the R-load cell at the 4 Mb SRAM density level, whereas the others looked for an alternative cell which would reduce the drawbacks of the R-load cell but retain its small size. An alternative approach in the 4 Mb generation was the development of a six-transistor cell with four NMOS transistors in the silicon substrate and the two PMOS load transistors formed in the thin-film polysilicon layers above the cell, in the same manner as the resistor loads were stacked in the R-load cell. A 1 Mb SRAM with stacked transistors-in-poly cell in 0.8 μm technology was developed by NEC in 1988 [19]. This SRAM employed the p-channel polysilicon resistors as cell load devices with offset gate–drain structure in the second-level poly with the gate below it in the first-level poly. It was stacked on the n-channel driver transistors. This vertical cell offered the size benefit of the R-load resistor cell and the noise margin of the six-transistor cell. Subsequently, various process enhancements were made by many companies to reduce the cell size and the standby current.

In 4 Mb and higher density SRAMs with increasing cell density and reduced supply voltage to eliminate hot carrier degradation, it becomes difficult to maintain low standby current without degrading the noise immunity. A

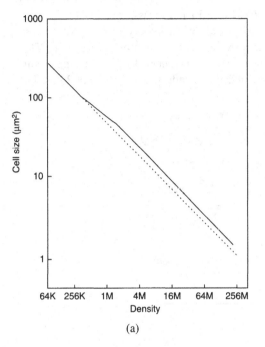

(a)

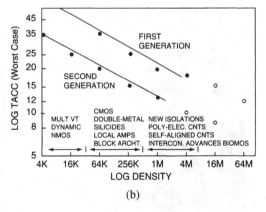

(b)

Figure 2-17. (a) SRAM cell size plotted against density [4] (source: various *ISSC Proceedings,* 1976–1990). (b) Evolution of MOS SRAM technology. (From [1], with permission of IEEE.)

stacked-CMOS SRAM cell with polysilicon thin-film transistor (TFT) load has been developed that leads not only to an increase of the cell static noise margin (SNM), but also to reduction of the cell area [20]. In this cell configuration, the off-current and on-current of the TFT load are the important parameters to reduce standby current and increase noise immunity, respectively. In this approach, an improvement of the electrical characteristics of the TFT load has been achieved by using a special solid phase growth (SPG) technique. The fabrication of this cell is based on three-level polysilicon technology: the first polysilicon

layer is used for the gates of the access and driver NMOS transistors and for the *Vss* line, the second polysilicon layer for the gate of the TFT load, and the third polysilicon layer for the active layer of the TFT load and the *Vcc* line. Figure 2-18 shows the stacked-CMOS SRAM cell (a) schematic cross-section, (b) circuit diagram, and (c) layout [20].

Hitachi has developed a 4 Mb CMOS SRAM in 0.5 μm process, four-level poly, two-level metal, with a polysilicon PMOS load memory cell with an area of 17 μm^2 [21]. This had a standby current of 0.2 μA at a supply voltage of 3 V and address access time of 23 ns. The

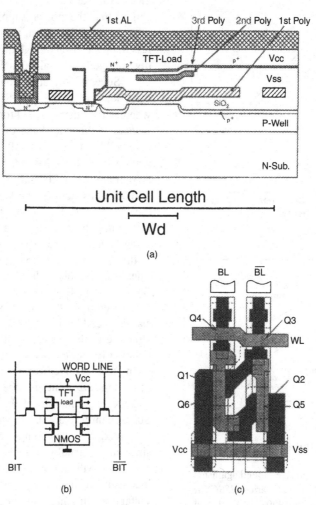

Figure 2-18. A stacked-CMOS SRAM cell. (a) Schematic cross-section. (b) Circuit diagram. (c) Layout. (From [20], with permission of IEEE.)

memory cell has excellent soft-error immunity and low-voltage operating capability. Fast sense amplifier circuits were used to reduce the access time. A noise immune data-latch circuit was used to get a low active current of 5 mA at an operating cycle of 1 μs. A quadruple-array word-decoder architecture was employed in which the word decoders for four memory sub-arrays are laid out in one region. This architec-

ture drastically reduced the word-decoder area, achieving a small chip area of 122 mm².

In 1990, Mitsubishi introduced a 4 Mb CMOS SRAM fabricated with 0.6 μm process technology which utilized quadruple-polysili-con (including polycide), double-metal, twin-well CMOS technology [22]. Figure 2-19(a) shows a memory cell cross-section, as well as the process and device parameters. The memory

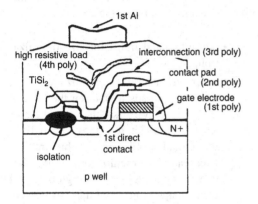

Process and Device Parameters

TECHNOLOGY	0.6μm CMOS
	N-SUB, TWIN-WELL QUADRUPLE POLY-Si DOUBLE METAL
DESIGN RULE	0.6μm (minimum)
TRANSISTOR	L = 0.6μm (NMOS) 0.8μm (PMOS) Tox = 15nm
MEMORY CELL	HIGH RESISTIVE LOAD R = 10TΩ

(a)

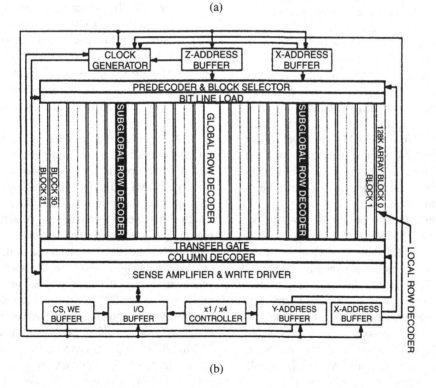

(b)

Figure 2-19. A 4 Mb CMOS SRAM. (a) Memory cell cross-section, process and device parameters. (b) Block diagram. (From [22], with permission of IEEE.)

cell size is $3.5 \times 5.3 \ \mu m^2$. The first polysilicon (polycide) is used for the gate electrode of the transistors. The second polysilicon layer, which is used to form a "contact pad," is electrically isolated from the first polycide and laid over it; this results in cell size compaction. The second polysilicon and the diffusion area are silicided with a self-alignment technique to reduce the sheet resistance. The third polysilicon is used for the interconnection and power supply line of the memory cell. The fourth polysilicon forms the high resistive load of $10 \ T\Omega$ to achieve a low standby current. The bit lines are formed by the first metal with a barrier metal layer. This SRAM has a 20 ns access time at a supply voltage of 3.3 V, achieved by a new word-decoding scheme and high-speed sense amplifiers. It also has a fast address mode which is suitable for cache memory applications. Figure 2-19(b) shows a block diagram of the SRAM.

2.2.6.2 16–64 Mb SRAM Development.
For 16 Mb and higher density SRAMs, improvements in the MOSFET characteristics alone cannot provide the high performance necessary for advanced processing requirements. As the cell size is reduced through scaling, it becomes more difficult to maintain stable operation at a low supply voltage of 3 V. Therefore, it is necessary to incorporate improved process and circuit techniques and optimization of the chip architecture. Currently, the 16 Mb SRAM is in development stages. In the ISSC Conferences during the last few years, several advanced 16 Mb CMOS SRAM designs and architectural developments have been presented.

In 1992, Fujitsu presented development plans on a 16 Mb CMOS SRAM, organized as 4M words $\times$ 4 b [23]. Low power dissipation and access time of 15 ns are obtained by a reduced voltage amplitude data bus connected to a latched cascaded sense amplifier and a current sense amplifier. Double-gate pMOS thin-film-transistor (TFT) load-cell technology reduces the standby current. A "split word-line" memory cell layout pattern, quintuple polysilicon layers, double metal layers, and a 0.4 μm photolithography result in a $2.1 \times 4.15 \ \mu m^2$ memory cell. This SRAM has a 16 b parallel test mode. Figure 2-

20(a) shows a memory cell layout, conventional versus split word-line cell. Figure 2-20(b) shows a block diagram of the 16 Mb CMOS SRAM.

NEC has developed a 3.3 V, 16 Mb CMOS SRAM with an access time of 12 ns and typical standby current of 0.8 μA by using a 0.4 μm quadruple-polysilicon, double-metal technology with thin-film transistor (TFT) load memory cells [24]. The fast access time (12 ns) was achieved through the various design optimization techniques, including:

- An automated transistor size optimization.
- Chip architectural optimization that used center pads to shorten the bit and data-bus delays. Also, the rotated memory cell architecture halves the length of the global word lines. A hierarchical data read-bus structure with a buffer between the subglobal and global read buses helps reduce the data bus delays by 33%. To further reduce the read bus transmission delay, presetting the bus lines to the midlevel called read-bus midlevel preset scheme (RBMIPS) is used.
- This SRAM has three test modes: redundancy row test mode, redundancy column test mode, and 16 b parallel test mode.

A single-bit-line cross-point cell activation (SCPA) architecture has been developed by Mitsubishi to reduce the size of 16 Mb and higher density SRAMs [25]. The supply voltage of this experimental device is 3 V. The column current which constitutes a major portion of active current in the SRAMs refers to the current flowing from the bit line to the storage node of the memory cell through the access transistor when the access transistor is in the on-state. The conventional divided word-line (DWL) structure to reduce the column current requires a large number of block divisions, proportional to the memory density in 16 Mb SRAM chips, whereas the SCPA architecture realizes a smaller column current with a 10% smaller memory core versus the DWL structure. The experimental device with $2.2 \times 3.9 \ \mu m^2$ cell was fabricated using a five-layer polysilicon, sin-

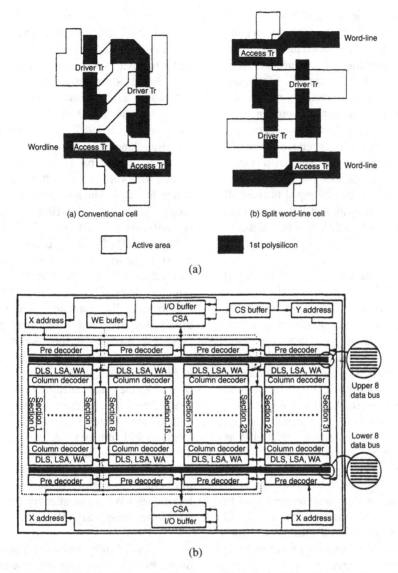

(a)

(b)

Figure 2-20. A 16 Mb CMOS SRAM Fujitsu. (a) Memory cell layout. (b) Block diagram. (From [23], with permission of IEEE.)

gle-aluminum, 0.4 μm CMOS wafer process technology. These are some of the major features of the SCPA architecture:

- A new PMOS precharging boost circuit which is used to boost the word line and activated only during the write cycle.
- A modified data-sensing scheme utilizing a dummy cell since the conventional differential amplifier cannot be used in

the SCPA architecture which has a different memory cell configuration.

A 9 ns 16 Mb CMOS SRAM has been developed by Sony Corporation using a 0.35 μm process with quadruple-polysilicon and double-aluminum technology [26]. This design utilizes a current-mode, fully nonequalized data path by using a stabilized feedback current sense amplifier (SFCA) that provides a small input resistance and an offset compensating effect. The chip is organized as 4M words × 4 b, and the memory cell array is divided into 4 Mb quadrants with 1 b of the 4 b data word residing in each quadrant. Also, each quadrant is further divided into 32 sections of 1024 rows × 128 columns. By placing the sense amplifiers adjacent to the I/O pins and dedicating each quadrant to one data bit, the *RC* delay of the read data path is reduced. It uses a line mode parallel test circuit which is a current mode EXCLUSIVE-OR gate consisting of two SFCAs with reference current equal to one-half of the memory cell current. A parallel access occurs in write and read cycles. As a result, the 16 Mb SRAM is compressed to 256 kb memory in the test mode.

A novel architecture has been developed for 16 Mb CMOS SRAM by Hitachi that enables fast write/read in poly-PMOS load or high-resistance polyload, single-bit-line cell [27]. The write operation involves alternate twin word activation (ATWA) and pulsing of a single bit line. A dummy cell is used to obtain a reference voltage for reading. To obtain the required balance between a normal cell signal line and a dummy cell signal line, a balanced common data-line architecture is used. A novel self-bias control (SBC) amplifier provides excellent stability and fast sensing performance for input voltages close to V_{cc} at a low power supply of 2.5 V. This single-bit-line architecture, used to fabricate a 16 Mb SRAM in 0.25 μm CMOS technology, reduced the cell area to 2.3 μm^2, which is two-thirds of a conventional two-bit-line cell with the same process.

In the last few years, in addition to process enhancements in the planar transistors, several advanced structures have been developed. In 1988, Toshiba proposed the "surrounding gate transistor" (SGT) which had a gate electrode surrounded by a pillar silicon island and reduced the occupied area of the CMOS inverter by 50% [28]. The source, gate, and drain were arranged vertically using the sidewalls of the pillar silicon island as the channel region. The pillar sidewalls were surrounded by the gate electrode, and the gate length was adjusted by the pillar height. This SGT structure poten-

tially reduces serious problems from both the short channel effects (gate length) and the narrow channel effects (gate width) of the planar transistors with submicron geometries.

Some other possibilities for the 64 Mb generations and beyond include combinations of the vertical and stacked transistor and cell technologies. In 1989, Texas Instruments proposed an innovative six-transistor cell with a thin-film polysilicon PMOS transistor and a vertical sidewall trench NMOS transistor in 0.5 μm technology, which resulted in a 23 μm^2 cell area [29]. Since the thin-film transistor may be stacked over the vertical NMOS, this cell has the potential for future scaling. Another interesting combination structure investigated for 0.2 μm technologies was the "delta transistor" proposed by Hitachi in 1989 [30]. This combines the vertical pillar technology with the fully depleted SOI technology using only bulk silicon process techniques. In this process, a pillar of silicon was formed using the standard trench etching technique, and silicon nitride was deposited as the gate dielectric over it. Then beneath the pillars, a selective oxidation of the bulk silicon isolates the silicon of the pillar and forms an SOI vertical transistor above the oxide. This process eliminated the need for isolation, permitting the larger gate length and width for the vertical transistor. These vertical stacking techniques have a potential for future 256 Mb and higher density SRAMs.

2.2.6.3 Gallium Arsenide (GaAs) SRAMs. Some experimental gallium arsenide (GaAs) SRAMs have been developed for very high-speed applications that have 2–3 ns cycle time and densities of 1–16 kb. However, there are a number of technological and processing problems with this technology. The normal processing is more difficult for a GaAs substrate because, unlike SiO_2, As_2O_3 is conductive, and Ga and As tend to separate above 650 °C. The cost of GaAs wafers is also higher than the silicon wafers, the wafer size is smaller, and yield is low. Also, the voltages that could be supported in a sub-5 ns GaAs SRAM are much lower than the standard interface circuitry, and thus special design considerations are required to interface the RAM to an external ECL system environment.

A schematic cross-section of a GaAs memory cell is shown in Figure 2-21 [31]. This is fabricated with 1.0 μm tungsten self-aligned gate field-effect transistors using double-level metal interconnect technology. The ohmic metal is AuGe–NiAu and the interconnect metals are Ti–Au. The dielectric insulator is phosphorous-doped CVD deposited Si_3N_4. An ion implant was used to form the source and the drain regions.

The process improvements underway in the GaAs technology including 0.3–0.5 μm gates may permit the development of 64 kb SRAMs with access times near 1 ns. These ultra-high-speed devices can be used in cache applications for the next generation supercomputers and supercomputing workstations.

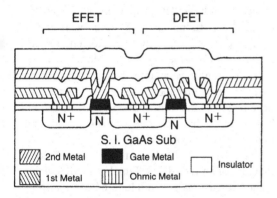

Figure 2-21. Schematic cross-section of a GaAs memory cell. (From [31], with permission of IEEE.)

2.2.7 Application-Specific SRAMs

Application-specific SRAMs include some extra logic circuitry added to make them compatible for a specific task. There is no clear-cut distinction between the application-specific SRAMs and the SRAM arrays embedded in the logic. Usually, the application-specific SRAMs are made in the high-density, optimized processes which include customized features such as the buried contacts and straps to reduce the memory cell size. A few application-specific SRAMs are described in the following subsections.

2.2.7.1 Serially Accessed Memory (Line Buffers). The first-in first-out (FIFO) buffer is an example of the shift register memory architecture through which the data are transferred in and out serially. The serial or sequential architecture requires that the memory cell containing the information be arranged so as to receive and retrieve one bit at a time. The data stored in each cell are retrieved strictly in accordance with their position in the sequence. The data may be retrieved on a first-in first-out (FIFO), first-in last-out (FILO), or last-in first-out (LIFO) basis. The FIFOs are widely used in telecommunication types of applications. These are generally made using the SRAM cells if there is a requirement for the data to be maintained in the FIFO, and from the DRAM cells if the data simply pass through the FIFO. The data are transferred from cell to cell by the sequential operation of the clock.

Many FIFOs currently available are RAM-based designs with self-incrementing internal read and write pointers, which results in very low fall-through times compared to the older shift-register-based designs. The fall-through time of a FIFO is the time elapsing between the end of the first write to the FIFO and the time the first read may begin. Also, the serial registers are not shift registers, but bit wide memory arrays with self-incrementing pointers. These FIFOs can be expanded in any depth to any level by cascading the multiple devices. The FIFOs can range in depth from 64 × 4 b and 64 × 5 b to 16K × 9 b and higher devices.

Figure 2-22 shows the functional block diagram of an IDT 2K × 9 b parallel CMOS FIFO [24]. It is a dual-port memory that loads and empties data on a first-in first-out basis. This device uses Full and Empty flags to prevent the data overflow and underflow, and expansion logic to allow for unlimited expansion capability in both the word size and depth. The reads and writes are internally sequential through the use of ring pointers, with no address

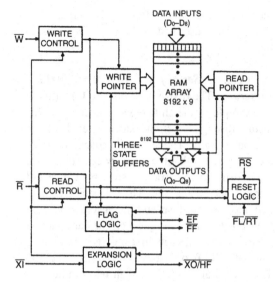

Figure 2-22. Functional block diagram of IDT7205 FIFO (2K × 9 b). (From [32], with permission of Integrated Device Technology, Inc.)

information required to load and unload the data. It is toggled in and out of the device using Write (*W*) and Read (*R*) pins. It utilizes a 9 b wide data array to allow for control and parity bits at the user's option. It also features a retransmit (*RT*) capability that allows for the reset of the read pointer to its initial position. A Half-Full flag is available in the single device and width expansion modes. The device has a read/write cycle time of 30 ns (33 MHz).

2.2.7.2 Dual-Port RAMs.

The dual-port RAMS allow two independent devices to have simultaneous read and write access to the same memory. This allows the two devices to communicate with each other by passing data through the common memory. These devices might be a CPU and a disk controller or two CPUs working on different but related tasks. The dual-port memory approach is useful since it allows the same memory to be used for both working storage and communication by both devices without using any special data communication hardware. A dual-port memory has sets of addresses, data, and read/write control signals, each of which accesses the same set of

memory cells. Each set of memory controls can independently and simultaneously access any word in the memory, including the case where both sides are accessing the same location at the same time.

Figure 2-23(a) shows an example of a dual-port memory cell [32]. In this cell, both the left- and right-hand select lines can independently and simultaneously select the cell for a readout. In addition, either side can write data into the cell independently of the other side. The problem occurs when both sides try to write into the same cell at the same time, and the result can be a random combination of both data words rather than the data word from one side. This problem is solved by the hardware arbitration Busy Logic design by detecting when both sides are using the same location at the same time, and then causing one side to wait until the other side is done. It consists of a common address detection logic and a cross-coupled arbitration latch. A common problem in dual-port RAMs is the need to temporarily assign a block of memory to one side. Some devices currently available have "semaphoric logic," which provides a set of flags specially designed for the block assignment function. This ensures that only one side has permission to use that block of memory.

Figure 2-23(b) shows the functional block diagram of an IDT 8K × 8 dual-port SRAM designed for use as a standalone 64 kb dual-port RAM or as a combination MASTER/SLAVE for 16 b or more word systems [32]. This device provides two independent ports with separate control, address, and I/O pins, which permit asynchronous access for reads or writes to any location in the memory. These are rated for a maximum access time of 35 ns.

A family of modular multiport SRAMs with a built-in self-test (BIST) interface has been developed using a synchronous self-timed architecture. The basic port design is self-contained and extensible to any number of ports sharing access to a common-core cell array. The same design was used to implement modular one-port, two-port, and four-port SRAMs and a one-port DRAM based on a four-transistor (4-T)

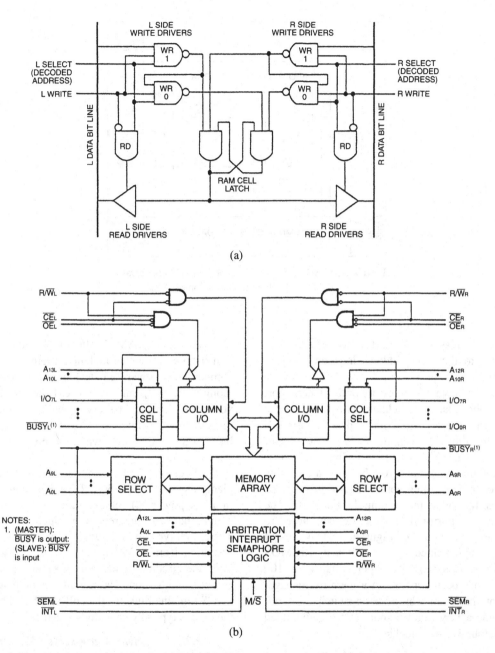

Figure 2-23. (a) A dual-port RAM cell. (b) Functional block diagram of IDT7005 dual-port SRAM. (From [32], with permission of Integrated Device Technology, Inc.)

cell. Figure 2-24 shows the schematic of a four-port SRAM core array as an example [33]. One access transistor pair is added per core cell to provide fully independent read/write capability for each port. This provides symmetry in design, layout, simulation, and BIST circuitry. Each port functions independently with its own control block and model signal path. Also, each core cell is provided with adequate margin for stability to permit simultaneous reads from the

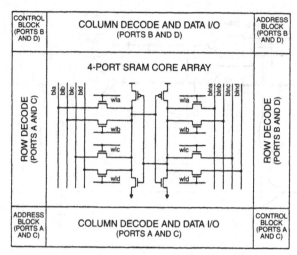

Figure 2-24. Schematic of a four-port SRAM core array. (From [34], with permission of IEEE.)

same address on all ports. No simultaneous writes to the same address location are permitted.

The layout architecture of the generic port periphery places the address block in one corner, with the row and column decoders extending from this block along the adjacent sides of core cell array. The data I/O circuitry is integrated into the same layout block as the column decoder. The timing control circuitry is located at the far end of the column decode block. The ports are mirrored around the cell array to achieve multiport configuration. This design was implemented in a 0.8 μm BiCMOS process. A system level performance of 100 MHz has been verified on a memory management chip from a broad-band switch core containing a 12K × 8 two-port SRAM subdivided into four 3K × 8 blocks.

2.2.7.3 Nonvolatile SRAMs The battery-operated power systems require not only very low operating power and battery-backup data retention current levels from the memory, but also logic circuitry which will protect the stored data if the battery voltage gets low. This logic circuitry also contains protection against accidental writes during power-down. In the early 1980s, Mostek (later SGS-Thomson Microelectronics) introduced byte-wide SRAMs called

the "zero power RAMs," which included a lithium battery. They also included a temperature-compensated power fail detection circuitry to monitor the V_{dd}, and switched from the line voltage to back-up battery when the voltage drops below a specified level, and then switched back to the line voltage when the power is restored.

Figure 2-25 shows the functional block diagram of an IDT 71256 (32K × 8 b) available in a low-power (L) version which also offers a battery backup data retention capability [32]. The circuit typically consumes only 5 μW when operating from a 2 V battery. When $\overline{CS}$ goes high, the circuit automatically goes into a low-power standby mode and remains there as long as $\overline{CS}$ remains high. In the full standby mode, the device typically consumes less than 15 μW.

2.2.7.4 Content-Addressable Memories (CAMs). The content-addressable memories (CAMs) are designed and used both as embedded modules on the larger VLSI chips and as a standalone memory for specific systems applications. Unlike the standard memories which associate data with an address, the CAM associates an address with data. The data are presented to the inputs of CAM which searches for a match to those data in the CAM, regardless of the address. When a match is found, the CAM

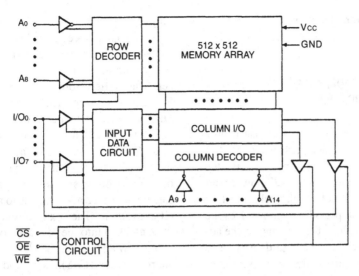

Figure 2-25. Functional block diagram of IDT71256 (32K × 8 b). (From [32], with permission of Integrated Device Technology, Inc.)

identifies the address location of the data. The applications using CAMs include database management, disk caching, pattern and image recognition, and artificial intelligence.

Figure 2-26 shows the block diagram of Advanced MicroDevices AM99C10A, a CMOS CAM with a capacity of 256 words (each consisting of a 48 b comparator and a 48 b register)

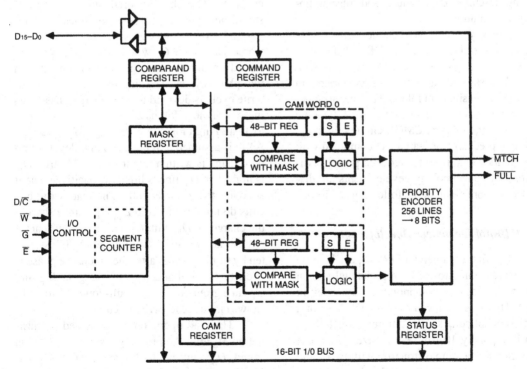

Figure 2-26. A block diagram of Advanced MicroDevices AM99C10A CMOS CAM. (From [35], with permission of Advanced MicroDevices, Inc.)

and a user-programmable word width of 16 or 48 b [35].

2.3 DYNAMIC RANDOM ACCESS MEMORIES (DRAMs)

Dynamic random access memories use charge storage on a capacitor to represent binary data values. These are called dynamic because the stored charge leaks away, even with power continuously applied. Therefore, the cells must be read and refreshed at periodic intervals. Despite this apparently complex operating mode, the advantages of cost per bit and high density have made DRAMs the most widely used semiconductor memories in commercial applications. In the early 1970s, the DRAM cell designs ranging from a six-transistor to a one-transistor configuration were proposed. In 1973, the one-transistor (1-T) cell became standard for the 4K DRAMs [36]. In the following years, the density of DRAMs increased exponentially with rapid improvement in the cell design, its supporting circuit technologies, and photolithographic techniques.

The key circuit technologies which have contributed to the progress in DRAM development are advanced vertical cell structures with trench and stacked capacitors; improvements in differential sensing and folded data-line arrangement for noise reduction; and the use of dynamic amplifiers and drivers, CMOS circuits, half V_{dd} data-line precharge, a shared I/O combined with multidivided data lines for reduced power dissipation. These technology developments are discussed in more detail in the following sections.

2.3.1 DRAM Technology Development

The first commercial DRAM, a 1 kb chip using three transistor cells in p-channel silicon gate technology, was announced by Intel in 1970. This was followed by 4 kb DRAMs in a single-polysilicon, single-aluminum metallization in typically 10 μm feature size. The sense amplifier was a simple structure with two static inverters and two switches with a bootstrapped output stage which resulted in fast charging of the load capacitances. It had nonmultiplexed addressing. A shift register generated the three regulated timing signals needed to operate the sense amplifier from a single external clock. Subsequently, the first address-multiplexed DRAMs were introduced, also at the 4 kb density level. With time-multiplexed addressing, these 4 kb DRAMs used six address pins per device to select one of the 64 rows, followed by one of 64 columns to select a particular bit instead of using 12 address pins to access one specific location. The six multiplexed addresses were clocked into the row and column decoders by the row address strobe ($\overline{RAS}$) and the column address strobe ($\overline{CAS}$). Input and output latches were also used, and the data to be written into a selected storage cell were first stored in an on-chip latch. This simplified the memory system design.

The first generation DRAMs were manufactured using a three-transistor (3-T) cell as shown in Figure 2-27(a) [37]. The transistors M_2, M_3, and M_4 are made small to minimize the cell area. The charge stored on C_1 represents stored binary data. The selection lines for reading and writing must be separated because the stored charge on C_1 would be lost if M_3 were turned on during reading. The cell operates in two-phase cycles. The first half of each read or write cycle is devoted to a precharge phase during which the columns D_{in} and D_{out} are charged to a valid high logic level through M_{y1} and M_{y2}. A "1" is assumed to represent a high level stored on C_1, and is written by turning on M_3 after D_{in} is high. The D_{in} line is highly capacitive since it is connected to many cells. The read–write circuits do not need to hold D_{in} high since sharing the charge on D_{in} with C_1 does not significantly reduce the precharged high level. A "0" is written by turning on M_3 after the precharge phase is over, then simultaneously discharging D_{in} and C_1 via a grounded-source pull-down device (not shown) in the read–write circuit.

The read operation is performed by turning on M_4 after the precharge is over. If a "1" is stored, D_{out} will be discharged through M_2 and M_4. If a "0" is stored, there will be no conducting

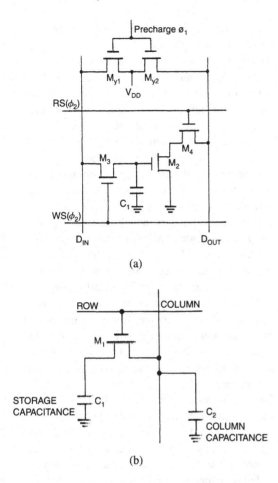

(a)

(b)

Figure 2-27. (a) Three-transistor (3-T) DRAM cell. (From [37], with permission of McGraw-Hill Inc., New York.) (b) One-transistor (1-T) DRAM cell.

current drain would be required for D_{out} recovery to a high level after reading a "1." Similarly, fast changes of logic states on D_{in} would require excessive power if static drivers were used. For this 3-T cell, the output data are inverted compared to the input data. Therefore, to have memory data input and output the same logic polarity, an extra inversion is included in either the read or the write data path. There is some cost to using dynamic techniques arising from the need to provide several clocking and timing signals. However, for large memories, the cost per bit is lowest for the dynamic designs.

A significant improvement in the DRAM evolution was the switch from three-transistor designs to one-transistor (1-T) cell design, as shown in Figure 2-27(b). This cell has been used to fabricate dynamic memories to 1 Mb densities and higher with some variations on the basic configuration, such as the number of polysilicon layers, methods of capacitor formation, different materials used for the word and bit lines, etc. The storage capacitor C_1 is made very small, and consists of a polysilicon plate over a thin oxide film, with the semiconductor region under the oxide serving as the other capacitor plate.

For the 1-T DRAM cell in Figure 2-27(b), the selection for reading or writing is accomplished by turning on M_1 with the single row line. Data are stored as high or low level on C_1. Data are written into the cell by forcing a high or low level on the column when the cell is selected. To understand operation of this 1-T cell, assume that the substrate is grounded and that 5 V power is applied to the polysilicon top plate of the storage capacitor (referred to as the plate electrode of the capacitor as opposed to the storage electrode on which the charge is stored). The semiconductor region under the polysilicon plate serves as the other capacitor electrode, and in the NMOS cell, this p-type region is normally inverted by the 5 V bias. As a result, a layer of electrons is formed at the surface of the semiconductor and a depleted region is created below the surface.

To write a "1" into the cell, 5 V is applied to the bit line and 5 V pulse is simultaneously

path through M_2, so that the precharged high level on D_{out} will not change. The cell may be read repeatedly without disturbing the charge on C_1. The drain junction leakage of M_3 depletes the stored charge in a few milliseconds. Therefore, a refresh operation has to be performed by reading the stored data, inverting the result, and writing back into the same location. This is done simultaneously for all the bits in a row, once every 2–4 ms. The use of dynamic precharge reduces power consumption compared to the static NMOS/CMOS operation. For example, if the D_{out} high level were to be established only through a static pull-up device, a higher average applied to the word line. The access transistor is turned ON by this pulse since its threshold voltage is about 1 V. The source of the access transistor is biased to 5 V since it is connected to the bit line. However, the electrostatic potential of the channel beneath the access-transistor gate and the polysilicon plate of the storage capacitor is less than 5 V because some of the applied voltage is dropped across the gate oxide. As a result, any electrons present in the inversion layer of the storage capacitor will flow to the lower potential region of the source, causing the storage electrode to become the depletion region which is emptied of any inversion-layer charge. When the word-line pulse returns to 0 V, an empty potential well which represents a binary "1" remains under the storage gate.

To write a "0," the bit-line voltage is returned to 0 V, and the word line is again pulsed to 5 V. The access transistor is turned ON, and the electrons from the n^+ source region (whose potential has been returned to 0 V) have access to the empty potential well (whose potential is now lower than that of the source region). Hence, the electrons from the source region move to fill it, thus restoring the inversion layer beneath the poly plate. As the word-line voltage is returned to zero, the inversion-layer charge present on the storage capacitor is isolated beneath the storage gate. This condition represents a stored binary "0."

It should be noted that for a stored "1," an empty potential well exists, which is not an equilibrium condition since the electrons that should be present in the inversion layer have been removed. As a result, thermally generated electrons within and nearby the depletion region surrounding the well will move to recreate the inversion layer, so that a stored "1" gradually becomes a "0" as electrons refill the empty well. To prevent this charge loss, each cell must be periodically refreshed so that the correct data remain stored at each bit location. The time interval between the refresh cycles is called the "refresh time." The 16 kb DRAMs retained the read operation refresh common to all single-transistor cell memories, which means that a read operation always refreshes the data in the cell being read. It also had $\overline{RAS}$ only refresh, which became standard on the address-multiplexed DRAMs. This operation permitted refresh using only the $\overline{RAS}$ clock without activating the $\overline{CAS}$ clock. As a result, any sequence in which all row addresses were selected was sufficient to refresh the entire memory.

The development of 16 kb DRAMs was followed by scaling of the basic cell to smaller geometries and developing the operational controls needed for these smaller cells. Also, a significant amount of work was done on sense amplifier architectures in making them sensitive enough to support the smaller amount of signal charge stored in the capacitor. It should be noted that a sense amplifier is an extremely sensitive comparator, and its design is critically important to the success of DRAM manufacture. Figure 2-28(a) shows the basic one-transistor storage cell structure with reference and a cross-coupled latch sense amplifier, and (b) shows the associated timing required for its operation. This was the refresh structure used in the first 16 kb DRAMs introduced in 1976 by Intel [38]. Three clock signals are required for the operation of this cell, one of which generates the other two on the chip. In order to read and restore the data stored in the cell, the bit lines are normally divided on either side of the centrally located and balanced differential amplifier.

During the sense amplifiers' evolutionary phase, the idea of a "dummy cell" was conceived to aid in sensing the low data signal from the cell by providing a more sensitive reference level [39]. In this approach, only one memory cell is accessed along with a dummy cell on the opposite side of the latch, with the dummy cell storing a voltage midway between the two voltage levels allowed in the memory cell, so that the latch is unbalanced in opposite directions, depending on the voltage stored in the memory cell. For example, in the sense amplifier arrangement shown in Figure 2-28(c), the bit and $\overline{bit}$ lines are either shorted together, or provided a common reference level before attempting to read the cell [40]. During this time, the reference voltage level is written into the dummy cell structures on either side of the

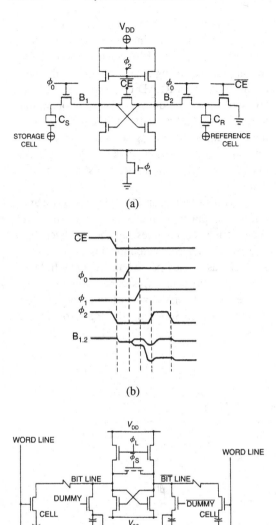

(a)

(b)

(c)

Figure 2-28. (a) Basic one-transistor storage cell with cross-coupled latch sense amplifier [38]. (b) The associated timing diagram [38]. (c) A DRAM differential sense amplifier showing dummy cell structure [40]. (From [38] and [40], with permission of IEEE.)

inverter. When the memory is selected by the select clock, the dummy cells are unshorted. As the word line is selected, the data from a storage cell are dumped to one side of the sense ampli-

fier, which has a reference level from the dummy cell provided on the other side. This causes a voltage differential to be set up across the two nodes of the sense amplifier.

The load clock now turns on the two load transistors, causing both nodes to rise together until the higher side starts to turn on the opposite transistor. This regenerative action then results in the amplification of the signal to full logic levels and the restoration of a logic "0" or "1" back into the storage cell. During the resettling period, both bit and $\overline{\text{bit}}$ are precharged by the clock to the applied power supply voltage level. Such a sense amplifier can provide very fast signal detection since it is sensitive to low-voltage difference signals.

Further improvements were made in the sense amplifier designs which could compensate for differences in threshold voltages between the two cross-coupled latch devices such that smaller sense signals could be used. This was accomplished in double cross-coupled sense amplifiers used in IBM's first 64 kb RAM [41]. As in the case of 16 kb DRAMs, for scaling to 64 kb density level architecture, there were two options available. One of them used two blocks of 128 rows × 256 columns, such that within a block, there were 64 cells on a bit line on each side of a row of 256 sense amplifiers [42]. Both blocks of 128 rows could be refreshed simultaneously in 2 ms, thereby retaining compatibility with the 16 kb DRAMs that had 2 ms refresh time. The second option used an architecture with one block of 256 rows × 256 columns having 128 cells on a bit line on each side of 256 sense amplifiers [43]. Since the cycle time was the same on both parts, it took 4 ms to refresh the 256 rows. However, in the second scheme, the number of sense amplifiers is halved, thereby reducing the power dissipation.

In the early RAMs, the word-line noise was a common cause of pattern sensitivity bit failures, and the sources of noise included: imbalance of dynamic sense amplifiers from process variations, array noise caused by capacitive effects between the bit line and word line during row access, and the noise from peripheral circuitry. The open bit lines use a row of sense amplifiers in the middle and memory arrays on each side. In this layout, the noise injected will not be balanced since the bit lines go through the column decoder area where a great deal of coupling normally takes place. As a result, an error could be introduced in the cell due to wrong sensing. One solution commonly used was a grounded-word-line technique in which an additional dc low-conductance path to ground was applied to all word lines. This had some disadvantages. A better approach was developed for the 64 kb DRAMs with the "folded bit-line" arrangement in which the column decoders are at one end of the array and sense amplifiers at the other [44]. A single pair of bit lines is folded back on itself between the two, so that any fluctuations on the column decoders will appear as common mode noise on the line and will not be transferred to the differential signal on the bit lines.

Figure 2-29(a) shows the schematic for a sense-refresh amplifier implemented in an NSC/Fairchild 64 kb DRAM, which combined the two approaches using an architecture with open bit lines during a READ operation and folded-"twisted" bit lines during a WRITE operation [45]. Figure 2-29(b) shows the associated timing diagram and operating sequence. This sense-refresh amplifier was capable of detecting a 30 mV differential signal under worst case timing conditions.

In this scheme, an ultrasensitive sense amplifier was placed in the center of two pairs of cross-connected bit lines, as shown. Each bit-line section was connected to 64 cells instead of 128 cells as in the open-bit-line approach. During a read, write, or refresh operation, only one of the two bit-line pairs was initially selected during sensing. Subsequently, the two bit-line

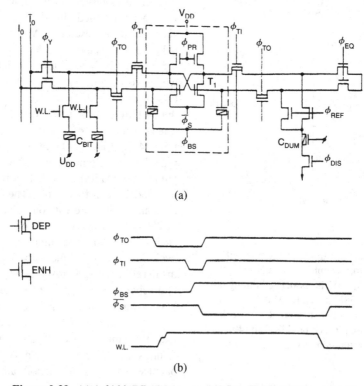

(a)

(b)

Figure 2-29. (a) A 64 kb DRAM sense amplifier circuit showing combination open and folded bit lines. (b) The associated timing diagram and operating sequence. (From [45], with permission of IEEE.)

pairs were reconnected on a single-folded metal bit line. A full size dummy capacitor C_{DUM} is used by taking a "half charge" from the two bit lines. The circuit contained preamplification capacitors to restore full V_{dd} level, and depletion mode devices were used to isolate nonselected bit line halves during the sensing operation. A bit-line equalizing device is used between the folded-"twisted" bit-line pair to ensure bit-line balance with a minimum precharge time. This scheme was effective in noise reduction for 64K DRAMs, but it was not very useful at the next higher density level of 256K.

The first redundancy scheme was developed to enhance the memory yield through error-correction circuitry. The single-bit error correction using redundant bits at the system level worked fairly well as long as the defects were single-cell failures [46]. However, it appeared that some defects could produce open or short circuits on the word or bit lines, which could cause a large number of cells to malfunction. Therefore, a redundancy scheme was later developed to add some extra word and bit lines on chip to replace the faulty ones. This memory design for redundancy and fault tolerance will be discussed in more detail in Chapter 5.

The first generation 256K NMOS technology ranged from the experimental 1.0 μm design rules to the 2.5 μm parts built on the 64K DRAM lines. The slightly larger chips encouraged the development and use of redundancy techniques for yield enhancements. Several of these DRAMs used new operational modes such as "nibble" and "$\overline{CAS}$-before-$\overline{RAS}$" that were developed for 64K DRAMs. The nibble mode allows up to four bits of data available to read serially in a continuous stream before another full access must be made. This enables several parts to be interleaved to give a continuous stream of data at the fast nibble rate. In the $\overline{CAS}$-before-$\overline{RAS}$ refresh mode, the refresh address is generated internally. It is triggered by activating the column address strobe ($\overline{CAS}$) prior to the row address strobe ($\overline{RAS}$). It made DRAMs more convenient to use in systems with a smaller number of memories.

2.3.2 CMOS DRAMs

With the introduction of 256 kb dense memories, the DRAM circuit designs began to change from NMOS to a mixed NMOS/CMOS technology. The cells of the mixed-technology memory array are all built in a common well of a CMOS wafer. The access transistor and storage capacitor of each cell are usually fabricated using the NMOS technology, whereas the peripheral circuits are designed and fabricated in CMOS. The advantages of CMOS designs over the pure NMOS include lower power dissipation, reduction in noise sensitivity, and smaller soft error rates (see Section 2.3.7). One result of the reduced sensitivity of the cell transistor to noise is that a shorter effective channel length can be used in the transfer gate. The CMOS designs can improve refresh characteristics over those of the NMOS DRAMs which tend to be limited by thermally generated carriers from defect sites in the substrate within a diffusion length of the memory cells. Also, a circuit technique known as "static column decoding" which significantly reduces the memory access time can be successfully implemented in the CMOS rather than in the NMOS designs.

A main disadvantage of the earlier CMOS DRAMs was the greater number of mask steps used relative to the NMOS process. Another potential reliability problem was the latchup which can occur as a result of both internal and external noise sources. A substrate bias generator can help prevent latchup during operation by improving the isolation characteristics of parasitic transistors and reducing the possibility of minority carrier injection. As the DRAM cell has been scaled down in size, the minimum amount of stored charge needed to maintain reliable memory operation has remained the same. Therefore, for scaling to higher densities, it is necessary that the cell's storage capacity be increased. Although the planar storage capacitors were used in some 1 Mb designs, stacked and trench capacitors were developed to achieve the DRAM cells with larger capacitance values without increasing the area these cells occupy on the chip surface of the 4–64 Mb designs.

These advanced cell structures will be discussed in more detail in Section 2.3.3.

Table 2-1(a) shows the relative die size trends for 1–64 Mb DRAMs [114]. It shows that although the memory capacity from 1–64 Mb has increased by a factor of 64, the die area has increased only by an estimated 5.2 times, which implies a factor of 12 times improvement in efficiency. However, this improvement in efficiency has been attained at the cost of process complexity going up from six mask alignments (100 process steps) to 24 mask alignments (560 process steps). These product and process complexity trends with feature size are shown in Table 2-1(b).

The increasing length of the bit lines in the higher density DRAMs which resulted in higher capacitance ratios and increasing RC delay times required new sensing architectures. A Mitsubishi 1 Mb DRAM introduced an alternative to the conventional shared sense amplifier, which was referred to as the distributed sense and restore amplifier [47]. This was designed to minimize the access time penalty of the shared sense amplifiers. In this case, the bit line is also divided into two segments: the NMOS sense amplifier and the PMOS restore circuit are connected to each segment. The column decoder and I/O bus lines are shared by eight 64K memory arrays. The sense circuit halves the bit-line length, thus improving the C_b/C_s ratio.

In another approach, the sensing circuitry used bootstrapped word lines to allow a full supply voltage level to be written into the memory cell. In 1984, IBM introduced a half-V_{CC} bit-line sensing scheme in the n-well CMOS DRAMs [48]. This scheme resulted in a reduction of peak current at both sensing and bit-line precharge by almost a factor of two and increased the chip reliability because of the IR drop. Also, the resulting narrower metal lines decreased the parasitic wiring capacitance, thus allowing higher speed. A $V_{CC}/2$ bias on the cell

TABLE 2-1 (a) Relative Die Size Trends for 1–64 Mb DRAMs. (b) DRAM Product and Process Complexity Trends and Feature Size [114].

DRAM Size	Die Size (mm)	Die Area (mm²)	Relative Size
1 Mb	5 × 10	50	1.0
4 Mb	6 × 15	90	1.8
16 Mb	7 × 20	140	2.8
64 Mb	12.7 × 20.4	260	5.2

(a)

DRAM Size	Feature Size (μm)	Process Steps	Mask Steps
16K	4	100	6
64K	3	140	7
256K	2	200	8
1M	1.2	360	15
4M	0.8	400	17
16M	0.5	450	20
64M	0.35	560	24

(b)

plate reduces the electric field stress across the thinner oxide of the megabit generation storage capacitor and minimizes the influence of supply voltage variations. Some 1 Mb DRAM designs used clock circuitry with address transition detection (ATD) for high-speed operation. There was also improvement in test modes to reduce the memory test time. Several types of parallel test modes were developed which allowed mul-

tiple bits (four or eight) to be tested simultaneously instead of one bit at a time.

2.3.2.1 1 Mb DRAM (Example).

Figure 2-30(a) shows the block diagram of an NEC 1 Mb DRAM [49] organized as 1,048,576 words × 1 b and designed to operate from a single +5 V supply. This uses advanced polycide technology with trench capacitors which minimizes

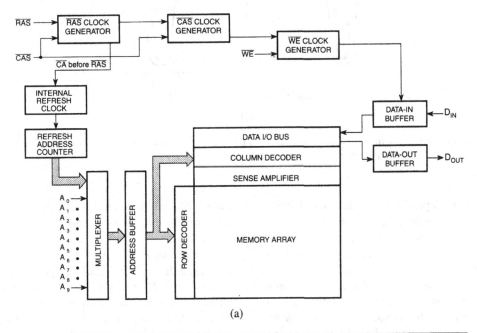

(a)

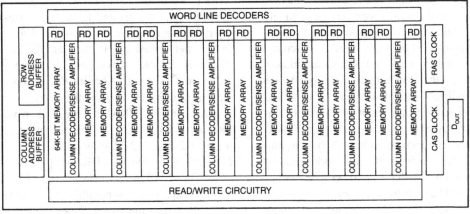

Notes:
[1] The memory is divided into sixteen 64-kbit memory cell arrays.
[2] RD = row decoder/word driver.

(b)

Figure 2-30. (a) Block diagram of 1 Mb DRAM (NEC μPD421000). (b) Chip layout.

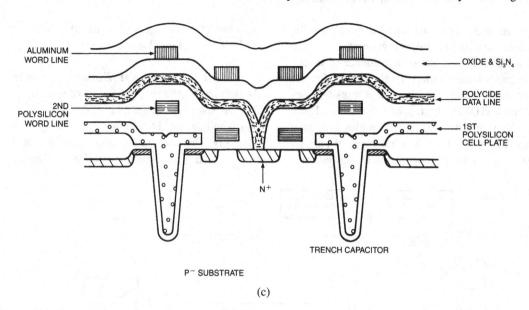

(c)

Figure 2-30 (cont.).   (c) Cross-section of one-transistor memory cell. (From [49], with permission of NEC
Electronics Inc.)

silicon area and provides high storage cell capacity with a single-transistor dynamic storage cell.
Figure 2-30(b) shows the chip layout, and (c)
shows the cross-section of a one-transistor memory cell.

Two inputs to the device are row address
strobe $\overline{RAS}$ and column address strobe $\overline{CAS}$ (or
chip select $\overline{CS}$). In addition to reading row addresses A_0–A_9, selecting the relevant word line,
and activating the sense amplifiers for read and
write operations, the $\overline{RAS}$ input also refreshes
the 2048 b selected by row addressing circuitry.
The $\overline{CAS}$ input latches in the column address
and connects the chip's internal I/O bus to the
sense amplifiers activated by the $\overline{RAS}$ clock,
thereby executing data input or output operations. This device features an address multiplexing method in which an address is divided into
two parts, the upper ten bits (row address) and
the lower ten bits (column address). The row address is latched into the memory at the falling
edge of the $\overline{RAS}$ clock. After an internal timing
delay, the column address input circuits become
active. Flow-through latches (voltage level activated) for column addresses are enabled, and
the column address begins propagating through

the latches to the column decoders. A column
address is held in the latches by the falling edge
of $\overline{CAS}$.

The read and write cycles are executed by
activating the $\overline{RAS}$ and $\overline{CAS}$ (or $\overline{CS}$) inputs and
controlling $\overline{WE}$. An early write cycle is exe-
cuted if $\overline{WE}$ is activated before the falling edge
of $\overline{CAS}$ (or $\overline{CS}$) during a write cycle, and a late
write (read–modify–write) cycle is executed if
the $\overline{WE}$ input is activated later. The process of
rewriting data held in a memory cell, refreshing, is performed by a sense amplifier. This
device is capable of executing the same $\overline{RAS}$-
only and $\overline{CAS}$ (or $\overline{CS}$)-before-$\overline{RAS}$ refresh cycles as are executed by other conventional,
general-purpose DRAMs. All 512 rows of
memory cells must be refreshed within any 8
ms period. Figure 2-31 shows access timing for
NEC μPD421000-series DRAMs [49].

These DRAMs also have "fast-page
mode" which makes it possible to randomly access data in the same row address. The 1024 b of
memory are obtained from the combinations of
column address inputs A_0–A_9 within one row address. In another operational mode called the
"nibble mode," the first data location is specified

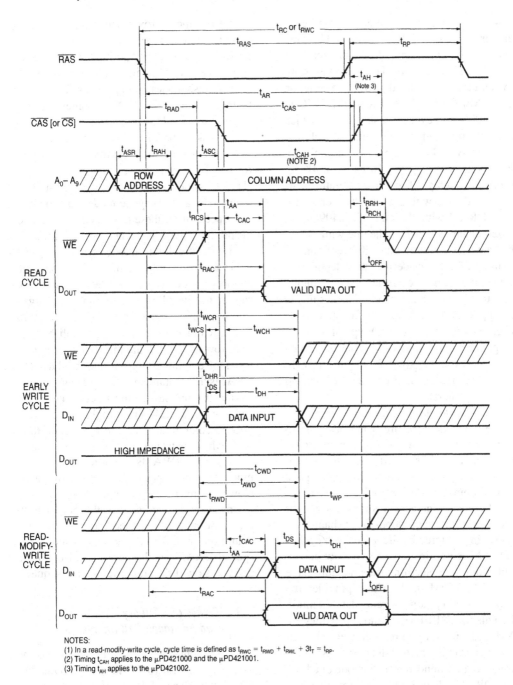

Figure 2-31. Access timing for a 1 Mb DRAM (NEC μPD421000-series). (From [49], with permission of NEC Electronics Inc.)

by row and column addresses A_0–A_9 during a read or write cycle. When the DRAM (NEC μPD 421001) internally sequences the two high-est order addresses (A_9) during the next $\overline{CAS}$ clock cycle, read and write cycles can be executed in less time than in the fast-page operation.

Many 4 Mb DRAMs (e.g., Micron Semi-conductors MT4C4M4B1) utilize 11 row-address bits and 11 column-address bits for 2048 (2K) cycle refresh in 32 ms [116]. However, industry demands for reduced power consumption at the 16 Mb DRAM level have led JEDEC to approve 4096 (4K) cycle refresh in 64 ms, which requires 12 row-address bits and 10 column-address bits. A 4 Mb × 4 DRAM at a minimum random cycle time (t_{RC} = 110 ns) with 4K refresh draws roughly 30 mA less operating current than a device with 2K refresh cycles. A 4 Mb × 4 DRAM with 4K refresh has 4096 rows and 1024 columns, whereas the one with 2K refresh cycles has 2048 rows and 2048 columns. If the number of columns defines the "depth" of a page, then the 4K device has half the page depth. In order to select a refresh standard most suitable for an application, the following factors have to be taken into consideration: (1) type of addressing supported by the DRAM controller, i.e., 11 row/11 column, 12 row/12 column, or both; (2) frequency and length of the page access; and (3) average cycle rates.

Many newer DRAM controllers, including some 3.3 V devices, are being designed to support both of the standards. In a memory design, the choice of 2K or 4K refresh also depends upon power consumption versus page depth requirements. For example, a system requiring frequent page accesses may not benefit by sacrificing the page depth in exchange for power savings offered by the 4K refresh cycles. The DRAM memories targeted for low-power, portable consumer electronics applications have a SELF REFRESH mode which provides the user with a very low-current data-retention mode. This SELF REFRESH mode provides the DRAM with an ability to refresh itself while in an extended standby mode (sleep or suspend). It is similar to the extended refresh mode of a low-power DRAM (LPDRAM), except that the SELF REFRESH DRAM utilizes an internally generated refresh clock while in the SELF REFRESH mode.

Instead of refreshing the DRAM at selected intervals as in the conventional scheme, a burst mode refreshes all the rows in one burst. Al-though this mode provides the CPU with a longer period of uninterrupted access time, it also produces a certain latency period if CPU needs access to the DRAM during a refresh burst [117]. In some refresh modes for high-reliability and large-memory systems, the DRAM's contents are occasionally scrubbed to remove any soft errors.

In addition to various refresh modes, the new generation DRAMs provide several modes for accessing data in the storage cells, such as: (1) page mode, (2) enhanced page, (3) static-column mode, and (4) cyclic mode. To enhance access speed, all of these modes take advantage of the principle that it is not necessary to strobe both the row and column addresses into the DRAM while accessing small portions of the consecutive memory space. Therefore, these access modes shorten the overall access time and precharge period by eliminating some of the address strobes. The page mode memory access establishes a constant row address, while the controller strobes a series of column addresses into the DRAM. In DRAMs that feature an enhanced page mode operation, the internal column address latch is transparent when the $\overline{CAS}$ line is high, which gives the DRAM column decoder direct access to the address when the latch is in its transparent state.

Some DRAMs feature static-column mode for high-speed read and write access. However, this mode consumes more power than the page mode access since the output buffers are continuously enabled during the static-column access. The cyclic access mode needs special DRAMs with sophisticated controllers, and it is slow if the addresses are not sequential.

2.3.3 DRAM Cell Theory and Advanced Cell Structures

The one-transistor cell which has been used from 16 kb to 1 Mb density DRAMs is called the planar DRAM cell. The capacitor used in the DRAM cell has usually been placed beside a single transistor and has occupied typically 30% of the cell area. A cross-section of the conventional double-poly planar cell structure, along with an equivalent circuit representation, is shown in Figure 2-32.

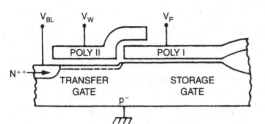

Figure 2-32. Schematic cross-section of a planar DRAM cell with equivalent circuit representation. (From [50], with permission of IEEE.)

The charge storage in the cell takes place on both the oxide and the depletion capacitances. The amount of charge in the storage region of a linear capacitor corresponding to a full charge well $Q(0)$ is given by the following equation:

$$Q(0) = C_{01}(V_p - P_1) - K_1\sqrt{P_1}.$$

In this equation, C_{01} is the oxide capacitance per unit area for the storage region, K_1 is the bulk doping concentration, and P_1 is the surface potential corresponding to a full charge well. The amount of charge in the storage region corresponding to an empty well is given by

$$Q(1) = C_{01}(V_p - P_3) - K_1\sqrt{P_3}$$

where P_3 is the surface potential in the transfer gate region for a gate voltage of V_p.

The signal charge is defined as the difference between $Q(1)$ and $Q(0)$ and is given by

$$\Delta Q = Q(0) - Q(1)$$

$$\Delta Q = \int_{Q(0)}^{Q(1)} dQ = \int_{P_1}^{P_3} C_{01}\left(\frac{1}{V_{t1}}, \frac{1}{P_1}\right) dP$$

$$+ \int_{P_1}^{P_3} C_d(K_1, P_3) dP.$$

Therefore, in order to increase the storage charge capacity, the following options are available: (1) increasing the oxide capacitance C_{01}

by thinning the oxide in the storage gate region, (2) reducing threshold voltage V_{t1} and P_1 by reducing the doping concentration and oxide thickness in the storage area, (3) increasing P_3 by thinning oxide in the transfer area, and (4) increasing the depletion capacitance C_d by increasing K_1. The first three methods, out of several options listed above, have been used extensively as scaling procedures on the planar DRAM cell.

The upper limit on the storage charge of these two-dimensional cells was reached in the mid-1980s with 1 Mb DRAMs. For higher density DRAMs, three-dimensional structures such as trenches etched into the silicon and stacked layers rising above the surface were developed to achieve higher cell capacitance without increasing the cell surface area on the chip. References [51] and [52] describe the cell structure designs before 1979, whereas the recent developments in three-dimensional cell structures are given in [53] and [54]. In order to optimize a cell structure, understanding the basic criteria for a cell design is important [55]. The following equation summarizes the cell design basic requirements:

$$V_{sen} + \Delta V \cong \frac{C_s V_s}{2(C_s + C_{BL})} = \frac{\epsilon A V_s}{2d(C_s + C_{BL})}$$

where V_{sen} is the sensing signal; ΔV is leakage noise due to reference cell mismatch, supply voltage variations, etc.; C_s and C_{BL} are the storage and bit-line capacitances (including sense amplifier loading); V_s is the storage signal; ϵ is the dielectric constant; A is capacitor storage area; and d is the thickness of the capacitor insulator. It can be seen from the equation that V_{sen} can be increased by reducing ΔV, C_{BL}, and d, as well as enlarging ϵ, A, and V_s. Therefore, some of the basic requirements for a good cell and sensing signal enhancement can be summarized as follows:

- low leakage current
- high noise immunity due to word-line and bit-line swings, etc.
- low soft-error rate (SER)

• low R and C parasitics for fast charge transfer

• process simplicity, scalability, and manufacturing ease.

The reduction of dielectric thickness helped to increase bit densities from 16 kb to 1 Mb. However, there are some physical limitations which prevent further dielectric thinning for 4 Mb and higher density DRAMs. As a result of these limitations, further increases in densities have been achieved, mainly through cell design innovations. This has led to the development of 3-D cell capacitors, such as the trench and stacked structures.

An approach has been presented to increase the charge stored in poly-to-poly (interpoly) capacitors by more than 30% without increasing cell area or incurring additional masking steps [33]. It involves texturing the bottom polysilicon electrode before depositing an adequate ultrathin dielectric, thus creating asperities on the polysilicon surface which increase the effective storage area. This technique can be used in all stacked capacitor cells and in all trench capacitor cells that have a bottom polysilicon electrode, such as the stacked-trench capacitor cells.

2.3.3.1 Trench Capacitor Cells. Most of the early 4 Mb DRAMs used trench capacitors. The main advantage of the trench concept was that the capacitance of the cell could be increased by increasing the depth of the trench without increasing the surface area of silicon occupied by the cell. The insulator formed on the trench walls serves as the capacitor dielectric and should be as thin as possible. The material that refills the trench serves as one plate of the capacitor and consists of highly doped polysilicon. In order to obtain increased capacitance through an increase in trench depth and for reliable refilling of the trenches, the sidewalls must be suitably sloped. To obtain Hi-C capacitor structures, the trench walls may need to be selectively doped. Many techniques have been developed to achieve a dielectric film that is thin enough to provide both high capacitance and

breakdown voltage equivalent to those for the planar capacitor cells. The composite dielectric films (e.g., thermally grown oxide and CVD nitride) are used. Since the nitride has a higher dielectric constant than SiO_2, a composite film yields higher capacitance compared to single SiO_2 film of the same thickness. The oxide–nitride film also prevents capacitor leakage due to oxide breakdown. The quality of thermal oxide film is also critical; thinner film growth at the bottom and top corners will produce higher electric fields across these regions, which may cause trench capacitors to exhibit higher leakage currents. This problem can be avoided for bottom corners by ensuring that the etch produces a trench with rounded bottom corners. The use of rapid thermal processing (RTP) to grow the trench capacitor oxide has provided good results.

In the first generation of trench-capacitor-based cells, the plate electrode of the storage capacitor is inside the trench and the storage electrode is in the substrate. The access transistor is a planar MOS transistor fabricated beside the trench capacitor, and the trenches are 3–4 μm deep. The cell size required about 20 μm² of surface area, making them suitable for 1 Mb DRAM designs. Figure 2-33 shows a simple trench cell structure [56]. The first generation cells exhibited some disadvantages. Since the charge is stored in the potential wells, high leakage currents can arise between adjacent cells due to the punchthrough or surface conduction. One solution is to increase doping between the regions of the cells to narrow the depletion region [57]. However, that may lead to avalanche breakdown of the reverse-biased junctions of the access transistors at spacings of 0.8 μm or less. Another option used is to make deeper and narrower trenches which are difficult to fabricate. Also, since the storage node is in the substrate, it is susceptible to charge collection from alpha particles and soft errors.

Several design modifications have been made to increase the capacitance without making the trenches deeper or increasing the cell size. In one approach, the plate electrode is folded around the sides of the storage electrode

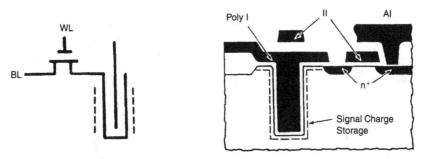

Figure 2-33. A simple trench cell structure 1–4 Mb DRAM. (From [56], with permission of IEEE.)

to create a folded capacitor cell structure (FCC) [58], [59]. A shallow trench is etched around most of the perimeter of the storage electrode, and the plate electrode is deposited over this trench. This cell apparently utilizes both the planar and trench capacitor concepts. Also, the cell's storage plate edges are electrically isolated from those of the adjacent cells by utilizing a BOX-type isolation structure.

In another approach called isolation vertical cell (or IVEC), the walls of the storage electrode were made to follow the outside edges of the cell perimeter, and the access transistor was placed inside [60]. Since the cell does not use all four sidewalls and the bottom of trench for storage, the capacitance can be limited and a more advanced variation called FASIC (folded-bit-line adaptive sidewall-isolated cell) has been developed [61]. This cell uses selective doping of certain trench walls and several advanced processing steps, such as oblique ion implantation and deep isolation with groove refill. It has been demonstrated using 0.8 μm design rules in a 4 Mb chip to produce a cell area of 10.9 μm^2 that has a grid trench only 2 μm deep.

The trench capacitor structures with the storage electrode inside the trench (also called inverted trench cell) were developed to reduce punchthrough and soft-error problems by placing the plate electrode on the outside of the trench and the storage electrode inside. Since the charge is stored inside the oxide isolated trench, it can only leak through the capacitor oxide or laterally diffused contact. A few examples of cell designs using this approach are the

buried-storage-electrode cell or BSE [62], the substrate-plate-trench capacitor cell or SPT [63], and the stacked-transistor-capacitor cell [64]. In the first two approaches, the plate electrode is heavily p-doped and connected to power supply, while the inside storage plate is heavily n-doped. Since the substrate is maintained at an equipotential, the punchthrough problem exists only around the region through which the charge is introduced into the trench. The heavy doping of the plates helps maximize the cell capacitance. A major problem with this type of cell (BSE) is that the gated-diode structure can cause sufficient leakage current to flow into the storage node, which affects the cell's data retention time.

Figure 2-34(a), shown for the IBM SPT cell, overcomes this problem by using the PMOS access transistors and p-type doped inner storage electrodes, and then creating the SPT cells in an n-well on a p-substrate [65]. In some advanced cell designs of this type, the substrate is not used as the storage electrode; both the plate and storage electrode are fabricated inside the trench opening, which allows both electrodes to be completely oxide isolated. This provides the cells with punchthrough protection and reduces the soft-error rate relative to other inverted trench cells. Several cells of this type have been reported, such as the dielectrically encapsulated trench (DIET) [66], or half-V_{cc} sheath plate capacitor (HSPC) [67], and the double-stacked capacitor [68].

In addition to the inverted trench cell structures, another key development has been to

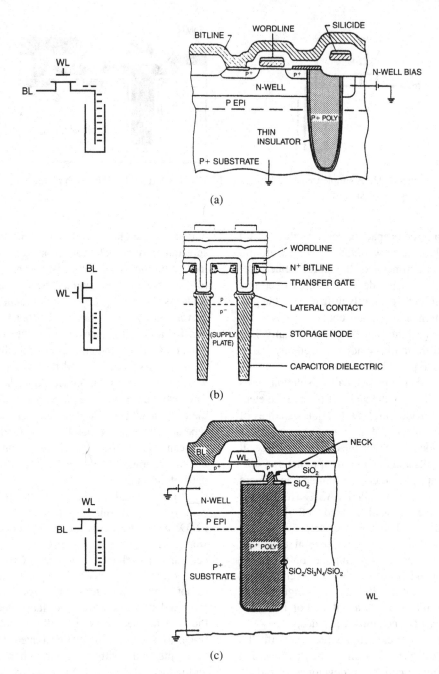

Figure 2-34. (a) Substrate-plate-trench (SPT) capacitor cell [65]. (b) Trench transistor cell (TTC) [71]. (c) Epitaxy-over-trench cell (EOT) [63], [72], with permission of IEEE.

build trench capacitor cells with the access transistor stacked above the trench capacitor. In trench transistor cell designs, the access transistor occupies a significant fraction of the cell area. For planar access transistors which are placed alongside the trench capacitors, the surface area must be devoted to both structures. A more efficient use of space would be to stack the transistor above the trench capacitor to form a vertical cell. This was initially tried in the VMOS DRAM cell [69] in which the sharp edges at the bottom of the V-groove resulted in severely degraded oxide integrity, causing manufacturability problems. Recent advances in process technology have made this a viable approach to realize an ultrahigh-density 3-D buried trench cell for scaling to 64 Mb and higher densities. Two examples of this cell are the trench transistor cell with self-aligned contact (TSAC) [70] and the selective epitaxy-over-trench (EOT) cell [63].

Figure 2-34(b) shows the cross-section of a trench transistor cell (TTC) in which the vertical access transistor is above the trench capacitor in a single deep trench. Its source is connected to the n^+ polysilicon storage electrode of the capacitor by a lateral contact which is made by an oxide undercut etch and polysilicon refill [71]. The drain, gate, and source of the trench transistor are formed by a diffused buried n^+ bit line, an n^+ polysilicon word line, and a lateral contact, respectively. The transistor width is determined by the perimeter of the trench. This cell stores charge on the poly inside the trench. Another version of this cell design is the trench transistor cell with self-aligned contacts, or TSAC. The TSAC cell uses a shallow trench forming the access device channel to improve both the short and narrow channel defects.

Figure 2-34(c) shows the cross-section of an epitaxy-over-trench (EOT) cell in which the storage electrode is first completely isolated from the substrate, and then selective epitaxy is grown [63], [72]. A single-crystal silicon layer is grown over the trench which is surrounded by an exposed Si area. When the epitaxy growth is stopped before the lateral epitaxial film has grown completely over the trench, a self-aligned window is formed on the top of the trench. Then the oxide on top of the trench surface is etched and a second epitaxial film is grown. A pyramidal window of polysilicon is formed on top of the exposed polysilicon in the trench; the mater-

ial surrounding this pyramid is single-crystal silicon formed by means of lateral epitaxy. Then the isolation structure and MOS transistor are fabricated.

An advanced gigabit-DRAM capacitor-over-bitline (COB) structure with a straight-line trench isolation and trench gate transistor (SLIT) has been developed which has the advantages of a simple layout and precise pattern delineation [73]. It uses a simple pattern-shrinking technique called sidewall aided fine pattern etching (SAFE) to provide manufacturable alignment margins by reducing the trench width to less than the design rule. The SLIT uses two kinds of trenches: an isolation and a gate trench. They are arranged orthogonally and define a rectangular active area. The cell size is given by $6 F^2$, where F is the feature size. A sub-0.1 μm SLIT transistor has been fabricated to investigate potential use in ULSI DRAMs.

2.3.3.2 Stacked Capacitor Cells (STC).

In contrast to the trench capacitor cells, another approach that allows cell size reduction without loss in storage capacity is stacking the storage capacitor on top of the access transistor rather than trenching it down into the silicon. The stacked capacitor is formed by layers of deposited dielectric such as silicon dioxide sandwiched between two layers of deposited polysilicon. Figure 2-35(a) shows a comparison among the double-polysilicon planar cell, basic trench capacitor cell, and simple stacked cell. The effective capacitance of a stacked cell is greater than that of the planar cell because of the increased surface area as well as the curvature and sidewall effects, as shown in Figure 2-35(b) [74]. The commercial 4 Mb DRAMs are currently made with both the trench or stacked cell structures. However, for 16 Mb and higher densities, STC cell structures are favored because of the ease of fabrication, immunity against the soft errors, and relative insensitivity to various leakage mechanisms.

Figure 2-35(c) shows a single-layer storage node in which, after formation of the word lines and bit lines, capacitor oxide is followed by polysilicon for the storage node, deposited

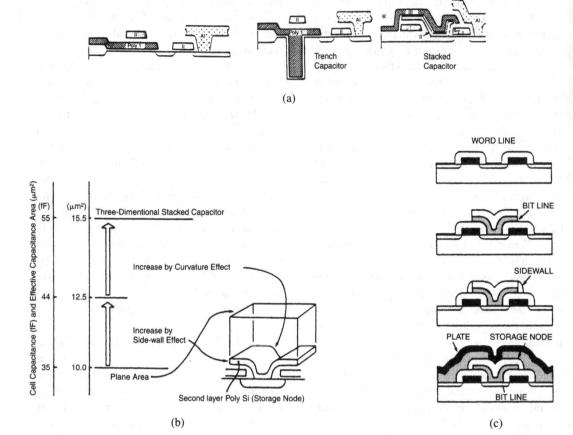

(a)

(b)   (c)

Figure 2-35. (a) Comparison among planar, trench, and stacked capacitor cells. (b) Planar versus trench capacitor cell showing components of effective capacitance [74]. (c) STC cell single-layer [75], with permission of IEEE.

over the bit lines [75]. Then the electrode plate is formed. The storage node of the capacitor is between the oxide and the polysilicon layer, which results in a single-layer capacitor stacked over the bit lines instead of occupying surface area of the chip. A number of layers can be stacked to increase the capacitance. Several advanced stacked capacitor designs have been introduced in the last few years. Based on the morphology of the polysilicon storage electrode used, these are usually categorized as smooth or rough capacitors.

The polysilicon fins or membranes which expand vertically, horizontally, or in both directions are representative designs of the smooth capacitors. The storage area is increased by using both sides of the storage electrodes which are surrounded by a thin oxide–nitride (ON) dielectric and the top capacitor plate. The horizontal fin structures include double-stacked [76], fins [77], and spread-stacked [78]. The schematics for these are shown in Figure 2-36. All of these structures require deposition of one or more polysilicon layers, and the storage node must completely overlap the storage node contact. Also, the double- and spread-stacked designs have two additional masking steps in their process flows.

The vertical fin structures include the cylindrical [79], crown [80], [81], and ring [82] capacitor structures. The schematics for these

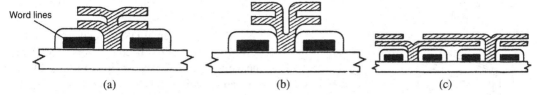

Figure 2-36. Schematics of (a) double-stacked [76], (b) fins [77], and (c) spread-stacked [78] structures. (With permission of IEEE.)

are shown in Figure 2-37. All three vertical structures require additional topography relative to the standard single-layer STCs, but no extra masks are needed. The cylindrical and ring designs require an additional polysilicon layer. The crown capacitor structure is potentially scalable to 256 Mb DRAMs.

In contrast to smooth capacitors, the rough capacitors use microscopic surface granularities (roughness) on the storage electrode to enhance the surface area without additional topography or increased masking steps. Their fabrication process may require an additional polysilicon layer, but the biggest advantage is the ease of manufacturing.

A novel capacitor structure called the modulated stacked (MOST) capacitor has been developed to increase the cell DRAM capacitance by using surface modulation technology [83]. A hemispherical-shaped polysilicon deposited on the top of a thick polysilicon storage electrode is used as a mask to selectively form a deep groove in the electrode, which increases the surface area by as much as a factor of 8 compared to that for a conventional STC occupying the same chip area. A variation on this technique

is the "engraved storage electrode" process developed by NEC for use in 64 Mb DRAMs [115]. In this approach, the electrode is entirely covered with an uneven, hemispherically grained Si surface which increases the storage capacitance value without increasing the cell area or the storage electrode height.

Another innovative 3-D stacked capacitor cell, the spread-vertical-capacitor cell (SVC), has been investigated for the 64 and 256 Mb ULSI DRAMs [84]. The basic structure of SVC consists of a double cylindrical storage capacitor which is spread to the neighboring memory cell area. Figure 2-38(a) shows the cross-section of SVC and an exploded view [84]. Each storage electrode A and B of the two adjacent memory cells A, B is connected via polysilicon (poly-Si) to the memory cells A and B, respectively. The storage electrode A forms an inner wall loop, while the storage electrode B forms an outer wall loop enclosing the inner wall loop. The storage electrode A extends over to the cell areas of memory cells B and C. Also, the storage electrode B extends to the cell areas of memory cells A and C. The storage capacitance of SVC is proportional to the peripheral length of the

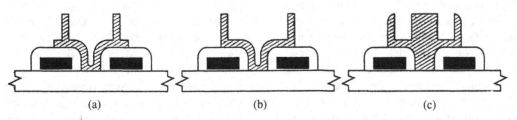

Figure 2-37. Schematics of vertical fin structures. (a) Cylindrical [79]. (b) Crown [81]. (c) Ring [82]. (With permission of IEEE.)

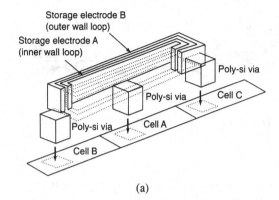

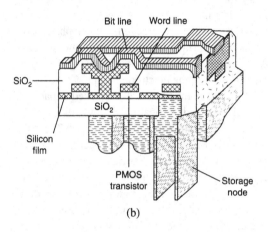

Figure 2-38. (a) A 3-D spread-vertical-capacitor cell (SVC) cross-section and exploded view (From [84], with permission of IEEE). (b) Cross-sectional view of buried capacitor (BC) cell [118].

capacitor footprint, including two adjacent memory cells, and is three times larger than the conventional STC.

Most memory cells currently use SiO_2/Si_3N_4 films as insulation layers. As DRAM densities increase and consequently the cell area shrinks, SiO_2/Si_3N_4 thin layers are being used to obtain the required capacitance. However, these films have practical limitations because of the leakage current problems. One approach has been to use stronger dielectric films. In 1990, Hitachi announced a 64 Mb DRAM crown-type memory cell using a Ta_2O_5 film. NEC has announced the development of two 256 Mb

DRAM cells, one using a Ta_2O_5 film and the other a $Ba_{0.5}Sr_{0.5}$ film.

An experimental DRAM cell structure for 256 Mb and 1 Gb has been developed in which the cell capacitors are flexibly formed like a stacked capacitor (STC), but are buried under a thin silicon layer of a bonded SOI substrate. Figure 2-38(b) shows the cross-sectional view of this buried capacitor (BC) cell structure in which the cell capacitors are completely buried under the thin silicon layer and are connected to the back surface of the silicon film [118]. Therefore, they can have a large area and depth, which means a large capacitance (C_S). Since the cell capacitors do not occupy the space over the silicon layer, the cell size is determined simply by the pitch of the bit and word lines. Thus, the cell size can be reduced to 8 F^2 (where F is the minimum feature size) for a folded bit-line structure. The use of an SOI structure significantly reduces the conventional wide junction area device leakage path to the bulk substrate and the intercell leakage path. These BC cells of 1.28 μm^2 size with a C_s of 47 fF that were fabricated using 0.4 μm design rules exhibited leakage currents of less than 1 fA per cell, resulting in good data retention characteristics.

In gigabit DRAM fabrication, Hitachi Corporation has reported the development of a 0.29 μm^2 metal/insulator/metal crown-shaped capacitor (MIM-CROWN) using 0.16 μm process technologies [119].

2.3.4 BiCMOS DRAMs

The BiCMOS technology has certain advantages over the pure CMOS designs, particularly in access time. Also, it is convenient for achieving ECL interface DRAMs which makes higher performance systems integration possible. The high speed of these DRAMs results mainly from the BiCMOS drivers and the use of a highly sensitive BiCMOS amplifier for direct sensing techniques. The BiCMOS driver provides a larger load current compared with a CMOS driver for a given input capacitance. This is due to the inherently smaller input gate capacitance of the BiCMOS driver and the smaller number of inverter stages required to make it. A good design of a BiCMOS driver is very important because the input gate capacitance is closely related to circuit power dissipation and speed. In conventional DRAMs, the timing margins required among the word pulse, rewrite amplifier activation pulses, and the Y control pulse, connecting the data line to the I/O line, slow up the access time. The direct sensing scheme employed in the BiCMOS designs eliminates these timing margins, allowing for high-speed operation. It has been reported that the BiCMOS circuitry is less sensitive to fabrication process deviations and temperature changes relative to the CMOS

circuitry [85]. Therefore, the BiCMOS DRAMs should have an advantage, especially under the worst case operating conditions.

In 1990, a 1 Mb BiCMOS DRAM using 1.3 μm design rules was described as having 23 ns access time [86]. This used a nonaddress-multiplexing TTL interface and advanced circuit developments such as high-speed direct sensing, a bipolar main amplifier with bipolar cascade connection and data latch functions, an output buffer to prevent noise, a new BiCMOS driver connection to reduce the substrate current, and a static CMOS word driver. Figure 2-39(a) shows

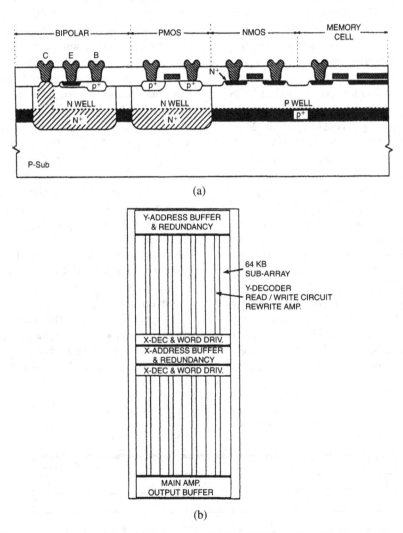

(a)

(b)

Figure 2-39. (a) Cross-sectional view of 1 Mb BiCMOS DRAM. (b) Chip architecture [86]. (With permission of IEEE.)

a cross-sectional view of the chip which has only two more masking steps than the conventional CMOS process [86]. All nMOSTs, including memory cell transistors, have an underlying p⁺ layer to eliminate one masking step. The memory cell capacitance is increased by reducing the oxide thickness. Figure 2-39(b) shows the chip architecture, which includes four distributed Y-decoder columns. The 1 Mb memory cell array is divided into 16 subarrays of 64 kb, and then each of these subarrays consists of 128 words × 512 data-line pairs.

Hitachi has introduced a sub-10 ns ECL 4 Mb BiCMOS DRAM with 0.3 μm as the minimum feature size [87]. It includes a combination of a MOS DRAM core including major periphery and bipolar ECL I/O with a V_{CC} connected limiter, a double-current mirror address buffer with reset and level conversion functions providing high speed and low power against MOS device parameter variations, and an overdrive rewrite amplifier for a fast cycle time. The MOS DRAM core consists of nMOS memory cell arrays, and CMOS drivers, decoders, and level converters. The bipolar circuits are used for ECL interface input/output buffers, highly sensitive amplifiers, and voltage limiters. The memory core incorporates a 1.3 μm² stacked capacitor cell, and the chip size is 47 mm². Figure 2-40 shows a cross-sectional view of the 4 Mb BiCMOS DRAM [87].

A new BiCMOS circuit technology features a novel bit-line sense amplifier that can reduce the access time to half that of a conventional CMOS DRAM access time [88]. The bit-line sense amplifier consists of a BiCMOS differential amplifier, an impedance-converting means featured by a CMOS current mirror circuit or the clocked CMOS inverter between the bit line and the base node of the BiCMOS differential amplifier, and a conventional CMOS flip-flop. This technology applied to a 1 kb test chip fabrication demonstrated that a fast access time can be successfully achieved for megabit DRAMs with only 10% area penalty.

2.3.5 Soft-Error Failures in DRAMs

The DRAM soft errors are single-bit read errors occurring on single bits of stored information in a memory array. A soft error is a not a permanent error; it is correctable by clearing the memory array, writing in the correct data, followed by a read operation. A change in the logic stage of a storage element such as a memory cell is also referred to as an "upset." The common usage tends to use the terms "error" and "upset" interchangeably in the case of single-event-induced logic state reversals (e.g., logic "1" changing to a logic "0" and vice versa). Although soft errors can be caused by circuit-related problems such as supply voltage fluctuations, inadequate noise margins, and sense amplifier imbalance, the source of a particular failure mode was identified in 1978 by May and Woods as being the alpha particles [89]. These alpha particles were produced by the decay of trace amounts of uranium and thorium present in some packaging material, causing random errors.

Figure 2-41 shows the various stages of soft-error creation by an alpha-particle hit in a DRAM [89]. The "critical charge" Q_c is defined as the charge differential between the logic "1"

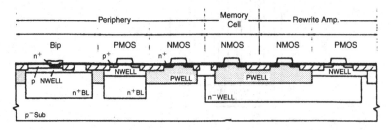

Figure 2-40. Cross-sectional view of the 4 Mb BiCMOS DRAM. (From [87], with permission of IEEE.)

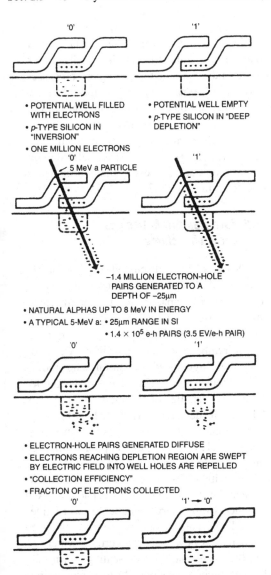

- POTENTIAL WELL FILLED
  WITH ELECTRONS
- *p*-TYPE SILICON IN
  "INVERSION"
- ONE MILLION ELECTRONS

- POTENTIAL WELL EMPTY
- *p*-TYPE SILICON IN "DEEP
  DEPLETION"

'0'
'1'
5 MeV a PARTICLE

−1.4 MILLION ELECTRON-HOLE
PAIRS GENERATED TO A
DEPTH OF −25μm

- NATURAL ALPHAS UP TO 8 MeV IN ENERGY
- A TYPICAL 5-MeV a: • 25μm RANGE IN SI
  • 1.4×10^5 e-h PAIRS (3.5 EV/e-h PAIR)

'0'
'1'

- ELECTRON-HOLE PAIRS GENERATED DIFFUSE
- ELECTRONS REACHING DEPLETION REGION ARE SWEPT
  BY ELECTRIC FIELD INTO WELL HOLES ARE REPELLED
- "COLLECTION EFFICIENCY"
- FRACTION OF ELECTRONS COLLECTED

'0'
'1' → '0'

- POTENTIAL WELL REMAINS
  FILLED

- POTENTIAL WELL NOW
  FILLED

- NO APPRECIABLE COLLECTION
- IF (COLLECTION EFF.) × (a ELECTRONS GENERATED) >
  CRITICAL CHARGE, A "SOFT ERROR" RESULTS
- A SINGLE ALPHA CAN CAUSE AN ERROR
- NO PERMANENT DAMAGE RESULTS

Figure 2-41. The various stages of soft-error creation by an alpha-particle hit in a
DRAM. (From [89], with permission
of IEEE.)

and "0" states. For the conventional one-transistor DRAM cell described by May and Woods,
the logic state in which the storage node is filled

with minority-carrier charge (electrons) is designated as "0." The electron–hole pairs produced by an alpha particle first passing through
the passivation layers are not collected by the
empty potential well, which means that some
impact energy is lost before reaching the substrate. A typical 5 MeV alpha particle with a
penetration depth of 25 μm in the silicon substrate may generate roughly 1.4×10^6 e–h pairs.
The electrons reaching the depletion region are
swept by the electric field into the well, whereas
the holes are repelled. The electrons generated
in the bulk that diffuse to the edge of the well
are also collected by the storage nodes.

As illustrated in Figure 2-41, the charge
deposited by an alpha-particle track, which
reaches the storage node through drift and diffusion, does not alter the electrical state of the
nodes holding "0"s. However, it may "upset"
the states of the nodes holding "1"s. A soft error
or an "upset" occurs when a large enough fraction of electrons collected by the potential well
produces net charge > critical charge (Q_c).

Since the DRAMs are susceptible to upset
by charge collected through diffusion, several physically adjacent RAM cells may be affected by a single ionization track from an
alpha-particle hit. For high-density DRAMs, the
probability of multibit errors caused by an incidence of an alpha particle increases. To minimize
the probability of multiple upsets from a single-
particle strike, some application-specific DRAMs
are designed such that the adjacent cells tend to
carry the opposite charge states. In DRAMs,
where the information is stored as packets of
charges on isolated nodes (ideally), the critical
charge is defined as the absolute value of difference between the average charge content of a
node in logic "1" state from that in a "0" state.

The soft-error rate (SER), expressed in
number of errors per hours (or per day), is a
measure of the device susceptibility to soft errors. The soft errors first reported in 4 kb
DRAMs became a significant concern for 64 kb
and higher density memories. In the earlier
memory designs, an attempt was made to make
the SER comparable to the hard-error rate. This
was usually done by scaling the oxide thickness

to obtain adequate capacitance in the cell. For example, the Hi-C cell was introduced for increasing storage capacity without increasing capacitor size. Another approach was to use the dual dielectric films (such as SiO_2 and Si_3N_4) as the insulator for the charge storage capacitor, thus increasing the cell's capacitance.

In later NMOS/CMOS DRAM designs, various other techniques were used to reduce their susceptibility to the soft-error phenomenon. The entire array of NMOS DRAM cells was built in a p-well, so that the well–substrate junction acted as a reflecting barrier for the minority carriers produced outside the well, preventing their diffusion into the unfilled storage wells. The use of heavily doped substrate and lightly doped epitaxial layers also helped improve the SER rates. Several advances have been made in storage cell structures, such as trench capacitors and stacked capacitor cells discussed in Section 2.3.3. Trench capacitor structures with the storage electrode inside the trench (also called inverted trench cells) provided better SER rates than the cells in which the storage electrode is outside the trench. For the cells with very little or no isolation between the closely spaced trench walls, an alpha-particle track crossing through two cells can cause charge transfer between the two cells by forward-biasing the cell junctions. A paper was presented by Fujitsu making a comparison between several types of trench and stacked capacitor cell structures by simulating their soft-error rates [90]. According to the conclusions, a stacked capacitor cell was found more suitable for high-density, megabit-level designs than a trench capacitor cell.

Another method to reduce SER has been the use of thick coatings of radioactive contaminant-free polymer material on the die surface of memory chips. The purification of wafer fabrication materials and improvements in packaging materials with a much lower concentration of radioactive impurities has also helped reduce soft-error problems. More recently, the use of redundancy through the on-chip error-correcting codes (ECC) has been implemented for recovery from "soft errors" caused by alpha particles. So far, we have discussed soft errors in memories resulting from alpha particles released by the radioactive contaminant materials introduced during the fabrication and packaging processes. However, in the natural space environments, the soft errors may be caused by ionizing particles (such as MeV-energy protons) trapped within the radiation belts surrounding earth, or from heavy ion cosmic ray fluxes. These space-related memory soft errors and "upsets" will be discussed in more detail in Chapter 7.

2.3.6 Advanced DRAM Designs and Architectures

Section 2.3.3 discussed scaling of the CMOS DRAMs to 1 Mb densities by increasing the storage cell capacity, the use of improved sense amplifiers, and incorporation of advanced features such as the latchup and noise improvement circuitry, address transition detection (ATD) for high-speed operation, and redundancy schemes. The earliest DRAMs used open bit-line architecture for the memory array. For 64 kb–1 Mb densities, the folded bit-lines approach implemented was briefly discussed in Section 2.3.1. However, that approach imposed some restrictions on the array size. A major problem with 4 Mb and higher density DRAMs has been the increased parasitic capacitance and resistance from a large number of word lines and bit lines. This results in slow sensing speed and increased power supply current, which affects the noise margin. For scaling down to 4 Mb and higher densities, further innovations have been made in design and architectures.

One of the design improvements was the requirement for shorter channel length MOS transistors in the memory array than in the peripheral circuitry. These smaller geometry transistors were susceptible to hot electron effects if operated at a 5 V external supply voltage. Therefore, on-chip voltage converters were used to limit the voltage to less than 4 V for the memory array. For example, Hitachi, in their 16 Mb DRAM, used an internal voltage regulator to supply 3.3 V to 0.6 μm minimum channel length transistors in the memory array [91]. The transistors in the peripheral circuitry with channel length scaled only to 0.9 μm were operated with a full external supply voltage of 5 V. Several other options for improving the hot-electron immunity were investigated at the 4 Mb and higher density levels. One of them was the use of lightly doped drain (LDD) structures. Another approach involved the use of a dual-gate inverter called the "cascade" structure in which an additional n-channel transistor is connected in series to the n-channel switching transistor in the standard CMOS inverter [92]. The lifetime of this dual-gate transistor has been reported to be two orders of magnitude higher than the single-gate transistor.

The use of lower operating voltages at the 16 Mb DRAM level to avoid problems associated with hot-electron degradation affects the drive capability of the sense amplifiers. To reduce this problem, improved differential amplifier designs and common I/O schemes have been developed. The new sensing schemes have increased speed and reduced bit-line coupling noise. The asymmetries due to normal process variations can affect the performance of sense amplifiers. Several modifications to the conventional sense amplifiers have been proposed to minimize the effect of those asymmetries. In one of the schemes proposed, each of the paired transistors in a conventional sense amplifier is divided into two transistors which have reverse source and drain current flows [93]. This modification produces symmetrical current–voltage characteristics and a balanced circuit which enhances the sensitivity of the sense amplifiers.

For the 4 Mb and higher density DRAMs, attempts have been made to combine the higher density of the open bit-line architecture with improvements in the signal-to-noise (S/N) ratios of the folded bit-line approach. In one approach, a "cross-point" cell array with segmented bit-line architecture was proposed which has higher noise immunity than the conventional cell array [94]. For the sense circuit, it used the $V_{CC}/2$ sensing scheme, a conventional CMOS sense amplifier, and a full reference cell. This design technique allowed a cell to be placed at every intersection of a word line and segmented bit lines. Several other techniques have been suggested, such as the pseudo-open bit line [95] and a staggered bit-line arrangement [96].

In Section 2.3.5, the susceptibility of DRAMs to soft errors, particularly from alpha-particle interaction, was discussed along with the various design techniques used to reduce their soft-error rates (SERs). One approach being used on some advanced DRAM architectures is the implementation of redundancy and the on-chip error-correcting codes (ECCs). The use of this technique to overcome the manufacturing defects and increase reliability and yield has been proposed in several papers [97–99]. A scheme has been developed by IBM that makes use of redundant word and bit lines in conjunction with ECC as implemented on the 16 Mb chips [100]. The error-correcting circuits in this 16 Mb chip can be quite effective in correcting single-cell failures. However, any additional faults in an ECC word must be fixed with the on-chip redundancy provided.

2.3.6.1 A 16 Mb DRAM (Example). A 16 Mb DRAM (LUNA-C DD4) organized as 4 Mb × 4 with on-chip error-correction circuitry (ECC) and capable of operation in page mode has been designed and fabricated by the IBM Corporation. The chip is divided into four quadrants of 4 Mb each, with each quadrant dedicated to two of the four I/Os. Each quadrant is divided into 8 blocks of 512 kb each, and each block into two segments of 256 kb each. The word line is 1024 data bits plus 72 check bits, which is segmented into four driven segments. Figure 2-42 shows the LUNA-C DD4 block diagram [101]. A double-error-detection, single-error-correction (DED/SEC) Hamming code (odd weight) has been designed in each quadrant. All the bits being transferred from the page to the SRAM buffer are being checked and corrected within the ECC capability. This architecture results in a minimum delay for the first bits accessed and no delay for the subsequent bits accessed within the page. The chip is designed to operate at 3.6 V, and is provided with a power-on reset protection which causes the chip to remain unselected, with all I/Os maintained at

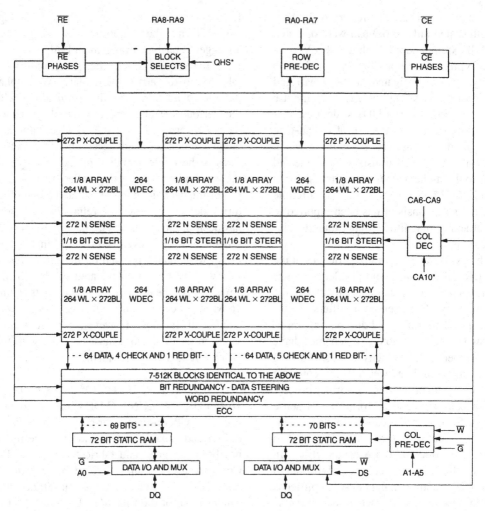

* In 11/11 addressing mode, CA10 is used to select 1 of 2 blocks
In 12/10 addressing mode, only one QHS (quadrant half select) is activated.

Figure 2-42. A 16 Mb DRAM LUNA-C DD4 Block Diagram (From [101], with permission of IBM Corporation)

high impedance until V_{cc} reaches approximately 1.5 V above V_{ss}.

The LUNA-C memory chip is fabricated in a CMOS V 0.5 μm/0.2 μm process on a 7.8 mm × 18 mm p-type Si-substrate with a p-epi layer, utilizing an n-well process, three-level metal interconnections, trench isolation, and LDD spacers. Figure 2-43 shows the chip cross-section [130].

2.3.6.2 ULSI DRAM Developments.

In ULSI (64–256 Mb) DRAM fabrication, the applied high-voltage stress field across the gate oxide increases with scaling. The operating voltage of the word line cannot be scaled satisfactorily due to the nonscaling threshold voltage of the transfer gate transistor. Therefore, for the ultrahigh-density 64 Mb DRAMs, the reliability of the thin gate oxide becomes a critical issue,

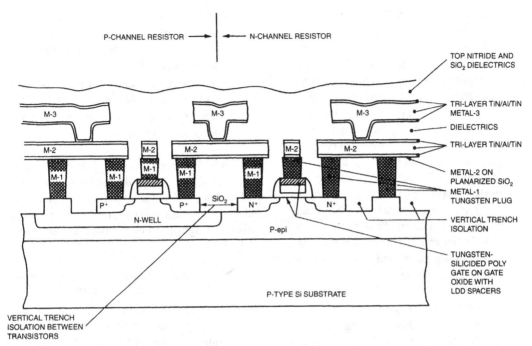

Figure 2-43. A cross-section of 16 Mb DRAM IBM LUNA-C chip [130].

especially because of the time-dependent dielectric breakdown (TDDB) phenomenon. This TDDB susceptibility is a function of gate oxide thickness (T_{ox}), variations in operating temperature, and supply voltage. Toshiba has proposed a word-line architecture for 64 Mb DRAMs, which can control and compensate for the reliability degradations from the gate oxide thickness variations and temperature fluctuations [102].

The ULSI 64 and 256 Mb DRAM designs require innovative process, fabrication, and assembly techniques such as fine-line patterning or thicker vertical topography; low-temperature deposition, oxidation, and planarization with very stringent uniformity specifications; and an efficient assembly process to encase the large die into ultrathin packages. The overall large die size of 64 Mb DRAMs ($\sim$ 200 mm^2) as well as the die length presents additional design constraints that may reach present reticle size limits [103]. To provide architectural flexibility while satisfying package and reticle size constraints, leads-over-chip (LOC) layouts for the die are being considered which allow bonding pads to

be arranged down the middle of the chip. The LOC layout has other advantages, e.g., the internal power bus routing is reduced and the output switching characteristics are improved. Self-refresh sleep modes internally generated to provide refresh controls and signals are being investigated for ULSI DRAMs to reduce the power required to maintain the memory contents while the devices are in the standby mode. The optimized test structures can provide up to 256:1 parallel expanded test mode capability.

The ULSI DRAMs are currently in developmental stages, and the following technology challenges must be overcome for high-yield successful production:

- A 0.35 μm resolution lithography for the 64 Mb DRAMs that may require successful development of X-ray, deep-UV, or i-line combined with phase-shift technology.

- A vertical capacitor structure to maintain sufficient cell capacitance. It is estimated that at 3.3 V, a 45 fF capacitor is needed for the 64 Mb DRAMs to provide good soft-error immunity, sense amplifier read capability, and noise immunity. Therefore, the 64 Mb DRAM challenge is to provide enough cell capacitor surface area for the required capacitance. The vertical structures can provide more surface area, but their use can result in several problems during photolithographic steps. Another way to achieve high cell capacitance is with alternate dielectrics such as tantalum pentaoxide (Ta_2O_5), yttrium trioxide (Y_2O_3), and PZT. These alternate dielectrics can provide higher charge density, but the technology is not mature and is still evolving. Another way to achieve higher capacitance is by texturing the capacitor by using several techniques available.

- Another major fabrication challenge for the 64 Mb DRAMs is the manufacture of highly reliable 0.35 μm transistors with shallow junction depths, low thermal budget requirements, and ultrathin oxides.

In the 64 Mb generation and beyond, the focus of DRAM core array design will be the elimination of array noise. The conventional open bit-line and folded bit-line architectures used in typical cross-point cell memory arrays have some fundamental limitations. A new DRAM signal-sensing scheme called a divided/shared bit line (DSB) has been developed that eliminates common-mode array noise [104]. The DSB approach has the following major features:

- It realizes a folded bit-line sensing operation in the cross-point cell arrangement which provides high-density memory core with noise immunity.
- In the DSB array architecture, the bit-line pitch is relaxed as compared to the conventional folded bit-line scheme, allowing the twisted bit-line technique to be implemented such that interbit-line coupling noise is eliminated.

- The reduced bit-line capacitance per sense amplifier due to the divided bit-line scheme contributes to high-speed sensing operation.
- A divided/pausing bit-line sensing (DIPS) scheme based on the DIPS design shows excellent interbit-line coupling noise immunity.

A 64 Mb, 1.5 V DRAM has been designed and fabricated with a goal of maintaining speed performance and stable operation in spite of the reduced operating voltage [105]. To achieve a stable and high-speed operation, these are some of the features used:

- A complementary current-sensing scheme for high-speed operation.
- A 1.28 μm^2 crown-shaped stacked capacitor (CROWN) cell to ensure an adequate storage charge and to minimize the data-line interference noise.
- A new speed-enhanced, half-V_{cc} voltage generator utilized to improve the response characteristics and voltage accuracy.
- A word-line driver circuit having a high boost ratio to store the complete data-line voltage swing in the memory cell.

Figure 2-44 shows the schematic of a CROWN cell (a) top view, and (b) cross-sectional view ($A - A'$). The capacitor area is maximized with this new capacitor structure featuring a cylindrical storage electrode formed on the data line [105]. A storage capacitance of more than 40 fF can be obtained even in a small area of 1.28 μm^2. An important feature of the CROWN cell is that the data line is shielded by either the plate or storage layer. Figure 2-44(c) shows a microphotograph of the chip based on the 0.3 μm design rule. This memory chip works with either a 1.5 or 3.3 V external power supply. A typical operating current is 29 mA at a cycle time of 180 ns.

An experimental 256 Mb DRAM with a NAND-structured cell has been fabricated using a 0.4 μm CMOS technology [106]. The NAND-

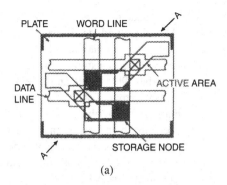

(a)

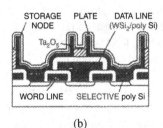

(b)

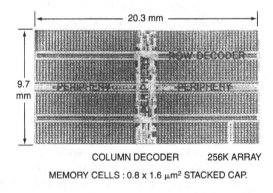

COLUMN DECODER   256K ARRAY

MEMORY CELLS : 0.8 x 1.6 μm² STACKED CAP.

(c)

Figure 2-44. A 64 Mb DRAM memory with CROWN cell. (a) Top view. (b) Cross-sectional view $(A - A')$. (c) Microphotograph. (From [105], with permission of IEEE.)

structured cell has four memory cells connected in series, which reduced the area of isolation between the adjacent cells, and also the bit-line contact area. The cell measures 0.962 μm²/b, which is 63% compared with that for a conventional cell. For temporary storage cells that are essential for restoring data to the NAND-structured cell, the RAM adopts the one-

transistor and one-capacitor cell to minimize the chip penalty. To reduce the die size, a new time-division multiplex sense amplifier (TMS) architecture has been introduced in which a sense amplifier is shared by four bit lines. The total chip area is 464 mm², which is 68% compared with the conventional DRAM. The data can be accessed by a fast block-access mode up to 512 bytes, as well as a random access mode. It has a typical 112 ns access time for first data in a block and 30 ns serial cycle time.

Figure 2-45 shows a comparison between the NAND-structured cell and a conventional cell [106].

Another experimental 256 Mb DRAM with a multidivided array structure has been developed and fabricated with 0.25 μm CMOS technology [107]. For memory cells, it uses the hemispherical gram (HSG) cylindrical stacked capacitors that achieve 30 fF capacitance and trench isolation transistors to reduce the diffusion layer capacitance for bit lines. The memory cell size is 0.72 μm². This DRAM has row and column redundancy subarrays with their own sense amplifiers and subword decoders at every 64 Mb block instead of at every 128 kb subarray. The layout with 256 cells per bit line reduces the total chip size by 16% compared to that with 128 cells per bit line. The chip size is 13.6 × 24.5 mm². The timing measurements have shown a typical access time of 30 ns. This experimental 256 Mb DRAM uses the following major circuit technologies:

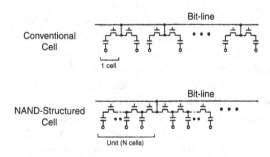

Figure 2-45. A comparison between the NAND-structured cell and a conventional cell. (From [106], with permission of IEEE.)

- A partial cell array activation scheme for reducing the power line voltage bounce and operating current.
- A selective pull-up data line architecture to increase I/O width and produce a low operating current.
- A time-sharing refresh scheme to maintain a conventional refresh period without increasing the power-line voltage bounce.

Future gigabit DRAM developments will have a wide range of applications in the new field of file storage and battery-operated portable devices for which low-voltage and low-power operation is essential. One of the most important design issues for DRAMs used in file storage applications is the reduction of data-retention current which consists of the refresh current (I_{REF}) and the standby current (I_{SB}). The I_{SB} consists of a subthreshold current of the MOS transistors and a dc current of the internal circuits which must be reduced. The threshold voltage V_T of the MOS transistors should be lowered in the low-voltage DRAMs to maintain a reasonable operating speed. A self-reverse biasing scheme for the word drivers and decoders has been proposed for suppressing the subthreshold current to 3% of that for the conventional scheme [108]. A subarray-replacement redundancy technique reduces the manufacturing cost by doubling the yield. Since this redundancy technique requires subarrays, it is not suitable for small-capacity memory with only a small number of subarrays.

As the density of DRAMs increases to gigabit levels, the cell area decreases exponentially. Therefore, the capacitance density, which is defined as the cell capacitance divided by the projection area of storage node, should increase with the DRAM density. For example, a capacitance density of 145 and 330 fF/μm^2 is required for the 1 and 4 Gb DRAM, respectively, for a target capacitance of 25 fF/cell. These are some of the new developments in the fabrication of gigabit density DRAMs that were reported during the 1994 International Electron Devices Meeting (IEDM):

- A Ta$_2$O$_5$ capacitor applicable to a 1 Gb DRAM and beyond, developed by Samsung Electronics Company Ltd. [126]. This capacitor has a storage capacitance of more than 90 fF/cell and a leakage current lower than 2×10^{-5} A/cell that is obtained by applying a WN/poly-Si top electrode and a 3.5 nm Ta$_2$O$_5$ capacitor dielectric to a cylindrical capacitor with a rugged poly-Si surface (projection area = 0.4 μm^2).
- A new technology called the low-temperature integrated (LTI) process for gigabit DRAMs with a thin Ta$_2$O$_5$ capacitor developed by NEC [127]. This uses two key technologies as follows: low-temperature interlayer planarization and low-temperature contact formation. An experimental device using 2.5 nm Ta$_2$O$_5$ capacitors by the LTI process showed ten times improvement in data retention time.
- Another development reported by NEC is a new capacitor-over-bit-line (COB) cell with an area of 0.375 μm^2, fabricated by 0.2 μm electron-beam direct writing lithography and a single-layer resist [128]. It uses a diagonal bit-line (DBL) configuration which reduces cell area down to 75% compared to the folded-bit-line cell. A low-temperature ($\leq 500°C$) fabrication process was developed to prevent leakage current increase due to Ta$_2$O$_5$ degradation. The obtained storage capacitance (C_s) was 28.5 fF for a 0.40 μm height capacitor. An edge operation MOS (EOS) design transfer gate with low body effects that are characteristic of SOI technology permits low-voltage operation.

In ULSI DRAMs beyond the 256 Mb generation, DRAMs fabricated in SOI substrate may be a more practical alternative. Some experimental work has been reported on the SOI DRAM memory cell structures and the DRAM on thick SOI film prepared by wafer bonding [120]. The expected advantages of the SOI

MOSFETs include an improved short-channel effect, reduced parasitic capacitance, reduction of body effect, and latchup immunity. In a study, a comparison was made between the SOI and bulk-Si DRAM memory cell data retention characteristics. In both memory cell structures, p–n junction leakage current is the dominant leakage mechanism. For 1.5 V V_{cc} operation, the lower limit of memory cell capacitance (C_s) from the data retention requirement is only 4.5 fF for SOI in comparison to 24 fF for bulk-Si. This is because the junction area of the SOI memory cell is reduced to 7.5% of the bulk-Si DRAM. Therefore, the thin-film SOI DRAM provides a greater advantage in cell data retention characteristics.

Advanced cell structures are under development, e.g., the buried capacitor (BC) cell discussed in Section 2.3.3.2 that uses bonded SOI for a 1 Gb DRAM. The implementation of 0.15–0.10 μm SOI technology will require the use of fully depleted accumulation mode transistors [121]. The key feature in the optimization of such devices is the silicon film thickness. The transistors fabricated with film thicknesses as low as 50 nm have shown better hot electron behavior, reduced short-channel effects, and higher breakdown voltage characteristics than for the bulk-Si devices.

2.3.7 Application-Specific DRAMs

The technical advances in multimegabit DRAMs have resulted in a greater demand for application-specific products incorporating specialized performance requirements. For example, the rapidly expanding video graphics market has spawned the need for high-speed serial interfaces which encouraged the development of dual-port video RAMs (VDRAMs) with one serial and one random port, or the frame buffer with two serial ports. The memory designers have made efforts to simplify the complex external control requirements for DRAM chips to make them more compatible with the SRAM systems. These design modifications of lower cost DRAMs for use in the SRAM-specific applications are called pseudostatic DRAMs (PSRAMs), or virtual-static DRAMs (VSRAMs). Some other advanced high-speed RAM architectures include the enhanced DRAM (EDRAM), cached DRAM (CDRAM), synchronous DRAM (SDRAM), Rambus DRAM (RDRAM), and 3-D RAM.

2.3.7.1 Pseudostatic DRAMs (PSRAMs).
A PSRAM (or PS-DRAM) is basically a one-transistor dynamic RAM cell with nonmultiplexed addresses, and an on-chip refresh address counter driven by an external "refresh pin" to refresh the memory cells. It uses dynamic storage cells, but contains all refresh logic on-chip, so that it is able to function similarly to a SRAM. It has synchronous operation since it operates from a system clock applied to the chip enable pin, and memory cell refresh occurs during the portion of cycle time when the memory is not being accessed. However, pseudostatic RAMs are not that easy to use since they must execute internal refresh cycles periodically, and there is a potential for conflict between an external access request and an internal cycle.

The main advantage of PSRAM over SRAM results from the general DRAM technology life cycle leading that of SRAM by several years. However, the use of PSRAMs adds complexity to the system with its requirements of synchronous operation and the need to clock the refresh control pin. An alternative developed was the virtual-static DRAMs (or VSRAMs) which have a refresh capability totally transparent to the user and are identical in a system to a SRAM. They are basically pseudostatics which have an on-chip refresh timer that generates the refresh request signal for the timing generator. Since the refresh operation is now transparent to the user, the cycle time and address access time must be equal as in a standard asynchronous SRAM. A disadvantage of the VSRAM is that the on-chip timer requires power, and therefore its standby current tends to be an order of magnitude higher than a comparable PSRAM.

2.3.7.2 Silicon File.
A technology gap has existed between the faster primary memories (CPU main memory/cache) and the slower secondary storage media (hard disk, magnetic tapes, etc.). The disk market requires nonvolatility and low operating power, but the data access

is serial and orders of magnitude slower than that of standard DRAMs. An alternative proposed is the "silicon file" as an economical mass storage device specially designed to replace the magnetic media in hard disks, solid-state recording, and for other system back-up applications. Figure 2-46 shows a block diagram of the NEC µPD42601 silicon file organized as 1,048,576 words × 1 b, and provides a battery back-up feature [49]. It is based on the trench cell technology of NEC's 1 Mb DRAMs and implements the same read and write cycles. However, the system bandwidth is optimized with a page cycle that repeatedly pulses $\overline{CAS}$ while maintaining $\overline{RAS}$ low. The silicon file also periodically executes standard $\overline{RAS}$-only and $\overline{CAS}$-before-$\overline{RAS}$ refresh cycles to refresh its cells within a specified interval of 32 ms, which is four times slower than the one for a 1 Mb DRAM.

An important feature of this silicon file is its ability to retain data while being powered by a back-up battery. The silicon file has a slower access time and lower active current than a comparable size DRAM.

2.3.7.3 Video DRAMs (VRAMs).
Another class of application-specific devices developed is VRAMs for use as multiport graphic buffers. The basic configuration of a dual-port VRAM includes a random access (RAM) port and a serial access (SAM) port. The chip architecture includes a RAM array coupled to serial data registers to provide unified text and graphics capabilities. This organization virtually eliminates the bus contention by decoupling the video data bus from the processor data bus. Figure 2-47 shows a block diagram of the NEC dual-port graphic buffer equipped with a 256K × 4 b random access port and a 512 × 4 b serial

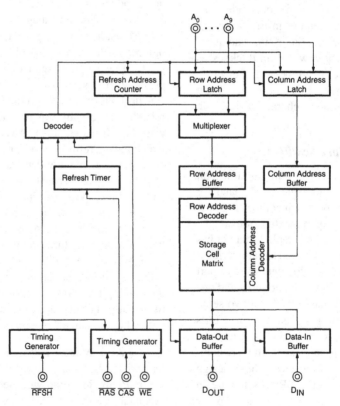

Figure 2-46. Block diagram of a silicon file NEC µPD42601. (From [49], with permission of NEC Electronics Inc.)

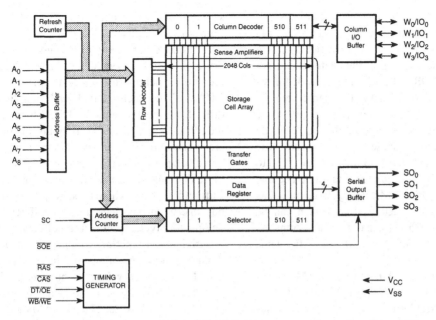

Figure 2-47. Block diagram of NEC μPD42273 dual-port graphics buffer. (From [49], with permission of NEC Electronics Inc.)

read port [49]. The serial read port is connected to an internal 2048 b data register through a 512 × 4 b serial read output circuit. The random access port is used by the host CPU to read or write addressed data in any desired order. A write-per-bit capability allows each of the four data bits to be individually selected or masked for a write cycle.

This device operates in fully asynchronous dual-access mode, except when transferring stored graphics data from a selected row of the storage array to the data register. During a data transfer, the random access port requires a special timing cycle using a transfer clock; the serial read port continues to operate normally. The refreshing operation is performed by means of $\overline{RAS}$-only refresh cycles or by normal read or write cycles on the 512 address combinations of A_0–A_8 during an 8 ms period. An automatic internal refresh mode is also available on these devices.

A variation on VRAM is a new two-port chip called the window RAM (WRAM). The WRAM is basically on 8 Mb VRAM with some special features. Instead of a 16 b wide random port, the WRAM has a 32 b random port, making the chip easier to use in moderate-sized

graphic frame buffers for multimedia applications. In addition to the 32 b random port, the WRAM has some significant internal architectural improvements based on a 256 b internal bus that can route data from the memory array through on-chip pixel processing logic and back to the memory array. The WRAM uses these functions to gain its speed advantage over VRAMs and wide DRAM arrays.

2.3.7.4 High-Speed DRAMs. The high-speed specialty SRAMs offer at best only one-fourth the density of DRAMs, and are costlier at per-bit equivalent storage capacity. Also, the secondary caches built from SRAMs require additional board space and power. The main disadvantage of the VRAMs is the higher cost of the chip, typically about 20% more required to implement the serial-output registers and control logic.

The DRAMs are traditionally used in large volume, slower speed main level systems. The DRAMs' performance can be enhanced with some circuit design techniques, as well as special on-chip operating features. For example, the standard page mode DRAMs primarily consist

of the memory cell array, row and column address decoders, sense amplifiers, and data I/O
buffers. In a way, the sense amplifier acts as a
cache to provide faster data than a direct data access from the DRAM array. The extended data-
out (EDO) DRAMs have the same internal
organization as the standard page mode
DRAMs, with the exception that a "D"-type
latch is added to the sense amplifier output. Due
to this output latch, the $\overline{CAS}$ line does not need to
remain low while waiting for data-out to become
valid. The EDO DRAMs can cycle at speeds as
fast as the address access time, which can be
quite low (e.g., 25 ns). Also, the burst transfers
can cycle up to 30% faster than the fast-page
DRAMs.

The enhanced DRAM (EDRAM) introduced by Ramtron International Corporation in
a 0.76 μm CMOS process features a new isolation technique to reduce parasitic capacitance
and increase transistor gain. As a result, the
EDRAM has a row access time of 35 ns, a
read/write cycle time of 65 ns, and a page-write
cycle time of 15 ns [122]. The EDRAM also includes an on-chip static RAM cache, write-post-
ing register, and additional control lines that
allow the SRAM cache and DRAM to operate
independently.

A high-speed DRAM (HSDRAM) has
been developed with random access time approaching that of the SRAMs while retaining
the density advantage of a one-transistor
DRAM. An example of this category of memories is the development of a 1 Mb HSDRAM
which has a nominal random access time of less
than 27 ns and a column access time of 12 ns
with address multiplexing [109]. This device
uses a double-polysilicon, double-metal CMOS
process having PMOS arrays inside the n-wells
and with an average 1.3 μm feature size. This 1
Mb HSDRAM is about three times faster than
the typical MOS DRAM, but with an increased
chip area of about 20–40%. The increased area
is mainly due to the large drivers, greater number of sense amplifiers due to shorter bit lines, a
larger planar capacitor cell, and one-quarter selection of the chip. However, this HSDRAM is
about two–three times denser than the typical

same-generation technology SRAMs using
four-transistor cells.

Some other examples of innovative memory architectures used for high-speed DRAMs
are the synchronous, cached, and Rambus
DRAMs. The synchronous DRAMs (SDRAMs)
differ from the conventional DRAMs mainly in
the following respects: (a) all inputs and outputs
are synchronized (referenced) to the rising edge
of a clock pulse; and (b) during a read operation,
more than one word is loaded into a high-speed
shift register from where these words are shifted
out, one word per clock cycle. In SDRAM development cycle, prefetch architectures were
one option, but pipelined architectures are more
dominant now. As a result, the synchronous
DRAMs can have very high burst rates. However, the access times are no faster than for the
conventional DRAMs. The SDRAMs are capable of read or write data bursts at synchronous
clock rates (up to 66 or 100 MHz) after the initial access latency period. The SDRAM developers originally proposed the following two
approaches: (1) a level-$\overline{RAS}$ signal with a sin-
gle-bank internal architecture, and (2) a pulsed-
$\overline{RAS}$ scheme with a dual-bank architecture
(banks are subdivisions of the internal memory
array into controllable blocks which are easier
to manipulate). Currently, the level-$\overline{RAS}$ device
architecture is no longer being pursued.

SDRAMs appear to be good candidates
for fast frame buffer multimedia applications.
However, the mainstream SDRAMs are considered too big and too narrow to fit well into
midrange graphics applications. A group of
memory suppliers have recently introduced synchronous graphics RAM (SGRAM) which combine 256 K × 32 b organization and some of the
internal functions of the VRAMs with the econ-
omy of a single port and the speed of a 100 MHz
synchronous interface. SGRAMs differ from
SDRAMs by providing an eight-column
BLOCK WRITE function and a MASKED
WRITE (or WRITE-PER-BIT) function to ac-
commodate high-performance graphics applications. SGRAMs offer substantial advances in
dynamic memory operating performance, including the ability to synchronously burst data

at a high data rate with automatic column address generation, the ability to interleave between internal banks in order to hide the precharge time, and other special functions.

Figure 2-48 shows the functional block diagram of micron 256K × 32 SGRAM, containing 8,388,608 bits [129]. It is internally configured as a dual 128K × 32 DRAM with a synchronous interface (all signals are registered on the positive edge of the clock signal, CLK). Each of the 128K × 32 b banks is organized as 512 rows by 256 columns by 32 b. Read and write accesses to the SGRAM are burst-oriented; accesses start at a selected location and

continue for a programmed number of locations in a programmed sequence. The accesses begin with the registration of an ACTIVE command, which is then followed by a READ or WRITE command. The address bits registered coincident with the ACTIVE command are used to select the bank and row to be accessed (BA selects the bank, A0–A8 select the row). Then the address bits registered coincident with the READ or WRITE command are used to select the starting column location for the burst access.

The SGRAM provides for programmable READ or WRITE burst lengths of two, four, or eight locations, or the full page, with a burst

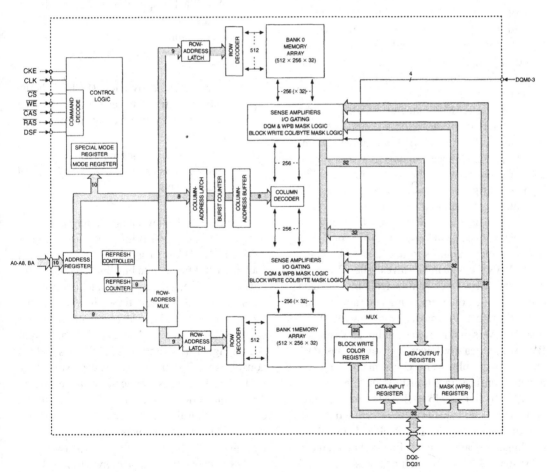

Figure 2-48. Functional block diagram of a 256K × 32 SGRAM. (From [129], with permission of Micron Technology Inc.)

terminate option. An AUTO PRECHARGE function may be enabled to provide a self-timed row precharge that is initiated at the end of the burst sequence. This SGRAM uses an internal pipelined architecture to achieve high-speed operation, and is designed to operate in 3.3 V, low-power memory systems.

Another proprietary scheme, called the Rambus DRAM (RDRAM), offers a complete solution to a system memory architecture by providing a specification that describes the protocol, electrical interface, clocking requirements, the register set, device packaging and pinout, and the board layout. It uses an interface called the Rambus channel which contains the I/O drivers, phase-locked loop, and other control logic. The core of RDRAM basically consists of a standard 8 or 16 Mb DRAM array divided into two independent, noninterleaved logical banks. Each of these banks has an associated high-speed row cache (sense amplifiers) that is roughly ten times larger than the row cache on standard DRAMs.

The RDRAM eliminates *RAS/CAS* at the system interface by handling all address transitions internally. The RAMBUS channel is composed of 13 high-speed signals as follows: nine data, two bus control pins, and two clocks. The high-speed signals are routed between a memory controller containing a RAMBUS ASIC cell (RAC) and up to 300 RDRAMs by using the RAMBUS expansion slots. The memory controller delivers requests to the RDRAMs which act as RAMBUS channel slaves. The RDRAMs use a 250 MHz clock to transfer 1 byte of data every 2 ns, i.e., on every rising and falling clock edge. Therefore, the maximum bandwidth (BW) of a DRAM is 500 Mbytes/s, although sustained BW is significantly lower. For higher bus BW, all signals are physically constrained using terminated, surface-trace transmission lines. An example is 16 and 18 Mb RDRAMs announced by NEC, which are capable of transferring data at a rate of 500 Mbytes/s. The RDRAM and RAC combination can be used in graphic memory subsystem design for high-end desktops, and embedded applications such as multimedia and asynchronous transfer mode (ATM) operations.

A cache DRAM (CDRAM) has a small SRAM cache inserted between its external pins and an internal 4 or 16 Mb DRAM array. This SRAM cache consists of a segmented 16 kb cache of 128 cache lines, each with eight 16 b words [124]. This 128 b data path between the SRAM and DRAM facilitates a high-speed data transfer, so that the access times within a location in the cache are much faster than the typical DRAM accesses. An example of this architecture is a 100 MHz, 4 Mb CDRAM which integrates a 16 kb SRAM as cache memory and a 4 Mb DRAM array into a monolithic circuit [110]. This 4 Mb CDRAM has the following main features:

- A 100 MHz operation of cache hit realized by an improved localized CDRAM architecture that results in only 7% area penalty over the conventional 4 Mb DRAM.
- A new fast copy-back (FCB) scheme with three times faster access time of a cache miss over the conventional copy-back method.
- A maximized mapping flexibility realized by independent DRAM and SRAM address inputs.

Figure 2-49 shows a block diagram of the 4 Mb CDRAM consisting of the following: a 1 Mb × 4 DRAM array, a 4 kb × 4 SRAM cache, 16 × 4 block data bus lines, data transfer buffers (DTBs), I/O circuits, and control circuits [110]. 16 b block times four are transferred between the DRAM block and a SRAM column. All input signals except for the output enable signal # G are registered and synchronously controlled by an external master clock K. The cache access time can be made fast by direct connection of the cache addresses to the CPU, avoiding the multiplexer for DRAM addresses. This CDRAM was fabricated in a 0.7 μm quad-polysilicon, double-metal, twin-well CMOS process used in a conventional 4 Mb DRAM process, and required only 7% additional chip area overhead. The system simula-

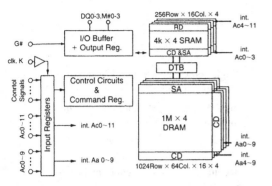

Figure 2-49. Block diagram of a 4 Mb CDRAM. (From [110], with permission of IEEE.)

tion indicated better performance than a conventional cache system with eight times the cache capacity.

A new development called the 3-D RAM is a frame buffer memory architecture which is optimized for high-performance 3-D graphics and imaging. This development is a result of joint collaboration between Sun Microsystems (Mountain View, CA) and Mitsubishi Electronics (Cypress, CA). A traditional bottleneck of the 3-D graphics hardware has been the rate at which the pixels can be rendered into a frame buffer. The 3-D RAM increases the speed of Z-buffer operations and allows complete data modification in one clock cycle, which is about ten times faster than the current generation of conventional VRAMs, at a comparable system cost.

2.3.7.5 Application-Specific RAM Glossary and Summary of Important Characteristics.

Table 2-2(a) lists a glossary of some commonly used application-specific RAMs, and (b) summarizes the application-specific RAMs important characteristics [125].

TABLE 2-2. (a) A Glossary of Some Commonly Used Application-Specific RAMs [125].

- 3-D RAM—Cache DRAM with on-board ALU for 3-D graphics functions.
- Burst EDO—EDO plus a counter to transfer a linearly addressed string of data.
- CDRAM—Cache DRAM: internal SRAM cache added to DRAM.
- EDORAM—Extended Data Out RAM, also called hyper page; a modification of fast page mode to hold data after CAS goes high, allowing faster CAS cycles.
- EDRAM—Enhanced DRAM: very fast DRAM cells directly mapped to SRAM cache.
- Fast Page Mode—a modification of the basic DRAM to allow multiple column accesses from a single row access.
- MDRAM—Multibank DRAM, a collection of smaller, fast blocks of DRAM with on-chip interleaving, pipelining.
- RDRAM—Rambus DRAM: specialized interface and 500 MHz 8 b wide bus controller plus on-chip interleaving.
- SBSRAM—Synchronous Burst Static RAM: SSRAM plus a burst counter.
- SDRAM—Synchronous DRAM: a standard DRAM with all functions referenced to the system clock, burst output mode.
- SGRAM—Synchronous Graphics DRAM: an SDRAM with block write and write per bit.
- SRAM—Static RAM.
- SSRAM—Synchronous Static RAM: addition of clock to synchronize RAM to system clock.
- VRAM—Video RAM: Dual Port or multi-port RAM.
- WDRAM—Window DRAM, a modification of VRAM to reduce internal complexity.

TABLE 2-2 (Cont.).   (b) A Summary of Application-Specific RAMs' Important Characteristics [125].

Description	Advantages	Disadvantages
3DRAM	On-chip cache, on-board ALU, optimized for 3-D operations.	Single source, nonstandard, requires complex controller.
Burst EDO DRAM	Addition of burst counter allows fast output of a string of linear addresses.	High latency if next address is not linearly consecutive.
CDRAM	On-chip cache, separate address decoders for DRAM and Cache.	Requires special off-chip cache controller.
DRAM	Industry standard parts, low cost.	Becoming too slow for either main memory or video.
EDO DRAM	Uses "standard" interfacing, makes more of inherent internal speed available to user.	May not be fast enough without external interleaving.
EDRAM	SRAM substitute; static buffer and cache controller on chip.	Single source.
MDRAM	High throughput, low latency, fine granularity.	Single source, nonstandard, requires complex controller.
RDRAM	High throughput, many suppliers.	Nonstandard, requires a separate complex controller.
SBSRAM	Addition of burst counter allows fast output of a string of linear addresses.	Potential for high latency if the next address is not linearly consecutive.
SDRAM	JEDEC standards exist, much higher throughput than DRAM.	Not all parts meet standards.
SGRAM	JEDEC standards exist, superset of SDRAM, single-port.	Not all parts meet standards.
SRAM	High-speed parts with standard interface.	Low density, high price/bit.
SSRAM	Addition of logic allows sync with system clock, allows faster timing.	High cost, smaller memory sizes.
VRAM	Dual port, standard interface requirements.	High cost, extra functions may not be used.
WRAM	High-speed, dual port; simpler interface than VRAM or RAMBUS.	Single source, nonstandard

REFERENCES

[1] S. Flanagan, "Future technology trends for static RAMs," in *Proc. 1988 IEDM.*

[2] K. Itoh, "Trends in megabit DRAM circuit design," *IEEE J. Solid-State Circuits,* vol. 25, June 1990.

[3] J. J. J. Bastiaens and W. C. H. Gubbels, "The 256k bit SRAM: An important step on the way to submicron IC technology," *Philips Tech. Rev.,* vol. 44, p. 33, Apr. 1988.

[4] B. Prince, *Semiconductor Memories: A Handbook of Design, Manufacture and Applications,* 2nd ed.   New York: Wiley.

[5] D. A. Hodges and H. Jackson, *Analysis and Design of Digital Integrated Circuits,* 1st ed.   New York: McGraw-Hill, 1983.

[6] T. Watanabe *et al.,* "Comparison of CMOS and BiCMOS 1-Mbit DRAM performance," *IEEE J. Solid-State Circuits,* vol. 24, June 1989.

[7] A. Alvarez *et al.,* "Tweaking BiCMOS circuits by optimizing device design," *Semiconductor Int.,* p. 226, May 1989.

[8] C. Lage, "BiCMOS memories: Increasing speed while minimizing process complexity," *Solid State Technol.,* pp. 31–34, Aug. 1992.

[9] Y. Okajima *et al.*, "7ns 4 Mb BiCMOS with parallel testing circuit," in *IEEE ISSC C Tech. Dig.*, 1991.

[10] J. W. Hayden *et al.*, "A high-performance 0.5 μm BiCMOS technology for fast 4-Mb SRAMs," *IEEE Trans. Electron Devices*, vol. 39, pp. 1669–1677, July 1992.

[11] K. Nakamura *et al.*, "A 6-ns ECL 100K I/O and 8-ns 3.3 V TTL I/O 4-Mb BiCMOS SRAM," *IEEE J. Solid-State Circuits*, vol. 27, pp. 1504–1510, Nov. 1992.

[12] J. P. Colinge, presented at the 5th Int. Workshop on Future Electron Devices Three Dimensional Integration, Miyagi, Zao, Japan, May 1988.

[13] T. D. Stanley, "The state-of-the-art in SOI technology," *IEEE Trans. Nucl. Sci.*, vol. 35, Dec. 1988.

[14] S. Wolf, *Silicon Processing for the VLSI Era, Vol. 2: Process Integration. Lattice Press, 1990.*

[15] J. P. Colinge, "Thin-film SOI technology: The solution to many submicron CMOS problems," in *Proc. 1989 IEDM*, p. 817.

[16] P. H. Woerlee *et al.*, "Half-micron CMOS on ultra-thin silicon on insulator," in *Proc. 1989 IEDM*, p. 821.

[17] *RCA High-Reliability Integrated Circuits Databook, CMOS/SOS Technology.*

[18] T. Yoshihiro *et al.*, "Low power and high-stability SRAM technology using a laser-recrystallized p-channel SOI MOSFET," *IEEE Trans. Electron Devices*, vol. 39, pp. 2147–2152, Sept. 1992.

[19] M. Ando *et al.*, "A 0.1 A standby current bouncing-noise-immune 1 Mb SRAM," in *Proc. 1988 Symp. VLSI Technol.*, p. 49.

[20] Y. Uemoto *et al.*, "A stacked-CMOS cell technology for high-density SRAMs," *IEEE Trans. Electron Devices*, vol. 39, pp. 2259–2263, Oct. 1992.

[21] K. Sasaki *et al.*, "A 23-ns 4-Mb CMOS SRAM with 0.2-μa standby current," *IEEE J. Solid-State Circuits*, pp. 1075–1081, Oct. 1990.

[22] T. Hirose *et al.*, "A 20-ns 4-Mb CMOS SRAM with hierarchical word decoding architecture," *IEEE J. Solid-State Circuits*, pp. 1068–1074, Oct. 1990.

[23] M. Matsumiya *et al.*, "A 15-ns 16 Mb CMOS SRAM with reduced voltage amplitude-data bus," presented at IEEE ISSCC, session 13, paper 13.5, 1992.

[24] H. Goto *et al.*, "A 3.3-V, 12-ns 16-Mb CMOS SRAM," *IEEE J. Solid-State Circuits*, vol. 27, pp. 1490–1496, Nov. 1992.

[25] M. Ukita *et al.*, "A single-bit-line cross-point cell activation (SCPA) architecture for ultra-low-power SRAMs," *IEEE J. Solid-State Circuits*, vol. 28, pp. 1114–1118, Nov. 1993.

[26] K. Seno *et al.*, "A 9-ns 16-Mb CMOS SRAM with offset-compensated current sense amplifiers," *IEEE J. Solid-State Circuits*, vol. 28, pp. 1119–1124, Nov. 1993.

[27] K. Sasaki *et al.*, "A 16-Mb CMOS SRAM with a 2.3-μm^2 single-bit-line memory cell," *IEEE J. Solid-State Circuits*, vol. 28, pp. 1125–1130, Nov. 1993.

[28] H. Takato *et al.*, "High performance CMOS surrounding gate transistor (SGT) for ultra high density LSIs," in *Proc. 1988 IEDM*, p. 223.

[29] R. Eklund *et al.*, "A 0.5 μm BiCMOS technology for logic and 4 Mbit-class SRAMs," in *Proc. 1989 IEDM*, p. 425.

[30] D. Hisamato *et al.*, "A fully developed lean-channel transistor (DELTA)—A novel vertical ultra thin SOI MOSFET," in *Proc. 1989 IEDM*, p. 833.

[31] S. Takano *et al.*, "A 16k GaAS SRAM," in *IEEE ISSCC Tech. Dig.*, Feb. 1987, p. 140.

[32] Integrated Devices Technology, Inc., *Specialized Memories Databook 1990–91.*

[33] P. C. Fazan *et al.*, "Thin nitride films on textured polysilicon to increase multimegabit DRAM cell storage capacity," *IEEE Electron Device Lett.*, vol. 11, pp. 279–281, July 1990.

[34] A. Silburt *et al.*, "A 180-MHz 0.8-μm BiCMOS modular memory family of DRAM and multiport SRAM," *IEEE J. Solid-State Circuits*, vol. 28, pp. 222–231, Mar. 1993.

[35] Advanced Micro Devices, Inc., *CMOS Memory Products 1991 Data Book/Handbook.*

[36] L. Boonstra *et al.*, "A 4096-b one-transistor per bit random-access memory with internal timing and low dissipation," *IEEE J. Solid-State Circuits*, vol. SC-8, p. 305, Oct. 1973.

[37] D. A. Hodges and H. Jackson, *Analysis and Design of Digital Integrated Circuits, Semiconductor Memories*, 2nd ed. New York: McGraw-Hill, 1988.

[38] C. N. Ahlquist *et al.*, "A 16K 384-bit dynamic RAM," *IEEE J. Solid-State Circuits*, vol. SC-11, Oct. 1976.

[39] K. U. Stein *et al.*, "Storage array and sense/re-fresh circuit for single-transistor memory cells," *IEEE J. Solid-State Circuits*, vol. SC-7, p. 336, 1972.

[40] R. C. Foss and R. Harland, "Peripheral circuits for one transistor cell MOS RAMs," *IEEE J. Solid-State Circuits*, vol. SC-10, Oct. 1975.

[41] L. G. Heller and D. P. Spampinato, "Cross-cou-pled charge transfer sense amplifier," U.S. Patent 4,039,861, Aug. 1977; also, L. Heller, "Cross-coupled charge transfer sense amplifier," in *IEEE ISSCC Tech. Dig.*, Feb. 1979, p. 20.

[42] T. Wada *et al.*, "A 150ns, 150mw, 64k dynamic MOS RAM," *IEEE J. Solid-State Circuits*, vol. SC-13, p. 607, Oct. 1978.

[43] T. Wada *et al.*, "A 64k × 1 bit dynamic ED-MOS RAM," *IEEE J. Solid-State Circuits*, vol. SC-13, p. 600, Oct. 1978.

[44] R. C. Foss, "The design of MOS dynamic RAMs," in *IEEE ISSCC Tech. Dig.*, Feb. 1979, p. 140.

[45] J. J. Barnes and J. Y. Chan, "A high performance sense amplifier for a 5V dynamic RAM," *IEEE J. Solid-State Circuits*, vol. SC-15, p. 831, Oct. 1980.

[46] R. H. Dennard, "Evolution of the MOSFET dy-namic RAM—A personal view," *IEEE Trans. Electron Devices*, pp. 1549–1555, Nov. 1984.

[47] H. Miyamoto *et al.*, "A fast 256k × 4 CMOS DRAM with a distributed sense and unique re-store circuit," *IEEE J. Solid-State Circuits*, vol. SC-22, Oct. 1987.

[48] N. C. Lu and H. H. Chao, "Half-VDD bit-line sensing scheme in CMOS DRAMs," *IEEE J. Solid-State Circuits*, vol. SC-19, p. 451, Aug. 1984.

[49] NEC Electronics Inc., *NEC 1991 Memory Prod-ucts Databook*.

[50] Y. A. El-Mansy and R. A. Burghard, "Design parameters of the Hi-C SRAM cell," *IEEE J. Solid-State Circuits*, vol. SC-17, Oct. 1982.

[51] V. Rideout, "One device cells for dynamic ran-dom-access memories: A tutorial," *IEEE Trans. Electron Devices*, vol. ED-26, pp. 839–852, June 1979.

[52] P. Chatterjee *et al.*, "A survey of high-density dy-namic RAM cell concepts," *IEEE Trans. Electron Devices*, vol. ED-26, pp. 827–839, June 1979.

[53] H. Sunami, "Cell structures for future DRAMs," in *Proc. 1985 IEDM*, pp. 694–697.

[54] P. Chatterjee *et al.*, "Trench and compact struc-tures for DRAMs," in *Proc. 1986 IEDM*, pp. 128–131.

[55] N. C. C. Lu, "Advanced cell structures for dy-namic RAMs," *IEEE Circuits and Devices Mag.*, pp. 27–36, Jan. 1989.

[56] H. Sunami *et al.*, "A corrugated capacitor cell (CCC) for megabit dynamic MOS memories," in *Proc. 1982 IEDM*, p. 806.

[57] T. Furuyama *et al.*, "An experimental 4Mb CMOS DRAM," in *IEEE ISSCC Tech. Dig.*, 1986, pp. 272–273; 1986; also, *IEEE J. Solid-State Circuits*, vol. SC-21, pp. 605–611, Oct. 1986.

[58] T. Furuyama and J. Frey, "A vertical capacitor cell for ULSI DRAMs," in *Symp. VLSI Tech-nol., Dig. Tech. Papers*, 1984, pp. 16–17.

[59] M. Wada *et al.*, "A folded capacitor cell (FCC) for future megabit DRAMs," in *Proc. 1984 IEDM*, pp. 244–247.

[60] S. Nakajima *et al.*, "An isolation-merged verti-cal capacitor cell for large capacity DRAMs," in *Proc. 1984 IEDM*, pp. 240–243.

[61] M. Mashiko *et al.*, "A 90 ns 4 Mb DRAM in a 300 mil DIP," in *IEEE ISSCC Tech. Dig.*, pp. 12–13.

[62] M. Sakamoto *et al.*, "Buried storage electrode (BSE) cell for megabit DRAMs," in *Proc. 1985 IEDM*, pp. 710–713.

[63] N. C. C. Lu *et al.*, "A buried-trench DRAM using a self-aligned epitaxy over trench technol-ogy," in *Proc. 1988 IEDM*.

[64] F. Horiguchi *et al.*, "Process technologies for high density, high speed 16 megabit dynamic RAM," in *Proc. 1987 IEDM*, pp. 324–327.

[65] N. C. C. Lu *et al.*, "The SPT cell—A new sub-strate-plate trench cell for DRAMs," in *Proc. 1985 IEDM*, pp. 771–772; also, *IEEE J. Solid-State Circuits*, vol. SC-21, pp. 627–634, 1985.

[66] M. Taguchi *et al.*, "Dielectrically encapsulated trench capacitor cell," in *Proc. 1986 IEDM*, pp. 136–139.

[67] T. Kaga *et al.*, "A 4.2μm^2 half-V_{CC} sheath-plate capacitor DRAM cell with self-aligned buried plate wiring," in *Proc. 1987 IEDM*, pp. 332–335.

[68] K. Tsukamoto *et al.*, "Double stacked capacitor with self-aligned poly source/drain transistor (DSP) cell for megabit DRAM," in *Proc. 1987 IEDM*, pp. 328–331.

[69] F. B. Jenne, U.S. Patent 4,003,036, 1977.

[70] M. Yanagisawa et al., "Trench transistor cell with self-aligned contact (TSAC) for megabit MOS DRAM," in Proc. 1986 IEDM, pp. 132–135.

[71] W. F. Richardson et al., "A trench transistor cross-point cell," in Proc. 1985 IEDM, pp. 714–717.

[72] G. Bronner et al., "Epitaxy over trench technology for ULSI DRAMs," in Symp. VLSI Technol., Dig. Tech. Papers, 1988, pp. 21–22.

[73] M. Sakao et al., "A straight-line trench isolation and trench-gate transistor (SLIT) cell for gigabit DRAMs," presented at the IEEE 1993 Symp. VLSI Technol., Kyoto, Japan, May 1993.

[74] Y. Takemae et al., "A 1Mb DRAM with a 3-dimensional stacked capacitor cells," in IEEE ISSCC Tech. Dig., Feb. 1985, p. 250.

[75] S. Kimura et al., "A new stacked capacitor DRAM cell with a transistor on a lateral epitaxial silicon layer," in Proc. 1988 IEDM, p. 596.

[76] T. Kisu et al., "A novel storage capacitance enlargement structure using a double stacked storage node in STC DRAM cell," in SSDM Ext. Abst., 1988, p. 581.

[77] T. Ema et al., "3-dimensional stacked capacitor cell for 16M and 64M DRAMs," in Proc. 1988 IEDM, p. 88.

[78] S. Inoue et al., "A spread stacked capacitor (SSC) cell for 64 Mb DRAMs," in Proc. 1989 IEDM, p. 31.

[79] W. Wakamiya et al., "Novel stacked capacitor cell for 64 Mb DRAM," in Symp. VLSI Technol., Dig. Tech. Papers, 1989, p. 31.

[80] Y. Kawamoto et al., "A 1.2 μm, 2-bit line shielded memory cell technology for 64 Mb DRAMs," in Symp. VLSI Technol., Dig. Tech. Papers, 1990, p. 13.

[81] T. Kaga et al., "Crown-shaped stacked capacitor cell for 1.5-V operation of 64 Mb DRAMs," IEEE Trans. Electron Devices, vol. 38, p. 255, 1991.

[82] N. Shinmura et al., "A stacked capacitor cell with ring structure," in SSDM Ext. Abst., 1990, p. 833.

[83] Y. K. Jun et al., "The fabrication and electrical properties of modulated stacked capacitors for advanced DRAM applications," IEEE Electron Device Lett., vol. 13, pp. 430–432, Aug. 1992.

[84] N. Matsuo et al., "Spread-vertical capacitor cell (SVC) for high-density DRAMs," IEEE Trans. Electron Devices, vol. 40, pp. 750–753, Apr. 1993.

[85] G. Kitsukawa et al., "A 1-Mbit BiCMOS DRAM using temperature compensation circuit techniques," in Proc. ESSCIRC'88, Sept. 1988.

[86] G. Kitsukawa et al., "A 23-ns 1-Mb BiCMOS DRAM," IEEE J. Solid-State Circuits, pp. 1102–1111, Oct. 1990.

[87] T. Kawahara et al., "A circuit technology for sub-10ns ECL 4 Mb BiCMOS DRAMs," presented at the 1990 IEEE Symp. VLSI Circuits.

[88] S. Watanabe et al., "BiCMOS circuit technology for high-speed DRAMs," IEEE J. Solid-State Circuits, vol. 26, pp. 1–9, Jan. 1993.

[89] T. C. May and M. H. Woods, "A new physical mechanism for soft errors in dynamic memories," in Proc. Rel. Phys. Symp., Apr. 1978, pp. 2–9.

[90] S. Ando et al., "Comparison of DRAM cells in the simulation of soft error rates," presented at the 1990 IEEE Symp. VLSI Circuits.

[91] M. Aoki et al., "A 60ns 16-Mbit CMOS DRAM with a transposed data-line structure," IEEE J. Solid-State Circuits, vol. 23, p. 1113, Oct. 1988.

[92] H. Terletzki and L. Risch, "Operating conditions of a dual gate inverter for hot carriers reduction," in Proc. ESSDERC, Sept. 1986, p. 191.

[93] H. Yamuchi et al., "A circuit design to suppress asymmetrical characteristics in 16 Mbit DRAM sense amplifier," presented at the VLSI Seminar, 1989.

[94] A. Shah et al., "A 4-Mbit DRAM with trench-transistor cell," IEEE J. Solid-State Circuits, vol. SC-21, p. 618, Oct. 1986.

[95] F. Horiguchi et al., "Process technologies for high density high speed 16 megabit dynamic RAM," in Proc. 1987 IEDM, p. 324.

[96] M. Inoue et al., "A 16-Mbit DRAM with a relaxed sense-amplifier-pitch open-bit-line architecture," IEEE J. Solid-State Circuits, vol. 23, p. 1104, Oct. 1988.

[97] J. Yamada, "Selector-line merged built-in ECC technique for DRAMs," IEEE J. Solid-State Circuits, vol. SC-22, pp. 868–873, Oct. 1987.

[98] L. M. Arzubi et al., "Fault tolerant techniques for memory components," in IEEE ISSCC Tech. Dig., Feb. 1985, p. 231.

[99] P. Mazumder, "Design of a fault-tolerant DRAM with new on-chip ECC," in Defect and Fault Tolerance in VLSI Systems, Vol. 1, I. Koren, Ed. New York: Plenum, 1989, pp. 85–92.

[100] H. Kalter *et al.*, "A 50-ns 16-Mb DRAM with a 10-ns data rate and on-chip ECC," *IEEE J. Solid-State Circuits*, pp. 1118–1128, Oct. 1990.

[101] IBM Preliminary Engineering Spec. for LUNA-C DD4 16M (4M × 4) DRAM with On-Chip ECC.

[102] T. Takeshima, "Word-line architecture for constant reliability 64Mb DRAM," in *Proc. 1991 Symp. VLSI Circuits*, June 1991, p. 57.

[103] T. Lowery *et al.*, "The 64 megabit DRAM challenge," *Semiconductor Int.*, pp. 47–52, May 1993.

[104] H. Hidaka *et al.*, "A divided/shared bit-line sensing scheme for ULSI DRAM cores," *IEEE J. Solid-State Circuits*, vol. 26, pp. 473–478, Apr. 1991.

[105] Y. Nakagome *et al.*, "An experimental 1.5V 64-Mb DRAM," *IEEE J. Solid-State Circuits*, vol. 26, pp. 465–472, Apr. 1991.

[106] T. Hasegawa *et al.*, "An experimental DRAM with a NAND-structured cell," *IEEE J. Solid-State Circuits*, vol. 28, pp. 1099–1104, Nov. 1993.

[107] T. Sugibayashi *et al.*, "A 30-ns 256-Mb DRAM with a multidivided array structure," *IEEE J. Solid-State Circuits*, vol. 28, pp. 1092–1098, Nov. 1993.

[108] G. Kitsukawa *et al.*, "256-Mb DRAM circuit technologies for file applications," *IEEE J. Solid-State Circuits*, vol. 28, pp. 1105–1111, Nov. 1993.

[109] N. C. C. Lu *et al.*, "A 22-ns 1-Mbit CMOS high-speed DRAM with address multiplexing," *IEEE J. Solid-State Circuits*, vol. 24, pp. 1198–1205, Oct. 1989.

[110] K. Dosaka *et al.*, "A 100-MHz 4-Mb cache DRAM with fast copy-back scheme," *IEEE J. Solid-State Circuits*, vol. 27, pp. 1534–1539, Nov. 1992.

[111] K. Nakamura *et al.*, "A 220 MHz pipelined 16 Mb CMOS SRAM with PPL proportional self-timing generator," in *IEEE ISSCC Tech. Dig.*, 1994, pp. 258–259.

[112] B. Dance, "European SOI comes of age," *Semiconductor Int.*, pp. 83–90, Nov. 1994.

[113] J. Alowersson, "A CMOS circuit technique for high-speed RAMs," in *Proc. IEEE ASIC Conf.*, 1993, pp. 243–246.

[114] M. Penn, "Economics of semiconductor production," *Microelectron. J.*, vol. 23, pp. 255–265, 1992.

[115] H. Watanabe *et al.*, "Hemispherical grain silicon for high density DRAMs," *Solid State Technol.*, pp. 29–33, July 1992.

[116] Micron Semiconductor Inc., *DRAM Data Book Technical Notes: 16 MEG DRAM—2K vs 4K Refresh Comparisons; and Self Refresh DRAMs.*

[117] S. Gumm and C. T. Dreher, "Unraveling the intricacies of dynamic RAMs," *EDN*, pp. 155–164, Mar. 30, 1989.

[118] T. Nishihara *et al.*, "A buried capacitor cell with bonded SOI for 256-Mbit and 1-Gbit DRAMs," *Solid State Technol.*, pp. 89–94, June 1994.

[119] T. Kaga *et al.*, "A 0.29-μm^2 MIM-CROWN cell and process technologies for 1-Gbit DRAMs," in *Proc. 1994 IEDM*, pp. 12.6.1–12.6.3.

[120] K. Suma *et al.*, "An SOI-DRAM with wide operating voltage range by CMOS/SIMOX technology," in *IEEE ISSCC Tech. Dig.*, 1994, pp. 138–139.

[121] T. Nishimura *et al.*, "SOI-based devices: Status overview," *Solid State Technol.*, pp. 89–96, July 1994.

[122] D. W. Bondurant, "High performance DRAMs," *Electron. Products*, pp. 47–51, June 1993.

[123] D. Bursky, "Fast DRAMs can be swapped for SRAM caches," *EDN*, pp. 55–68, July 22, 1993.

[124] M. Levy, "The dynamics of DRAM technology," *EDN*, pp. 46–56, Jan. 5, 1995.

[125] R. T. "Tets" Maniwa, "An alphabet soup of memory," *Integrated Syst. Design*, pp. 19–24, May 1995.

[126] K. W. Kwon *et al.*, "Ta_2O_5 capacitors for 1 Gbit DRAM and beyond," in *Proc. 1994 IEDM*, pp. 34.2.1–34.2.4.

[127] Y. Takaishi *et al.*, "Low-temperature integrated process below 500°C for thin Ta_2O_5 capacitor for giga-bit DRAMs," in *Proc. 1994 IEDM*, pp. 34.3.1–34.3.4.

[128] H. Mori *et al.*, "1 GDRAM cell with diagonal bit-line (DBL) configuration and edge operation MOS (EOS) FET," in *Proc. 1994 IEDM*, pp. 26.5.1–26.5.4.

[129] Micron Technology Inc., data sheets for 256K × 32 SGRAM.

[130] Jet Propulsion Laboratory (JPL), Construction Analysis Report for IBM 16 Mb DRAM LUNA-C chip.

3

Nonvolatile Memories

3.1 INTRODUCTION

It is often desirable to use memory devices that will retain information even when the power is temporarily interrupted, or when the device is left without applied power for indefinite periods of time. Magnetic and optical media offer such nonvolatile memory storage. However, a variety of semiconductor memories have also been developed with this characteristic, and are known as nonvolatile memories (NVMs). The ideal nonvolatile memory is one which would offer low cost per bit, high density, fast random access, read/write and cycle times of equal duration, low power consumption, operation over a wide temperature range, a single low-voltage power supply operation, and a high degree of radiation tolerance for military and space system applications. None of the NVMs currently available may have all the listed desirable features, and therefore in real-life applications, selection tradeoffs have to be made.

The first category of nonvolatile memories consists of read-only memories, also known as the masked ROMs (or simply ROMs) in which the data are permanently written during the manufacturing, and cannot be altered by the user. ROMs are usually customized devices procured from the manufacturer by supplying it with a truth table of the circuit logic functions. The manufacturer converts the truth table into

an appropriate ROM mask pattern, which implements the required logic function. Thus, the mask-programmed ROMs have either nonrecurring engineering charges (NRE) or minimum volume requirements, and sometimes both.

Another category of nonvolatile memories is those in which the data can be entered by the user (user-programmable ROMs). In the first example of this type, known as programmable ROMs (or PROMs), the data can be written only once; hence, they are also called OTPs (one-time programmables). The PROMs are manufactured with fusible links (usually made of nichrome, polysilicon, or titanium–tungsten) which can be blown by the user to provide connecting paths to the memory storage elements. The older fuse-link PROMs were generally bipolar technology, but currently, CMOS and BiCMOS devices are also available. Bipolar PROMs are relatively fast and provide good radiation immunity, but they have higher power dissipation and are limited in density. Bipolar and MOS PROMs using vertical polysilicon "antifuses" have been recently introduced, and the ones with amorphous "antifuses" are under development.

PROMs have historically been used for storing macros and programs, and have a much smaller market than the other user-programmable ROMs in which the data can be erased as well as entered. These devices are called erasable-programmable ROMs (EPROMs), electrically erasable PROMs (EEPROMs), and "flash" memories. In general, the PROMs are faster than the EPROMs. However, the PROM cells are much larger in size than the EPROM cells because the fuse element used is large and not merged with the select transistor. Also, the fuse-blowing operation requires higher currents and larger MOSFETs. Low-resistance fuses and antifuses may be less susceptible to the READ-disturb and data-retention problems than the EPROM or EEPROM elements. However, the partially blown fuses and fuse regrowth under severe environmental conditions are significant failure mechanisms for fusible link PROMs, and are discussed in more detail in Chapter 6.

In 1970, Frohman-Bentchkowsky developed a floating polysilicon-gate transistor. In this device, the hot electrons are injected into the floating gate and removed by either ultraviolet internal photoemission or by Fowler–Nordheim tunneling. This device was called EPROM (also known as UVEPROM), and is erased by removing it from the target system and exposing it to ultraviolet light for 20 min. Since the EPROM cell consists of a single transistor, these memories can be made in higher densities, comparable to the DRAMs. The EPROMs are usually supplied in the ceramic packages with quartz windows to provide access for UV light to erase the information programmed in the cells. Since the ceramic packages are expensive, often the EPROMs for commercial applications are programmed according to the user's configuration and supplied in plastic packages as OTP devices.

The user-programmable EPROMs compete with the mask-programmed ROMs for the same applications—mainly program code storage in computers or microcontrollers. For small-volume applications, the ROMs nonrecurring engineering cost is prohibitive, and the EPROMs are often chosen over the ROMs for ease of design change and inventory control. However, the EPROMs do have a few drawbacks. First, although the EPROMs are erasable and reprogrammable, this process is time consuming since the EPROM has to be removed

from the system and exposed to UV light. As previously mentioned, the UV-erase feature necessitates that the EPROMs be packaged in expensive ceramic packages with quartz windows which have to be obscured in the system to prevent an accidental erasure of stored information. Another limitation inherent in the use of EPROMs is the inability to selectively erase memory locations. The whole EPROM memory array has to be erased for reprogramming. This restriction makes the EPROMs ineffective memories for data storage. The limited endurance of EPROMs because of the hot carrier injection causing memory cell degradation is a potential failure mechanism discussed in more detail in Chapter 6.

In the late 1970s, there were many efforts to develop electrically erasable EPROMs using electron tunneling through a thin oxide, or the more conductive interpoly oxide. Most of these early laboratory devices combined tunneling-erase with EPROM's hot carrier programming method. Then, an EEPROM using hot carrier tunneling for the WRITE operation and Fowler–Nordheim tunneling for ERASE was developed. The EEPROM cell, consisting of two transistors and a tunnel oxide, is two or three times the size of an EPROM cell. This cell can be programmed and erased by appropriate polarity high-voltage pulses at the drain and control gates. The hot electrons (or holes) tunneling through the thin gate oxide are trapped at the floating gate. There were further EEPROM technology developments which included the following: (1) a metal–nitride–oxide–silicon (MNOS) combination for the gate region of a p-channel transistor storage cell, and (2) the silicon–oxide–nitride–oxide–semiconductor (SONOS) process. This technology used the Frenkel–Poole effect in which the electrons are trapped within the insulating silicon nitride instead of tunneling through an oxide to a conducting gate.

The floating-gate EEPROMs offer considerable advantages: 5 V-only programmability, no need for erasure before programming, byte and page mode write operations, moderate access times, moderate endurance (write cycles), low-power dissipation, moderate density, data protection lockout, full military temperature operation, and nonvolatility under severe environmental conditions. Many EEPROMs are specified to endure 100,000 or even 1,000,000 write cycles. But the most significant advantages of EEPROMs over the EPROMs are their in-circuit reprogrammability, general usage of a single 5 V supply, and their byte-erase capability.

The design techniques that provide a floating-gate EEPROM with its advantages are also responsible for its drawbacks. The EEPROMs are significantly larger than EPROMs at comparable bit densities because of the chip area taken up by the charge pumps that allow 5 V-only operation and the circuitry that permits byte erasure. Most full-featured EEPROMs have another shortcoming that complicates their use: EEPROMs do not respond to read cycles while performing a store operation since their contents become inaccessible for several milliseconds at a time. Other factors restricting applications of EEPROMs are higher access times and endurance limitations. The number of write cycles is limited due to the trapped charges in the oxide, causing drift in the transistor characteristics. The manufacturing process and device internal structures are complicated, leading to lower component yields, and hence relatively high cost.

In the mid-1980s, the combination of hot-carrier programming and tunnel ERASE was rediscovered as a means for achieving a single-transistor EEPROM cell, and this new technology was called "flash-EEPROM." This technology incorporates the programmability of a UV EPROM and the erasability of an EEPROM. The process of "hot-carrier" injection implemented on the UV EPROMs is used to program these devices, and the Fowler–Nordheim tunneling implemented on EEPROMs is used to remove the charge from the floating gate. Unlike the "full-featured" EEPROMs, flash EEPROMs do not have byte-erasure flexibility, and so they must be erased by the entire chip or large sectors of the chip (flash ERASE). However, the cell size and endurance reliability are improved. Each cell is made up of a single transistor, which simplifies the design and fabrication of 1 Mb and higher density flash-EEPROMs with moderate access time, and in-circuit erase and reprogrammability.

3.2 MASKED READ-ONLY MEMORIES (ROMs)

3.2.1 Technology Development and Cell Programming

Masked (or mask-programmed) ROMs are nonvolatile memories in which the information is permanently stored through the use of custom masks during fabrication. The user must provide the manufacturer with the desired memory bit pattern map, which is then used to generate a customized mask to personalize the ROMs for a specific application. Since the 1970s, these memory circuits have been implemented in bipolar, NMOS, and CMOS technologies in densities ranging from 1 kb to 4 Mb. The faster ROMs have been used to simplify the microprocessor interface by eliminating the wait states, which permits faster program execution. Besides interfacing with microprocessors, the ROMs have been used for various other applications, including look-up tables for mathematical calculations, character generators for alphanumeric display devices, and microcontrol storage.

Some of the early designs were implemented as a 1 kb CMOS ROM supplied in a 16-pin package. The memory matrix for these was organized as four 32×8 b sections. The address decoder flagged one bit (or specific location) in each of the four memory sections to produce a 4 b parallel output. In the early 1980s, the larger 16 and 64 kb ROMs were developed, and more recently, 4 Mb parts are being fabricated by some manufacturers. A ROM requires only one customized mask layer which contains a user-supplied bit pattern for permanent data storage of "1"s and "0"s. This customer-defined mask can be applied either late in the fabrication cycle to give a fast turnaround time, or introduced earlier in the production process for some other tradeoffs, such as small chip size, etc. Figure 3-1 shows a typical mask ROM development flowchart used by the manufacturers.

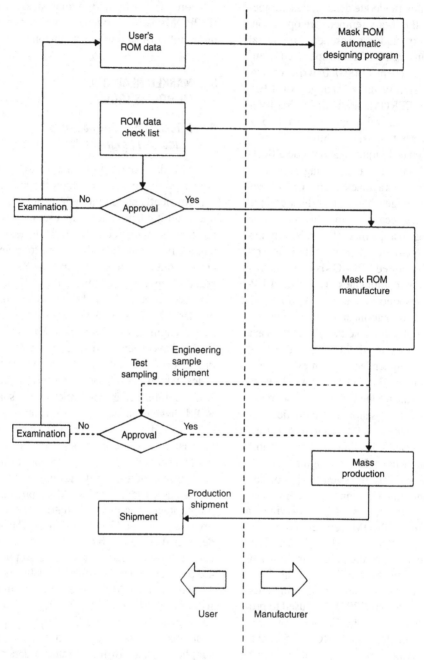

Figure 3-1. A typical mask ROM development flowchart.

Each information bit is stored in a ROM by the presence or absence of a data path from the word access line to a bit (sense) line. The data path is created or eliminated by cell programming during the fabrication process. For example, when the word line of a ROM is activated, the presence of a signal on the bit line implies that a "1" is stored, whereas the absence of a signal will indicate that a "0" is stored. There are three general methods of custom programming or mask programming of a cell: through-hole or contact programming, field oxide programming, and threshold voltage programming.

The through-hole or contact method is the historical ROM programming technique in which the contacts to the cells are opened just before the final interconnect mask. This is done by selectively opening the contact holes for each transistor to the drain. The main disadvantage is that it requires contact for every cell, which increases the size of the cell array and its cost. However, this has a very fast turnaround time since the contact programming is performed on the wafers during the final stages of the process.

Field oxide programming is performed during the early phases of the fabrication process, and uses different thicknesses of the gate oxides to produce different threshold voltages for the "off" and "on" transistors. For example, in a programmed cell, the gate oxide is just the thickness of the field oxide, so that the transistor is permanently "off" or in a logic "0" state. In an unprogrammed cell, a normal gate oxide is used so that the transistor is "on" or in a logic "1" state. Advantages of this technique are the large threshold voltage differential between the programmed and unprogrammed cells, no extra masks, and higher density arrays. The major disadvantage is the higher turnaround time after programming.

Figure 3-2(a) shows the memory cell layout of 128 kb of ROM by Nippon Telegraph and Telephone (NTT) Public Corporation, which has a polysilicon word line and a molybdenum bit line [2]. This ROM uses the field oxide programming method in which a patterned nitride layer is used to define the active areas for transistor channel gate oxides and nonconducting cell field oxides. Figure 3-2(b) shows the cross-section of two ROM cells: the one on the left is a cell with thin gate oxide, corresponding to a logic "1," and the one on the right-hand side is a cell with a thick field oxide corresponding to a logic "0."

The threshold voltage method of programming involves the use of ion implantation to change the threshold voltages of some transistor gates to create conducting and nonconducting cells. Typically, a heavy dose of boron implant is used in the n-channel transistors to raise their threshold voltage and force them permanently into an "off" state. This method has a shorter turnaround time than field oxide programming since the implants can be made through the polysilicon gate just before the contact etch. It also gives a high-density array by eliminating the need for individual cell contacts. However, the heavy implant dose has some side effects which could adversely affect the device reliability.

3.2.2 ROM Cell Structures

A ROM array is usually formed with transistors at every row–column intersection. ROM cell arrays generally have two types of structures: (1) conventional NOR gate type of parallel

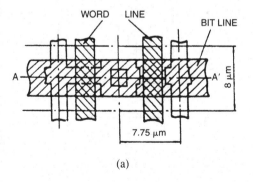

(a)

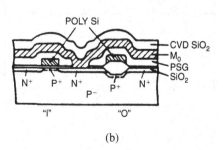

(b)

Figure 3-2. Example of a field oxide threshold programmed ROM cell. (a) Memory cell layout. (b) Cross-section of two ROM cells, corresponding to logical "1" and "0." (From [1], with permission of IEEE.)

structures, and (2) serial NAND gate type structures. Figure 3-3(a) shows NOR gate ROM array implementation in which a set of MOS transistors is connected in parallel to a bit line [2]. Using a positive logic, a stored "1" is defined as the absence of a conducting path to that cell transistor, which means omitting the drain or a source connection, or the gate electrode. In normal operation, all except one row line *is* held low. When a selected row line is raised to V_{DD}, all transistors with the gates connected to that line turn on, so that the columns to which they are connected are pulled low. The remaining columns are held high by the pull-up or load devices. This parallel cell structure has the advantage of high-speed operation from low series resistance, but low density because of large cell size resulting from 1 to 1.5 contact holes for every cell.

Figure 3-3(b) shows NAND ROMs in which the column output goes low only when all series-connected bit locations provide a conducting path toward the ground [2]. In this array structure, all except one of the row conductors are held at V_{DD}. When a selected row line is pulled low, all transistors with gates connected to that line are turned off and columns to which

they are connected are pulled high by the pull-up devices. The data output is taken at the top, as shown in Figure 3-3(b). In positive logic, a stored "1" is defined as the presence of a transistor. A stored "0" is achieved by shorting out the source–drain paths at the desired "0" locations by a diffusion or an implant.

A combination of serial–parallel ROM array structures also has been implemented for speed and bit density tradeoffs. The requirements for high-bit-density ROMs led to several innovative cell structures in the 1980s. For example, Motorola introduced a four-state ROM cell which stored 2 b per cell [3]. This four-state cell was made with varying polysilicon dimensions to vary the *W/L* ratio of the transistor by having an active channel area which was wider at one end than the other. Another cell structure was developed for speed rather than density by using complementary n- and p-channel transistors in the ROM cells [4]. A logic "1" was programmed by a p-channel transistor connected to the positive supply, while a logic "0" was programmed by an n-channel transistor connected to the ground. This cell structure required two access lines for each individual word, one for p-channel and the other for n-channel transistors.

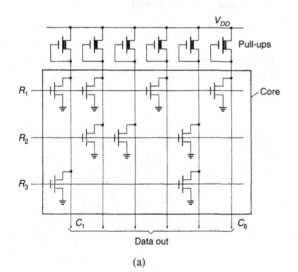

(a)

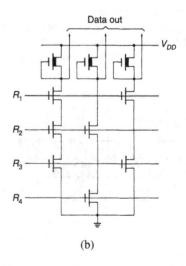

(b)

Figure 3-3. ROM cell structures. (a) NOR array. (b) NAND array. (From [2], with permission of McGraw-Hill Inc., New York.)

3.2.3 High-Density (Multimegabit) ROMs

Mask-programmable high-density ROMs are being manufactured from 1 to 16 Mb densities. Two examples are: (1) Hitachi 4 Mb CMOS ROM organized as 262,144 words $\times$ 16 b or as 524,288 words $\times$ 8 b with 200 ns maximum access time [5], and (2) NEC 16 Mb ROM with 250 ns access time, which can be configured either as 1M $\times$ 16 b or 2M $\times$ 8b. Figure 3-4 shows a block diagram of an NEC 16 Mb CMOS ROM [6].

3.3 PROGRAMMABLE READ-ONLY MEMORIES (PROMs)

This section discusses PROMs, a type of ROMs in which the information can be programmed only once by the user and then cannot be erased. In these PROMs, after the completion of the fabrication process, a data path exists between every word and bit line through a "fuse," usu-

ally corresponding to a stored logic "1" in every cell location. The storage cells are then selectively altered by the user to store a "0" by electrically blowing the "fuse" to open the appropriate word-to-bit connection paths. The user programming is a destructive process, and once a "0" has been programmed into a bit location, it cannot be changed back to "1." The PROMs were originally manufactured in bipolar technology, but are now available in MOS technology as well, including the radiation-hardened CMOS devices.

3.3.1 Bipolar PROMs

Bipolar PROMs have been manufactured using a number of memory cell designs and programming techniques. Two techniques are most commonly used to open the word-to-bit line paths in the desired bit locations. In the first technique, the cells composed of the n-p-n or p-n-p transistor emitter–followers, base–collector diodes, or Schottky diodes are programmed by blowing nichrome, titanium–tungsten, or

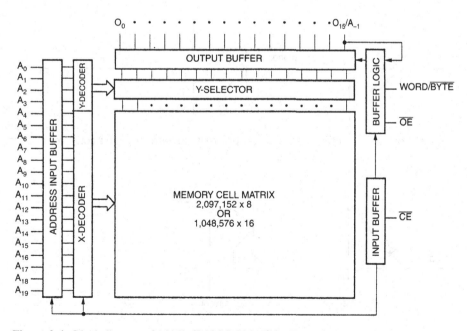

Figure 3-4. Block diagram of 16 Mb CMOS ROM NEC. (From [6], with permission of NEC Electronics Inc.)

polysilicon fuses. Figure 3-5 shows the schematics of a few of these cell designs [7]. For example, in the emitter–follower configuration PROMs, a small fusible link is placed in series with the emitter. These fusible link PROMs are designed to operate with a 5 V supply for normal operation, but need higher voltages (10–15 V) to produce the required current for fusing. The PROM programmers are used to "burn" in the program by providing the current necessary to open the fuses.

In the earlier implementation of bipolar PROMs, the fuse links were made of nichrome thin films which exhibited a "growback" phenomenon such that the blown-apart (disconnected) fuse filaments grew back together (reconnected). This regrowth process that was accelerated for partially blown fuses and under certain environmental test conditions posed a reliability problem, and is discussed in more detail in Chapter 6. The nichrome fusible links were replaced by the polysilicon resistor fuses which are more commonly used now. The investigation of polysilicon resistor programming characteristics have shown them to be more reli-

able. It has been found that an open circuit occurs only after the fuse makes a transition to a second-breakdown state in which the current flow is mainly through a molten filament [8]. This reference paper describes a model which predicts the I–V characteristics of polysilicon fuses in second breakdown, and open polysilicon resistors are guaranteed if the available fusing current is high enough.

For 4–16 kb density PROMs, stacked-fuse bipolar technology has been used. It uses polysilicon resistor fuses and the base–emitter diode for the memory cell which makes direct contact with the emitter region. This diode, in an emitter–follower configuration, provides current gain, relatively good conductance, and self-isolation. A double-level metallization system is used instead of a single to get denser arrays and higher speed. It also helps in reducing the cell size and ensures higher fusing current for a clean and permanent break of the polysilicon fuse. Figure 3-6(a) shows the schematic of 4 b in a stacked-fuse array, and Figure 3-6(b) shows the cross-section of a standard diffused-isolation Schottky bipolar technology used for

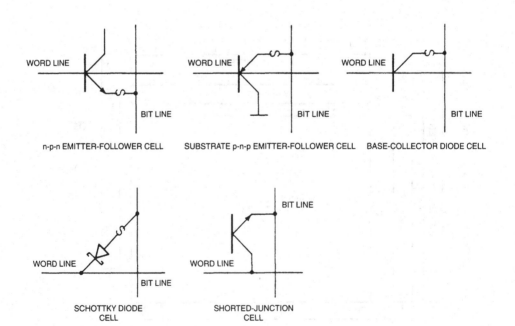

Figure 3-5. Schematics of various PROM cell designs. (From [7], with permission of Penton Publishing, Inc.)

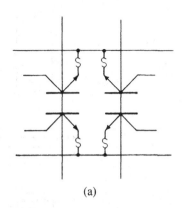

(a)

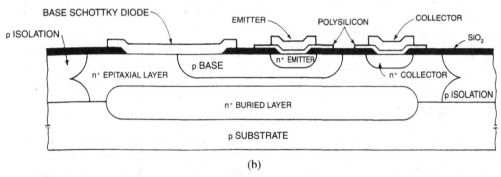

(b)

Figure 3-6. (a) Schematic of 4 b in a stacked-fuse array. (b) Cross-section of a standard diffused-isolation Schottky bipolar technology. (From [7], with permission of Penton Publishing, Inc.)

stacked-fuse memories [7]. The word lines are of bottom-level metal and the bit lines are of top-level metal.

This cell has two main features. First, the bit line (top-level metal) makes direct contact with the bit-line side of the polysilicon fuse instead of using the traditional top-metal-to-bottom-metal contact. Second, the other end of the fuse makes direct contact to the emitter of the emitter–follower circuit element which makes possible the cell size reductions. Also, each bit has its own bit- and word-line contact that virtually eliminates bit-to-bit nonuniformities.

The second popular technique for fabrication of PROMs uses junction shorting for programming instead of blowing fuses. In this technique, the p-n junction diode is diffused using the conventional Schottky wafer fabrication process, and the junction shorting is performed within the silicon bulk by using Si–Al eutectic. Figure 3-7(a) shows the schematic of a memory cell with the junction shorting programmable element before and after programming [9]. A memory cell consists of a programmable element of a p-n junction diode and a vertically connected p-n-p transistor which is used to reduce the output load of the decoder–driver and block the reverse current flow toward the programmable element. The aluminum electrode connected to the cathode of the p-n junction diode is a bit line. The common base of the p-n-p transistors acts as a word line. During programming, the selected word line is pulled down to sink the program current, and reverse current pulses are applied to the selected memory cell through the single bit line. This results in a temperature increase at the junction, and induces shorting of the junction using the diffused Al–Si eutectic. The current-blocking state of the reverse-biased diode represents a

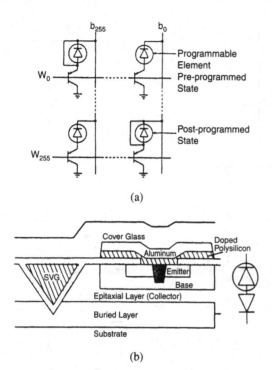

(a)

(b)

Figure 3-7. (a) Schematic of a memory cell with junction-shorting programmable element in the pre- and postprogrammed state. (b) Cross-section of a junction-shorting PROM with a cell in the programmed state. (From [9], with permission of IEEE.)

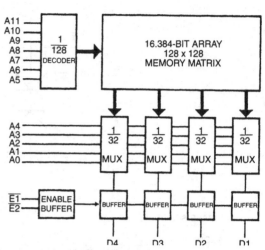

Figure 3-8. Block diagram of NSC DM77/87S195 (16,384 b) TTL PROM. (From [10], with permission of National Semiconductor Corporation.)

logical "0," while the current-conducting state of the shorted junction represents a logical "1." A schematic cross-section of a junction-shorting PROM with the cell in its programmed state is shown in Figure 3-7(b).

Figure 3-8 shows an example of a block diagram of an NSC 16,384 b Schottky-clamped TTL PROM organized as a 4096 word × 4 b configuration [10]. Memory enable inputs are provided to control the output states. When the device is enabled, the outputs represent the contents of the selected word. When disabled, the eight outputs are "OFF" (high-impedance state). It uses titanium–tungsten (Ti:W) fuse links which are designed to be programmed with 10.5 V applied. An advantage with these is that the low programming voltage requirement of 10.5 V eliminates the need for guard-ring de-

vices and wide spacings required for other fuse technologies. It is manufactured with all bits containing logical "0"s. Any bit can be programmed selectively to a logical "1" by following the programming algorithm. This Schottky PROM includes extra rows and columns of fusible links for testing the programmability of each chip. These test fuses are placed at the worst case chip locations to provide the highest possible confidence in the programming test's integrity.

In the last few years, process optimization, programming element (fuse or antifuse) advances, and innovative circuit design techniques have enabled faster and higher density fusible bipolar PROMs. An example is a 128K bipolar PROM organized as 16K × 8 b with 35 ns access time developed by Advanced Micro Devices [11]. It uses stepper technology and dry etch to produce a minimum feature size of 1.2 μm and a cell size of 188.5 μm² (IMDX III process). This process and design includes slot isolation, as well as other circuit techniques such as column current multiplex, to reduce power supply current consumption without penalizing the device performance. The programming element used is platinum silicide, which has demonstrated reliable fusing characteristics. Figure 3-9 shows a block diagram of this 128K PROM.

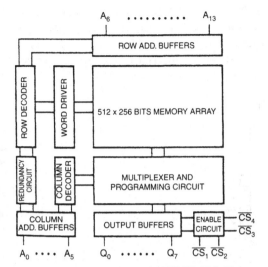

A_6 .... A_{13}

ROW ADD. BUFFERS

ROW DECODER

WORD DRIVER

512 x 256 BITS MEMORY ARRAY

REDUNDANCY CIRCUIT

COLUMN DECODER

MULTIPLEXER AND PROGRAMMING CIRCUIT

COLUMN ADD. BUFFERS

OUTPUT BUFFERS

ENABLE CIRCUIT — $\overline{CS_4}$ — $\overline{CS_3}$

A_0 •••• A_5 Q_0 •••••• Q_7 $\overline{CS_1}$ $\overline{CS_2}$

Figure 3-9. Block diagram of 128K PROM. (From [10], with permission of IEEE.)

A programmable low-impedance circuit element (PLICE), which is a dielectric-based "antifuse" developed by Actel Corporation, has also been used as a programming element for PROMs [12]. This antifuse consists of a dielectric between an n^+ diffusion and polysilicon, which is compatible with bipolar, CMOS, and BiMOS technologies. Each antifuse is incorporated into a cell structure such that when the antifuse is programmed, it connects a metal 1 and metal 2 line. It can be programmed within 1 ms, and has a tight resistance distribution centered around 500 Ω. However, this antifuse-based programming element for PROMs has found very limited usage. Another antifuse technology has been investigated by Texas Instruments (TI), which utilizes a thin film of undoped amorphous silicon formed in a contact window [13]. This can be programmed from a normally open state to a conductive state by the application of a bias with a potential high enough to cause a destructive breakdown, but at a significantly reduced programming current compared to the existing fusible link technology and the avalanche-induced migration vertical fuse technology.

A new programmable cell, described as the breakdown-of-insulator-for-conduction (BIC) cell, has been developed by Fujitsu which uti-

lizes the electrical breakdown of an insulator for programming [14]. A refined thin insulator with a delta-function type of breakdown voltage is used in the BIC cell. This cell can be programmed within 1 μs, and has a programmed cell resistance of about 100 Ω. Since a BIC cell has stacked-cell structure, it can be merged with a MOSFET to produce a PROM cell and form arrays.

3.3.2 CMOS PROMs

In CMOS technologies, there are three different approaches available to implement nonvolatile programmable devices. The first approach is similar to those for bipolar devices, and uses polysilicon fuses (blown or diffused) for cell programming. However, the densities are limited because of the scalability problems associated with these PROM cells. The second approach, which uses erasable and reprogrammable cells (EPROM and EEPROM) that are scalable to 1 Mb and higher densities, are discussed in Sections 3.4 and 3.5.

The PROMs that have latched address registers are called registered PROMs. An example of a blown polysilicon fusible link nonvolatile memory using a parasitic bipolar transistor in the fuse element is the Harris 16K CMOS registered PROM [15]. For military and space applications, the PROMs are required to have high radiation tolerance to prevent functional failure from accumulated total dose or heavy ion induced latchup. Manufacturers use various techniques for radiation hardening, some of which are proprietary designs. For example, Harris Semiconductors has used a self-aligned poly-gate, junction-isolated process by the insertion of a highly doped guardband around the edge of the p-well. The presence of this guardband greatly increases the amount of charge trapped at the interface necessary to invert the silicon surface and allow leakage currents to flow. The disadvantage of this technique is that it consumes valuable chip real estate. Some other methods include the use of radiation-hardened field oxides and gate oxides. An epitaxial layer of appropriate thickness is used to prevent latchup from heavy ions present in the space environment. These techniques will be discussed in more detail in Chapter 7.

An example of a radiation-hardened non-volatile memory used in space applications is the Harris 16K CMOS PROM, organized in a 2K word × 8 b format which uses nichrome fuses as the programmable elements [16]. Figure 3-10 shows the Harris 16K (organized as

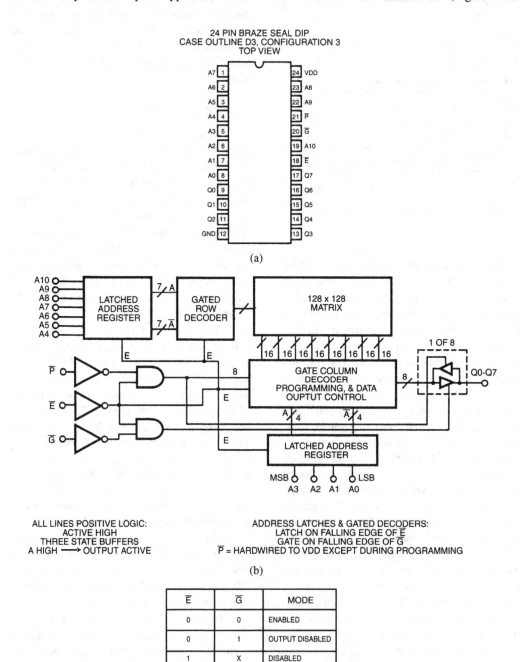

Figure 3-10. Harris 16K CMOS PROM (HS-6617RH). (a) Pinouts. (b) Functional diagram. (c) Truth table. (From [16], with permission of Harris Semiconductor.)

2K × 8) CMOS PROM (a) pinouts, (b) functional diagram, and (c) truth table. This PROM uses synchronous circuit design techniques combined with CMOS processing to give high speed (100 ns access time) and low power dissipation.

3.4 ERASABLE (UV)-PROGRAMMABLE READ-ONLY MEMORIES (EPROMs)

This section describes EPROM, which is a ROM that can be electrically programmed in a system by using a special 12 V programming pin, but has to be erased by removing from the system and exposing the ceramic package with a quartz window to ultraviolet light for about 20 min. The UV EPROM has a one-transistor cell which enables its fabrication in densities comparable to DRAMs, but the expensive package and reprogrammability constraints keep it from being a low-cost nonvolatile memory. The part damage can occur from extra handling problems involved in the erase process. Also, the test time becomes significantly longer due to a lengthy erase procedure. Another problem is the data-retention sensitivity of the EPROMs to long exposures to daylight or fluorescent lights. In commercial applications, a low-cost alternative offered by the manufacturers is the one-time programmable (OTP), which is a plastic-packaged EPROM that is functionally a PROM. It can be programmed only once, and cannot be erased.

In EPROMs, the charge is stored on a floating polysilicon gate (the term "floating" implies that no electrical connection exists to this gate) of a MOS device, and the charge is transferred from the silicon substrate through an insulator. The physical mechanism for the charge transfer is called the Fowler–Nordheim electron tunneling, a quantum–mechanical effect in which the electrons (pass) tunnel through the energy barrier of a very thin dielectric such as silicon dioxide. The following subsections discuss the floating-gate MOS transistor theory, EPROM cell structures, technology developments, OTPs, and high-density EPROMs.

3.4.1 Floating-Gate EPROM Cell

The EPROM transistor resembles an ordinary MOS transistor, except for the addition of a floating gate buried in the insulator between the substrate and the ordinary select-gate electrode (control gate). Figure 3-11(a) shows the basic EPROM transistor in which two layers of polysilicon form a double-gate structure with Gate F acting as a floating gate and Gate C for the cell selection as a control gate. Figure 3-11(b) shows the circuit symbol with drain (D), source (S), and body (B). The two gate capacitances in series are represented by

$$C_F = \text{floating-gate capacitance}$$
$$C_C = \text{control-gate capacitance.}$$

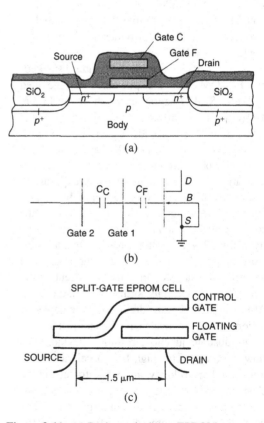

Figure 3-11. (a) Basic stacked-gate EPROM transistor. (b) Circuit symbol with equivalent control- and floating-gate capacitances. (c) Split-gate EPROM cell by Wafer Scale Integration [17].

The operation of this EPROM depends on its ability to store charge on the floating gate which is surrounded by the insulator (SiO_2). The charge on the floating gate can remain stored for a long time. However, it can be erased by exposing the cells to ultraviolet light (UV), which slightly enhances the conductivity of SiO_2 by generating electron–hole pairs in this material.

To explain the operation of the cell, let us assume that there is no charge on the floating gate (C_F), so that with the control gate, drain, and source all grounded, the potential of the floating gate is 0 V. When the voltage on the control gate (C_C) is increased, the floating-gate voltage also increases, but at a lower rate given by the capacitive differential of $C_C - C_F$. This creates an effect of raising the threshold voltage of this transistor with respect to the control gate. When the threshold voltage is high enough (roughly twice the normal V_T), a channel is formed and the device provides a positive logic stored "0" in a NOR array.

Another variation on the standard stacked-gate EPROM cell is the split-gate cell developed by Wafer Scale Integration, as shown in Figure 3-11(c) [17]. According to the manufacturer, this split-gate cell can provide higher read currents, and therefore access time as low as 35 ns.

The programming of EPROMs begins by discharging all the floating gates by exposure to UV radiation so that every cell initially stores a logic "0." To write a "1" into this cell, both the control gate and drain are raised to a high voltage (typically 12 V or higher for a few hundred microseconds), while the source and substrate are at ground potential. A large drain current flows through the device in its normal conduction mode. Also, a high field in the drain–substrate depletion region causes an avalanche breakdown of the drain–substrate junction, resulting in additional flow of current. This high field accelerates electrons to a high velocity, and while traversing the channel, they enter the substrate region where the electric field is about 10^5 V/cm or greater. At this point, the rate of energy gained from the electric field can no longer be attributed to the silicon temperature; hence, the term "hot electrons" is used.

Once the "hot electrons" gain sufficient energy, they can surmount the energy barrier of about 3.2 eV existing between the silicon substrate and SiO_2 insulator. A small fraction of electrons traversing the thin oxide becomes trapped on the floating gate which, due to capacitive coupling, has a more positive potential than the drain. If the floating gate is charged with a sufficient number of electrons, a channel inversion under the gate will occur. When the voltage on the control gate and drain are returned to 0 V, these electrons remain trapped on the floating gate, and will cause its potential to be about -5 V. Therefore, during the EPROM read operation, when a signal of only 5 V is applied to the control gate, no channel would be formed in the transistor. This denotes a logic "1" stored in the cell. Therefore, in a programmed device, the presence of a logic "1" or "0" in each bit location is determined by the absence or presence of a conducting channel.

The charge-transfer mechanism of EPROMs is based upon the avalanche breakdown of a reverse-biased drain–substrate p-n junction, causing the hot electrons to be injected into the floating polysilicon gate. As such, the early EPROM devices were called the floating-gate, avalanche-injection MOS transistors (FAMOS) [18], [19]. The operation of this FAMOS memory structure depends on the charge transport to the floating gate by avalanche injection of electrons from either the source or drain p-n junctions, as explained in the basic EPROM cell operation.

The amount of charge transferred to the floating gate is a function of the amplitude and duration of the applied junction voltage. After the applied junction voltage is removed, no discharge path is available for the electrons accumulated on the floating gate which is surrounded by high-resistivity thermal oxide. The remaining electric field is due only to the accumulated electron charge, which is insufficient to cause charge transport across the polysilicon-thermal-oxide energy barrier. This induced field on the floating gate then modifies the conductivity of the underlying channel region. The capacitive coupling of the floating gate can strongly distort the I–V characteristics of a floating-gate MOS transistor [20]. The device equations for a floating-gate transistor are derived in the reference paper by Wang [20].

At the beginning of the hot electron injection process, the inversion layer in the substrate beneath the floating polysilicon gate extends almost all the way to the drain. As the floating gate charges up, the floating gate-to-source voltage drops and the drain pinch-off region moves closer toward the source. The electron injection process is self-limiting. As the floating gate becomes fully charged, the injection current drops close to zero since the oxide field now begins to repel the electrons injected into the high-field region. After programming, the electrons are trapped in the polysilicon floating gate, as shown in Figure 3-12, the energy-band diagram of the FAMOS device after programming.

The EPROMs are programmed by using the manufacturer's supplied algorithms to assure that the cells are programmed correctly with the right amount of charge. A major concern with the high-voltage EPROM programming process is the electron trapping in the oxide after several programming cycles. These electrons can remain trapped in the oxide, even after UV exposure of the floating gate. As a result, they can potentially raise the threshold voltage of the channel to a point where the cell states may be sensed incorrectly. This condition is avoided by a proper growth of gate oxide underneath the floating polysilicon gate. Another programming problem may occur when the control-gate voltage is raised high for a device in a row that contains another device which has already been programmed. During operation, this would tend to increase the field strength in the upper gate oxide (between the control gate and floating polysilicon gate) where the

average field strength is roughly 1 MV/cm. A substantial increase in local electric field from surface asperities in the first polysilicon layer may cause partial erasure of the floating-gate charge. This field emission effect can be minimized by the use of specially optimized processing steps. The EPROM data-retention reliability issues will be discussed in more detail in Chapter 6.

An EPROM has to be erased completely before it can be reprogrammed. This is done by exposing the entire memory array to a UV light source such as a mercury arc or a mercury vapor lamp which emits radiation of wavelength 2537 Å (4.9 eV). The incident photons are absorbed by the electrons in the conduction and valence bands of the floating polysilicon gate. Most of the absorption occurs within 50 Å of the oxide interface. The electrons excited from the polysilicon gate entering the oxide are swept away to the control gate or substrate by the local field. During the erase process, the control gate, source, drain, and substrate are all held near ground potential. The UV erase process is a photoelectric process. The quantum yield of this process is defined as the ratio of the number of electrons transferred through the oxide per unit time to the number of photons absorbed by the floating gate in the same time interval.

A physical model has been presented to explain the various features of the UV erase process in FAMOS EPROM devices [21]. This model defines the erase sensitivity factor as being proportional to the floating-gate photoinjecting area, and inversely proportional to the oxide thickness and total capacitance of the floating gate. The photoinjection of electrons from the thin strips on the floating-gate edges to other available terminals is responsible for the charge removal from the floating gate. With an n-type polysilicon floating gate, electrons can be excited from the conduction band or the valence band. For an excitation from the conduction band, the required activation energy is 3.2 eV, whereas from the valence band it is 4.3 eV. An inherent problem with EPROMs is the accidental erasure from exposures to fluorescent and incandescent lamps emitting a wavelength of about 3000 Å, which corresponds to 4.1 eV. Hence, the programmed EPROMs must be shielded from long exposures

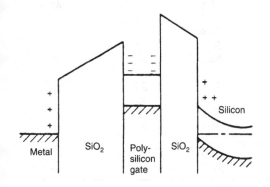

Figure 3-12. Energy-band diagram of a FAMOS device after programming.

to fluorescent lighting or sunlight. EPROMs are also affected by the gamma radiation and space environments consisting of protons and other heavy ions. These radiation effects will be discussed in more detail in Chapter 7.

3.4.2 EPROM Technology Developments

The first generation of UV EPROMs was fabricated in NMOS technology in densities ranging from 2 to 16 kb, and often used several power supplies for various read and program functions. The first popular industry standard 16 kb EPROM was introduced by Intel (2716), which used a single +5 V supply for the read function and a +25 V supply for programming. At the 64 kb level density, the 28-pin series EPROM was accepted by the JEDEC Memory Committee, and became the industry standard for densities ranging up to 1 Mb. In the early 1980s, several manufacturers introduced CMOS EPROMs which had a smaller chip size and lower power consumption than their equivalent NMOS counterparts.

The first 16 kb CMOS EPROMs introduced by Intersil had a dual polysilicon gate, single n-channel transistor cell in a p-well [22]. All following generations of EPROMs used n-well technology. Several innovative circuit design techniques were introduced, such as the use of differential sense amplifiers whose differential inputs were precharged before sensing to raise their sensitivity. Because of concerns with CMOS latchup during high-voltage programming, the voltage was clamped to minimize p-well injection current. An internal synchronous circuitry was used to reduce the standby power consumption. A 288 kb CMOS EPROM manufactured by Fujitsu had 150 ns access time and used n-well technology with CMOS peripheral circuitry [23]. It provided an organization flexibility of $\times 8$ or $\times 9$ (an extra parity bit added for redundancy). The channel lengths were 2.0 μm for NMOS and 2.5 μm for PMOS transistors, with a total cell size of 54 μm^2. Table 3-1 shows the technological advances and process characteristics of EPROMs ranging in densities from 16 kb to 16 Mb, as reported at various ISSCC Symposia.

TABLE 3-1. Technological Advances and Process Characteristics of EPROMs Published in Various *IEEE ISSCC Technical Digests*

Date	Company	Density	Channel (μm)	Cell (μm^2)	Chip (mm^2)	Gateox (nm)	Interox (nm)
1981	Intersil	16K	5	—	—	65	80
1982	Fujitsu	64K	2.5	133.6	20.8	—	—
1983	Signetics	64K	—	—	18.8	65	70
1983	Fujitsu	288K	2.0	54	34.6	50	—
1984	SEEQ	256K	1.5	37.5	20.9	45	65
1984	Toshiba	256K	—	—	34.7	—	—
1985	Hitachi	256K	—	36	21.4	—	—
1985	Intel	256K	1.0	36	25.6	35	45
1985	Toshiba	1 Mb	1.2	28.6	60.2	28	45
1985	Hitachi	1 Mb	1.0	19.3	39.4	—	—
1986	AMD	1 Mb	1.5	20.25	50.9	25	30
1987	T.I.	1 Mb	1.2	13.5	53.5	35	—
1987	Fujitsu	1 Mb	2.0	18.9	45.4	—	—
1988	SGS-T	1 Mb	1.0	18.9	45.7	25	30
1988	Hitachi	1 Mb	1.2	18.5	42.7	35	—
1988	WSI	256K	1.2	42.2	30.1	27	—
1989	Cypress	256K	0.8	39.7	25.3	24.5	30
1990	Toshiba	1 Mb	0.8	30.24	77.3	—	—
1990	Toshiba	4 Mb	0.9*	8.9	85.9	24†	—
1991	Toshiba	16 Mb	0.6*	3.8	124.9	20†	—

Notes: *Represents cell transistor gate length.
　　　†Represents peripheral transistors gate oxide thickness.

3.4.2.1 1 Mb EPROM (Example). Figure 3-13(a) shows a block diagram of a Fujitsu 1 Mb CMOS UV EPROM, organized in a 131,072 word/8 b or 65,536 word/16 b format [24]. This is fabricated using CMOS double-polysilicon technology with stacked single-transistor gate cells. Once programmed, the device requires a single +5 V supply for operation, and has a specified access time of 150 ns maximum. It has multiplexed address and data pins which permit the device to reduce the number of pin-counts for portable systems where a compact circuit layout is required. It requires a programming voltage of 12.5 V. Figure 3-13(b) shows the programming waveforms.

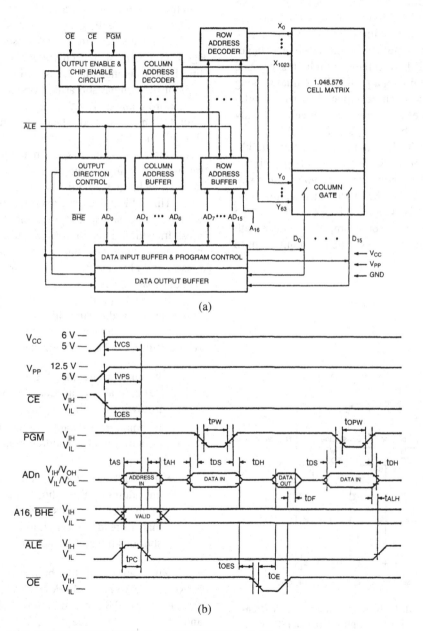

(a)

(b)

Figure 3-13. 1 Mb CMOS UV EPROM Fujitsu MBM27C1028. (a) Block diagram. (b) Programming waveforms. (From [24], with permission of Fujitsu Limited.)

3.4.3 Advanced EPROM Architectures

The data sensing in early EPROMs was performed with a single-ended sensing scheme which sensed the potential difference between a bit sense line and a reference line. The bit sense line potential depends on the state of a selected EPROM cell, whether it is erased or programmed. Other conventional sensing schemes have used fully differential techniques that require a pair of bit lines (two transistors/cell), and therefore result in speed delays and larger chip sizes [25]. To achieve both high speed and a smaller chip size, a pseudodifferential sensing technique has been developed that uses single-ended bit lines (one transistor per cell) and only two reference bit lines [26]. This technique has been used by Toshiba to fabricate in 0.9 μm lithography a 1 Mb EPROM with double-polysilicon CMOS process, polycide technology, and a single metal layer that has a 36 ns access time.

Figure 3-14 shows a block diagram of the pseudodifferential sense amplifier circuit with "0" data reference cells which are equivalent to

the programmed memory cells, and "1" data reference cells equivalent to erased memory cells. Data sensing is performed by comparing results between the selected memory cell with both "0" and "1" reference cells. However, if the memory cell is in the erased state ("1"), data sensing is performed by comparing the "1" memory cell with the "0" reference cell. This sensing scheme, which uses one transistor/cell, provides high-speed access time and stable sensing. At the first stage sense circuit, the sense line voltage is compared with each voltage of reference line 1 and reference line 2, respectively. The results of the comparison at this first stage sense circuit are compared at the second stage sense circuit. These sensed data are transferred through the data transfer circuit to the output buffer circuit. For a high-speed read operation, a bit line equalizing technique and a sensed data latching technique are utilized.

This EPROM uses a data transfer circuit whose data transfer speed is controlled by an address transition detection (ATD) pulse to get a high noise immunity against the power line noise caused by charging and discharging of the output capacitances.

For high-speed access EPROMs, techniques such as address transition detection (ATD) are used. The main requirements for the design of ATD circuitry can be summarized as follows:

- It must ensure that the shortest clock pulse created is adequate to trigger the following circuits.
- It should incorporate a valid sensing period after the final address transition occurs.
- It should have the capability to screen out address pulses shorter than a minimum specified pulse width to ensure that no read cycle is initiated from the spurious inputs (transients).

An example of the high-speed CMOS EPROM is a 16 ns, 1 Mb device developed by Toshiba using 0.8 μm minimum design rules and a double-metal process [27]. For high-speed sensing operations, it uses a differential sensing

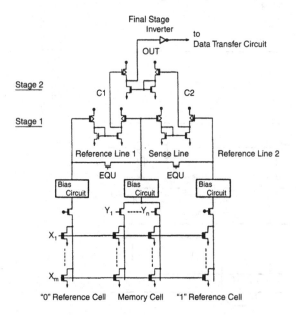

Figure 3-14. Block diagram of the pseudodifferential sense amplifier. (From [26], with permission of IEEE.)

scheme with ATD. To reduce the *RC* word-line delay inherent in large conventional EPROMs, a double word-line structure is used in the double-metal process. This device has a 16 b wide data bus and large noise generated in the output buffers. For sense amplifier high noise immunity, a bit-line bias circuit with a depletion load transistor is used. The data-out latch circuit is implemented to prevent the output data glitches which can induce ground bounce noise. Another technique used to achieve fast access time is by guaranteeing sufficient threshold voltage shift for the memory transistors. This is done by a threshold voltage monitoring program (TMP) scheme consisting of a single-ended sense amplifier parallel to a normal differential sense amplifier and transfer gates, which select a right or left cell to verify.

Figure 3-15(a) shows the 1 Mb CMOS EPROM sense amplifier circuit-structure in which complement data bits are programmed into a pair of stacked gate transistors that form one memory cell (differential) [27]. The differential sense amplifier is composed of three-stage differential sense circuits, each of which consists of paired current-mirror circuits to obtain highly symmetrical outputs. The individual complementary nodes are equalized by the ATD pulse to increase the sensing speed. Figure 3-15(b) shows the memory cell array structure that is used to reduce the word-line and bit-line delays. It consists of 512 rows × 2048 columns. The column lines are divided into eight sections, and each section word line contains only 128 cells, which reduces the word-line delay to less than 2 ns. As only 512 cells are connected to each bit line, the bit-line capacitance (and hence the bit-line delay) is also significantly reduced. Figure 3-15(c) shows the threshold monitor program (TMP) circuit.

Another approach to improve the read access time and programming characteristics has been to use new transistor structures for memory cells. Cypress Semiconductor, for the manufacturing of their low-density, high-speed EPROMs, has used a four-transistor differential memory cell [28], [29]. This cell was optimized for high read current and fast programmability

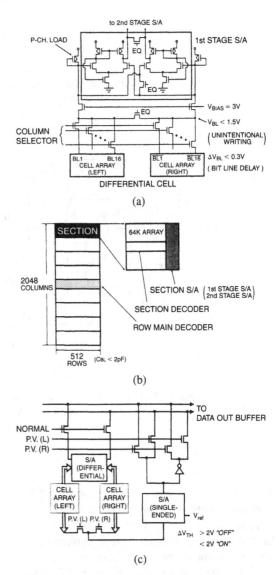

Figure 3-15. A 1 Mb CMOS EPROM. (a) Sense-amplifier circuit structure. (b) Memory cell array structure. (c) Threshold monitoring program (TMP) circuit. (From [27], with permission of IEEE.)

by using separate read and program transistors, as shown in Figure 3-16(a). The program transistor had a separate implant to maximize the generation and collection of hot electrons, while the read transistor implantation dose was chosen to provide large read currents. The two sets of

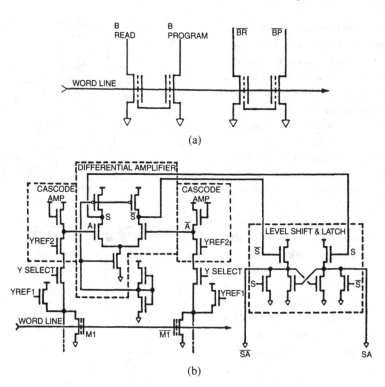

Figure 3-16. (a) A four-transistor memory cell with read and program transistors implanted separately. (b) Four-transistor memory cell
with differential sensing scheme. (From [28], with permission
of IEEE.)

read and program transistors were used for differential sensing, along with a three-stage sense
amplifier as shown in Figure 3-16(b). These
EPROMs have applications in high-performance
microprocessor systems with "zero-wait states."

A novel source-side injection EPROM
(SIEPROM) structure has been introduced
which has several advantages over the conventional EPROM cell [30]. The cell is an asymmetrical n-channel stacked gate MOSFET, with a
short weak gate-control region introduced close
to the source. Under a high gate bias, a strong
channel electric field is created in this local
region, even at a relatively low drain voltage
of 5 V. The programming speed for these
SIEPROMs, even at the low programming
voltage of 5 V, is much faster than that of the
drain-side injection PROMs. This high-speed
programming is desirable for the EPROMs

greater than 1 Mb density. Another advantage of
this structure is that the low drain programming
voltage is well below the device breakdown,
which gives a larger tolerance for the design and
process control variations.

The EPROMs are traditionally used with
high-performance microprocessors which require
fast read and write, and "byte-wide" organizations. The demand for high-density EPROMs has
increased. Advances in cell structures, scaling,
and process enhancements have made possible
the fabrication of 4–16 Mb EPROMs. As described earlier, the standard EPROM cell uses locally oxidized silicon (LOCOS) to isolate the
individual bits. A half contact per cell is required
to connect the drain diffusions to the metal bit
lines. An example of the new cell development is
the true cross-point EPROM cell which uses
buried n^+ diffusions self-aligned to the floating-

gate avalanche-injection MOS (FAMOS) transistor for the bit lines [31]. These diffusions are covered with a planarized low-temperature CVD oxide which isolates them from the orthogonal set of polysilicon word lines. This new cell, called SPEAR for self-aligned planar EPROM array, eliminates the bit-line contacts and the LOCOS isolation used in the standard EPROM cells.

Figure 3-17(b) shows the top view of this cross-point SPEAR cell (4 μm²) which is half the size of the conventional EPROM cell (9 μm²), as shown in Figure 3-17(a), both drawn by using the 1.0 μm design rule. Figure 3-17(c) shows a three-dimensional diagram of the EPROM array using the SPEAR cell. A 9 μm²

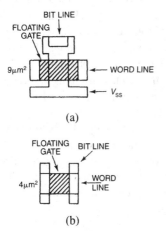

(a)

(b)

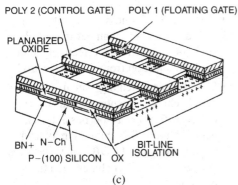

(c)

Figure 3-17. (a) Conventional 9 μm² EPROM cell.
(b) Cross-point 4 μm² SPEAR cell.
(c) Three-dimensional diagram of EPROM array using cell shown in (b).
(From [31], with permission of IEEE.)

SPEAR cell has been used to fabricate a 120 ns, 4 Mb CMOS EPROM [32]. Toshiba has fabricated 4 Mb EPROMs using innovative NAND gates instead of conventional NOR gate structures [33]. This NAND gate cell can be programmed by hot electron injection to the floating gate, and can be erased either by UV irradiation or by electric field emission of the electrons from the floating gate (flash-EEPROM mode).

The major requirements for 4 Mb and higher density EPROMs have been short programming time, fast access times, low power dissipation, and low cost. In the case of 4 Mb devices, these requirements have translated into smaller chip sizes with a cell size of 8 μm² or smaller. For 16 Mb generation of EPROMs with cost-effective die sizes, the cell area would have to be less than 4 μm². A new scaling guideline has been proposed for 16 Mb EPROMs using 0.6 μm design rules [34]. This method is based upon the investigation of scaling effects on cell reliability and performance. It introduces two scaling factors "k" (for lateral dimensions: L, W, x_j) and "h" (for vertical dimensions: tox_1, tox_2), and then derives the relationship between the two scaling factors to satisfy the reliability and performance constraints simultaneously.

Another high-performance process and device technology has been proposed for 16 Mb and higher density EPROMs using 0.6 μm design rules and a novel cell structure called the LAP cell [35]. It utilizes the following advanced processes and technologies:

- A large-tilt-angle implanted p-pocket (LAP) cell structure with 0.6 μm gate length
- A self-aligned source (SAS) technology
- A poly-Si plugged contact technology for CMOS devices by a novel multistep poly-Si deposition method
- A 0.8 μm poly-Si shield isolation structure for high-voltage circuits
- Advanced lithography techniques.

Figure 3-18 shows the LAP cell cross-section and key process sequence. This advanced

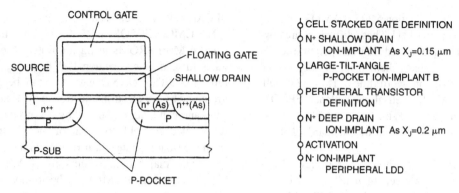

Figure 3-18. Large-tilt-angle implanted p-pocket (LAP) cell cross-section and key process sequence. (From [35], with permission of IEEE.)

EPROM process has the potential of scaling to 64 Mb and higher densities. Another 16 Mb EPROM process developed features 0.2 μm "bird's beak" isolation, lightly doped drain (LDD) n-channel and p-channel transistors, 20 nm interpoly dielectrics, and rapid thermal annealing [36]. This process resulted in a standard T-shaped cell less than 4.5 μm² in area and an even smaller cross-point type cell.

The scaling down of interpoly dielectric thickness is one of the most important issues for nonvolatile memories such as EPROMs. A study of thin interpoly dielectrics in stacked-gate structures has shown some intrinsic reliability problems. In polyoxide grown on poly 1 with a high concentration of phosphorous, an increase in the oxide defects and degradation of oxide quality is observed [36]. For polyoxides grown on low phosphorous-doped polysilicon, relatively reliable oxide can be formed on unpatterned polysilicon. However, a drastic increase in leakage current is observed when oxide is formed on patterned polysilicon.

It has been shown that oxide–nitride–oxide (ONO) structures have several advantages, such as smaller dependence on phosphorous concentration and lower defect densities. However, there are also some limitations on the use of ONO interpoly dielectrics for EPROM cells, particularly in the UV-erase speed which becomes slower compared to that for the polyoxide interpoly dielectrics. The degradation in UV speed for ONO cells is mainly caused by the decrease in photocurrent through the ONO interpoly dielectric. This phenomenon is closely related to the electron trapping and carrier transport mechanisms in the ONO films under UV light irradiation.

The relationship between the charge retention characteristics and ONO thickness composition have been investigated. For charge retention considerations, the ONO scaling methodology can be summarized as follows [37]:

- Top oxide thickness should be maintained at more than critical values (~3 nm), which can effectively block the hole injection from the top electrode.

- SiN thickness scaling can usually improve charge retention, and thinning down to 5 nm may be possible. However, too much SiN thinning may lead to degradation in TDDB characteristics and adversely affect charge retention capabilities.

- Thinning of the bottom oxide layer down to 10 nm does not lead to degradation, provided the bottom oxide quality can be well controlled.

3.4.4 One-Time Programmable (OTP) EPROMs

One-time programmable EPROMs have been introduced by the manufacturers to make them cost-competitive in the high-volume applications ROM market. The high cost of EPROMs results from their use of ceramic packages with quartz windows which permit the UV light to reach the chip during the erase operation. These ceramic packages are substantially more expensive than the molded plastic packages. A cost-effective alternative has been the OTP EPROMs, which are plastic-packaged parts programmed by the manufacturer to the customer-supplied data pattern (code). The parts are then tested like ordinary ROMs before shipment. A disadvantage of OTP EPROMs is their testing limitations since the part cannot be erased after assembly. This makes it difficult to verify the integrity of the programmed cells before shipment, or to perform final erase in case testing has partially programmed some of the cells. One of the solutions has been to use redundancy by including additional rows of cells on the chip. Faster EPROMs have used two or four transistor cells

with differential sensing techniques that allow the working cells to be checked without actually reading them.

Advanced Micro Devices offers another version of EPROMs, called the ExpressROM™ family. These are manufactured with the same process as AMD's standard UVEPROM equivalent, with the topside passivation layer for plastic encapsulation. This is identical in architecture, pinout, and density with both their current and future generation CMOS EPROM devices. These ExpressROM devices have shorter lead times, and can even cost less than the OTP EPROMs.

Figure 3-19(a) shows a block diagram of Advanced Micro Devices 4 Mb UV EPROM organized as 256K words × 16 b/word, which is available in both a windowed ceramic-packaged part, as well as in a plastic-packaged OTP version [38]. It operates from a single +5 V supply, and bit locations may be programmed singly, in blocks or at random. The separate Output Enable ($\overline{OE}$) and Chip Enable ($\overline{CE}$) controls are used to eliminate bus contention in a multiple bus microprocessor environment. Figure 3-19(b) shows the Flashrite™ programming algorithm (100 μs pulses) which results in typical programming time of less than 2 min.

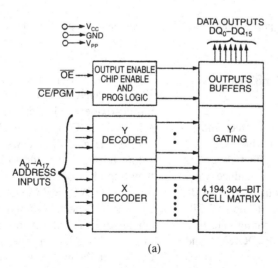

(a)

Figure 3-19. (a) Block diagram of Advanced Micro Devices 4 Mb CMOS EPROM Am27C4096.

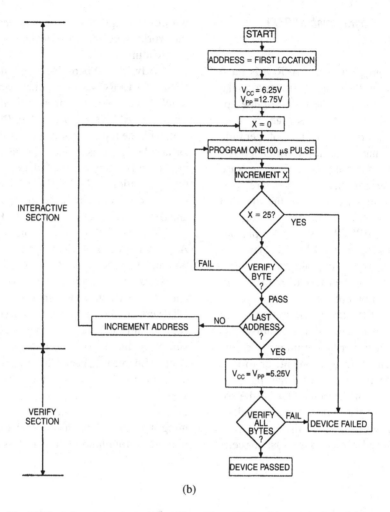

(b)

Figure 3-19 (cont.). (b) Flashrite programming algorithm. (From [38], with permission of Advanced Micro Devices.)

3.5 ELECTRICALLY ERASABLE PROMs (EEPROMs)

EEPROM takes advantage of an effect known as Fowler–Nordheim tunneling which was referenced in Section 3.4. This effect describes how a strong electric field can induce low-energy electrons to tunnel through a thin oxide into the floating gate. In the best known EEPROM technology, known as floating-gate tunneling oxide (FLOTOX), the PROGRAM and ERASE operations are carried out by electrons tunneling through the thin oxide (roughly 100 Å). Another EEPROM technology that was developed earlier is the metal–nitride–oxide–silicon (MNOS) and (poly)silicon–oxide–nitride–oxide–semiconductor (SONOS). Both MNOS and SONOS use techniques similar to the floating-gate EEPROMs, but instead of tunneling through an oxide to a conducting gate, they trap electrons within the insulating silicon nitride [41]. A typical MNOS memory transistor has a gate dielectric of about 30 Å of thermal silicon–dioxide and several hundred angstroms of deposited silicon nitride. The electric charge is stored in the nitride, whereas charge injection and removal are effected through tunneling via the very thin oxide. As mentioned earlier, a popular MNOS (or SONOS) technology implementation is the nonvolatile RAM (NVRAM), which combines SRAM and EEPROM on a single monolithic substrate. The MNOS technology has better endurance reliability and relatively higher radiation tolerance, but provides little margin in data retention due to the very thin oxide used. Also, the MNOS write speed is rather slow.

Another alternative to the tunneling oxide type of devices is the textured-polysilicon EEP-ROMs, which are also based on the floating-gate MOS technology. In this technology, the cell consists of three layers of polysilicon that partially overlap to create a storage structure which behaves like three MOS transistors in series. Textured-poly EEPROMs depend on the tunneling process whose physical mechanisms are not as well understood as the thin oxide devices and require tighter control over the process parameters. Also, three polysilicon layers require a more complex and costlier fabrication sequence. Therefore, this approach has not found any wide commercial usage. Three main EEPROM technologies, that is, MNOS (and SONOS), FLOTOX, and textured-polysilicon, are discussed in more detail in the following sections.

3.5.1 EEPROM Technologies

3.5.1.1 Metal–Nitride–Oxide–Silicon (MNOS) Memories.
In the early 1970s, nonvolatile devices called EAROMs (electrically alterable ROMs) utilized a metal–nitride–oxide–silicon (MNOS) combination for the gate region of a p-channel transistor storage cell. A schematic cross-section of the p-channel MNOS transistor is shown in Figure 3-20, along with the symbolic electrical representation [39]. It is a conventional MOSFET in which the thermal oxide is replaced by a double-dielectric layer in the gate area. The gate dielectric consists of a thin layer (15–500 Å) of thermally grown silicon dioxide, over which a silicon nitride layer (500–1000 Å) is deposited. The thin (~15 Å) silicon dioxide is thermally grown at 920 °C (dry O_2). For low

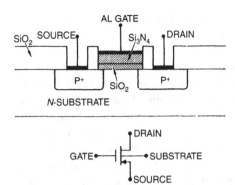

Figure 3-20. MNOS transistor schematic cross-section and its electrical representation. (From [39], with permission of IEEE.)

negative voltages applied to the gate, this MNOS transistor behaves like a conventional p-channel MOSFET. On the application of a sufficiently high positive charging voltage V_c to the gate, the electrons tunnel from the silicon conduction band to reach traps in the nitride oxide interface and nitride layer.

The movement of carriers in the silicon nitride is governed by a phenomenon known as Frenkel–Poole transport, which is different from Fowler–Nordheim tunneling. In this case, the electrons passing through the bulk of nitride are captured and emitted by numerous traps existing there. If J_n is the current density and $\mathscr{E}_n$ is the field in the nitride, then the current–field relationship is given by the following Frenkel–Poole transport equation [41]:

$$J_n = C_2 \mathscr{E}_n \, exp\left(-\frac{\phi_B - \sqrt{\dfrac{q\mathscr{E}_n}{\pi K_n \epsilon_o}}}{\phi_T} \right)$$

where ϕ_B represents the trap depth (or barrier height) in energy and C_2 is a constant. The experimental values of ϕ_B and C_2 are estimated to be 0.825 eV and 1.1×10^{-9} $(\Omega \cdot cm)^{-1}$. The barrier height is determined by the current versus $1/T$ plot. The equation above is applicable to transport through bulk nitride, and values are obtained at an electric field strength of 5.3×10^6 V/cm.

Most of the MNOS devices have an oxide conduction current (J_o) much larger than the nitride conduction current (J_n), which means that the electrons tunneling through the oxide accumulate in the interface traps until the electric field in the oxide is reduced to zero. It should be noted that while the tunneling current through an oxide is quite high during high electric fields for write and erase operations, there is essentially very little current conduction through the oxide during the read cycles and storage. It has been determined experimentally that the switching speed increases as the oxide thickness is reduced or the nitride trap density is increased. However, with an increase in speed, the ability to keep the stored charge, i.e., data-retentivity period, decreases.

The basic carrier transport properties in metal–oxide (top oxide)–nitride–oxide (tunnel oxide)–silicon (MONOS) memory structures have been investigated under steady-state conditions and negative bias voltage [42]. This has been done by measuring the hole and electron current simultaneously and separately. The experimental results showed that the dominant carriers are holes injected from the Si. The relatively thick top oxide layer acts as a potential barrier to the holes injected from the Si into thin nitride. Also, a portion of the electrons injected from the gate under negative polarity recombine with the holes injected from the Si, even in a thin nitride and/or at the top oxide–nitride interface.

The first EAROMs developed were low-performance devices. As a result, the MNOS memories using PMOS one-transistor cells had slow read speeds, used high-voltage power supplies, and nonstandard pin configurations. In the late 1970s, some manufacturers developed MNOS technology using the n-channel process, which improved performance and density. Also, these provided higher immunity to read disturbance because of the lower read voltage used, 4 V instead of a typical 9 V. Westinghouse Electric Corporation developed 8K and 16K MNOS EEPROMs for military applications. Hitachi

Corporation introduced a 16 kb device that used a two-transistor cell consisting of an MNOS storage transistor and an NMOS read transistor [43]. This cell configuration reduced data loss caused by the direct reading of the MNOS cell in a one-transistor structure. In addition, it significantly improved the read access time relative to earlier three-transistor cells due to reduced lead capacitance and an increase in the read current. It required a read voltage of $+5$ V applied externally and a program-erase voltage of $+25$ V. To counter the effect of a high program-erase voltage, a guard-ring structure and high-voltage DMOS transistors were used as the load devices.

A $+5$ V-only MNOS EAROM was developed by Inmos which used double-polysilicon gates, as shown in the schematic cross-section of Figure 3-21(a) [44]. The first polysilicon was used as an isolation layer to allow for a common source diffusion to all memory cells in the array. This poly layer, which also extended over the channel region, was used as a select gate. The storage part of the transistor consisted of a thin SiO_2 layer (~ 2 nm), followed by a thicker Si_3N_4 layer (~ 50 nm) which was topped by a second polysilicon storage gate. All high voltages required are generated on the chip. The programming is accomplished by applying high voltage to the storage (top) gate, and the erasing operation is carried out by grounding the top gate and raising the well to a high potential.

Figure 3-21(b) shows the functional block diagram of this $8K \times 8$ b array with 128 rows of 64 bytes each and a single row (64 bytes) of column latches [44]. These latches store in parallel all the data used for programming a row, or before erasure of that row. This results in 64 times programming time reduction compared to the single-byte programming.

Another example of high-performance MNOS (Hi-MNOS II) technology is the development of a 64 kb, byte-erasable, 5 V-only EEPROM by Hitachi [45]. It had a minimum feature size of 2 μm, a cell size of 180 μm^2, and a low programming voltage of 16 V generated on-chip. Its specifications include 150 ns access time, first write/erase time of less than 1 ms, en-

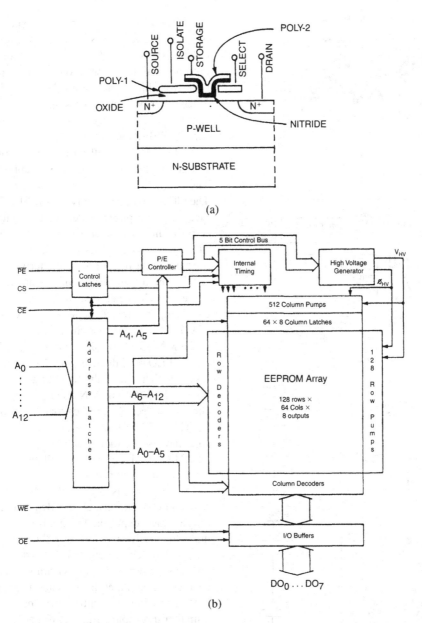

Figure 3-21. A +5 V-only Inmos EAROM. (a) Schematic cross-section. (b) Functional block diagram. (From [44], with permission of IEEE.)

durance greater than 10^4 write/erase cycles, and a ten-year data retention at 85°C. Figure 3-22(a) shows the memory cell layout. A memory cell array is formed in 32 wells, and each well consists of 8 × 256 memory cells. 8 b placed in the direction of the word line in each well correspond to 1 byte, or eight sense amplifiers and

eight input/output buffers. The high-voltage system for on-chip generation of programming voltage is regulated with an accuracy of ±1 V using a Zener diode formed in a p-type well. This high-voltage system and the memory cell array are shown in the functional block diagram of Figure 3-22(b).

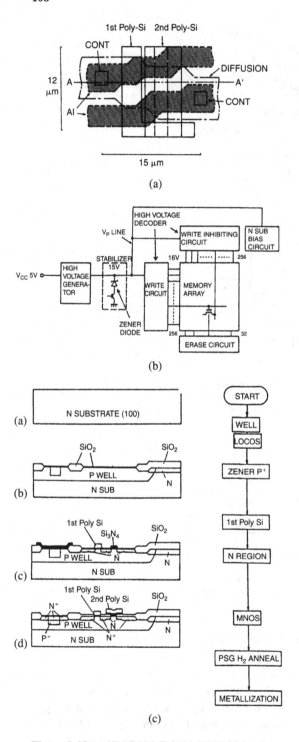

(a)

(b)

(c)

Figure 3-22. A Hi-MNOS II 64 kb EEPROM. (a) Memory cell layout. (b) High-voltage system and memory array block diagram. (c) Major fabrication steps. (From [45], with permission of IEEE.)

The Hi-MNOS II key processing features include: an effective use of Si_3N_4 at various processing steps, a high-voltage structure MNOS optimization, the Zener diode formation in a p-type well, high-temperature H_2 annealing, and 2 μm fine-pattern processing. Figure 3-22(c) shows the major processing steps for this 64 kb MNOS EEPROM.

In MNOS cell-based technology, the charge retention is a major reliability issue. The charge loss mechanism is time-dependent. The ultrathin oxide (relative to the FLOTOX devices) is susceptible to degradation with an increasing number of program-erase cycles. The MNOS cell-based parts are usually guaranteed to have data-retention characteristics at ten years over the military temperature range and endurance characteristics of 10^4 write–erase cycles compared to 10^6 write–erase cycles for the FLOTOX devices. The charge-carrier tunneling in the MNOS EEPROMs is more complex than for the FLOTOX devices, and the rate of charge loss is higher due to the tunneling through very thin oxides [46]. The MNOS technology is often chosen for certain applications instead of the FLOTOX EEPROMs because of its superior cumulative-dose radiation hardness characteristics. These radiation effects will be discussed in more detail in Chapter 7.

A study has shown that MNOS device write-state sensitivity is less dependent on the programmed depth, and is improved by reducing the silicon nitride thickness [47]. Erase/Write characteristics depend on nitride thickness, even if a constant amount of stored charge is maintained in the same programming field. The thinning of nitride thickness produces a smaller threshold voltage shift because the capacitance between the gate and stored charge increases. The initial threshold voltage (V_{thi}) of an MNOS memory device that has never been programmed corresponds to a thermal equilibrium state of the traps. Then, the required V_{th} shift from V_{thi} should be determined in order to meet ten-year data-retention requirements. This V_{th} shift should be ensured with the minimum programming time (t_{pMIN}) worst case. Figure 3-23 illustrates typical Erase/Write characteristics of an MNOS memory and the energy-band dia-

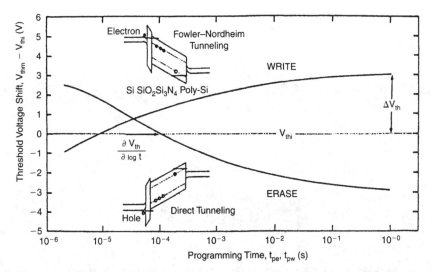

Figure 3-23. Erase/Write characteristics of a MNOS memory and energy-band diagrams during programming of an n-channel MNOS device. (From [47], with permission of IEEE.)

grams during programming of an n-channel MNOS device [47].

In the write mode, electrons are injected from the channel into the Si_3N_4 layer by modified F–N tunneling through SiO_2 and a portion of Si_3N_4, and stored as traps in Si_3N_4 [48]. Although fewer holes are also injected from the poly-Si gate, they decay more quickly and have no effect on V_{th}. However, in the erase mode, the holes are injected from the Si substrate into the Si_3N_4 layer by direct tunneling through SiO_2, and stored in the traps in Si_3N_4. Some electrons are also injected from the poly-Si gate, but they have very little effect on V_{th}.

A scaling guidelines investigation was performed on samples fabricated using n-channel Si-gate MNOS technology with tunnel oxide (t_{ox}) thickness of about 1.6 nm, but with various Si_3N_4 thicknesses. The optimum t_{ox} thickness improves: (1) the write-state retentivity and minimizes the programming voltage (V_p) lower limit, and (2) improves the erase-state retentivity and maximizes the V_p upper limit. The test results showed that while the erase-state decay rates increased with programmed depth, the write-state decay rate showed a minor dependence on the programmed depth, and write-state retention characteristics improved as the Si_3N_4

thickness (t_N) is reduced. These write-state charges are distributed rectangularly, while the erase-state charges are distributed exponentially. This indicates that t_N scaling down is effective for MNOS memory devices since the lower limit of V_p can be reduced and the V_p margin widened. It is estimated that MNOS structures have the scaling potential for fabrication of 16 Mb EEPROM devices.

3.5.1.2 Silicon–Oxide–Nitride–Oxide–Semiconductor (SONOS) Memories.
Silicon–oxide–nitride–oxide–semiconductor (SONOS) is considered to be a suitable candidate for military and space applications because of its good data-retention characteristics, high endurance, and radiation hardness. This technology has been used to fabricate 64K EEPROMs that were demonstrated to have retention time better than ten years at 80°C after 10,000 programming cycles [49]. The radiation-hardened process uses n-channel SONOS (NSONOS) memory devices on n-epi with n^+ starting material. A self-aligned p^+ guardband is used for total dose hardening of the NMOS transistors. EEPROMs have also been fabricated using p-channel SONOS (PSONOS) transistors technology. Figure 3-24(a) shows the cross-section of a SONOS process

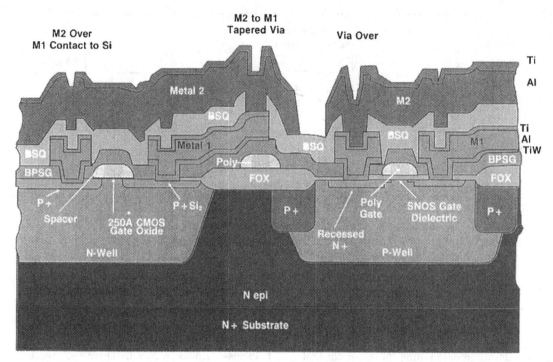

Figure 3-24. (a) WEC/SNL silicon–oxide–nitride–oxide–semiconductor (SONOS) process cross-section. (From [50], courtesy of WEC.)

jointly developed by Westinghouse Electric Corporation (WEC) and Sandia National Laboratories (SNL) [50].

This SONOS process has been used to fabricate 64K EEPROMs which use differential memory design with two memory transistors per bit for improved data-retention characteristics. Figure 3-24(b) shows a photomicrograph of WEC/SNL 64K EEPROM and its specifications [50]. It uses a 5 V power supply, and features a charge pump for internal generation of memory programming voltage. The writing and clearing of the SONOS memory arrays requires 2.5 ms/ 7.5 ms of ± 10 V pulses. Some other features supported by this part include the following: military operating temperature range (−55 to +125°C), self-timed programming, combined erase/write, auto program start, asynchronous addressing, 64 word page, and data polling.

3.5.1.3 *Floating-Gate Tunneling Oxide (FLOTOX) Technology.* The floating-gate tunneling oxide (FLOTOX) technology, most

commonly used for fabrication of the EEP-ROMs, is the derivative of UV EPROM technology developed by Intel, and was discussed earlier in Section 3.4. The major difference between the EPROM cell and the FLOTOX EEP-ROM structure is the small area of thin oxide fabricated between the floating gate and the drain which permits carrier tunneling. This transport mechanism follows the Fowler–Nordheim tunneling equation. There have been many variations of the floating-gate technology developed over the years. The original floating-gate memory device consisted of a p-channel IGFET on a thin-film transistor with a metal–insulator–metal–insulator–semiconductor structure (MIMIS). In this floating-gate IGFET, the charge was stored in the conduction band of the floating-gate metal M-1, instead of the insulator bulk traps as in the MNOS devices [46].

To avoid the leakage problems existing in an MIMIS structure, the floating-gate avalanche-injection MOS device (FAMOS) described in Section 3.4 was developed by Frohman-

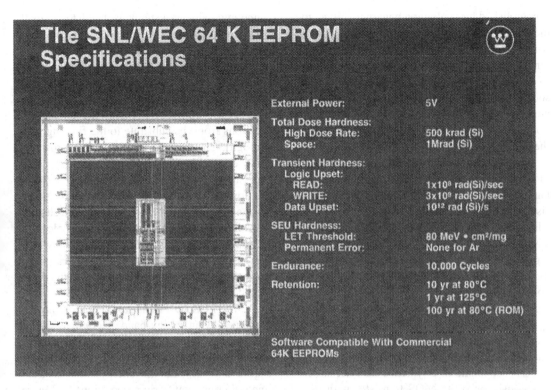

Figure 3-24. (Cont.). (b) WEC/SNL 64K SONOS EEPROM photomicrograph and specifications. (From [50], courtesy of WEC.)

Bentchkowsky. One of the advantages of the FAMOS device is its avalanche injection, which can be achieved at a lower oxide field rate than the Fowler–Nordheim injection (or direct tunneling). However, a major limitation is that for the erase operation, the use of an X-ray or UV source is required. An alternate way of extracting electrons from the floating gate isolated by thick oxide was developed by using the metal–insulator–semiconductor–insulator–semiconductor (MISIS) structure. In this approach, silicon–dioxide is used as an insulator for its low trap densities, ease of fabrication, and low Si–SiO$_2$ interface states [51]. Then, for a floating gate made of silicon, an electric field can be easily established throughout the entire thickness of the floating gate by application of voltage between the top gate and the silicon substrate. The electrons that are stored in the floating gate are accelerated by this electric field, and some of

them acquire enough energy to escape into the oxide and drift into the silicon substrate. For a pulsed operation (alternating voltage), the acceleration process is repeated, and the floating gate is discharged electronically.

Another alternative structure proposed for solving the erasure problem associated with the FAMOS-type devices was the channel-injection floating gate FET [52], [53]. As in the FAMOS and the MISIS structures, the floating gate in this device is isolated by a relatively thick oxide. A top gate is added (as in MISIS) to control the charging and discharging of the floating gate, which may be a semiconductor or a metal. This device uses another FET in series for its memory cell. The switching speed is slow since both the channel injection of electrons and the avalanche injection of holes are inefficient processes. A variation of this was the plane injection structure in which hole injection is accomplished by applying a negative voltage to the top gate and a positive voltage to the drain which causes avalanche breakdown at the drain junction similar to the FAMOS operation [54]. In an n-channel FET structure of this type, the source and drain should be open-circuited or positively biased in order to create deep depletion under the gate region by positively biasing the gate.

In one approach, called floating Si-gate tunnel injection MIS (FTMIS), small silicon islands have been used as the floating gate. It consists of a thin-oxide layer (20–40 Å) grown on a silicon substrate and a high-resistivity polycrystalline silicon layer (200–1000 Å) as the floating gate [55]. A gate oxide layer (800–1500 Å thick) is obtained by thermal oxidation of the polycrystalline layer. By using this approach, the devices have been fabricated with switching speeds of less than 20 ns. However, these devices may not have the same data-retentivity characteristics as the MNOS structures.

A schematic of a typical FLOTOX structure consisting of an MOS transistor with two polysilicon gates is shown in Figure 3-25 [53]. A thin (8–12 nm) gate oxide region is formed near the drain. The lower polysilicon layer is the floating gate, while the one on top is the control gate. The remainder of the floating-gate oxide, as well as the interpoly oxide, is about 50 nm thick.

As mentioned earlier, there are variations in thin-oxide floating-gate process technology from manufacturer to manufacturer. The thin

tunneling oxide is usually isolated to a small area over the drain region (e.g., in the Intel process). However, it can also be a small area over the channel. In an early Motorola FET-MOS process, the tunneling oxide extended over the entire gate region. The programming of a FLOTOX transistor involves the transfer of electrons from the substrate to the floating gate through the thin-oxide layer as the control-gate voltage is raised to a sufficiently high value (e.g., 12 V). This causes charge transfer through the Fowler–Nordheim tunneling process. As the charge builds up on the floating gate, the electric field is reduced, which decreases the electron flow. This tunneling process is reversible, which means that the floating gate can be erased by grounding the control gate and raising the drain voltage.

The FLOTOX transistor operation may be explained by an energy-band diagram, as shown in Figure 3-26 [57]. A cross-section of the device with thin tunneling oxide is represented by Figure 3-26(a), and the energy-band diagram under thermal equilibrium conditions by Figure 3-26(b). The cell programming means charging up the floating gate with the electrons. This is done by applying a high positive gate-bias voltage to the poly 2 (control gate) while the source and drain are at ground potential. As a result, a fraction of the applied gate voltage is developed in the floating gate through capacitive coupling, which establishes an electric field across the thin tunnel oxide. The energy-band diagram is shifted, as shown in Figure 3-26(c), so that electrons in the conduction band of the n^+ drain can tunnel through the oxide to reach the floating gate. The I–V characteristics vary for different oxide thicknesses, and the current density (J_o) is given by the Fowler–Nordheim equation. As the charge builds up on the floating gate, it lowers the electric field and reduces the electron flow.

Figure 3-26(d) shows the energy-band diagram in the storage mode after the gate voltage is removed. In this case, the Fermi levels do not coincide since the system is not in equilibrium. In the floating gate, the shift of the Fermi level closer to the conduction band indicates an increase of electron concentration. During the

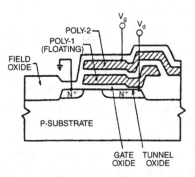

Figure 3-25. A schematic of a typical FLOTOX transistor structure. (From [56], with permission of IEEE.)

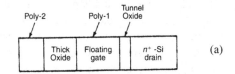
(a)

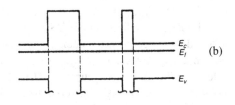
(b)

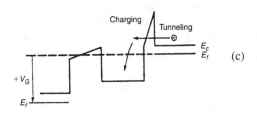
(c)

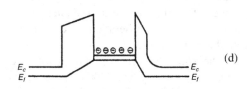
(d)

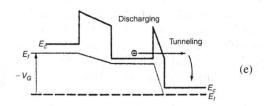
(e)

Figure 3-26. The FLOTOX transistor. (a) Cross-section with tunnel oxide. (b) Equilibrium energy-band diagram. (c) Conduction band with $+V_G$ programming mode. (d) Storage mode. (e) Erase mode with $-V_G$. (From [57], with permission of Academic Press, NY.)

erase operation, the poly 2 (control gate) is grounded and the source is left floating. As a high positive voltage is applied to the drain, a

fraction of that voltage is capacitively coupled to the floating gate, which enables the tunneling of electrons from the floating gate to the drain, as shown in Figure 3-26(e).

During the programming and erasing operations, the FLOTOX transistor may be considered a network of capacitors tied to the floating gate, as shown in the equivalent circuit representation by Figure 3-27 [57]. Now, the floating-gate voltage for this circuit can be written as

$$V_{FG} = \frac{Q_{FG}}{C_T} + V_G\frac{C_{pp}}{C_T} + V_B\frac{C_{fd}}{C_T} + V_S\frac{C_{gs}}{C_T}$$
$$+ V_D\frac{C_{gd}}{C_T} + V_{DO}\frac{C_{tox}}{C_T} + (V_{FB} + \psi_s)\frac{C_g}{C_T}$$

where

$$C_T = C_{pp} + C_{fd} + C_{gs} + C_{gd} + C_{tox} + C_g$$

and C_{pp} is the poly-to-poly capacitance, C_{fd} is the floating-gate over field-oxide capacitance, C_{gs} is the floating-gate-to-source capacitance, C_{gd} is the floating-gate-to-drain capacitance excluding the tunnel area, C_{tox} is the tunnel-oxide capacitance, and C_g is the floating-gate over gate-oxide capacitance. Also, Q_{FG} is the floating-gate charge, whereas V_B, V_S, V_D, and V_{DO} are the voltages of the body, source, drain, and effective drain under the tunnel oxide, respectively. The inversion layer charge can be written as

$$Q_I = C_g(V_{FG} - V_{FB} - \psi_s).$$

These equations have been used to predict fairly accurately the device characteristics, except

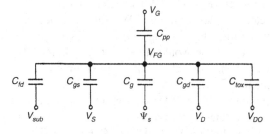

Figure 3-27. Equivalent circuit representing capacitances for charge-coupling calculations. (From [57], with permission of Academic Press, NY.)

during the erasure when the surface charge changes from the accumulation to inversion mode.

Another important design parameter is the program coupling ratio, which is defined as the ratio between the floating-gate voltage (V_{FG}) and the gate oxide voltage (V_g), expressed as a percentage:

$$C.R. = \left(\frac{V_{FG}}{V_g}\right) \times 100\%.$$

For a given threshold voltage ($\Delta\ VT$), a lower coupling ratio is required for a thinner oxide. This means a smaller cell area, but then the thinner oxide has reliability concerns. Therefore, in practice, the coupling ratio is determined by a tradeoff between the oxide thickness and cell area. The WRITE operation has been adequately modeled, based upon the simple concept of coupling ratios [58]. By using this model, the simulated values of the threshold voltage as a function of WRITE time (or WRITE pulse amplitude) agree closely with the measured data. However, this simple model is not adequate for accurate simulation of the ERASE operation. In certain cell structures, a hole current can flow from the drain into the substrate during the erase operation, through an effect associated with the positive charge trapping in the tunnel oxide. This hole current flow into the substrate can be avoided by a proper cell design.

The basic EEPROM cell consists of the FLOTOX transistor as a storage cell, along with an access (or select) transistor. This access transistor acts as an ON/OFF switch to select or deselect a particular cell. Without this access transistor, a high voltage applied to the drain of one transistor would appear at the drains of other transistors, also in the same column of the memory array. However, this large-sized (needed because of large programming and erasure voltages) access transistor increases the area of the FLOTOX EEPROM cell relative to EPROM cell. This has somewhat limited the density of the FLOTOX EEPROMs. As the memory size is increased, the failure rate due to

defect-related oxide breakdown problems also goes up [59]. One approach to overcome this limitation has been the use of on-chip error-detection and correction (EDAC) schemes which result in a larger chip size.

The scaling of nonvolatile memories has led to the need for memory cell analysis and accurate device simulation for design verification and reliability prediction. Two-dimensional (2-D) simulators such as HFIELDS have been developed that incorporate models for band-to-band tunneling, hot electron injection, and the Fowler–Nordheim tunnel current [60]. This simulator allows floating electrodes to be defined as the charge boundaries, so that their charge can be continuously updated with time. This simulator was used to study an anomalous spike in the tunnel current observed during experimental characterization of the erase operation of an EEPROM device [61]. The erase and program operations were simulated for two cells with different doping densities. The test data indicated that the anomalous spike observed was due to the relaxation of deep-depletion condition band-to-band generation of carriers beneath the tunnel oxide.

The simulators can be helpful tools for investigating process variation effects in the floating-gate EEPROMs. An EEPROM I–V model was developed and used in the simulation of an EEPROM structure [62]. This model helped predict the drain current for the above-threshold and subthreshold regions, and provided a theoretical understanding of the effects of the floating-gate and tunnel-oxide area on channel potential and total channel charge. It has been observed that the traditional 1-D and 2-D simulators may not be adequate for device performance characterization. A 3-D device simulator MAGENTA has been used to investigate the 3-D effects in small EEPROM devices in order to develop a device parametrization and design methodology for improving the cell performance [63]. The 3-D effects analyzed include short-channel and narrow-channel effects, enhancement, and reduction of the tunneling currents. A 3-D simulation study was performed with generic n-channel FLOTOX EEPROM which uses a storage transistor and a select transistor for each cell in the array [64]. The test results of this study can be summarized as follows:

- Threshold voltage for smaller devices require a higher charge density for programming. This effect, which depends on the cell aspect ratio and size, tends to degrade the performance of small devices, and could be significant for cell designs with larger floating gate capacitances.

- The 3-D effects can reduce tunnel current in the write mode due to 3-D spreading of field lines near the substrate. The strong enhancement of the erase current indicates sensitivity to the

shape of the top electrode, and therefore a good process control is required for the erase time consistency.

3.5.1.4 Textured-Polysilicon Technology.

Textured-polysilicon EEPROMs consist of three polysilicon layers that are partially overlapped to create a cell with three transistors in series. Figure 3-28 shows: (a) a top view of the cell with 1.2 μm design rules, (b) the cell cross-section, and (c) the cell equivalent circuit [65]. The floating gate is the poly 2 layer in the middle which is encapsulated by silicon dioxide for high charge retention. The charge is still transferred to the floating gate by Fowler–Nordheim tunneling, although the tunneling takes place from one polysilicon structure to another, in contrast to the FLOTOX technology where the

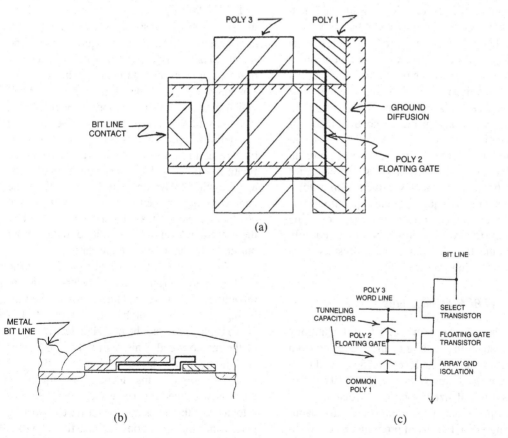

Figure 3-28. A textured-polysilicon EEPROM. (a) A cell top view. (b) Cell cross-section. (c) Cell equivalent circuit. (From [65], with permission of IEEE.)

tunneling takes place from the substrate to the floating gate. Also, the interpoly oxides through which the tunneling takes place can be significantly thicker (60–100 nm) than those for the FLOTOX devices since the electric field is enhanced by geometrical effects of fine texture at the surface of polysilicon structures.

Textured polysilicon cells are programmed by causing the electrons to tunnel from poly 1 to floating gate (poly 2), and erasure by the electron tunneling from poly 2 to poly 3. The poly 3 voltage is taken high in both the programming and erase operations. However, it is the drain voltage which determines whether the tunneling takes place from poly 1 to the floating gate, or from the floating gate to poly 3. This implies that the state of the drain voltage determines the final state of the memory cell. This has the advantage of being a "direct write cell" with no need to clear all the cells before write, as is the case with the FLOTOX-cell-based EEPROMs.

The manufacturing process for textured-polysilicon devices is basically an extension of the EPROM process with the addition of an extra polysilicon layer. The three polysilicon layers are vertically integrated, which results in a compact cell layout that is about a factor of two smaller than a FLOTOX cell for a given generation of technology. This gives them an edge over the FLOTOX devices for memories larger than 256 kb. However, the textured-polysilicon memory requires higher operating voltage (> 20 V) to support its operation. The dominant failure mechanism in a textured poly cell is electron trapping, which results in a memory window closure [66].

3.5.2 EEPROM Architectures

The advances in EEPROM technology have enabled the scaling of these devices from the older 16 and 64 kb levels to 256 kb and multimegabit generation. For high-density EEPROMs, the cell size is the most critical element in determining the die size, and hence the manufacturing yield. For given program/erase threshold performance requirements, the scaling of an EEPROM cell usually requires a thinner dielec-

tric. However, a thinner dielectric may result in higher infant mortality failures and a lower cell endurance life cycle. The oxynitride films of 90–100 Å thickness for the EEPROM cells have proven to be quite reliable because of their low charge trapping, lower defect density, and higher endurance characteristics. An n-well double-polysilicon gate CMOS technology was developed by SEEQ Technology, Inc., that uses oxynitride film 70–90 Å thick to produce a 54 μm^2 cell size for a 256 kb density EEPROM [67]. This cell has been shown to operate with a programming voltage as low as 17 V, and programming times as short as 0.1 ms. The endurance characteristics of this device (Q-cell design) are specified as over 1 million cycles/byte.

Figure 3-29(a) shows the Q-cell floating-gate memory transistor cross-section [68]. The SEEQ EEPROM memory cell consists of a floating-gate memory transistor and a select transistor, as shown in the schematic of Figure 3-29(b). The select transistor is used to isolate the memory transistor to prevent data disturb. The other peripheral logic is combined to form the Q-cell which incorporates the memory error-correction technique transparent to the user. The memory cell defines the logic states as either a "1" or a "0" by storing a negative or a positive charge on the floating gate (Poly 1). When the reference voltage is applied to the top control gate (Poly 2), the addressed memory cell will either conduct or not conduct a current. This cell current is detected by the sense amplifier and transmitted to the output buffers as the appropriate logic state.

The charge is transferred to and from the floating gate through the Fowler–Nordheim tunneling process, as explained earlier. The tunneling occurs when a high voltage (typically 17–20 V) is placed across the tunnel dielectric region of the memory cell. This high voltage is generated internally, on-chip. For a logic "1," the electrons are stored on the floating gate, using the conditions defined for the "erase" operation. For a logic "0," the holes are stored on the floating gate, using the conditions defined for a "write" operation. The Q-cell thresholds for a logic "1" and "0" are shown in Figure 3-29(c) [68].

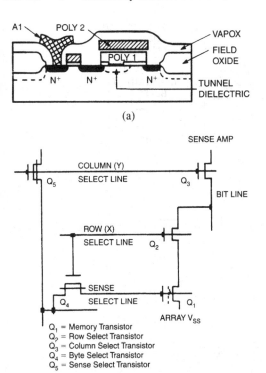

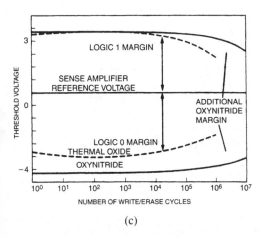

Figure 3-29. SEEQ EEPROM "Q-cell." (a) Floating-gate transistor cross-section. (b) Memory cell schematic. (c) Cell margin characteristics. (From [68], with permission of SEEQ Technology, Inc.)

For 256 kb EEPROMs, JEDEC Standard 21C requires some standard features, such as: operation with a primary power supply voltage

of 5 V, input levels between 0 and 5 V, read and write timing cycles consistent with the standard timing diagrams, operation in conformance with the standard truth table, etc. [69]. This standard also defines the implementation protocols and requirements for some optional features, including the page write mode, DATA-polling, software data protection, and write protect.

An example is the ATMEL 256 kb CMOS EEPROM organized as 32,768 words $\times$ 8 b, and offers a read access time of 70 ns, an active current of 80 mA, and a standby current of 3 mA [70]. Figure 3-30(a) shows the pin configurations and block diagram of this device which is accessed like a RAM for the read or write cycle without the need for any external components. It contains a 64 byte page register to allow for writing up to 64 bytes simultaneously using the page write mode. During a write cycle, the address and 1–64 bytes of data are internally latched, which frees the addresses and data bus for other operations. Following the initiation of a write cycle, the device automatically writes the latched data using an internal control timer. The end of a write cycle can be detected by DATA-polling of the I/O7 pin. Once the end of a write cycle has been detected, a new access for read or write can begin. In addition to the DATA-polling, there is another method for determining the end of the write cycle. During a write operation, successive attempts to read data from the device will result in I/O6 toggling between 1 and 0. However, once the write operation is over, I/O6 will stop toggling and valid data will be read. Figure 3-30(b) shows the AC Read waveforms, and (c) shows the Page Mode Write waveforms.

A hardware data protection feature protects against inadvertent writes. An optional software-controlled data protection feature is also available and is user-controlled. Once it is enabled, a software algorithm must be issued to the device before a write operation can be performed. This device utilizes internal error correction for the extended endurance and improved data-retention characteristics. The endurance is specified at 10^4 or 10^5 cycles, and the data retention at ten years.

A 1 Mb EEPROM using dual-gate MONOS (metal–oxide–nitride–oxide–semiconductor)

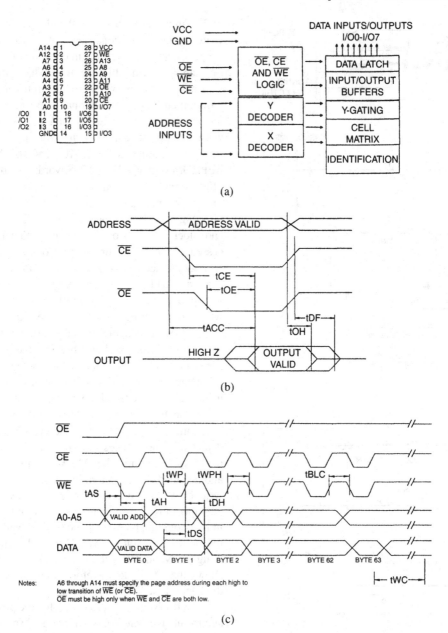

Figure 3-30. ATMEL 256 kb CMOS EEPROM AT28HC256/L. (a) Pin configurations
and block diagram. (b) AC Read waveforms. (c) Page Mode Write waveforms. (From [70], with permission of ATMEL Corporation.)

memory cells has been developed for semiconductor disk applications [71]. This has a low
program voltage of 5 V, a block erase/page programming organization, and a high endurance to
the WRITE/ERASE cycles. It has a program-

ming speed up to 140 ms/chip (1.1 μs/byte
equivalent).

The conventional, full functional EEP-
ROMs have several advantages including the
byte erase, byte program, and random access read

capabilities. As discussed earlier, the memory cell for these devices consists of two transistors per bit: a memory transistor and a select transistor. This results in a relatively large cell size and a lower level of integration. A new cell concept has been proposed, called a NAND-structured cell which dramatically reduces the number of cell components as well as the cell size, allows for block erasing, successive programming, and reading [72]. This new cell structure has been used to build experimental, 5 V-only 4 Mb EEPROMs with 1.0 μm design rules. The main features of this include: (1) a tight threshold voltage (V_t) distribution which is controlled by a program verify technique, and (2) successive program/erase operation capability [73], [74].

Figure 3-31(a) shows the NAND-structured cell equivalent circuit and its layout [75]. It has eight memory transistors in series which are sandwiched between the two select gates, $SG1$ and $SG2$. The first gate $SG1$ ensures

selectivity, whereas the second gate $SG2$ prevents the cell current from passing during the programming operation. The eight-NAND structures yield a cell size of 12.9 μm², which is only 44% of the area required by a conventional NOR-structured EEPROM cell. This cell can be programmed and erased by the Fowler–Nordheim tunneling process. A three-step program-verify algorithm used consists of initial-program, verify-read, and reprogram. Initial programming is performed by applying V_{pp} (20 V) to the selected control gate, while the intermediate voltage V_m (10 V) is applied to the unselected control gates. For a cell programming of "1," the bit lines are grounded. As a result, the electrons are injected into the floating gate and V_t becomes positive. For a cell programming of "0," the bit lines are raised to V_m (10 V) to prevent "1" programming. In a read operation, 5 V is applied to the control gates of the unselected cells, and the control gate of the

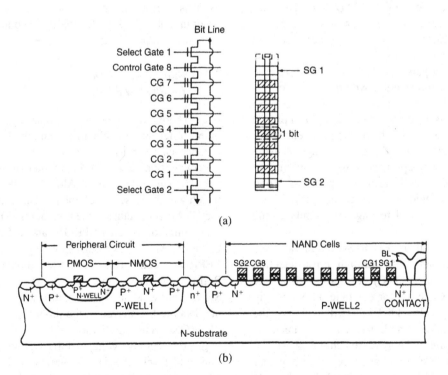

(a)

(b)

Figure 3-31. A 4 Mb CMOS EEPROM. (a) A NAND-structured cell equivalent circuit and its layout. (b) Cross-sectional view of twin p-well structure. (From [73] and [74], with permission of IEEE.)

selected cell is grounded. The unselected cells act as transfer gates.

Figure 3-31(b) shows a cross-sectional view of the NAND-structured cell made with the self-aligned, double-polysilicon, floating-gate, n-channel process. The floating gates are made of first (lower)-level polysilicon, whereas the control gates, which are the word lines in an array, are made of second-level polysilicon. The NAND cell array and peripheral CMOS circuitry are located in separate p-wells (1 and 2), which are electrically isolated from each other. Both the channel length and width of the memory cell transistors are 1.0 μm. The tunnel oxide thickness under the floating gate is about 100 Å.

A dual-mode sensing scheme (DMS) of a capacitor-coupled EEPROM cell utilizing both the charge-sensing mode and the current-sensing mode has been proposed for low-voltage operation and high endurance [76]. The yield and endurance of the EEPROMs have been increased by the use of error-correction techniques, such as in SEEQ "Q-cell." This cell, discussed earlier, used a pair of standard two-transistor EEPROM cells which were read out through separate sense amplifiers. These sense amplifiers outputs were fed into a NOR gate, so that correct information would be read out as long as at least one of the two cells was not defective. This was considered a valid approach since the probability of two adjacent cells wearing out is quite low. The disadvantage of using this redundancy approach was the doubling up of array size, resulting in a larger chip. However, this technique increased the yield by reducing the required testing for outputs of the Q-cell.

Some other EEPROM manufacturers have improved yield through error-detection and correction (EDAC) schemes such as the Hamming Code as, for example, in Microchip Technology's 256 kb EEPROM [77]. For the implementation of an EDAC circuitry, a penalty has to be paid in the increased chip area.

IBM designers have developed a planar EEPROM which can be embedded in the gate arrays, standard-cell blocks, microprocessors, or neural-net chips [102]. The planar cell, which consists of adjacently placed NMOS and PMOS transistors that share a polysilicon gate, also has a third element which is electrically isolated from the two transistors. The common polysilicon gate acts as a floating gate, while the n-well of the PMOS transistor serves as a control node. The EEPROM cell operates in either of the two read/write modes: (a) Mode I, in which the hot electrons are used to charge the floating gate, and (b) Mode II, in which the Fowler–Nordheim tunneling charges the gate.

The EEPROMs have also been used in combinations with other technologies, such as ROMs on the same chip. In some devices, for example, microcontrollers, the embedded EEPROM has been used as part of the standard logic process with 5 V power supply. An example of technology mix is the development of an embedded two-transistor EEPROM cell processed along with a single-transistor flash EEPROM cell by Philips Semiconductor [78]. The serial EEPROMs (like serial PROMs) are also used as peripheral devices to store the user-programmable features for the microcontrollers typically equipped with a serial interface.

3.5.3 Nonvolatile SRAM (or Shadow RAM)

There are some applications which require faster write-cycle times and higher endurance characteristics than those provided by the conventional EEPROMs. An alternative is to use the battery-backed SRAMs (or DRAMs), which was discussed in Chapter 2. One interesting NVM architecture is the nonvolatile SRAM, a combination of an EEPROM and a SRAM. Each SRAM cell has a corresponding "shadow" EEPROM cell. This NVRAM (nonvolatile RAM) appears to the system as a normal SRAM, but when a STORE command is issued, the data are transferred to the on-board shadow EEPROM. When used for critical data storage (e.g., the supply voltage falls below some threshold level), the data are transferred to the nonvolatile storage mode. A RECALL command transfers the data from EEPROM to static RAM. All NVSRAMs perform an automatic

RECALL operation on power-up. The NVS-RAMs are finding greater acceptance in military applications because of their reliability, unlimited read and writes, high nonvolatile endurance, and retention characteristics.

The first-generation NVRAMs included development of a 4 kb device with on-chip high-voltage generation using the HMOS I, dual-polysilicon FLOTOX technology by Intel Corporation [79]. The SNOS is a popular technology for NVRAM implementation because of some desirable reliability characteristics such as log-linear decay of the SNOS transistor threshold voltages, oxide pin hole immunity, and the absence of dielectric rupture. An example of the SNOS-based NVRAM technology is the n-channel, double-level polysilicon process 1K × 8 b devices that have internally 8 kb of static RAM and 8 kb of EEPROM for a total of 16 kb of usable memory [80].

This 8K NVRAM is word organized in 1K bytes, with a memory matrix organized as 64 rows × 128 columns. Figure 3-32(a) shows the schematic of this 8 kb NVRAM cell, which consists of a standard six-element static RAM cell with polysilicon resistors for loads. In addition, it contains two EEPROM elements, each made up of elements C_1, M_1, and M_2, as shown in the schematic. C_1 is a variable threshold capacitor used to retain the last stored state of the RAM cell. M_1, which shares a common gate with C_1, is a Poly II depletion transistor. M_2 is a Poly I depletion transistor with its gate strapped to V_{SS}, and serves as a decoupling transistor during the STORE operation. Figure 3-32(b) shows the cross-section of two EEPROM memory elements in a single cell, and below that, the surface potential profiles for different regions of the elements.

Currently, NVRAMs of 64 and 256 kb are available, and higher density devices are in development. Figure 3-33 shows a block diagram of the Simtek 256 kb NVSRAM [81]. A hardware store is initiated with a single pin ($\overline{SE}$), and the store busy pin ($\overline{SB}$) will indicate when it is completed. Similarly, a hardware recall may be initiated by asserting the $\overline{RE}$ pin. Both nonvolatile cycles may also be initiated with software sequences that are clocked with the

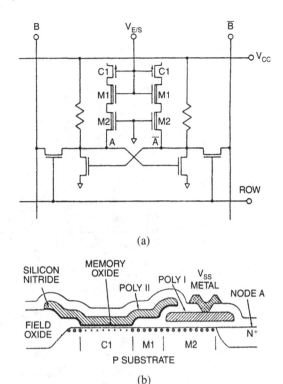

(a)

(b)

Figure 3-32. SNOS 8 kb static NVRAM. (a) A schematic of the cell. (b) Cross-section of EEPROM memory element. (From [80], with permission of IEEE.)

software strobe pin ($\overline{SS}$), or one of the chip enable pins. The chip is provided with several chip enable pins ($\overline{E}$, $\overline{E}1$, S, $S1$) and write enable pins ($\overline{W}$, $\overline{W}1$) to allow easy system level integration. For power-sensitive systems, a hardware-initiated sleep mode reduces the supply current well below the values expected for fast SRAMs in the standby mode.

The conventional NVRAM, which combines a six-transistor SRAM with EEPROM elements, requires a large cell size, making it difficult to achieve high density. To reduce the cell size, the NV-DRAM has been developed which, instead of SRAM, combines a DRAM with an EEPROM. An example is the stacked storage capacitor on the FLOTOX (SCF) cell for a megabit NV-DRAM [82]. It has a flash STORE/RECALL (DRAM to EEPROM/EEPROM to DRAM) operation capability. The store

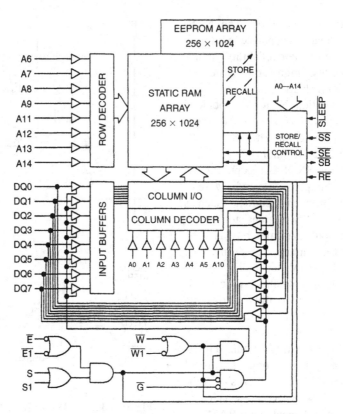

Figure 3-33. Block diagram of Simtek 256 kb NVSRAM. (From [81], with permission of IEEE.)

operation for this is completed in less than 10 ms, and the store endurance for a single cell is characterized as greater than 10^6 cycles.

The SCF cell has been successfully implemented into a 1 Mb NVRAM using a twin-well, triple-polysilicon, and double-metal CMOS process based on 0.8 μm design rules. This SCF cell had a chip area of 30.94 μm² and a DRAM storage capacitance of 50 fF. Figure 3-34 shows the SCF cell structure (a) schematic view, (b) SEM view, and (c) equivalent circuit [83]. The cell consists of three transistors (*T*1, *T*2, *MT*) and a storage capacitor (*C*). *T*1 is the word-select transistor, and its drain (*D*) is connected to the bit line. *T*2 is the recall transistor. *MT* is the stacked gate memory transistor which has a floating gate (*FG*) with tunnel oxide (*TO*) on the source (*S*). The DRAM data are stored on the gate (*MG*) of *MT*, which is connected to the sub-

strate through the memory contact (*MC*) between the select gate (*SG*) and the recall gate (*RG*). DRAM storage capacitor (*C*) is formed between the *MG* and the capacitor gate (*CG*) over the EEPROM element.

3.6 FLASH MEMORIES (EPROMs OR EEPROMs)

Flash memories are the devices for which the contents of all memory array cells can be erased simultaneously through the use of an electrical erase signal. They are based on either the EPROM or EEPROM technology, and their selection for a particular application requires making tradeoffs between the higher density of the EPROM technology versus the in-circuit programming flexibility of the EEPROM technol-

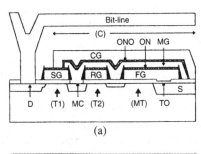

(a)

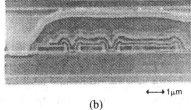

(b)

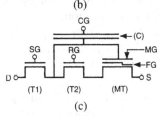

(c)

Figure 3-34. A 1 Mb NVRAM SCF cell. (a) Schematic view. (b) SEM view. (c) Equivalent circuit. (From [83], with permission of IEEE.)

programming, per chip for erasing [84]. The flash memories, because of their bulk erase characteristics, are unlike the floating-gate EEP-ROMs which have select transistors incorporated in each cell to allow for the individual byte erasure. Therefore, the flash EEPROM cells can be roughly made two or three times smaller than the floating-gate EEPROM cells fabricated using the same design rules.

The flash memory endurance or number of possible program/erase cycles is a key reliability issue related to the technology, cell architecture, process lithography, and program/erase voltage. In relatively low program/erase voltage technologies (e.g., Intel's ETOX™), the major effect of cycling is the degradation of program and erase time. In the EEPROM-based flash technologies, the high voltages generated on-chip may induce a short circuit across the defect sites in thin oxide separating the floating gate and the substrate. To overcome the inherent limitations of EEPROM-based flash memories, often techniques such as cell redundancy (two transistors per cell) and the EDAC are used. The following sections describe flash memory cell structure developments and advanced architectures.

ogy. The current generation flash memories in the EEPROM technology need a 5 V-only supply, whereas those in the EPROM technology require an additional 12 V supply and a multistep algorithm to verify erasure. Flash memories are particularly suitable for applications which require not only more memory capacity than the EEPROMs can provide, but also need faster and more frequent programming than can be accomplished with the EPROMs.

Most flash memories are programmed with the EPROMs' hot electron injection techniques. The erasing mechanism in flash EEP-ROMs is the Fowler–Nordheim tunneling off the floating gate to the drain region. However, there are some 5 V-only flash memories which depend upon tunneling for both the write and erase mechanisms. Typically, the hot electron programming takes less than 10 μs/byte, while the tunneling takes between 5–20 ms/page for

3.6.1 Flash Memory Cells and Technology Developments

The flash memory cell is structurally like the EPROM cell, although slightly larger and with a thinner gate oxide layer, usually 10–20 nm deep. However, each of the manufacturers in fabricating these devices has taken a slightly different approach. For example, Intel's ETOX™ (EPROM tunnel oxide) NOR-based flash technology, as the name implies, was an evolution from its EPROM technology. The ETOX memories have the same UV-erasure requirements and Fowler–Nordheim tunneling process as the EPROMs, but they use the channel hot electron injection (CHE) programming method [85]. The flash memory cell structure is similar to that of EPROM, except for the use of a high-quality tunneling oxide under the floating polysilicon gate. In flash erase, all cells are erased simultaneously by cold electron tunneling from the floating gate to the source on the application of 12 V to the source junctions and grounding the select gates.

Figure 3-35(a) shows the schematic of a cell Read operation during which the decoded address generates supply voltage levels on the CMOS FET select gate and drain, while the source is grounded [86]. In an erased cell, the select-gate voltage is sufficient to overcome the transistor turn-on threshold voltage (V_T), and the drain-to-source (I_{DS}) current detected by the sense amplifier produces a logic "1." For a programmed cell which has a higher V_T because

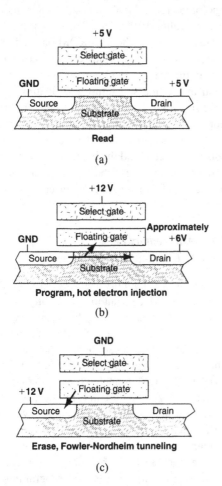

Read

(a)

Program, hot electron injection

(b)

Erase, Fowler-Nordheim tunneling

(c)

Figure 3-35. An ETOX cell. (a) Read operation. (b) Program operation hot electron injection. (c) Erase, Fowler–Nordheim tunneling [86].

of the additional electrons stored on the floating gate, the applied voltage on the select gate is not sufficient to turn it on. This absence of current results in a logic "0" at the flash memory output.

Figure 3-35(b) shows a schematic of the Program operation in which +12 V is applied between the source and select gate (capacitively coupled through the floating gate), and roughly 6 V between the source and drain. The source–drain voltage generates hot electrons which are swept across the channel from the source to drain. The higher voltage on the select gate overcomes the oxide energy barrier and attracts the electrons across the thin oxide where they accumulate on the floating gate. When enough electrons accumulate on the floating gate, the cell switches from an erased state (logic "1") to the programmed state (logic "0"). This programming technique is called hot electron injection. There are faster programming techniques for flash memories that use tunneling from the substrate to the floating gate and require higher internal voltages of up to 25 V.

Figure 3-35(c) shows a schematic of the Erase operation which uses Fowler–Nordheim tunneling in which the drain is left floating, the select gate is grounded, and 12 V is applied to the source. This creates an electric field across the thin oxide between the floating gate and the source which pulls electrons off the floating gate. There are some negative-gate erase flash memories which create essentially the same conditions by using voltages of +5 and −10.5 V on the source and select gate, respectively.

There are three distinct flash architectures: NOR EPROM, NOR EEPROM-based, and NAND EEPROM-based. All of them are nonvolatile and reprogrammable, but differ in such secondary characteristics as density, access time, and block size (the number of cells erased at one time). The NOR architecture is simplest, but it requires a dual power supply and has a large block size which creates problems for applications such as the hard drive emulation. The EEPROM flash device, which has a small block size of 64 bytes and a single external power supply, is more expensive. The NAND EEPROM flash has an intermediate block size of 4 kbytes and includes the EDAC circuitry. The random access capability of NOR-based flash makes it quite suitable for use as an EPROM replacement, although it lacks features necessary for the solid-state file storage applications. In comparison, the NAND flash architecture includes page programming and address register operation, which are required to emulate the industry standard file transfer operations. The NAND-based flash systems are capable of read, write, and erase operations at high speed and high density such as 16 and 32 Mb, which are suitable for data storage applications in the PCMCIA cards.

Figure 3-36(a) shows a schematic of the NOR-based flash memory design which requires one transistor for every contact [87]. Figure 3-36(b) shows a schematic of the NAND EEPROM cell structure in which the memory is organized in 16-page blocks, and each block has 264 cell structures consisting of 16 serially connected transistors. The NAND EEPROM is designed to transfer the data to and from its memory array in 264-byte pages, including 8 extra bytes that can be directly accessed by the system. This approach has the advantage of loading data quickly to the memory data register by using only a single programming instruction for each 264-byte page, which increases the effective programming speed compared to byte-at-a-time programming.

During the NAND EEPROM programming, the drain and source select gates are turned on to connect the cell structure to the bit line and turned off to isolate it from the ground. Then, 10 V is applied to each word line in the block, except the page to be programmed, which is set to 20 V. Then, the voltage level on

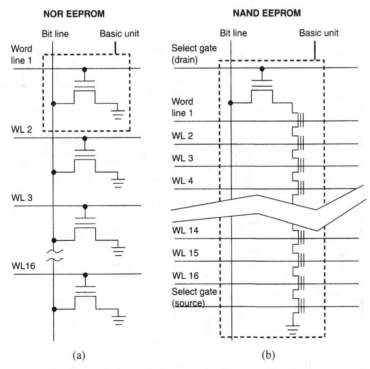

(a)   (b)

Figure 3-36. Flash memory cell structures. (a) NOR EEPROM that requires one transistor for every contact. (b) NAND EEPROM with 16 serially connected memory transistors per bit-line contact [87].

the bit line determines the data logic value ("1" or "0") to be programmed. For read operation, the select gates are both turned on, connecting the cell structure to a precharged bit line and to the ground. All word lines, except for the page to be read, are set at high voltage, turning on the deselected transistors. A logic "1" stored on the selected transistor will turn on and discharge the bit line and vice versa if logic "0" is stored. The program operation depends on the Fowler–Nordheim tunneling, where electrons migrate from the transistor substrate to the floating gate, driven by a potential difference of 20 V applied between the transistor's control gate and the substrate.

Intel developed a 256K flash memory using double-polysilicon, single-transistor 1.5 μm EPROM-based technology (ETOX™), which is electrically erasable and programmable from a 12 V power supply [88]. This device was optimized for microprocessor-controlled applications for reprogramming by providing a command port interface, address and data latches, internal erase, and program margin-voltage generation. The instructions for this include erase, erase-verify, program, program-verify, and READ. The READ performance of this device is equivalent to comparable density CMOS EPROM devices with a chip-enable access time of 110 ns at 30 mA active current consumption. The electrical byte programming is achieved through hot electron injection from the cell drain junction to the floating gate. The entire array electrical erase occurs in 200 ms typically, through Fowler–Nordheim tunneling.

Figure 3-37 shows a simplified device block diagram which includes the common-port architecture consisting of the command register, the command decoder and state latch, the data-in register, and the address register. This architecture allows easy microprocessor control for execution of various instructions. The cell is erased by placing a high voltage on the source

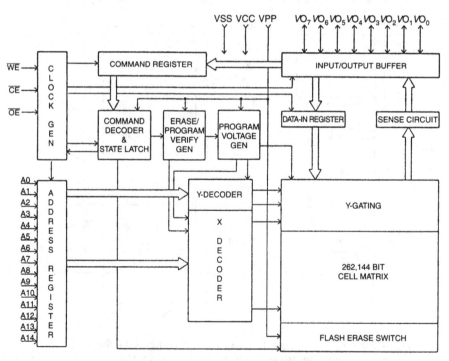

Figure 3-37. A 256K CMOS flash memory Intel chip block diagram. (From [88], with permission of IEEE.)

while the gate is grounded. The voltage across the thin first gate oxide causes tunneling of the electrons off the floating-gate oxide, thus reducing the cell threshold. The higher the voltage on the source, the faster the erase time. However, in choosing to relax the erasure time to 200 ms for this device, the electric-field-induced stress across the tunnel oxide is reduced, which yields better reliability. Since the tunneling occurs in a small area at the gate–source junction, this flash memory ETOX cell can tolerate a higher oxide defect density.

Toshiba has developed a 256 kb flash EEPROM with a single transistor per bit by utilizing triple-polysilicon technology resulting in a compact cell size of 8 × 8 μm² using conservative 2.0 μm design rules [89]. This device is pin compatible with a 256 kb UV-EPROM without increasing the number of input pins for erasing by introducing a new programming and erasing scheme. Figure 3-38(a)–(c) shows the basic cell structure, which consists of three polysilicon layers including the selection transistor. The erase gate is composed of the first polysilicon layer which is located on the field oxide. The floating gate is composed of the second polysilicon layer which partially overlaps the erase gate. The control gate at the top is composed of the third polysilicon layer, and functions as a word selection line during the programming and read operations. The bit line consists of a metal and ground line of n+ diffusion. The memory cell area is represented by a dashed line labeled as A, B, C, and D. This cell is programmed using a 21 V supply voltage, and the mechanism is channel hot carrier injection used for UV-EPROMs. Flash erasing is achieved by using the field emission of electrons from the floating gate to the erase gate. The programming time for this device is as fast as 200 μs/byte, and the erasing time is less than 100 ms/chip.

Atmel has developed a 5 V-only, in-system CMOS programmable and erasable read only memory (PEROM) with fast read access time (typically 120 ns), such as AT29C256 of 256 kb density organized as 32,768 words × 8 b [90]. The reprogramming for this can be per-

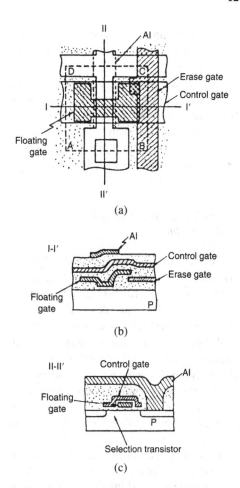

(a)

(b)

(c)

Figure 3-38. A 256 kb flash EEPROM Toshiba cell structure. (a) Top view of the cell. (b) Cross-section along I–I' line as shown in (a). (c) Cross-section along II–II' line. (From [89], with permission of IEEE.)

formed in a page mode in which 64 bytes of data are loaded into the device and simultaneously programmed. The contents of the entire memory array can be erased by using a 6 byte software code, although erasure before programming is not required. During the reprogramming cycle, the address locations and 64 bytes of data are internally latched, thus freeing the address and data bus for other operations. The initiation of a program cycle automatically erases the page and then programs the latched data by using an internal control timer. The end of a program cycle can be detected by data polling. Once the end of a program cycle has been detected, a new access for a read, program, or chip erase can begin. These PEROMs have write cycle times of 10 ms maximum per page or 5 s for full chip. They incorporate a hardware feature to protect against inadvertent programs, and a software-controlled data protection feature is also available.

3.6.2 Advanced Flash Memory Architectures

Channel hot electron injection (CHE) is widely used as a programming technique for EPROMs and flash EEPROMs. The major drawback for stacked gate devices which use this technique is that only a fraction of electrons injected into the gate oxide actually reach the gate, resulting in a low programming efficiency. An alternative investigated has been the use of a dual-transistor gate configuration where high efficiency can be obtained when a sufficiently high voltage is applied to the gate at the drain side of the device (called the D-gate), while a low voltage is applied to the gate at the source side of the device (called the S-gate). A study has shown that increased injection efficiency of the sidewall-gate concept can be achieved with a relatively simple split-gate transistor configuration which is compatible with the standard CMOS process. In order for the floating-gate potential to be higher than the potential of both the drain and control gate, the floating gate should be capacitively coupled to a high "program gate" voltage which can be generated on-chip by means of a high-voltage generator.

One of the key innovations in flash memory architecture is the "blocking" or partitioning of a flash memory chip array for different applications. For example, Intel's Boot Block flash memory line uses asymmetrical blocking in which large main blocks are reserved for the majority of code storage, very small blocks for parameter storage and/or back-up, and a small hardware-lockable boot block for the system's kernel start-up code. Symmetrical blocking is useful in mass storage applications, where equal size erase blocks emulate the sector and cluster arrangement of a disk drive. Several recent advances in photolithography, improvements in flash memory cell structures, and developments such as the high-speed page mode sensing scheme and divided bit line architecture have made possible the introduction of faster flash memories ranging in density from 1 to 64 Mb.

Hitachi introduced a 1 Mb flash memory with 80 ns read time which utilizes a one-transistor type cell with a stacked gate structure, a tunnel oxide of 10 nm thickness, interpoly dielectric 30 nm thick, and a cell area of 10.4 μm^2 [91]. The memory is fabricated in a 0.8 μm, one-poly, one-polysilicide, single-metal CMOS process similar to the conventional EPROM process. An advanced sense amplifier with a diode-like load, as well as scaled periphery transistors and polycided word lines are used.

Figure 3-39(a) shows a schematic block diagram of a 1 Mb flash memory [91]. The on-chip erase and erase-verify control system features a control signal latch, a sequence controller, and a verify voltage generator. The function of the control signal latch is to lock out any control signals and address signals except status polling instructions, the sequence controller performs an erase and erase–verify operation, and the verify voltage generator supplies low voltage to decoders and sense amplifiers during the verify operation in order to guarantee an adequate read operational margin. This memory can be programmed at the rate of 50 μs/byte, and the erase time is 1 μs. To achieve higher endurance, the peak electric field applied to the tunnel oxide during an erase operation is reduced by using a transistor with reduced channel width in a source driver circuit. This transistor limits the gate-induced junction breakdown current which occurs on the sources of memory array, and forces a slow rise time in the erase pulse.

In order to guarantee the read characteristics after an erasure, an appropriate control of the erase operation is required. An improvement has been made for in-system applications with a new automatic erase technique using an internal

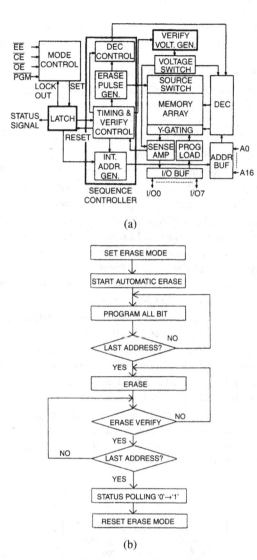

(a)

(b)

Figure 3-39. 1 Mb flash memory. (a) Schematic block diagram. (b) Internal automatic erase algorithm. (From [91], [92], with permission of IEEE.)

con layers and a self-aligned structure which allows for a smaller cell area of $4.0 \times 3.5 \ \mu m^2$ with 1.0 μm design rules. A select gate is used to prevent the undesirable leakage current due to overerasing. Figure 3-40 shows a cross-sectional view of the flash EEPROM SISOS cell. This cell is programmed by source-side injection of channel hot electrons, which means the injection occurs from the weak gate-controlled channel region under the oxide between the select gate and the stacked gate, as shown in the schematic cross-section. Since the hot electrons for programming operation are generated at the source side, the drain junction can be optimized independently to achieve erasure with no degradation in programmability. The cell erasure mechanism is the Fowler–Nordheim tunneling of electrons from the floating gate to the drain.

In the new high-density generation flash EEPROMs, TI has introduced a 4 Mb device which features 5 V-only operation and with full chip or sector-erase capabilities [94]. This device is fabricated in 0.8 μm CMOS technology which utilizes a unique, single-transistor advanced contactless EEPROM (ACEE) storage cell. This device can be programmed 1 byte at a time, or up to a full page of 256 bytes can be loaded into on-chip latches and programmed in a single cycle. Both the programming and erase operations are accomplished through Fowler–Nordheim

voltage generator [92]. Figure 3-39(b) shows the internal automatic erase algorithm.

The improvements in flash EEPROM cell structures have resulted in the fabrication of 4 Mb and higher density devices. An example is the stacked-gate MOSFET with a sidewall select-gate on the source side of the FET (SISOS cell) [93]. This cell uses three polysili-

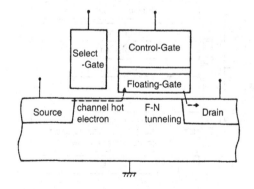

Figure 3-40. Schematic cross-section of flash EEP-ROM (SISOS cell). (From [93], with permission of IEEE.)

tunneling. Figure 3-41(a) shows the logical organization of this device. The memory array is divided into 64 segments, each consisting of 32 word lines. The corresponding word lines from each segment are connected in parallel and driven by a set of 32 row decoders that also provide on-chip generated high-voltage handling for the program mode and separate negative voltage generators for the erase mode.

Figure 3-41(b) shows the physical organization of the TI 4 Mb flash EEPROM chip. The read mode row decoders are located at the top of the array, whereas the programming portion and negative high-voltage generation circuitry are located at the bottom. The periodic breaks in the memory array are used for the segment select decoders which are required to enable the selected segment, and to provide routing channels for connecting the row decoders and high-voltage circuitry to the word lines.

Toshiba has developed a 5 V-only, 4 Mb (512K × 8) NAND EEPROM which incorporates a new program verify technique enabling a tight V_t distribution width of 0.8 V for the entire 4 Mb cell array [95]. The key element of the program verify technique is that the successfully data programmed cells with logic "1" are not reprogrammed again in order to prevent unnecessary overprogramming. The modified program responsible for tight V_t distribution consists of three steps: initial program, verify read, and reprogram. The cell size per bit is 11.7 μm². Both the channel length and width of the memory cell transistors are 1.0 μm.

Figure 3-42 shows a cross-sectional view of the 4 Mb NAND EEPROM for which the NAND-structured cell array is formed in the p-well region of the n-substrate [95]. The peripheral CMOS circuitry and NAND cell array are located in different p-wells, i.e., p-well 1 and p-well 2, respectively, which are electrically isolated from each other. The p-well depth is about 12 μm. In an erase operation, p-well 2 is raised to 20 V. The p-well 1 is always at ground potential. The peripheral circuitry is fabricated in a 2 μm triple-polysilicon, single-aluminum, and n-well CMOS process.

An innovation in flash memory designs has been the use of negative gate erase technology. According to Advanced Micro Devices (AMD), their 5 V-only negative gate erase technology provides designed-in reliability that is equal to their 12 V flash memory devices [96]. The 5 V-only technology with negative gate erase uses special design techniques to provide 10–20 mA (peak) current for band-to-band tunneling that comes directly from the system

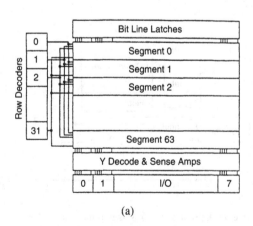

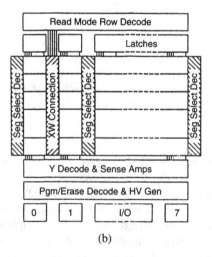

(a)   (b)

Figure 3-41. A 4 Mb TI flash EEPROM. (a) Logical organization. (b) Physical organization of the chip. (From [94], with permission of IEEE.)

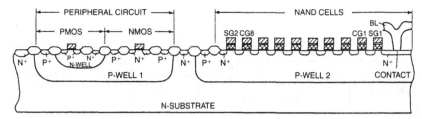

Figure 3-42. Cross-sectional view of Toshiba 4 Mb NAND EEPROM. (From [95], with permission of IEEE.)

V_{cc} through the Array Ground terminal. Figure 3-43(a) shows a circuit diagram of the negative gate erase implementation. The erase operation is performed by applying V_{cc} voltage to the source terminal while the gate is pumped to a negative voltage. It requires less than 10 μA of current on the gate for Fowler–Nordheim tunneling to occur. The erase time depends on the electric field across the floating gate and thickness of the tunnel oxide (T_{ox}).

The 5 V-only programming technique provides the same electric fields to the memory cells and the same programming mechanisms as the ones used by 12 V devices. Figure 3-43(b)

shows the 5 V-only drain programming circuitry. The drain is pumped from 5.0 to 6.7 V, and supplies approximately 0.5 mA current per cell. Since the charge pump delivers only a 34% voltage increase above the base supply voltage of V_{cc}, the electron injection mechanism is minimized. The current required by the gate voltage charge pump is less than 10 μA, which results in lower power dissipation.

An example of the flash memory architecture which uses negative gate erase and 5 V-only drain programming is the AMD 4 Mb device organized as 256K bytes of 16 b each. Figure 3-44 shows a block diagram of this flash memory

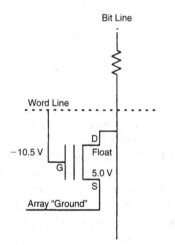

Notes:
1. Gate terminal is pumped to −10.5 V @ <10 μA current
2. 10-20 mA (peak) erase current is provided to the Source terminal by the system's Vcc supply

(a)

Bit Line

Word Line

+10.5 V

D
6.7 V

G
0.0 V

S

Array "Ground"

Notes:
1. Gate terminal is pumped to +10.5 V @ <10 μA current
2. Drain terminal is pumped to 6.7 V from 5.0 V Vcc supply @ 0.5 mA

(b)

Figure 3-43. AMD 5 V-only flash memory technology. (a) Negative gate erase circuitry. (b) Drain programming circuitry. (From [96], with permission of Advanced Micro Devices.)

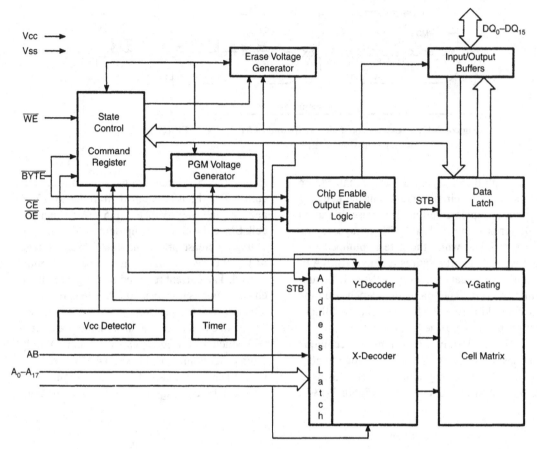

Figure 3-44. A block diagram of the AMD 4 Mb 5 V-only flash memory. (From [96], with permission of Advanced Micro Devices.)

which has a specified maximum access time of 70 ns. This device uses a single 5 V power supply for the read, write, and erase operations. It is programmed by executing the program command sequence which implements an embedded program algorithm that automatically times the program pulse widths and verifies the proper cell margin. The erasure is carried out by executing the erase command sequence which invokes an embedded erase algorithm and automatically preprograms the array if it is not already programmed before executing the erase operation. The device also features a sector erase architecture which allows for blocks of memory to be erased and reprogrammed without affecting the other blocks.

The conventional 12 V flash memories and 5 V-only technology produce approximately half the electric field on the tunnel oxide as compared with EEPROM technology. This lowering of tunnel oxide fields in the flash memories extends the life of oxides by several orders of magnitude. In addition, the flash memories which use source-side tunneling stress only a small region of the oxide relative to the EEPROM technologies that employ uniform channel tunneling stresses over the entire oxide region.

For very high-density, nonvolatile memory storage applications, Intel has developed a second generation of flash memories such as the 8 Mb device (28F008SA) organized as 1,048,576 $\times$ 8 b in a symmetrical blocking architecture (16 blocks × 64 kbytes) [97]. This memory is designed for 10,000 byte write/block erase cycles on each of the 16 64-kbyte blocks. It has features such as very high cycling endurance, symmetrical block erase, automation of byte write and block erase, erase suspend for data read, a reset/power-down mode, a write erase Status Register, and a dedicated RY/BY# status pin. The flash memory is fabricated with Intel ETOX™ III (EPROM tunnel oxide) which uses 0.8 μm double-polysilicon n-well/p-well CMOS technology that produces a 2.5 × 2.9 μm² single-transistor cell.

Figure 3-45 shows a block diagram of the Intel 8 Mb flash EPROM [97]. The command user interface (CUI) and Status Register interface to the power-up/down protection, address/data latches, and the write state machine (WSM). The WSM controls internal byte write, block erase, cell-margin circuits, and the dedicated READY/BUSY status outputs. The WSM-controlled block erasure, including pre-

programming, typically requires 1.6 s. All cells within the selected block are simultaneously erased via Fowler–Nordheim tunneling. Byte write is accomplished with the standard EPROM mechanism of channel hot electron injection from the cell drain junction to the floating gate. The programming mode is initiated by raising both the control gate and the cell drain to high voltage.

The CUI itself does not occupy an addressable memory location. The interface register is a latch used to store the command and address and data information needed to execute the command. After receiving the "Erase Setup" and "Erase Confirm" commands, the state machine controls block preconditioning and erase, monitoring progress via the Status Register and RY/BY# output. The "Erase Suspend" command allows block erase interruption in order to read data from another block of memory. Once the erase process starts, writing the Erase Suspend command to CUI requests that the WSM

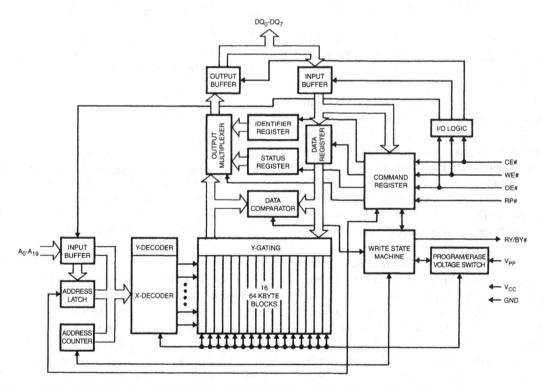

Figure 3-45. Block diagram of Intel 8 Mb flash EPROM. (From [97], with permission of Intel Corporation.)

suspend the erase sequence at a predetermined point in the erase algorithm. The Erase Suspend status and WSM status bits of the Status Register will be automatically cleared and RY/BY# will return to V_{OL}. After the "Erase Resume" command is sent, the device automatically outputs Status Register data when read.

A 5 V-only 16 Mb CMOS flash memory has been developed by NEC which utilizes an optimized memory cell with a diffusion self-aligned (DSA) drain structure and new types of row decoders to apply negative bias that allows 512-word sector-erase [98]. This flash memory has been fabricated using a 0.6 μm, double-layer metal, and triple-well CMOS process, as shown in the schematic cross-section of Figure 3-46(a). A tungsten silicide interconnection layer is used to minimize row decoder and column selector areas. Figure 3-46(b) shows the stacked-gate type memory cell cross-sectional view and typical operating conditions. The memory cell measures 1.7×2.0 μm². Its gate channel length and width are 0.6 and 1.0 μm, respectively. The tunnel gate oxide (SiO₂) thickness is 11 nm.

For the memory cell, a positive voltage (V_{cc}) is applied to both the p-well and deep n-well during the erase operation. This improves the channel erase capability since the electric field between the floating gate and p-well (channel area) is increased. For the negative voltage NMOS, a negative voltage can be applied to the p-well, source, and drain. The peripheral circuitry uses two types of transistors, each having different gate oxide thickness, i.e., 20 nm for normal transistors and 30 nm for high-voltage transistors. The minimum gate channel lengths for nMOS and pMOS transistors are 0.8 and 1.0 μm, respectively.

In order to achieve selective erase in one sector (one word line) with negative-gate bias channel erase, a new row decoding architecture was developed which consists of two row-main decoders and nine row-subdecoders. The row-main decoders are placed on both sides of the cell array and select 8 out of 2048 word lines. The row-subdecoders are placed on each side of the cell array blocks and select one out of eight

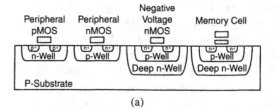

(a)

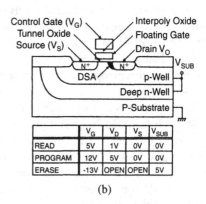

	V_G	V_D	V_S	V_{SUB}
READ	5V	1V	0V	0V
PROGRAM	12V	5V	0V	0V
ERASE	-13V	OPEN	OPEN	5V

(b)

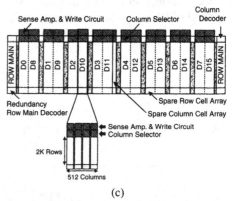

(c)

Figure 3-46. A 16 Mb flash memory NEC. (a) Schematic cross-section of triple-well CMOS structure. (b) Memory cell cross-sectional view and typical operating conditions. (c) Block diagram. (From [98], with permission of IEEE.)

word lines. This row decoder organization reduces word-line delay and occupies 12% of the chip area. For large-capacity flash memories, preprogramming and erase verify make auto chip erase time longer. This problem is overcome by using 64 b simultaneous operation and an improved auto chip erase sequence. Figure 3-46(c) shows a block diagram of this memory, in which the I/O block has four sense amplifiers and four write circuits which allow simultaneous 64 b preprogramming and erase verify. Also, this sense amplifier and write circuit organization permits four-word or 8 byte simultaneous operations, such as the page-mode read and page programming.

An ultrahigh-density NAND-structured memory cell has been developed using a new self-aligned shallow trench isolation (SA–STI) technology [99]. This SA–STI technology has been used to fabricate devices in 0.35 μm technology that have a very small cell size of 0.67 μm^2/bit. These are the key process features used to realize small cell size: (1) 0.4 μm width shallow trench isolation (STI) to isolate the neighboring bits, and (2) a floating gate that is self-aligned with the STI, eliminating the floating-gate wings. In this approach, the floating gate does not overlap the trench corners, so that enhanced tunneling at the trench corner is avoided, which increases the tunnel oxide reliability. This SA–STI cell has shown fast programming characteristics (0.2 μs/byte), fast erasing (2 ms), good write/erase endurance (10^6 cycles), and a high read-disturb tolerance (> ten years).

There are two key aspects in further developments of high-density flash memories: low-voltage operation and high-erase/write cycling endurance. The conventional flash memories described use the Fowler–Nordheim (FN) tunneling effect to inject or emit electrons through the tunnel oxide. The FN erasure uses the entire region of the channel, including the gate/n$^+$ overlapping region, which means a high electric field across the tunnel oxide affecting the erase/write cycling endurance characteristics. A new erase scheme called divided bitline NOR (DINOR) has been proposed which utilizes substrate hot electron (SHE) injection instead of FN tunneling [100]. In this approach, the cell is formed in a triple-well structure, and the bottom n-well layer surrounding the p-well is used as an electron supply source.

Figure 3-47 shows a schematic drawing for SHE erasure. In this erasure scheme, the

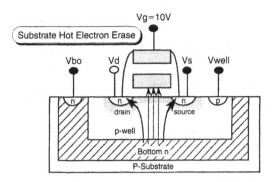

Figure 3-47. A schematic drawing of the substrate hot electron (SHE) erasure scheme. (From [100], with permission of IEEE.)

electrons pass only through the central region of the channel, and the electric field required is roughly reduced to one-fourth as compared to the FN erasure. The DINOR operation needs less than 1 s for erasure. The analysis shows that the SHE erasure scheme can extend erase/write cycling endurance by more than an order of magnitude.

A new write/erase method has been suggested to improve the read disturb characteristics by drastically reducing the stress-induced oxide leakage [101]. The analysis for this proposed write/erase method shows that the degradation of the read disturb lifetime after 10^6 write/erase cycles can be reduced to 50% as compared to the conventional bipolarity write/erase method. This write/erase method has the following key features: (1) application of an additional pulse to the control gate just after completion of the write/erase operation, (2) the voltage of this additional pulse is higher than that of a control gate during a read operation and lower than that of a control gate during a write operation, and (3) the polarity of the voltage is the same as that of the control gate voltage in the read operation. This proposed write/erase method based on the deactivation mechanism of the leakage current appears to be a promising approach for the realization of 256 Mb and higher density flash memories.

REFERENCES

[1] K. Kiuchi et al., "A 65mW 128k EB-ROM," *IEEE J. Solid-State Circuits*, vol. SC-14, p. 855, Oct. 1979.

[2] D. A. Hodges and H. G. Jackson, *Analysis and Design of Digital Integrated Circuits.* New York: McGraw-Hill, 1983.

[3] B. Donoghue et al., "Variable geometry packs 2 bits into every ROM cell," *Electronics*, p. 121, Mar. 24, 1983.

[4] T. P. Haraszti, "Novel circuits for high speed ROMs," *IEEE J. Solid-State Circuits*, vol. SC-19, p. 180, Apr. 1984.

[5] *Hitachi IC Memory Data Book,* 1990.

[6] *NEC Memory Products Data Book,* 1991.

[7] R. K. Wallace and A. Learn, "Simple process propels bipolar PROMs to 16-K density and beyond," *Electronics*, 1980.

[8] D. W. Greve, "Programming mechanism of polysilicon resistor fuses," *IEEE Trans. Electron Devices*, vol. ED-29, pp. 719–724, Apr. 1982.

[9] T. Fukushima et al., "A 40-ns 64 kbit junction-shorting PROM," *IEEE J. Solid-State Circuits*, vol. SC-19, pp. 187–194, Apr. 1984.

[10] *National Semiconductor Corporation Non-Volatile Memory Databook,* 1987.

[11] P. Thai et al., "A 35 ns 128K fusible bipolar PROM," in *IEEE ISSCC Tech. Dig.*, 1986, pp. 44–45, 299.

[12] E. Hamdy et al., "Dielectric based antifuse for logic and memory ICs," in *Proc. 1988 IEDM*, pp. 786–789.

[13] B. Cook and S. Keller, "Amorphous silicon antifuse technology for bipolar PROMs," in *Proc. IEEE Bipolar Circuits Tech. Meet.*, Oct. 1986, pp. 99–100.

[14] N. Sato et al., "A new programmable cell utilizing insulator breakdown," in *Proc. 1985 IEDM*, pp. 639–643.

[15] L. R. Metzger, "A 16k CMOS PROM with polysilicon fusible links," *IEEE J. Solid-State Circuits*, vol. SC-18, p. 562, Oct. 1983.

[16] *Harris Semiconductor Rad-Hard/Hi-Rel Data Book,* 1990.

[17] M. Wright, "High speed EPROMs," *EDN*, Sept. 1987.

[18] D. Frohman-Bentchkowsky, *Solid State Electron.*, vol. 17, p. 517, 1974.

[19] D. Frohman-Bentchkowsky, "A fully decoded 2048-bit electrically programmable FAMOS read-only memory," *IEEE J. Solid-State Circuits*, vol. SC-6, pp. 301–306, Oct. 1971.

[20] S. T. Wang, "On the *I-V* characteristics of floating-gate MOS transistors," *IEEE Trans. Electron Devices*, vol. ED-26, pp. 1292–1294, Sept. 1979.

[21] R. D. Katznelson et al., "An erase model for FAMOS EPROM devices," *IEEE Trans. Electron Devices*, vol. ED-27, pp. 1744–1752, Sept. 1980.

[22] P. Y. Cheng et al., "A 16k CMOS EPROM," in *IEEE ISSCC Tech. Dig.*, Feb. 1981, p. 160.

[23] M. Yoshida et al., "A 288K CMOS EPROM with redundancy," *IEEE J. Solid-State Circuits*, vol. SC-18, p. 544, Oct. 1983.

[24] *Fujitsu CMOS 1,048,576 bit UV Erasable Read Only Memory Data Specification Sheets*, Sept. 1988.

[25] D. Hoff et al., "A 23ns 256K EPROM with double-layer metal and address transition detection," in *IEEE ISSCC Tech. Dig.*, Feb. 1989, p. 130.

[26] H. Nakai et al., "A 36 ns 1 Mbit CMOS EPROM with new data sensing techniques," presented at the IEEE 1990 Symp. VLSI Circuits.

[27] M. Kuriyama et al., "A 16-ns 1-Mb CMOS EPROM," *IEEE J. Solid-State Circuits*, pp. 1141–1146, Oct. 1990.

[28] S. Pathak et al., "A 25 ns 16K CMOS PROM using a 4-transistor cell," in *IEEE ISSCC Tech. Dig.*, Feb. 1985, p. 162.

[29] S. Pathak et al., "A 25 ns 16k CMOS PROM using a 4-transistor cell and differential design techniques," *IEEE J. Solid-State Circuits*, vol. SC-20, p. 964, Oct. 1985.

[30] A. T. Wu et al., "A novel high speed 5-V programming EPROM structure with source-side injection," in *Proc. 1986 IEDM*, pp. 584–587.

[31] A. T. Mitchell et al., "A self-aligned planar array cell for ultra high density EPROMs," in *Proc. 1987 IEDM*, pp. 548–553.

[32] S. Atsumi et al., "A 120 ns 4 Mb CMOS EPROM," in *IEEE ISSCC Tech. Dig.*, 1987, p. 74.

[33] F. Masuoka et al., "New ultra high density EPROM and flash EEPROM with NAND structure cell," in *Proc. 1987 IEDM*, p. 552.

[34] K. Yoshikawa *et al.*, "0.6 μm EPROM cell design based on a new scaling scenario," in *Proc. 1989 IEDM*, pp. 587–590.

[35] Y. Oshima *et al.*, "Process and device technologies for 16Mbit EPROMs with large-tilt-angle implanted P-pocket cell," in *Proc. 1990 IEDM*, p. 95.

[36] S. Mori, "Polyoxide thinning limitations and superior ONO interpoly dielectric for nonvolatile memory devices," *IEEE Trans. Electron Devices*, vol. 38, pp. 270–276, Feb. 1991.

[37] S. Mori *et al.*, "ONO interpoly dielectric scaling for nonvolatile memory applications," *IEEE Trans. Electron Devices*, vol. 38, pp. 386–391, Feb. 1991.

[38] *Advanced Micro Devices CMOS Memory Products 1991 Databook/Handbook.*

[39] D. Frohman-Bentchkowsky, "The metal-nitride-oxide-silicon (MNOS) transistor characteristics," *Proc. IEEE*, vol. 58, Aug. 1970.

[40] E. S. Yang, *Microelectronic Devices.* New York: McGraw-Hill, 1988.

[41] S. M. Sze, "Current transport and maximum dielectric strength of silicon nitride films," *J. Appl. Phys.*, vol. 38, June 1967.

[42] E. Suzuki *et al.*, "Hole and electron current transport in metal-oxide-nitride-oxide-silicon memory structures," *IEEE Trans. Electron Devices*, vol. 36, pp. 1145–1149, June 1989.

[43] T. Hagiwara *et al.*, "A 16k bit electrically erasable PROM using N-channel Si-gate MNOS technology," *IEEE J. Solid-State Circuits*, vol. SC-15, p. 346, June 1980.

[44] A. Lancaster *et al.*, "A 5 V-only EEPROM with internal program/erase control," in *IEEE ISSCC Tech. Dig.*, Feb. 1983, p. 164.

[45] Y. Yatsuda *et al.*, "Hi-MNOS II technology for a 64k bit byte-erasable 5 V only EEPROM," *IEEE J. Solid-State Circuits*, vol. SC-20, p. 144, Feb. 1985.

[46] C. C. Chao and M. H. White, "Characterization of charge injection and trapping in scaled SONOS/MOMOS memory devices," *Solid State Electron.*, vol. 30, p. 307, Mar. 1987.

[47] S. Minami, "New scaling guidelines for MNOS nonvolatile memory devices," *IEEE Trans. Electron Devices*, vol. 38, pp. 2519–2526, Nov. 1991.

[48] Y. Yatsude *et al.*, "Scaling down MNOS nonvolatile memory devices," *Japan J. Appl. Phys.*, vol. 21, suppl. 21-1, pp. 85–90, 1982.

[49] D. Adams (Westinghouse Electric Corp.), J. Jakubczak (Sandia National Labs.) *et al.*, "SONOS technology for commercial and military nonvolatile memory applications," in *IEEE Nonvolatile Memory Technol. Rev., Proc. 1993 Conf.*

[50] SONOS technology report and photographs provided by Westinghouse Electric Corp., Baltimore, MD.

[51] J. J. Chang, "Nonvolatile semiconductor memory devices," *Proc. IEEE*, vol. 64, July 1976.

[52] Y. Tarui *et al.*, "Electrically reprogrammable nonvolatile semiconductor memory," in *IEEE ISSCC Tech. Dig.*, pp. 52–53, Feb. 1972.

[53] Y. Tarui *et al.*, "Electrically reprogrammable nonvolatile semiconductor memory," *IEEE J. Solid-State Circuits*, vol. SC-7, pp. 369–375, Oct. 1972.

[54] Y. Tarui *et al.*, "Electrically reprogrammable nonvolatile semiconductor memory," in *Proc. 5th Conf. Solid State Devices*, Tokyo, Japan, 1973, pp. 348–355.

[55] M. Horiuchi, "High-speed alterable, nonvolatile MIS memory," in *Proc. 1972 IEDM*.

[56] A. Kolodony *et al.*, "Analysis and modeling of floating-gate EEPROM cells," *IEEE Trans. Electron Devices*, vol. ED-33, pp. 835–844, June 1986.

[57] S. K. Lai and V. K. Dham, "VLSI electrically erasable programmable read only memory (EEPROM)," in N. G. Einspruch, Ed., *VLSI Handbook, Vol. 13.* New York: Academic, 1985.

[58] A. Kolodony *et al.*, "Analysis and modeling of floating-gate EEPROM cells," *IEEE Trans. Electron Devices*, vol. ED-33, pp. 835–844, June 1986.

[59] A. Bagles, "Characteristics and reliability of 10 nm oxides," in *Proc. Int. Reliability Phys. Symp.*, 1983, p. 152.

[60] G. Baccarani *et al.*, "HFIELDS: A highly flexible 2-D semiconductor device analysis program," in *Proc. 4th Int. Conf. Numerical Anal. of Semiconductor Devices and Integrated Circuits (NASECODE IV)*, June 1985, pp. 3–12.

[61] A. Concannon *et al.*, "Two-dimensional numerical analysis of floating-gate EEPROM devices," *IEEE Trans. Electron Devices*, vol. 40, pp. 1258–1261, July 1993.

[62] L. C. Liong *et al.*, "A theoretical model for the current-voltage characteristics of a floating-gate

EEPROM cell," *IEEE Trans. Electron Devices,* vol. 40, pp. 146–151, Jan. 1993.

[63] T. D. Linton, Jr. *et al.,* "A fast, general three-dimensional device simulator and its applications in a submicron EPROM design study," *IEEE Trans. Computer-Aided Des.,* vol. 8, pp. 508–515, May 1989.

[64] T. D. Linton *et al.,* "The impact of three-dimensional effects on EEPROM cell performance," *IEEE Trans. Electron Devices,* vol. 39, pp. 843–850, Apr. 1992.

[65] D. Guterman *et al.,* "New ultra-high density textured poly-Si floating gate EEPROM cell," in *Proc. 1985 IEDM,* pp. 620–623.

[66] H. A. R. Wegener, "Endurance model for textured-poly floating gate memories," in *Proc. 1984 IEDM,* p. 480.

[67] L. Chen *et al.,* "A 256K high performance CMOS EEPROM technology," in *Proc. 1985 IEDM,* pp. 620–623.

[68] *Seeq Technology, Inc. Databook.*

[69] Electronic Industries Association (EIA) JEDEC Standard 21C, "Configurations for solid state memories."

[70] *ATMEL CMOS Databook,* 1991–1992.

[71] T. Nozaki *et al.,* "A 1 Mbit EEPROM with MONOS memory cell for semiconductor disk application," presented at the IEEE 1990 Symp. VLSI Circuits.

[72] F. Masouka *et al.,* "New ultra high density EPROM and flash EEPROM with NAND structure cell," in *Proc. 1987 IEDM,* pp. 552–555.

[73] M. Momodami *et al.,* "New device technologies for 5 V-only 4 Mb EEPROM with NAND structured cell," in *Proc. 1988 IEDM,* pp. 412–415.

[74] T. Tanaka *et al.,* "A 4-Mbit NAND-EEPROM with tight programmed V_t distribution," presented at the IEEE 1990 Symp. VLSI Circuits.

[75] M. Momodami *et al.,* "An experimental 4-Mbit CMOS EEPROM with a NAND-structured cell," *IEEE J. Solid-State Circuits,* pp. 1238–1243, Oct. 1989.

[76] M. Hayashikoshi *et al.,* "A dual-mode sensing scheme of capacitor coupled EEPROM cell for super high endurance," presented at the IEEE 1992 Symp. VLSI Circuits.

[77] T. J. Ting *et al.,* "A 50 ns CMOS 256k EEPROM," *IEEE J. Solid-State Circuits,* vol. 23, p. 1164, Oct. 1988.

[78] F. Vollebregt *et al.,* "A new EEPROM technology with a TiSi$_2$ control gate," in *Proc. 1989 IEDM,* p. 607.

[79] N. Becker *et al.,* "A 5 V-only 4K nonvolatile static RAM," in *IEEE ISSCC Tech. Dig.,* 1983, pp. 170–171.

[80] D. D. Donaldson *et al.,* "SNOS 1K × 8 static nonvolatile RAM," *IEEE J. Solid-State Circuits,* vol. SC-17, pp. 847–851, Oct. 1982.

[81] C. E. Herdt, "Nonvolatile SRAM—The next generation," in *IEEE Nonvolatile Memory Technol. Rev., Proc. 1993 Conf.,* pp. 28–31.

[82] Y. Yamauchi *et al.,* "A versatile stacked storage capacitor on FLOTOX cell for megabit NVRAM applications," in *Proc. 1989 IEDM,* pp. 595–598.

[83] Y. Yamauchi *et al.,* "A versatile stacked storage capacitor on FLOTOX cell for megabit NVRAM applications," *IEEE Trans. Electron Devices,* vol. 39, pp. 2791–2795, Dec. 1992.

[84] R. D. Pashley *et al.,* "Flash memories: The best of two worlds," *IEEE Spectrum,* p. 30, Dec. 1989.

[85] K. Robinson, "Endurance brightens the future of flash," *Electron. Component News, Technol. Horizons,* Nov. 1989.

[86] B. Diapert, "Flash memory goes mainstream," *IEEE Spectrum,* pp. 48–52, Oct. 1993.

[87] J. Eldridge, "Filling in a flash," *IEEE Spectrum,* pp. 53–54, Oct. 1993.

[88] V. N. Kynett *et al.,* "An in-system reprogrammable 32K × 8 CMOS flash memory," *IEEE J. Solid-State Circuits,* vol. 23, pp. 1157–1162, Oct. 1988.

[89] F. Masouka *et al.,* "A 256k bit flash EEPROM using triple-polysilicon technology," *IEEE J. Solid-State Circuits,* vol. SC-22, p. 548, Aug. 1987.

[90] *ATMEL CMOS Databook,* 1993.

[91] K. Seki *et al.,* "An 80-ns 1-Mb flash memory with on-chip erase/erase-verify controller," *IEEE J. Solid-State Circuits,* vol. 25, pp. 1147–1152, Oct. 1990.

[92] K. Shohji *et al.,* "A novel automatic erase technique using an internal voltage generator for 1 Mbit flash EEPROM," in *Proc. IEEE 1990 Symp. VLSI Circuits,* p. 99–100.

[93] K. Naruke *et al.,* "A new flash-erase EEPROM cell with a sidewall select-gate on its source side," in *Proc. 1989 IEDM,* pp. 603–606.

[94] H. Stiegler *et al.,* "A 4Mb 5 V-only flash EEP-ROM with sector erase," presented at the IEEE 1990 Symp. VLSI Circuits.

[95] M. Momodomi *et al.,* "A 4-Mb NAND EEP-ROM with tight programmed V_t distribution," *IEEE J. Solid-State Circuits,* vol. 26, pp. 492–496, Apr. 1991.

[96] *Advanced Micro Devices Flash Memory Products 1992/1993 Data Book/Handbook.*

[97] *Intel 28F008SA-L 8-MBIT (1 MBIT × 8)* Flash Memory Advance Information Sheets; *and* ER-27 Eng. Rep.

[98] T. Jinbo *et al.,* "A 5V-only 16-Mb flash memory with sector erase mode," *IEEE J. Solid-State Circuits,* vol. 27, pp. 1547–1553, Nov. 1992.

[99] S. Aritome *et al.,* "A 0.67 μm^2 self-aligned shallow trench isolation cell (SA-STI cell) for 3V-only 256 Mbit NAND EEPROMs," in *Proc. 1994 IEDM,* pp. 3.6.1–3.6.4.

[100] N. Tsuji *et al.,* "New erase scheme for DINOR flash memory enhancing erase/write cycling endurance characteristics," in *Proc. 1994 IEDM,* pp. 3.4.1–3.4.4.

[101] T. Endoh *et al.,* "A new write/erase method for the reduction of the stress-induced leakage current based on the deactivation of step tunneling sites flash memories," in *Proc. 1994 IEDM,* pp. 3.4.1–3.4.4.

[102] K. Oshaki et al., "A planar type EEPROM cell structure by standard CMOS process for integration with gate array, standard cell, microprocessor and for neural chips," IEEE 1993 Custom IC Conf., pp. 23.6.1–23.6.4.

4

Memory Fault Modeling and Testing

4.1 INTRODUCTION

Memory device failures are usually represented by a bathtub curve, and are typically grouped into three categories, depending upon the product's operating life cycle stage where the failures occur. These categories are as follows: infant mortality, useful life, and wear-out failures (see Chapter 6). These failures are usually caused by design errors, materials and process defects, operational environment extremes, and aging effects. To analyze the faulty memory circuit behavior and development of techniques for the detection of failures, abstract fault models are commonly used. These fault models allow the cost-effective development of test stimuli that will identify the failed chips, and if possible, diagnose the failure mode. The fault models can be based upon different levels of abstraction such as the behavioral model, the functional model, the logical model, the electrical model, and the geometrical model which are briefly described as follows [1]:

- The behavioral model is totally based on the system specification and is the highest level of abstraction.
- The functional model is based upon functional specification of the system which requires certain assumptions to

be made about the internal structure of the system.

- The logical model has the properties similar to the electrical model which can allow faults to be detected and localized at the gate level.
- The electrical model is based upon both the functional specification of the system and complete knowledge of the internal structure at the electrical level.
- The geometrical model assumes complete knowledge of the chip layout.

In general, memory faults can be divided into two classes. In memory fault modeling, the physical faults that can appear on a chip are often modeled as logical faults. This modeling approach is based upon the assumption that given any fault, it is always possible, at least in principle to determine its effect on the logical behavior of the circuit. Some advantages of modeling the physical faults as the logical faults are: (1) physical examination of the chip for the location of actual physical faults is often impossible, and hence tests must be used which are based on the logical comparison with the known good units; and (2) logical fault modeling allows the test approach to become technology (e.g., TTL, NMOS, CMOS)-independent, and hence more generally applicable.

Semiconductor memories differ from the random logic, both combinational and sequential, which is implemented in typical VLSI devices. The submicron memory devices are more densely packed, and have to be tested thoroughly for multiple faults of various types. However, since memories perform simple and well-defined tasks, they can be easily modeled at the functional levels and tested for parametric and functional faults. The parametric faults include excessive leakage currents, unacceptable input and output voltage levels, inadequate drive and fan-out performance, low noise margins, and low data retention time. Functional faults are based upon the various fault models such as stuck-at and transition faults, bridging faults, coupling faults, and pattern-sensitive faults, which are discussed in Section 4.2. The RAM electrical testing including various functional algorithms is described in Section 4.3, and pseudo-random test techniques in Section 4.4.

The RAM and nonvolatile memory technologies, cell structures, parametric, and functional performance characteristics were discussed in Chapters 2 and 3. The most common memory device from the modeling point of view is a RAM, and the RAM model can be easily modified into a ROM, EPROM, or an EEP-ROM chip model. Figure 4-1 shows the functional block diagram of a RAM chip, whose basic function consists of writing data, storing data, and providing data readout access from any arbitrary cell location [2]. For DRAMs, this would also include memory cell array refresh circuitry. Some of the faults that can occur in this chip can be listed as follows:

- Cells stuck
- Read/write line stuck
- Chip select line stuck
- Data line stuck or open
- Short and/or crosstalk between two data lines
- Address line stuck or open
- Short between address lines
- Wrong access or multiple access
- Pattern-sensitive interaction between cells.

Since a RAM is designed to store any arbitrary data pattern in its cells, ideally the functioning of each cell should be verified for all possible data patterns in the remaining cells. However, for large memory cell arrays, this can become very time consuming, and in practice, many of these faults are unlikely to occur. Therefore, the reduced functional fault model sets are used, which are discussed in the following sections.

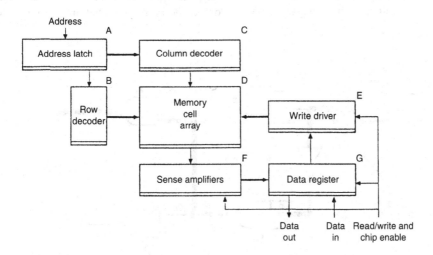

Figure 4-1. Functional block diagram of a RAM chip [2].

4.2 RAM FAULT MODELING

The most commonly used RAM memory fault models are for stuck-at or transition faults, bridging faults, coupling faults, pattern-sensitive faults, and dynamic (or delay) faults, which are discussed in this section. In general, the fault models and fault detection algorithms discussed here are based upon one bit per word memory organizations, but can be extended to word (or byte)-oriented memories with appropriate modifications.

4.2.1 Stuck-At Fault Model

The single-stuck fault model (SSF), because of its wide usage, is also referred to as the classical or a standard fault model. In this fault model, only one logical connection in a circuit may be permanently stuck at logical one (s-a-1) or stuck at logical zero (s-a-0). Therefore, in this fault model, the logical value of a stuck-at (SA) cell or a line is always "1" or a "0," and cannot be changed to the opposite state. The test generation for SSF is relatively easy. Several test generation methods such as the D-algorithm and LASAR use a technique called path sensitization, as illustrated in Figure 4-2. Assuming that for an SSF, the output of G_0 which is stuck-at-1 (s-a-1) is to be detected. Then it is sufficient to apply a pattern of input signals to the circuit which creates a path (shown by the heavy line) over which an error signal can propagate from

the faulty line to an observable output node (e.g., in this case, the output line Z_1). Then, this path is considered "sensitized" since any change in the signal applied to the input end propagates to the output node.

By using this standard path-sensitization technique, the tests for many SSFs can be generated simultaneously, thereby reducing the test generation costs. Considering the circuit in Figure 4-2, let us say it is desired to obtain the test for the SSF fault G_0 which is stuck-at-1 (s-a-1). Let $E/\overline{E}$ denote the signal on a line such that the state of the line is E if no fault is present and $\overline{E}$ if G_0 is (s-a-1). Thus, a test for this fault must apply 0/1 to the output line of G_0, and must also apply a signal of the form $E/\overline{E}$ to every line along the sensitized path in order for the desired error-signal propagation to take place. This implies that the given test pattern will detect an SSF of the (s-a-$\overline{E}$) type associated with every line on the sensitized path.

In comparison to other fault models, the number of SSFs is relatively small, and the number of faults to be explicitly analyzed can be reduced by the fault-collapsing techniques. For example, if the circuit contains N distinct logic lines, then there are $2N$ possible SSFs. This number can be further reduced by using the equivalence and dominance properties to eliminate faults that will be detected by tests generated for other faults. SSF test sets can also detect the fault types that are not equivalent to

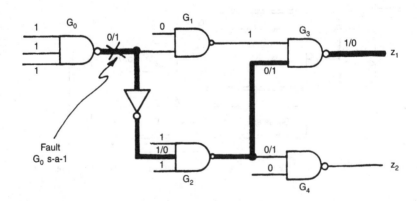

Figure 4-2. SSF detection via a sensitized path (heavy line).

SSFs, such as the shorts between logic signals and multiple stuck-line faults. However, the SSF model has some limitations since there are many physical lines in a chip (e.g., power and ground lines) which do not appear in the standard logic diagrams, and conversely, not all lines in logic diagrams correspond to physical connections. In addition, there are many fault situations where one or more of the basic assumptions underlying the SSF may not be valid. For example, "soft faults," unlike the SSFs, are transient and more likely to escape detection by SSF tests which are applied periodically under the assumption that the fault remains after its first occurrence.

A circuit under test can have more than one fault, i.e., multiple stuck line faults, or the MSF model. In this case, as for the SSF model, the faults are assumed to be permanent, and the circuit components and operating speed are assumed to be fault-free. These MSFs are of interest for several reasons, as follows:

- They can be used to model more physical faults than the SSF model. For example, if a circuit contains N distinct logic lines, then there are $2N$ possible SSFs, but 3^N-1 possible MSFs. As the component density increases, physical failures are more likely to resemble the MSFs than SSFs.
- The use of an MSF model can often simplify the SSF analysis. In a sequential circuit, an SSF may propagate faulty signals to many different parts of the circuit, and therefore, the MSF model may be the most appropriate model for analyzing subsequent behavior of the circuit [3].

It is usually recognized that the SSFs tend to dominate the MSFs, i.e., a test "T" which detects all the SSFs in a circuit is likely to detect most, if not all, MSFs. However, this test set "T" may fail to detect some MSFs due to the phenomenon of "fault masking." In general, an SSF can fail to detect a multiple fault $(f_1, f_2 \cdots f_k)$ if the detection of each component fault f_i is masked by the remaining faults. The SSFs can be present in the memory data and address registers, address decoders, memory cell arrays, and read/write logic. The memory scan (MSCAN) is a simple test that can detect any SSF in the memory array, in the memory data register, or in the read/write logic, but will not detect any SSF in the memory address register or decoder. The MSCAN test writes each word, first with a 0 and then with a 1, and verifies the write operation (denoted by W_i) with two read statements. The MSCAN test complexity is $4n$, and the algorithm can be represented as follows [4]:

$$\text{For } i = 0, 1, 2, \cdots n - 1 \quad \text{do:}$$
$$\text{Write: } W_i \leftarrow 0$$
$$\text{Read : } W_i (=0)$$
$$\text{Write: } W_i \leftarrow 1$$
$$\text{Read : } (W_i = 1).$$

A more complex procedure called the algorithmic test sequence (ATS) has been developed that detects any combination of stuck-at-0 or stuck-at-1 multiple faults in a RAM, which is assumed to consist of a memory address register (MAR), a memory data register (MDR), an address decoder, and a memory cell array [5]. It is assumed that a single fault within the decoder does not create a new memory address to be accessed without also accessing the programmed address, and that the RAM has a wired-OR behavior. The complexity of the ATS algorithm is $4n$ since $4n$ memory accesses are needed. A discussion of the ATS algorithm follows.

Assume that the memory contains n words, each of them denoted as W_i for $0 \le i \le n - 1$, and that each word is m bits long. The memory words are grouped into three partitions, G_0, G_1, and G_2, such that

$$G_0 = \{ W_i \,|\, i = 0 \text{ (modulo 3)} \}$$
$$G_1 = \{ W_i \,|\, i = 1 \text{ (modulo 3)} \}$$
$$G_2 = \{ W_i \,|\, i = 2 \text{ (modulo 3)} \}.$$

ATS Procedure:

1. For all $W_i \in G_1 \cup G_2$
   Write: $W_i \leftarrow 0$ "0 denotes the
   all 0 word"

2. For all $W_i \in G_0$
   Write: $W_i \leftarrow 1$ "1 denotes the
                              all 1 word"
3. For all $W_i \in G_1$
   Read: $W_i (=0)$
4. For all $W_i \in G_1$
   Write: $W_i \leftarrow 1$
5. For all $W_i \in G_2$
   Read: $W_i (=0)$
6. For all $W_i \in G_0 \cup G_1$
   Read: $W_i (=1)$
7. For all $W_i \in G_0$
   Write: $W_i \leftarrow 0$
   Read: $W_i (=0)$
8. For all $W_i \in G_2$
   Write: $W_i \leftarrow 1$
   Read: $W_i (=1)$
END

Several modified versions of ATS have been developed such as ATS+ and MATS. The MATS algorithm modified the ATS test sequence to create a simple structure which is eas-

ier to implement [6]. It detects any combination of SSFs in a RAM, irrespective of the design of the decoder. There are other versions of MATS called Complemented MATS and MATS+. Table 4-1 lists a summary of some commonly used stuck-at fault algorithms, their relative test complexities, and the fault coverages [4].

A special case of stuck-at faults is the transition fault (TF) in which a cell or a line fails to undergo a $0 \rightarrow 1$ transition or a $1 \rightarrow 0$ transition when it is written into. A test that has to detect and locate all the TFs should satisfy the following requirement [1]: each cell must undergo a $0 \rightarrow 1$ transition and a $1 \rightarrow 0$ transition, and be read after each transition before undergoing any further transitions.

The SAF and TF detection procedures can also be extended to the word-oriented memories which contain more than one bit per word, i.e., $B \geq 2$, where B represents the number of bits per word and is usually expressed by a power of 2. Since the SAFs and TFs involve only single cells, the memory width of B bits does not affect the detectability of these faults. However, an important difference exists in the test procedures (or algorithms) of bit- versus word-oriented memories.

TABLE 4-1. Summary of Some Commonly Used Stuck-at Fault Algorithms, Their Relative Complexities, and Fault Coverages. (From [4], with Permission of Association for Computing Machinery [ACM], Inc.)

Test Procedure	Complexity	Fault Coverage	
		Stuck-at Faults	Coupling and PSFs
MSCAN	$4n$	Covers memory array faults, but does not cover decoder or MAR faults	No
ATS	$4n$	Covers all, for RAMs with wired-OR logic behavior and noncreative decoder design	No
Complemented ATS	$4n$	Same as ATS, but for RAMs with wired-AND logic behavior	No
ATS+	$13n/3$	Covers all, for RAMs with arbitrary wired logic behavior and noncreative decoder design	No
MATS	$4n$	Covers all, for RAMs with wired-OR logic behavior and arbitrary decoder design	No
Complemented MATS	$4n$	Same as MATS, but for RAMs with wired-AND logic behavior	No
MATS+	$5n - 2$	Covers all, for RAMs with arbitrary wired logic behavior and arbitrary decoder design	No

In bit-oriented tests, the read and write operations for 0s and 1s (e.g., "$r\,0$," "$w1$") are applied to a single bit, whereas in the word-oriented memories, an entire word of B bits has to be read (or written). This word level read or write mode is called the "data background." Therefore, the word-oriented memory SAF and TF test procedures have to be modified from the bit-level operations to the word-level operations (read and write) with a data background of B bits.

4.2.2 Bridging Faults

A major shortcoming of the stuck-at fault model is that the fault simulation using a model is no longer an accurate indicator of ICs like memory chips. A large percentage of the physical faults occurring in ICs can be considered as bridging faults (BFs), consisting of a short between two or more cells or lines. These are also significantly more complex than stuck-at faults since, in general, the signals associated with shorted lines are variables rather than constants. For example, in an N-line circuit, the number of single BFs, i.e., faults involving just one pair of shorted lines, is given by

$$\binom{N}{2} = \frac{N(N-1)}{2}.$$

A bridging fault, unlike an SSF or an MSF, can introduce unexpected fault modes, de-pending upon the circuit technology being used. Figure 4-3 shows a bridging (SC) fault involving two signal lines z_1 and z_2. As in the case of MSFs, most of the bridging faults of a certain type can be detected by a complete test set for the SSFs. The research into BF models has generally been restricted to faults that result in well-defined wired-AND or wired-OR behavior. However, recent work has shown that the bridging faults cause more complex circuit behavior than the one predicted by permanent wired-logic fault models [7]. The two major aspects of bridging faults are: (a) fault simulation, (b) fault diagnosis and detection. The bridging faults can be modeled at the gate, switch, and circuit level. At the gate level, they can be modeled as permanent reconfigurations which alter the logical structure of the circuit [8]. The switch level fault simulators such as COSMOS [9], CHAMP [10], or IDSIM3 [11] can mimic a bridging defect by inserting a large, permanently conducting transistor between the nodes, or by simply joining the two nodes to form more complex pull-up and pull-down networks. The circuit level simulators such as SPICE2 [12] and SPICE3 can realistically model a bridging defect by insertion of a resistor between the two circuit nodes.

A study was performed to identify the key factors involved in obtaining fast and accurate CMOS bridging fault simulation results [8]. It was shown that the key steps to model physical

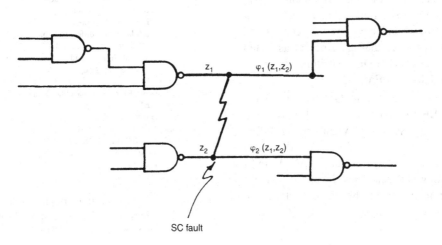

Figure 4-3. A typical bridging fault involving two signal lines z_1 and z_2.

mechanisms must be taken as follows: (1) determination of the voltage at the bridging site by resolving the drive contention between the bridged gates, and (2) interpretation of that voltage by comparison to the logic threshold voltage of each gate input connected to the bridged node. The BRIDGESIM program used precomputed values for these voltages to obtain accurate values for feeding into a gate-level simulator. This technique provided excellent throughput and accuracy, but is limited by the scalability of the table look-up procedures and the assumptions made during the circuit-level characterization.

Figure 4-4 shows an example of the test for locating bridging faults in data lines, the TLBFDL algorithm [13]. It uses the upper right half of a Boolean matrix BRIDGE $[0 \cdots B - 0 \cdots B - 1]$. If the entry BRIDGE$[d1, d2]$ ($d1 < d2$) is FALSE, then data lines d_{d1} and d_{d2} cannot be members of the same bridge. The variable MASK contains one bit for every data line which is stuck-at, as located by the test for locating stuck-at faults in data lines (TLSAFDL) [14]. $M0$ and $M1$ are the results from the TLSAFDL algorithm; $M0$ is a B-bit vector which contains a "1" for every corresponding SA1 data line, and $M1$ contains a "0" for every corresponding SA0 data line. SA data lines cannot be tested for bridging faults. The TLBFDL algorithm steps are briefly summarized below:

- Step 1: Initializes variable $w1$ to an arbitrary word within a row, and the upper right half of the Boolean matrix BRIDGE is initialized to TRUE, except for rows and columns that correspond to data lines that are stuck-at as tested by the TLSAFDL algorithm; these are initialized to FALSE.

- Step 2: Locates all OR-bridging faults (OBFs) by writing 2^{d1} to $A[r1, w1]$ and reading that word. When a bit $d1$ differs from a bit $d2$ ($d2 > d1$), the entry BRIDGE$[d1, d2]$ is set to FALSE, indicating that data lines d_{d1} and d_{d2} cannot be members of the same bridge. The variable c is used to store the data from $A[r1, w1]$.

Step 1: {Initialize variables $w1$, *MASK* and matrix *BRIDGE*}
  $w1 := 0$;{ Arbitrary word within a row }
  $MASK := M0$ **OR NOT** $(M1)$;
  { *MASK* now contains ones for every SAF }
  for $d1 := 0$ **to** $B - 1$ **do**
  { For all rows in *BRIDGE* }
    for $d2 := d1 + 1$ **to** $B - 1$ **do**
    { For all columns in *BRIDGE* }
      if ($MASK$ **AND** $(2^{d1}$ **OR** $2^{d2})) = 0$ **then**
      { Check if one of data lines d_{d1} or d_{d2} is SA }
        $BRIDGE[d1,d2] := $ TRUE;
        { Data lines d_{d1} and d_{d2} can be member
          of the same bridge}
      else
        $BRIDGE[d1,d2] := $ FALSE;
        { Data lines d_{d1} and d_{d2} cannot be member
          of the same bridge}
Step 2: { Locate OBFs }
  for $r1 := 0$ **to** q-1) **do** { Repeat for all q rows }
    for $d1 := 0$ **to** (B-2) **do** { Test only $B - 1$ data lines }
      if ($MASK$ **AND** $2^{d1}) = 0$ **then**
      { If data line $d1$ is not SA }
      begin
        $A[r1,w1] := 2^{d1}$; { Write $0 \ldots 010 \ldots 0$ }
        $c := A[r1,w1]$ { Read the same address }
        for $d2 := d1 + 1$ **to** $B - 1$ **do**
        { Check bits $d1 + 1 \ldots B - 1$ in c }
          if ($MASK$ **AND** $(2^{d2}) = 0$ **then**
          { If data line $d2$ is not SA }
            if $c[d1] \neq c[d2]$ **then**
            { If bit $d1$ in c differs from bit $d2$ }
              $BRIDGE[d1,d2] := $ FALSE;
              { Data lines d_{d1} and d_{d2} cannot be member
                of the same bridge }
      end;
Step 2: { Locate ABFs }
  for $r1 := 0$ **to** $q - 1$ **do** { Repeat for all q rows }
    for $d1 := 0$ **to** B - 2 **do** { Test only $B - 1$ data lines }
      if ($MASK$ **AND** $2^{d1}) = 0$ **then**
      { If data line $d1$ is not SA }
      begin
        $A[r1,w1] := -1 - 2^{d1}$; { Write $1 \ldots 101 \ldots 1$ }
        $c := A[r1,w1]$ { Read the same address }
        for $d2 := d1 + 1$ **to** $B - 1$ **do**
        { Check bits $d1 + 1 \ldots B - 1$ in c }
          if ($MASK$ **AND** $(2^{d2}) = 0$ **then**
          { If data line d_{d2} is not SA }
            if $c[d1] \neq c[d2]$ **then**
            { If bit $d1$ in c differs from bit $d2$ }
              $BRIDGE[d1,d2] := $ FALSE;
              { Data lines d_{d1} and d_{d2} cannot be member
                of the same bridge }
      end;
Step 4: { If there is a BF between data lines $d1$ and $d2$ }
    { $BRIDGE[d1,d2] = $ TRUE ($d2 > d1$)}
  for $d1 := 0$ **to** $B - 2$ **do**
    for $d2 := d1 + 1$ **to** $B - 1$ **do**
      if $BRIDGE[d1,d2]$ **AND** $(d1 \neq d2)$ **then**
        output("BF between data line",d_{d1},"
          and data line ",d_{d2});

Figure 4-4. Test for locating bridging faults in data lines (TLBFDL algorithm). (From [13], with permission of IEEE.)

- Step 3: Locates all AND-bridging faults (ABFs) by repeating Step 2 with inverse data.
- Step 4: Checks for the bridges and prints the members of each bridge.

In addition to robust TLBFDL, the complete set contains two other robust algorithms: the test for locating bridging faults in row select lines (TLBFRSL), and the test for locating BFs in word select lines (TLBFWSL) [13].

In the case of word-oriented memories, the detection of BFs and state coupling faults (SCFs) between cells in a word requires all states of two arbitrary memory cells i and j to be checked, i.e., the states

$$(i, j) \in \{(0,0), (0,1), (1,0), (1,1)\}.$$

Let B be the number of bits per word and D the required data background. A minimum set of data background for detecting the BFs and SCFs can be found based upon the following requirements:

- No two vectors Z_i and Z_j may be equal, i.e., $Z_i \neq Z_j$. This means that the maximum possible number of different vectors of D bits relationship to the word length can be expressed by

$$2^D = B.$$

- No two vectors may be each other's inverses, i.e., $Z_i \neq Z_j - 1$. Therefore, the number of maximum possible vectors is halved, or

$$2^D - 1 = B.$$

Therefore

$$D = \log_2 (B + 1).$$

A memory array may be considered as a collection of RAM chips interconnected via the data lines, row select lines, and word select lines. An array can be tested in the same way as a memory chip. For fault location in the memory arrays, it is necessary to store all addresses at which the tests indicate faults, together with the expected and actual data. Test algorithms have been devel-oped for locating the bridging faults in memory arrays using the k-bridging fault model [13]. It involves "k lines" which are supposed to be in close proximity on a printed circuit board, and are called "members of the bridge." As in the case of bit-oriented memories, two types of bridging faults are identified: (1) AND-bridging fault (ABF), in which the resulting state of the members of the bridge is the logical AND of the members; and (2) OR-bridging fault (OBF), in which the resulting state of the members is random or pseudorandom. Before the data lines can be tested for BFs, it is necessary to test them for stuck-at faults, which are considered to have precedence over bridging faults.

The memory array test algorithms are quite complicated because fault localization is required, and several faults may be present in the array simultaneously, including faults within the memory chips. The preferred test sequence for a memory array is to test the data lines first, then the row select lines, followed by the word select lines, and last of all, the memory chips. When only detection of the data line, row select line, word select line, and memory data stuck-at faults is required (such as for the power-on test), the MAT+ algorithm may be used.

A study has shown that the traditional approach to diagnosing stuck-at faults with the fault dictionaries generated for SAFs is not appropriate for diagnosing CMOS bridging faults [15]. A new technique was developed that takes advantage of the following relationship between detecting the bridging and SAFs: if a test vector detects a bridging fault, then it must also detect an SAF on one of the shorted nodes. This technique allows effective diagnosis of the bridging faults, even when the reduced SAF dictionaries are used. Another technique for detecting the bridging faults which is finding increasing usage for CMOS devices is based on the quiescent current (I_{DDQ}) monitoring that will be discussed in Section 4.7.

4.2.3 Coupling Faults

An important type of fault that can cause a semiconductor RAM cell to function erroneously is a coupling fault (CF). These faults can occur as the result of a short circuit or a circuit's parasitic effects, such as current leakages and stray capacitances. A CF caused by a transition write operation is a "2-coupling fault" which involves two cells. It is defined as the write operation which generates a $0 \rightarrow 1$ or a $1 \rightarrow 0$ transition in one cell that changes the content of the second cell. This coupling fault involving two cells is a special case of "k-coupling faults" which can become quite complicated if no restriction is placed on the location of k cells. Special forms of CFs are the bridging and the state coupling faults that may involve any number of lines or cells, and are caused by a logic level rather than a low-to-high or a high-to-low transition. Another special case is the dynamic coupling fault in which a read or a write operation on one cell forces the contents of a second cell to a certain value (0 or 1). The coupling faults can be "unlinked faults" because each of the coupled cells is coupled to one other cell in only one way, or they can be "linked faults" in which a cell is coupled to a single cell in more than one way.

In the inversion coupling fault (CF_{in}), a $0 \rightarrow 1$ (or $1 \rightarrow 0$) transition in one cell inverts the contents of a second cell. An idempotent coupling fault (CF_{id}) is the one in which a $0 \rightarrow 1$ (or $1 \rightarrow 0$) transition in one cell forces the contents of a second cell to a certain value (0 or 1). It should be noted that the tests for CF_{ids} will detect CF_{ins}. Since the tests for TFs and CFs include reading every cell in the expected state 0 and 1, the SAF is always detected, even when TFs or CFs are linked with the SAF. There is no problem detecting unlinked TFs and CFs through the March tests. However, in case of linked TFs and CFs, the requirements for detecting them may have to be combined into a new test algorithm.

A common approach for the detection of multiple occurrences of functional faults including coupling faults (noninteracting) can be summarized as follows:

- Identification of test elements necessary to detect faults under consideration, if one fault is present at a time.

- Grouping of test elements based on initial memory state which can either be a zero or one.
- Applying an optimal (or near optimal) collection of test elements that would cover the presence of multiple faults.

Let C_i and C_j be two memory cells in states x and y, respectively ($i \neq j$), and let s be a Boolean 0 or 1. Then C_i is said to be ($y \rightarrow \bar{y}$; s) coupled to C_j only when cell j undergoes a $y \rightarrow \bar{y}$ transition and $x = s$ [16]. This condition is defined as one-way asymmetric coupling. Now, for two memory cells C_i and C_j, let s and r be the two Booleans. Then C_i is ($y \rightarrow \bar{y}$; s) and ($\bar{y} \rightarrow y$; r) coupled to C_j if it is both ($y \rightarrow \bar{y}$; s) and ($\bar{y} \rightarrow y$; r) one-way asymmetrically coupled to C_j. This condition is known as two-way asymmetric coupling. C_i is ($x \rightarrow \bar{x}$) one-way symmetrically coupled to C_j if transition $x \rightarrow \bar{x}$ in C_j causes transition $x \rightarrow \bar{x}$ in C_i, independent of the state of the two other cells. Also, two-way symmetric coupling denoted by ($x \rightarrow \bar{x}$) and ($\bar{x} \rightarrow x$) means that both one-way symmetric couplings $x \rightarrow \bar{x}$ and $\bar{x} \rightarrow x$ occur between C_i and C_j.

As an example, to identify a two-coupling fault, it is first sensitized and detected by using a read operation which is assumed to be fault-free. Then the condition for detecting coupling fault ($x \rightarrow \bar{x}_j$; s) between cells C_i and C_j, assuming the memory is initialized in state s, involves the following basic operations in the given sequence: (1) read C_j, (2) make a transition in C_j to affect C_i, and (3) read C_i. It is important to note that for the coupling fault model or the pattern-sensitive faults (PSFs) discussed in Section 4.2.4, the memory bits cannot be treated as being functionally independent as they are considered for the stuck-at fault models [4]. The memory under test is an array of n cells (bits) rather than an array of multiple-bit words. The memory manufacturers use various test algorithms such as marching 1s and 0s (March) and galloping 1s and 0s (Galpat) for different classes of coupling faults. The complexities of these test algorithms range from $O(n)$ to $O(n^2)$, and will be discussed in more detail in Section 4.3, "RAM Electrical Testing." This section will discuss an analytical approach for coupling fault modeling and the various test strategies used.

The March tests, because of their reasonable test time and simplicity of algorithms, have been found to be quite effective for detecting the SAFs, TFs, and CFs. A March element consists of a sequence of operations as follows: writing a 0 into a cell ($w0$), writing a 1 into a cell ($w1$), reading a cell with expected value 0 as ($r0$), and reading a cell with expected value 1 as ($r1$). After all the operations of a March element have been applied to a given cell, they will be applied to the next cell given by the address order which can begin from the lowest address (e.g., cell 0) and proceed to the highest cell address (e.g., cell $n-1$), or it can be in the reverse order from the highest address to the lowest [17]. The March from the lowest to the highest address can be denoted by the symbol $\Uparrow$, and from the highest to the lowest by the symbol $\Downarrow$. For "do not care" address order, the symbol $\Updownarrow$ will be used. For example, the MATS+ March test can be written as: $\{\Uparrow (w0); \Uparrow(r0,w1); \Downarrow(r1,w0)\}$. It means that the test consists of three March elements: M_0, M_1, and M_2. March element M_1 uses the $\Uparrow$ address order, and performs an "$r0$" followed by a "$w1$" operation on a cell before moving to the next cell. Addressing faults (AFs) will be detected if the following two conditions are satisfied: (a) $\Uparrow(rx, \cdots, wx)$, and (b) $\Downarrow(r\overline{x}, \cdots, w\overline{x})$.

There are several variations of the March test such as MATS+, March C−, and March B, each of them optimized for a particular set of functional faults. Table 4-2 lists some of these algorithms [17]. The table shows that the March test should have at least two March elements which have reverse address orders, and may include other elements such as initializing March element $\Updownarrow$ ($w0$), etc. The notation "$\cdots$" in the March elements indicates the allowed presence of any number of operations, and read operations can occur anywhere. The MATS+ algorithm is represented by equation (1) in the table, which requires $5n$ operations and detects all the AFs and all the SAFs. The March C- algorithm, which is an improvement over March C and is represented by equation (2), requires $10n$ operations. It detects all the AFs, all the SAFs, all the TFs, all unlinked CF_{ins}, and all CF_{ids}. The March B algorithm is represented by equation (3), and requires $17n$ operations. It detects all the AFs, SAFs, TFs including those linked with CFs, CF_{ins} unlinked and some when linked with CF_{ids}, and linked CF_{ids}.

These March tests can also be used to test for the stuck-open faults (SOFs) and the data-retention faults (DRFs). An SOF is caused by an open word line which makes the cell inaccessible. For example, the MATS+ represented by equation (1) and March C- by equation (2) can

TABLE 4-2. Various March Test Algorithms. (1) MATS+. (2) March C−.
(3) March B. (4) March G. (From [17], with permission of IEEE.)

$$\{\Updownarrow(w0)\Uparrow(r0,w1)\Downarrow(r1,w0)\} \tag{1}$$
$$\phantom{\{}M_0 \quad M_1 \qquad M_2$$

$$\{\Updownarrow(w0);\Uparrow(r0,w1);\Uparrow(r1,w0);\Downarrow(r0,w1);\Downarrow(r1,w0);\Updownarrow(r0)\} \tag{2}$$
$$\phantom{\{}M_0 \quad M_1 \qquad M_2 \qquad M_4 \qquad M_5$$

$$\{\Updownarrow(w0);\Uparrow(r0,w1,r1,w0,w1);\Uparrow(r1,w0,w1);\Downarrow(r1,w0,w1,w0);\Downarrow(r0,w1,w0)\} \tag{3}$$
$$\phantom{\{}M_0 \quad M_1 \qquad\qquad M_2 \qquad\quad M_3 \qquad\quad M_4$$

$$\{\Updownarrow(w0);\Uparrow(r0,w1,r1,w0,r0,w1);\Uparrow(r1,w0,w1);\Downarrow(r1,w0,w1,w0);\Downarrow(r0,w1,w0);$$
$$\phantom{\{}M_0 \quad M_1 \qquad\qquad M_2 \qquad\quad M_3 \qquad\quad M_4$$

$$\text{Del; } \Updownarrow(r0,w1,r1);\text{Del;}\Updownarrow(r1,w0,r0)\} \tag{4}$$
$$\phantom{\{\text{Del; }}M_5 \qquad\qquad M_6$$

be modified to detect the SOFs by extending M_1 with an "$r1$" operation and M_2 with an "$r0$" operation. The detection of a data-retention fault requires that a memory cell be brought into one of its logic states, and then a certain time allowed before the contents of the cell are verified to detect leakage currents that have to discharge the open node of the SRAM cell. Then this test must be repeated with the inverse logic value stored into the cell to test for a DRF due to an open connection in the other node of the cell. The March G represented by equation (4) is the most general March test, and is obtained by extending the March B to cover SOFs and DRFs. It consists of seven March and two delay elements which requires $23n + 2del$, and is of interest only when this large test time is affordable.

A study was performed by comparing five test strategies for their abilities to detect the coupling faults in an n word × 1 b RAM [18]. In all five, the data-in line is driven randomly, and three of the five strategies used random selection of both the address lines and read/write (R/W) control. The other two strategies sequentially cycle through the address space with a deterministic setting of R/W control. These five test strategy models are listed below:

1. Explicit memory test with word operations (ETWO)
2. Explicit memory test with cycle operations (ETCO)
3. Random memory test
4. Segmented random memory test
5. Segmented random memory test with memory unload

In the fault model assumed, a coupling fault exists between a pair of cells such that a $0 \rightarrow 1$ transition in one cell causes a $0 \rightarrow 1$ transition in the other cell only for a fixed value of other cells G in the neighborhood, where G denotes some patterns in other cells of the memory. The comparison of five test strategies for various memory sizes, segments, and fault escape probabilities showed the ETWO as offer-

ing the best performance and ease of implementation. The relative merit of these five test strategies is measured by the average number of accesses per address needed to meet a standard test quality level. This three-step ETWO algorithm is briefly described as follows:

- An initial write cycle to every address to load fixed values is performed.
- A number of complete cycles through the address space with a read and then a write (R/W) to each address, i.e., a read followed by a write operation, is performed. For each cycle, the address space is covered in the same sequence.
- A final read cycle-through the addresses to detect the possibility that a fault has been triggered, but not read out in the last R/W cycle.

Another study was performed to find efficient tests for detecting classes of single and multiple write-triggered coupling faults in $n \times 1$ RAMs. The basic strategy used was: (1) to first precisely define the fault model, i.e., the set of single and/or multiple faults which are to be detected, (2) derive a lower bound on the length of any test that detects the fault model, and (3) find the shortest possible test that detects the fault model [19]. Two hierarchies of fault models were found. One of them had three classes of toggling faults detectable by the tests whose lengths grew by n^2, $n \log_2 n$, and n. The k-limited toggling faults imply a set of all multiple toggling faults consisting of $< k$ distinct single toggling faults, where $k \geq 1$; k-limited coupling is defined similarly in terms of component single coupling faults. A second hierarchy consisted of five classes of coupling faults with identified near optimal lengths of $10n - 4$, $12n - 8$, $14n - 12$, $4n [\log_2 n] + 17n$, and $2n^2 + 4$. For these fault models detectable by linear tests, the March tests were found to be near optimal.

The CF algorithms for the bit-oriented memories can be modified for word-oriented memory chips by substituting the bit-wide read and write operations with read/write operations consisting of a data background of B bits. By using this modified algorithm approach, the CFs in which the coupling cell and the coupled cell are in different words will be detected since the memory word (or byte)-wide organization does not influence the fault coverage [1]. However, if the coupling cell and the coupled cell exist in the same word, the CFs may not be detectable, depending upon the dominance of the write operation over the coupling fault. Under this condition of write operation dominance, the CF will have no effect, and does not have to be detected. An idempotent CF will not be detectable with the modified bit-wide algorithm, and may require a more rigorous test procedure.

4.2.4 Pattern-Sensitive Faults

The pattern-sensitive faults (PSFs) may be considered the most general case of k-coupling faults, for $k = n$, where n represents all cells in the memory. A PSF can be defined as the susceptibility of the contents of a cell to be influenced by the contents of all other cells in the memory. A group of cells that influences the base cell's behavior is called the neighborhood of the base cell. The PSFs are primarily caused by the high component densities of RAMs and unwanted interference (e.g., electromagnetic) between closely packed lines and signals. A variety of physical fault mechanisms are responsible for PSFs, making it difficult to define the logical fault models to represent them. Some other memory fault classes such as shorts, SAFs, and coupling faults can also be considered as special types of PSFs. The RAM testing for unrestricted PSFs would be very costly and impractical since it requires a test length of $(3N^2 + 2N)2^N$ [20]. As a result, the PSF models assume that interaction can occur only between the cells within certain physical proximity. Therefore, these are called the physical neighborhood pattern-sensitive faults (NPSFs).

The base cell and adjacent neighborhood cells are usually represented by a tiling configuration. Figure 4-5 shows five-cell and nine-cell physical neighborhood layouts [21]. Some of the definitions used in testing for PSFs are given below:

> *Base cell:* A cell in the memory array which is under test consideration.
>
> *Row (Column) Neighborhood:* Row or column containing the base cell, but excluding the base cell itself.
>
> *Three-Cell Row (Column) Neighborhood:* A set of three cells consisting of the base cell and any two cells from its row (column) neighborhood.
>
> *Five-Cell Row (Column) Neighborhood:* The union of a three-cell row neighborhood and a three-cell column neighborhood, both sharing the same base cell.

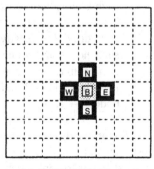

(i) five-cell Physical Neighborhood

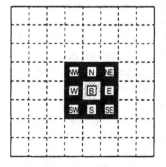

(ii) nine-cell Physical Neighborhood

Figure 4-5. A schematic representation of five-cell and nine-cell physical neighborhoods of a base cell. (From [21], with permission of IEEE.)

Five-Cell Logical Neighborhood: The set of five-cells consisting of the base cell, the two-cells in its row neighborhood whose column addresses differ from that of the base cell by unity, and the two-cells in its column neighborhood whose row addresses differ from that of the base cell by unity. This is also termed the "five-cell neighborhood" or "five-cell adjacent neighborhood."

Five-Cell Physical Neighborhood: A set of five-cells consisting of the base cell and the four-cells physically adjacent to the base cell (as shown in Figure 4-5).

The neighborhood pattern-sensitive faults (NPSFs) are often classified by the following three categories:

- Active NPSF (or ANPSF) in which the base cell changes its contents as a result of changes in the pattern of the neighborhood cells.
- Passive NPSF (or PNPSF) in which the contents of the base cell cannot be changed due to the influence of an existing pattern in the neighborhood cells.
- Static NPSF (or SNPSF) in which the contents of a base cell are forced to a certain value by the contents of an existing pattern in the neighborhood cells.

A general approach to testing the base cell for the ANPSF and PNPSF consists of: (1) sensitizing any fault using the appropriate neighborhood pattern, and (2) reading the state of the base cell after each sensitizing pattern to detect any fault [4]. The number of patterns needed to sensitize all the ANPSFs for a neighborhood of size k is given by $(k-1)2^k$ since each one of the $(k-1)$ neighbors of the base cell must exercise both the $1 \rightarrow 0$ and $0 \rightarrow 1$ transitions while $(k-1)$ neighbors take all possible binary values. A pattern that sensitizes one ANPSF is called the active neighborhood pattern (ANP). Table 4-3(a) shows all ANPs for a neighborhood size of 5 [22].

The number of patterns needed to sensitize all the PNPSFs in a k-cell neighborhood is

given by 2^k. These patterns are called the passive neighborhood patterns (PNPs). Table 4-3(b) shows all the PNPs for a neighborhood size of 5 [22]. As shown, each ANP or PNP contains exactly one transition write operation (i.e., a $0 \rightarrow 1$ or a $1 \rightarrow 0$ transition). For generation of these patterns, each neighborhood must be initialized to a known state, and then the contents of one cell must be changed at a time by a write operation. The sum of all the ANPs and PNPs is given by

$$(k - 1)2^k + 2^k = k\,2^k,$$

which represents all ANPSFs + PNPSFs.

Also, there are 2^k static NPSFs (or SNPSFs). Table 4-3(c) shows all the SNPs for a five-cell neighborhood. For an efficient sensitizing pattern generation, the test sequence length should be minimal, and to achieve that, following methodologies are used:

- When the patterns only contain 0s and 1s (e.g., SNPs), then a Hamiltonian sequence should be used. For a k-bit Hamiltonian sequence, each pattern differs by 1 b from the preceding pattern, which requires a minimum number of write operations to go from one pattern to the next.
- When the patterns contain only $0 \rightarrow 1$ or $1 \rightarrow 0$ transitions (e.g., ANPs and PNPs), then a Eulerian sequence should be used. This sequence is represented by a Eulerian graph which has a node for each k-bit pattern of 0s and 1s, and there is an arc between the two nodes only if they differ by exactly 1 b. When two such nodes are connected, the connection is made by two arcs, one going from node $1 \rightarrow 2$ and the second from node $2 \rightarrow 1$.

So far, the discussion has been to generate a test sequence that will sensitize all the NPSFs. The next step is to apply such a test sequence to every neighborhood in the memory, and this can be done with the "tiling method" in which the memory is totally covered by a group of neighborhoods which do not overlap. Such a group is

TABLE 4-3. Five-Cell Neighborhood Pattern. (a) ANPs. (b) PNPs. (From [22], with Permission of IEEE.)

```
A  ↑ ↑ ↑ ↑ ↑ ↑ ↑ ↑ ↓ ↓ ↓ ↓ ↓ ↓ ↓ ↓ ↑ ↑ ↑ ↑ ↑ ↑ ↑ ↑ ↓ ↓ ↓ ↓ ↓ ↓ ↓ ↓
B  0 0 0 0 1 1 1 1 0 0 0 0 1 1 1 1 0 0 0 0 1 1 1 1 0 0 0 0 1 1 1 1
C  0 0 1 1 0 0 1 1 0 0 1 1 0 0 1 1 0 0 1 1 0 0 1 1 0 0 1 1 0 0 1 1
D  0 1 0 1 0 1 0 1 0 1 0 1 0 1 0 1 0 1 0 1 0 1 0 1 0 1 0 1 0 1 0 1
E  0 0 0 0 0 0 0 0 0 0 0 0 0 0 0 0 1 1 1 1 1 1 1 1 1 1 1 1 1 1 1 1

A  0 0 0 0 1 1 1 1 0 0 0 0 1 1 1 1 0 0 0 0 1 1 1 1 0 0 0 0 1 1 1 1
B  ↑ ↑ ↑ ↑ ↑ ↑ ↑ ↑ ↓ ↓ ↓ ↓ ↓ ↓ ↓ ↓ ↑ ↑ ↑ ↑ ↑ ↑ ↑ ↑ ↓ ↓ ↓ ↓ ↓ ↓ ↓ ↓
C  0 0 1 1 0 0 1 1 0 0 1 1 0 0 1 1 0 0 1 1 0 0 1 1 0 0 1 1 0 0 1 1
D  0 1 0 1 0 1 0 1 0 1 0 1 0 1 0 1 0 1 0 1 0 1 0 1 0 1 0 1 0 1 0 1
E  0 0 0 0 0 0 0 0 0 0 0 0 0 0 0 0 1 1 1 1 1 1 1 1 1 1 1 1 1 1 1 1

A  0 0 0 0 1 1 1 1 0 0 0 0 1 1 1 1 0 0 0 0 1 1 1 1 0 0 0 0 1 1 1 1
B  0 0 1 1 0 0 1 1 0 0 1 1 0 0 1 1 0 0 1 1 0 0 1 1 0 0 1 1 0 0 1 1
C  ↑ ↑ ↑ ↑ ↑ ↑ ↑ ↑ ↓ ↓ ↓ ↓ ↓ ↓ ↓ ↓ ↑ ↑ ↑ ↑ ↑ ↑ ↑ ↑ ↓ ↓ ↓ ↓ ↓ ↓ ↓ ↓
D  0 1 0 1 0 1 0 1 0 1 0 1 0 1 0 1 0 1 0 1 0 1 0 1 0 1 0 1 0 1 0 1
E  0 0 0 0 0 0 0 0 0 0 0 0 0 0 0 0 1 1 1 1 1 1 1 1 1 1 1 1 1 1 1 1

A  0 0 0 0 1 1 1 1 0 0 0 0 1 1 1 1 0 0 0 0 1 1 1 1 0 0 0 0 1 1 1 1
B  0 0 1 1 0 0 1 1 0 0 1 1 0 0 1 1 0 0 1 1 0 0 1 1 0 0 1 1 0 0 1 1
C  0 1 0 1 0 1 0 1 0 1 0 1 0 1 0 1 0 1 0 1 0 1 0 1 0 1 0 1 0 1 0 1
D  ↑ ↑ ↑ ↑ ↑ ↑ ↑ ↑ ↓ ↓ ↓ ↓ ↓ ↓ ↓ ↓ ↑ ↑ ↑ ↑ ↑ ↑ ↑ ↑ ↓ ↓ ↓ ↓ ↓ ↓ ↓ ↓
E  0 0 0 0 0 0 0 0 0 0 0 0 0 0 0 0 1 1 1 1 1 1 1 1 1 1 1 1 1 1 1 1
```

*Notation is as follows:
  1. A = top cell; B = bottom cell; C = left cell; D = right cell; E = base cell.
  2. ↑ = 0-to-1 transition; ↓ = 1-to-0 transition.
  3. Each column with five entries denotes one ANP of a base cell.

(a)

Top	Bottom	Left	Right	Base	Top	Bottom	Left	Right	Base
0	0	0	0	↑	0	0	0	0	↓
0	0	0	1	↑	0	0	0	1	↓
0	0	1	0	↑	0	0	1	0	↓
0	0	1	1	↑	0	0	1	1	↓
0	1	0	0	↑	0	1	0	0	↓
0	1	0	1	↑	0	1	0	1	↓
0	1	1	0	↑	0	1	1	0	↓
0	1	1	1	↑	0	1	1	1	↓
1	0	0	0	↑	1	0	0	0	↓
1	0	0	1	↑	1	0	0	1	↓
1	0	1	0	↑	1	0	1	0	↓
1	0	1	1	↑	1	0	1	1	↓
1	1	0	0	↑	1	1	0	0	↓
1	1	0	1	↑	1	1	0	1	↓
1	1	1	0	↑	1	1	1	0	↓
1	1	1	1	↑	1	1	1	1	↓

*Notation is as follows:
  1. ↑ = 0-to-1 transition; ↓ = 1-to-0 transition.
  2. Each column with five entries denotes one p-n-p of a base cell.

(b)

TABLE 4-3 (cont.). (c) SNPs. (From [22], with Permission of IEEE.)

```
A  0 0 0 0 0 0 0 0 1 1 1 1 1 1 1 1 1 0 0 0 0 0 0 0 0 1 1 1 1 1 1 1 1
B  0 0 0 0 1 1 1 1 0 0 0 0 1 1 1 1 0 0 0 0 1 1 1 1 0 0 0 0 1 1 1 1
C  0 0 1 1 0 0 1 1 0 0 1 1 0 0 1 1 0 0 1 1 0 0 1 1 0 0 1 1 0 0 1 1
D  0 1 0 1 0 1 0 1 0 1 0 1 0 1 0 1 0 1 0 1 0 1 0 1 0 1 0 1 0 1 0 1
E  0 0 0 0 0 0 0 0 0 0 0 0 0 0 0 0 1 1 1 1 1 1 1 1 1 1 1 1 1 1 1 1
```

(c)

called the "tiling group," and the set of all neighborhoods in a tiling group is called the "tiling neighborhood." In this approach, every cell in the k-cell neighborhood is assigned an integer number from 0 to k-1 so that every cell in the memory will be marked with all numbers of $\{0,1,2, \cdots, k-1\}$ [23]. Figure 4-6 illustrates the procedure for a five-cell tiling neighborhood marking with a fixed pattern from $\{0,1,2,3,4\}$ assigned to every tile [23].

It can be shown that for optimal write sequence N_n for a memory n size with tiling neighborhoods of size k, the number of write operations is given by $n(2^k)$ [22]. This procedure produces the optimal transition write operations needed to sensitize the PSFs caused by transition write operations. Therefore, all PSFs (multiple) caused by the transition write operations can be detected and located. However, it

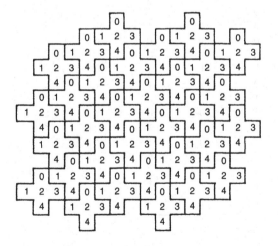

Figure 4-6. A procedure for a five-cell tiling neighborhood marking with a fixed pattern assigned to every tile. (From [23], with permission of IEEE.)

does not cover any PSFs caused by the nontransition writes, or any PSFs caused by the read operations. The complexity of this test procedure is $(k + 5)2^{k-1}n$. Another test procedure has been proposed that sensitizes single PSFs caused by the read operations, and single PSFs caused by the transition as well as nontransition write operations [23]. Its complexity is $(3k + 2)2^k n$, which is $O(n)$, meaning that the procedure can be implemented for large memories.

Many algorithms that have been proposed for detecting the NPSFs are based on the assumption that the memory array's physical and logical neighborhood are identical. However, in current high-density RAMs that are designed to minimize the die size and reduce critical path lengths, the physically adjacent rows (and columns) may not be logically adjacent. Also, for higher yields, state-of-the-art memory chips are being designed with the spare rows and columns for memory chip array reconfiguration, if needed. This means that in the reconfigured memory chip, the physically adjacent cells may no longer have consecutive addresses, and the logical-to-physical address mapping of the memory array is also lost.

A test algorithm has been developed to detect five-cell and nine-cell NPSFs in DRAMs, even if logical and physical addresses are different and the physical-to-logical address mapping is not available [21]. For N-bit RAMs, these algorithms have test lengths of $O(N[\log_3 N]^4)$, and can also detect the SAFs and coupling faults. The test procedure depends on the development of an efficient three-coloring algorithm which trichromatically colors all triplets among a number of n objects in at most $[\log_3 n]^2$ coloring steps.

It should be noted that for the word-oriented memories, the write operation involves

B cells ($B \geq 2$). Therefore, the word-oriented memory NPSF detection and/or location involves modification of the algorithms for the bit-oriented memories, which becomes quite complicated.

4.2.5 Miscellaneous Faults

The functional fault models indirectly cover only a small percentage of the parametric and dynamic faults. The parametric faults concern the external behavior of the memory, e.g., voltage and current levels, and input/output delay on the pins of a chip. The dynamic faults are electrical faults which are time-dependent and internal to the chip. Some important dynamic faults are sense amplifier recovery, write recovery, retention faults, and imbalance faults [24].

In classical RAMs (without BIST), autodiagnostic procedures require test transparency which assures that at the end of the test, the contents of RAM are the same as its contents before testing. These autodiagnostic procedures can cover parametric faults through functional testing. However, developing autodiagnostic procedures for the dynamic faults requires not only consideration of appropriate test patterns, but also time between subsequent operations performed within the checking memory. In the autodiagnostic mode covering static faults, the bit-oriented RAM tests can be transformed into transparent tests using special algorithms [24]. Many dynamic fault-testing algorithms have been proposed that were developed for the external memory testers. For in-system test of these, the following issues have to be considered:

- Identification of critical memory operation sequences which depend upon dynamic faults
- Mapping those sequences into equivalent operation sequences in the system
- Developing transparency of the test sequences

The dynamic tests which check memory behavior for long delays are easy to implement. However, checking minimal delays between the memory operations referring to the same or different memory cells requires more complex algorithms. In the case of cache memories, the faults have to be checked within the data memory array, directory memory arrays, cache control logic, and comparators. These memories can be adapted for in-system transparent testing by reserving a special area in the main RAM memory for testing purposes. A transparent testing of the dual-port memories can be performed using appropriate algorithms under the assumption that all application processes on both sides of the memory ports are inactive. Functional testing must be performed from both sides of the memories. In the dual-port memories, the most difficult problem is checking communication mechanisms and the arbitration circuitry.

4.2.6 GaAs SRAM Fault Modeling and Testing

The basic structure and processing sequence for the GaAs devices is different from that for silicon MOSFETs, and hence the failure mechanisms also vary. The process parameter variations, such as threshold voltage shifts across the wafer, are more significant in GaAs than in MOS devices. In GaAs circuits, failure modes such as ohmic contact degradation, interdiffusion of gate metal with GaAs, and interelectrode resistive path formation have been observed [25]. These failure modes in GaAs SRAMs can lead to different types of faults such as PSFs which may not be easily detected by normally used test procedures [23]. Also, the data-retention faults have a different failure mechanism compared to the stuck-open failures observed for silicon-based MOS RAMs.

A study was performed to analyze the fault mechanisms in GaAs memories by using the 1 kb SRAM as a test vehicle [26]. In this SRAM, which uses high electron mobility transistors (HEMTs), the memory circuits were simulated with different values of process and design parameters such as the HEMT widths and threshold voltages. The simulation results showed that process variations and design-related errors can lead to the following circuit faults:

- Stuck-open transistors
- Resistive shorts between transistor electrodes
- Bridging of metal lines causing shorts between two adjacent signals

The effects of shorts between the bit and word lines and between adjacent cells were also simulated. Test results showed that all these shorts could be caused by the interelectrode metal bridging and bridging between the metal lines. In a standard six-transistor SRAM cell, this can result in the following possible short mechanisms:

1. Bit line to word line
2. Word line to cell storage node
3. Cell storage node to power supply
4. Cell storage node to complementary cell storage node
5. Cell storage node to ground

A fault model for the GaAs SRAMs was identified based on an extensive analysis of the defect mechanisms. It was observed that the marginal design-related failures leading to simple read/write errors could be detected by standard SRAM test algorithms such as the March 13N [27]. The process- and layout-related faults can cause data retention, delay, and some specific types of PSFs. Row and column PSFs were observed that could not be detected by the conventional five-cell and nine-cell NPSF patterns.

4.2.7 Embedded DRAM Fault Modeling and Testing

The high-density embedded DRAM modules which are being increasingly used in the microcontroller and digital signal processing (DSP) environment are harder to test because of their limited observability and controllability. These are often subjected to limited testing through the scan chains, and are susceptible to catastrophic as well as soft defects. The RAM functional fault models discussed earlier in this section are usually not based on the actual manufacturing defects. However, the defect-oriented (inductive fault) analysis has been shown to be very useful in finding various defect mechanisms in a given layout and technology [28]. The catastrophic (or hard) defects in the process and layout are mapped onto the device schematic and classified into various fault categories. These fault categories are similar to that developed for the SRAM, such as stuck-at-0 or stuck-at-1, stuck open, coupling to another cell, multiple-access, and data-retention faults.

The occurrence of noncatastrophic defects is significantly high, degrading the circuit performance and/or causing coupling faults in the memory. The DRAMs, because of their dynamic operation, are more susceptible to potential coupling faults than the SRAMs. A pair of memory cells i and j are "coupled" if a transition from x to y in one cell of the pair (e.g., i) changes the state of the other cell j from 0 to 1 or 1 to 0 [4]. In a DRAM, a cell is driven when it is accessed, and not otherwise. Therefore, a read, write, or refresh operation on a DRAM cell causes it to be driven. This situation is different from that of a SRAM in which, at the time of coupling, both coupling and the coupled cell are driven [29]. Therefore, two DRAM cells, say i and j, are assumed to be dynamically coupled if the driven state of cell i with value $x(x \in \{0,1\})$ causes cell j to be in state $y(y \in \{0,1\})$. A test algorithm 8N has been developed for the bit-oriented DRAMs with a combinational R/W logic, and a data-retention test is added to cover the data-retention faults.

Figure 4-7 shows an 8N test algorithm including the data-retention test [29]. The basic test algorithm consists of an initialization step and a set of four March sequences (March 1–March 4). This is followed by a data-retention test sequence consisting of a wait state (disable DRAM), March 5, another wait state, and then March 6. There are 18 possible two-coupling faults modeled between cell i and cell j. The 8N test algorithm will detect all stuck-at-0 and stuck-at-1 faults, stuck-open faults, decoder multiple-access faults, and all two-coupling modeled faults. This algorithm can be easily modified to the 9N test algorithm if R/W logic is

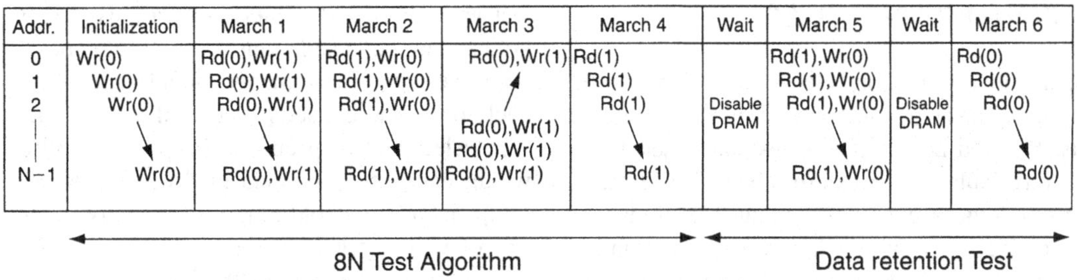

Addr.	Initialization	March 1	March 2	March 3	March 4	Wait	March 5	Wait	March 6
0	Wr(0)	Rd(0),Wr(1)	Rd(1),Wr(0)	Rd(0),Wr(1)	Rd(1)		Rd(1),Wr(0)		Rd(0)
1	Wr(0)	Rd(0),Wr(1)	Rd(1),Wr(0)	↑	Rd(1)		Rd(1),Wr(0)		Rd(0)
2	Wr(0)	Rd(0),Wr(1)	Rd(1),Wr(0)		Rd(1)	Disable DRAM	Rd(1),Wr(0)	Disable DRAM	Rd(0)
⋮				Rd(0),Wr(1)					
				Rd(0),Wr(1)					
N−1	Wr(0)	Rd(0),Wr(1)	Rd(1),Wr(0)	Rd(0),Wr(1)	Rd(1)		Rd(1),Wr(0)		Rd(0)

◄──────────────── 8N Test Algorithm ──────────────►◄──────── Data retention Test ────────►

Figure 4-7. March 8N test algorithm including data-retention test sequence. (From [29], with permission of IEEE.)

sequential. For the word-oriented DRAMs, different backgrounds can be utilized to cover intrafaults.

A fault model verification and validation study was performed on Philips 4K × 8 embedded DRAM modules. A large number of devices from 34 wafers were tested, out of which 579 failed the 9N algorithm. The performance of March 1 and March 2 steps was comparable to the most complex test element, March 3, which caught the maximum number of failures. Table 4-4 summarizes the test results, with the first and second columns showing the types and number of failed devices, respectively [29]. The third column shows the pass/fail response of the failed devices in terms of four March test ele-

ments, e.g., FFFP means that the device failed March 1, 2, and 3, but passed the March 4 test. The fourth column shows an explanation of the failure.

Most of the device failures (89%) appeared to be complete chip failures or memory stuck-at faults affecting a large number of bits. These failures could be explained by catastrophic bridging or open defects in the layout. A portion of the failed devices (7%) showed bit stuck-at behavior that could be explained with catastrophic defects in the cell. The remaining 4% of the failures could not be explained by the catastrophic defect model, but by the coupling fault model based on the noncatastrophic defects.

TABLE 4-4. March 9N Algorithm Test Results on Philips 4K × 8 Embedded DRAM Modules. (From [29], with Permission of IEEE.)

No.	Devices	Signature	Explanation
1	318	FFFF	Word line stuck-at, read/write failure, open
2	201	FFFP	Word, bit line failures, read/write failures
3	21	PFFF	Bit line stuck-at-0, bit stuck-at-0
4	18	FPFP	Bit line stuck-at-1, bit stuck-at-1
5	11	PFPF	Coupling 3
6	3	PFPP	Coupling 1
7	3	FPFF	Combination of stuck-at-1 and coupling 10
8	3	FFPF	Combination of coupling faults 9 and 10
9	1	FFPP	Coupling 9

4.3 RAM ELECTRICAL TESTING

In general, the memory electrical testing consists of dc and ac parametric tests and functional tests. DC and ac parametric tests are used for detecting faults in the external behavior of the memory devices, e.g., input/output (I/O) threshold voltages, I/O current levels, quiescent and dynamic power supply currents, and propagation delays and access times. They can be characterization tests which usually consist of a repetitive sequence of measurements to locate the operating limits of a device. They can be production tests to determine whether the devices meet their specifications and are performed by the manufacturer as an outgoing production test, or by the user as part of the incoming inspection. A production test can be the GO/NO GO testing to make pass/fail decisions, whereas the user-detailed testing may consist of data read and record on all the dc and ac parametric and functional testing. The functional tests are based upon the reduced functional fault models that were discussed in Section 4.2, such as the stuck-at, stuck-open, bridging, coupling, pattern-sensitive, and dynamic faults.

4.3.1 DC and AC Parametric Testing

During the memory electrical testing using automated test equipment (ATE), dc, functionals, and ac parametric tests are performed with a parametric measurement unit (PMU) which is capable of forcing the programming voltages and currents at the appropriate inputs/outputs and making corresponding measurements. For a typical SRAM, the dc electrical characteristics include quiescent and operating supply currents (I_{DD}), output voltages low and high (V_{OL}, V_{OH}), I/O pin leakage currents (I_{ILK}, I_{OLK}), data-retention voltage and current (V_{DR}, I_{DR}), input voltages low and high (V_{IL}, V_{IH}), etc. Functional tests consist of various algorithms such as March, Row/Column Gallop, Checkerboard, etc. (see Section 4.3.2).

The ac performance characteristics are measured for both the read and write cycles. For a read operation, the ac measurements include the read cycle time, address access time, chip se-

lect and chip enable times, and various output hold and output enable times. For a write operation, the ac measurements include the write cycle time, address setup to end of write cycle, chip select and chip enable times, write pulse width, data setup and data hold times, address setup and address hold times, output active after end of write cycle, write enable to output disable, and write pulse width. For the megabit DRAMs, additional ac measurements include the column and row address setup times, and various chip address select (*CAS*) and refresh address select (*RAS*) related timing specifications.

Table 4-5 shows Loral Federal System's/ 1Mb SRAM radiation-hardened (RH version) dc electrical characteristics, ac read and write cycle specifications, and the timing diagrams. These are supplied in two different access times (30, and 40 ns) versions. For a functional block diagram of this RH 1 Mb SRAM, technology and process features, and radiation hardness characteristics, see section 7.3.3.

4.3.2 Functional Testing and Some Commonly Used Algorithms

Section 4.2 reviewed an analytical approach to fault modeling, the most commonly used fault models, and the development of test algorithms for the detection of various faults. This section discusses some well-known algorithms used for the functional testing of RAMs by the manufacturers and end users. The development of various test patterns over the years has shown that no single pattern could exercise a RAM thoroughly enough to detect all the failure modes. The test patterns developed have been based upon the following common functional failure modes:

1. *Address Decoder Malfunction:* In this failure mode, an open address decode line internal to the device or a defective decoder inhibits proper addressing of portions of the memory array.

2. *Multiple Write Errors:* The data bits are written into a cell other than the one addressed because of capacitive coupling between the cells.

TABLE 4-5. (a) 1 Mb SRAM dc Electrical characteristics. (From [30], with permission of Loral Federal Systems.)

Electrical Performance Characteristics (see notes at the end of Table 4-5)

Test	Symbol	Test Conditions 1/ $-55 \le T_{case} \le +125°\,C$ $3.14V \le V_{DD} \le 3.46\,V$ unless otherwise specified	Device Type 2/	Limits Min.	Max.	Units
Supply Current (Cycling Selected)	I_{DD1}	$F = F_{max}$ 4/ $\bar{S} = V_{IL} = GND$ $E = V_{IH} = V_{DD}$ No output load	All		180	mA
Supply Current (Cycling Deselected)	I_{DD2}	$F = F_{max}$ 4/ $\bar{S} = V_{IH} = V_{DD}$ $E = V_{IL} = GND$	All		2.0	mA
Supply Current (Standby)	I_{DD3}	$F = 0$ MHz $\bar{S} = V_{IH} = V_{DD}$ $E = V_{IL} = GND$	All		2.0	mA
High-Level Output Voltage	V_{OH}	$I_{OH} = -4$ mA	All	2.4		V
High-Level Output Voltage	V_{OH}	$I_{OH} = -200\ \mu A$	All	$V_{DD} - 0.05V$		V
Low-Level Output Voltage	V_{OL}	$I_{OL} = 8$ mA	All		0.4	V
Low-Level Output Voltage	V_{OL}	$I_{OL} = 200\ \mu A$	All		0.05	V
High-Level Input Voltage	V_{IH}		All	2.2		V
Low-Level Input Voltage	V_{IL}		All		0.8	V
Input Leakage	I_{ILK}	$0\,V \le V_{IN} \le 3.46\,V$	All	-5	5	μA
Output Leakage	I_{OLK}	$0\,V \le V_{OUT} \le 3.46\,V$	All	-10	10	μA
C_{in}		3/	All		6	pF
C_{out}		3/	All		8	pF
Functional Tests			All			

3. *Pattern Sensitivity:* The cell storage characteristics (or responses) vary with the test pattern and the data pattern stored in the surrounding cells (neighborhood) or the rest of memory array.

4. *Refresh Sensitivity:* In DRAMs, the data can be lost during the specified minimum period between the refresh cycles due to excessive voltage or current leakage and other faults in the rewriting circuitry.

5. *Slow Access Time:* An above-normal capacitive load on the output driver circuit causes excessive time to sink

TABLE 4-5 (Cont.). (b) 1 Mb SRAM read cycle ac specifications and timing diagrams.

Electrical Performance Characteristics (see notes at the end of Table 4-5)

Test	Symbol	Test Conditions 1/ $-55 \le T_{case} \le +125°$ C $3.14V \le V_{DD} \le 3.46$ V unless otherwise specified	Device Type 2/	Limits Min.	Limits Max.	Units
Read Cycle ac Specifications						
Read Cycle Time	t_{AVAV}		×3×	30		ns
			×4×	40		ns
Address Access Time	t_{AVQV}		×3×		30	ns
			×4×		40	ns
Chip Select Access Time	t_{SLQV}		×3×		30	ns
			×4×		40	ns
Chip Enable Access Time	t_{EHQV}		×3×		30	ns
			×4×		40	ns
Output Enable Access Time	t_{GLQV}		×3×		12	ns
			×4×		15	ns
Chip Select to Output Active	t_{SLQX}		All	3		ns
Chip Enable to Output Active	t_{EHQX}		All	3		ns
Output Enable to Output Active	t_{GLQX}		All	3		ns
Output Hold after Address Change	t_{AXQX}		All	3		ns
Chip Select to Output Disable	t_{SHQZ}		×3×		12	ns
			×4×		15	ns
Chip Disable to Output Disable	t_{ELQZ}		×3×		12	ns
			×4×		15	ns
Output Enable to Output Disable	t_{GHQZ}		×3×		12	ns
			×4×		15	ns

Read Cycle Timing Diagram

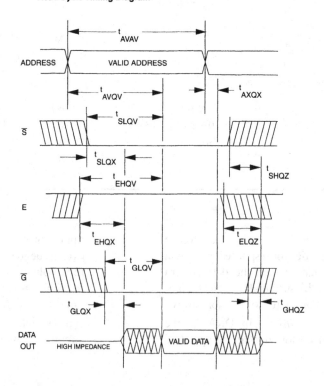

TABLE 4-5 (Cont.). (c) 1 Mb SRAM write cycle ac specifications and timing diagrams.

Electrical Performance Characteristics (see notes at the end of Table 4-5)

Test	Symbol	Test Conditions 1/ $-55 \le T_{case} \le +125°$ C 3.14V $\le V_{DD} \le 3.46$ V unless otherwise specified	Device Type 2/	Limits Min.	Limits Max.	Units
Write Cycle AC Specifications 5/ to 7/						
Write Cycle Time	t_{AVAV}		×3×	30		ns
			×4×	40		ns
Address Setup to End of Write	t_{AVWH}		×3×	25		ns
			×4×	35		ns
Chip Select to End of Write	t_{SLWH}		×3×	25		ns
			×4×	35		ns
Chip Enable to End of Write	t_{EHWH}		×3×	25		ns
			×4×	35		ns
Write Pulse Width Access Time	t_{WLWH}		×3×	25		ns
			×4×	35		ns
Data Setup to End of Write	t_{DVWH}		×3×	20		ns
			×4×	30		ns
Data Hold after End of Write	t_{WHDX}		×3×	3		ns
			×4×	5		ns
Address Setup to Start of Write	t_{AVWL}		All	0		ns
Address Hold after End of Write	t_{WHAX}		All	0		ns
Output Active after End of Write	t_{WHQX}		×4×	1		ns
Write Enable to Output Disable	t_{WLQZ}		×3×		12	ns
			×4×		15	ns
Write Disable Pulse Width	t_{WHWL}		All	5		ns

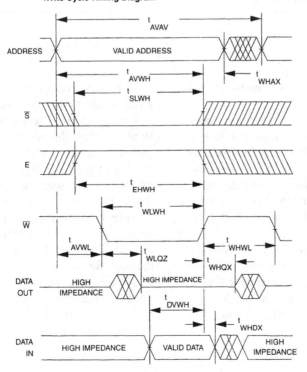

Write Cycle Timing Diagram

TABLE 4-5 (Cont.). (d) 1 Mb SRAM electrical characteristics notes and output load circuit.

Notes:

1) Test conditions for ac measurements are listed below.

Input Levels	0 V to V_{DD}
Input rise and fall time	≤ 2.0 ns/V
Input and output timing reference levels (except for tristate parameters)	2.5 V
Input and output timing reference levels for tristate parameters	$V_{OL} = 1.23$ V, $V_{OH} = 2.23$ V.

   Output load circuit as shown below.

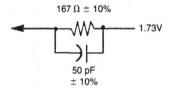

2) The device types are defined as follows: The delineation in this table is by device speed; thus the nomenclature: $\times 4 \times$ represents a 40 ns device, and $\times 3 \times$ represents a 30 ns device.

3) Guaranteed by design and verified by periodic characterization.

4) $F_{max} = 1 / t_{AVAV(min)}$.

5) $\overline{S}$ high, $\overline{W}$ high, or E low must occur while address transitions.

6) The worst case timing sequence of $t_{WLQZ} + t_{DVWH} + t_{WHWL} = t_{AVAV}$ (write cycle time).

7) $\overline{G}$ high will eliminate the I/O output from becoming active (t_{WLQZ}).

and source currents, thereby increasing access time.

6. *Slow Write Recovery:* Memory access time increases when a read follows immediately after a write operation that may have caused a sense amplifier to saturate, and thus have been unable to recover in time to allow detection of the differential voltage of the cell being read.

7. *Slow Sense Amplifier Recovery:* This failure mode is caused by a sense amplifier that requires excessive time to detect one logic state after a long period of detecting the opposite logic state.

Failure modes (1), (2), and (4) indicate a gross malfunction of the device due to the errors which are easily detected with specially designed patterns that have execution times proportional to the number of cells "n." For example, the MASEST and MARCH (C) algorithms have an execution time of $10n$ times the device cycle time, and each of these patterns provides reasonable assurance that the memory device does not exhibit failure modes (1), (2), and (4). However, these tests do not provide a thorough functional exercise of the memory device under test. Memory testing time is the square of the memory size (n^2), which means that even for RAMs of 16K and 64K size, typical execution time for the WAKPAT is

1–13 min, and for the GALPAT, it is 2–60 min. These are prohibitively long times in a memory production environment. Many of the addressing related failures in 16K and 64K RAMs have been eliminated by the design improvements such as internal address, dummy address codes, and internal resetting of the address lines.

A test pattern designed for a particular device may not be the best for another manufacturer's device with an identical pin configuration, timing, and voltage specifications. Therefore, in a testing environment where a RAM has not yet been fully characterized, its test failure modes may be established by applying the largest running test pattern such as the GALPAT to provide maximum coverage of all possible failure modes. After full characterization of its failure modes, for subsequent testing, certain test patterns can be selected to provide coverage for its specific failure modes. The following section describes various test patterns commonly used for a 64K SRAM, with "n" being the number of device cells, and the test time includes a pass for an initial background pattern and another pass for the complementary pattern. These are the patterns sent to the RAM, rather than the resulting patterns stored in the RAM.

4.3.2.1 ZERO–ONE.
This is a minimal test consisting of writing 0s and 1s to the memory cells followed by read operations. This algorithm is also known as Memory Scan, MSCAN (see Section 4.2.1 for the algorithm). The test length for this is $4n$ operations, and therefore it is an $O(n)$ test. It has limited usage for AFs, SAFs, and cannot detect all TFs. It does not cover CFs.

4.3.2.2 CHECKERBOARD.
This is another short and simple test in which the memory cells are divided into two groups, cells-1 and cells-2, forming a checkerboard pattern. The checkerboard algorithm can be represented as follows:

Step 1: Write 1 in all cells-1, and 0 in all cells-2.

Step 2: Read all cells (words).

Step 3: Write 0 in all cells-1, and 1 in all cells-2.

Step 4: Read all cells (words).

For this test, $2n$ operations (read or write) are needed for each step. Therefore, the test length for this is $4 \cdot 2n$ operations, and therefore it is an $O(n)$ test. It has limited coverage for the AFs, SAFs, TFs, and CFs.

4.3.2.3 MASEST.
This is an alternating multiple-address selection test which is used to detect faulty operation of the device address decoders. Test time is $10n$.

Pattern generation scheme:

Step 1:
Write a background pattern to the RAM of alternating 0s and 1s.

0	0
1	1
2	0
3	1
•	•
•	•
•	•
$n-1$	0
n	1

Step 2:
Read and test each of the n RAM cells.

Test Cell	Reads & Tests	Reads & Tests
0	0	n
1	1	$n-1$
2	2	$n-2$
•	•	•
n	n	0

The "read and test" procedure for test cell 0 is illustrated.

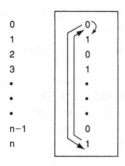

Step 3:
Read all RAM cells sequentially.

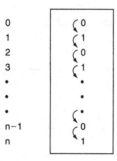

Step 4:
Write a complementary background pattern to the RAM of alternating 1s and 0s, and then repeat Steps 2 and 3.

4.3.2.4 MARCH C.   This is a single test word pattern that "marches" as a complement from cell to cell. It is used to test whether each RAM cell (bit) can be accessed and written into with a 0 and a 1, and to detect coupling faults. Test time is $10n$.

Pattern generation scheme:

Step 1:
Write a background pattern to the RAM of all 0s.

0	0	1
1	0	1
2	0	1
3	0	1
•	•	
•	•	
•	•	
n−1	0	1
n	0	1

Step 2:
Read, test, and write the complement into each of n RAM cells. The RAM now contains the complement of the background pattern.

0	0	1
1	0	1
2	0	1
3	0	1
•	•	
•	•	
•	•	
n−1	0	1
n	0	1

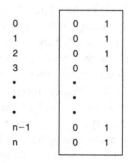

Test Cell	Reads & Tests Cell	Writes Compliment Into Cell
0	0	0
1	1	1
2	2	2
•	•	•
n−1	n−1	n−1
n	n	n

Step 3:
Repeat Step 2 in reverse, starting with cell n. The RAM now contains the original background pattern.

0	0
1	0
2	0
3	0
•	•
•	•
•	•
n−1	0
n	0

Test Cell	Reads & Tests Cell	Writes Compliment Into Cell
n	n	n
$n-1$	$n-1$	$n-1$
$n-2$	$n-2$	$n-2$
•	•	•
0	0	0

Step 4:

Write a background pattern to the RAM of all 1s and repeat Steps 2 and 3.

4.3.2.5 WAKPAT. This is a single test word walking pattern used to detect the internal multiple address selection failures, destruction of stored data due to noise coupling within a column, and slow sense amplifier recovery. The fault coverage is as follows: all AFs are detected and located; all SAFs and TFs are located. Test time is $2(n^2 + 3n)$.

Pattern generation scheme:

Step 1:

Write a background pattern to the RAM of all 0s.

```
0        0
1        0
2        0
3        0
•        •
•        •
•        •
n−1      0
n        0
```

Step 2:

For a test cell, perform the following sequence: write the complement of the test cell into the test cell. Read and test the remaining RAM cells.

Read the test cell; write in the complement so that the cell is in its original background state.

Repeat the sequence for each RAM cell.

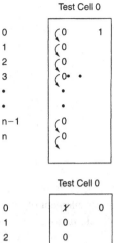

Test Cell 0

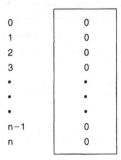

Test Cell 0

Step 3:

Write a background pattern to the RAM of all 1s and repeat Step 2.

4.3.2.6 GALPAT. It is a single test word galloping pattern which is used to detect unsatisfactory address transitions between each cell and every other cell, slow sense amplifier recovery, and destruction of stored data due to noise coupling between cells within a column. The fault coverage is the same as for the WAKPAT. Test time is $2(2n^2 + n)$.

Pattern generation scheme:

Step 1:

Write a background pattern to the RAM of all 0s.

```
0        0
1        0
2        0
3        0
•        •
•        •
•        •
n−1      0
n        0
```

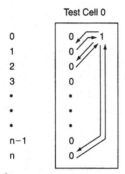

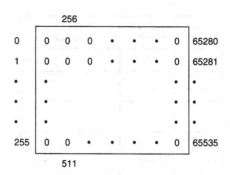

Step 2:

For a test cell, perform the following sequence: write the complement of the test cell into the test cell. To check the transitions between the test cell and each remaining RAM cell, alternately read the test cell and every other RAM cell.

Write the complement of the test cell into the test cell, thus restoring the test to its original background state.

Repeat the sequence for each RAM cell.

GALDIA test assumes a 256-element diagonal for each cell of a 64K RAM, counting the cell itself and 255 other cells whose locations are a multiple of 255 greater than the test cell location. For example: cell 1 diagonal includes cell 1 plus cells 256, 511, ⋯, 65281. Cell 65535 diagonal includes cell 65535 plus 255, 510, ⋯, 65280.

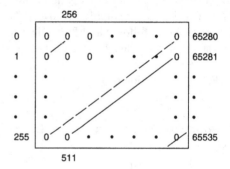

Step 2:

For a test cell, perform the following sequence: read the previous cell, then read the test cell. Write the complement of the test cell into the test cell.

Step 3:

Write a background pattern to the RAM of all 1s and repeat Step 2.

4.3.2.7 GALDIA. It is a galloping diagonal pattern which is used to detect unsatisfactory transitions between each cell and the position in the cell diagonal, slow sense amplifier recovery, and destruction of stored data due to noise coupling among cells in a column. Test time is $2(2n^{3/2} + 5n)$.

Pattern generation scheme:

Step 1:

Write a background pattern to the RAM of all 0s.

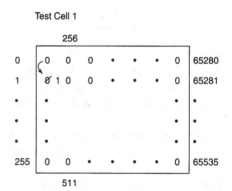

To check the transitions between the test cell and all other positions of its diagonal, alternately read the test cell and each cell of the test cell diagonal.

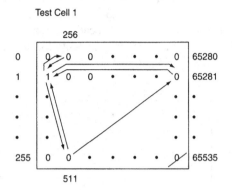

Test Cell 1

Write the complement of the test cell into the test cell, thus restoring it to its original background state.

Repeat sequence for each RAM cell.

Test Cell 1

Step 3:
Write a background pattern to the RAM of all 1s and repeat Step 2.

4.3.2.8 GALCOL. A galloping column pattern used to detect unsatisfactory address transitions between every cell and the cell–row position, destruction of stored data due to noise coupling among cells in the same column, and refresh sensitivity in the DRAMs. This test is often used to reduce the larger test time needed for the WAKPAT and the GALPAT, while accepting the loss of some fault detection and location capability. Test time is $2(2n^{3/2} + n)$. During the performance of a GALCOL test for the DRAMs, the pattern presented assumes that the test system has an asynchronous refresh mode. If that is not the case, then the GALCOL must be modified to be self-refreshing.

Pattern generation scheme:

Step 1:
Write a background pattern to the RAM of all 0s.

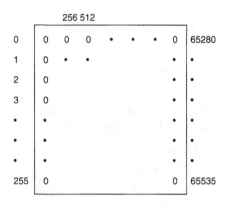

Step 2:
For a test cell of a current test column, perform the following sequence: write the complement of the test cell into the test cell. Check address transitions for the test cell row by alternately reading the test cell and locations (test cell + 256n), where $n = 1, 2, \cdots, 255$.

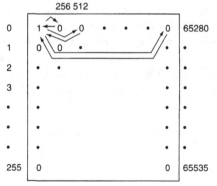

For example, for test column 0, test cell 0: read cell 0, read cell 255, then read cell 0, read cell 512 $\cdots$; read cell 0, read cell 65280.

Step 3:
Read the test cell, then write the complement into the test cell, so that it is in the original

background state. The test cell location is incremented by 1 and a 1 written into the test cell. Repeat Step 2 for all RAM columns.

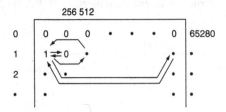

Step 4:

Write a background pattern to the RAM of all 1s and repeat Steps 2 and 3.

4.3.2.9 DIAPAT. A shifting diagonal pattern which detects internal multiple-address selection failures, destruction of stored data due to noise coupling, faulty sense amplifiers, or slow sense amplifier recovery. The test time is $2(2n^{3/2} + 2n)$. The DIAPAT tests every positively sloping diagonal varying in length from 1 to 256 elements, in contrast to the GALDIA diagonals which are all 257 elements long.

Pattern generation scheme:

Step 1:

Write a background pattern to the RAM of all 0s. For this memory array, the elements of positively sloping diagonals tested by the DIAPAT can be generated by the following formulas. For cells 0–255, if B represents the beginning element location, then each element location is defined by: $B + 255N$, where $N = 0, 1, 2 \cdots B$. Example: The diagonal beginning in cell location 3 contains elements located at 3, 258, 513, 768.

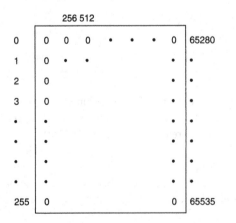

Similarly, for positively sloping diagonals beginning with cells 511–65535, the beginning cell can be defined by: $511 + 256K$, where $K = 0, 1, 2, \cdots 254$. Then, within each diagonal, a cell location can be defined by: $B + 255N$, where $N = 0, 1, 2, \cdots, (254-K)$. Example: The diagonal beginning in cell location 64767, i.e., $511 + 256(251)$, contains elements located at 64767, 65022, 65277, 65532.

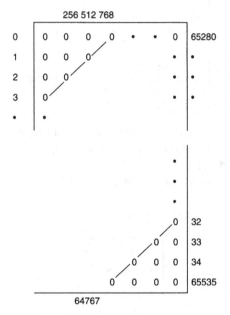

Step 2:

Write the complement of each diagonal element into the cell location. Read the nondiagonal cells, obtaining their locations by adding the multiples of 255 to the location of the last diagonal element.

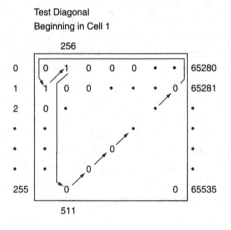

Read the elements of the test diagonal, then write the complement of each diagonal element to the cell location so that the test diagonal is restored to its original background state. Repeat the sequence for all positively sloping diagonals of the RAM.

Step 3:
Write a background pattern to the RAM of all 1s and repeat Step 2.

4.3.2.10 WAKCOL.

A single test word walking column pattern that detects the destruction of stored data due to noise coupling within a column and slow sense amplifier recovery. Test time is $2(2n^{3/2} + 3n)$.

Pattern generation scheme:

Step 1:
Write a background pattern to the RAM of all 0s.

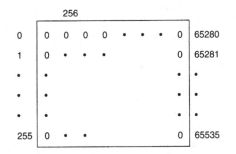

Step 2:
For a column, write the complement of each cell into the cell's location.

Read the remaining cells of the RAM sequentially.

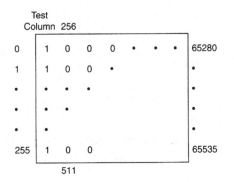

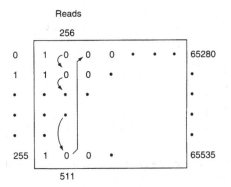

Read the test column, then write the complement of each cell into the cell's location so that the test column is restored to its original background state.

Repeat the sequence for every column of the RAM.

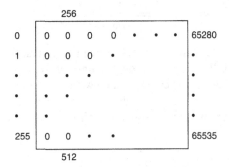

Step 3:
Write a background pattern of all 1s to the RAM and repeat Step 2.

4.3.2.11 DUAL WAKCOL.

It is a dual walking column pattern which detects destruction of the stored data due to noise coupling within a column and slow sense amplifier recovery. Test time is $2[\frac{1}{2}(n^{3/2}) + 3n]$.

Pattern generation scheme:

Step 1:
Write a background pattern to the RAM of all 0s.

Step 2:
For a pair of columns which are 128 columns apart, perform the following sequence:

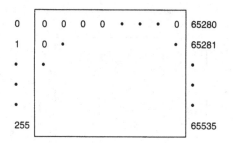

write the complement of each cell into the cell location.

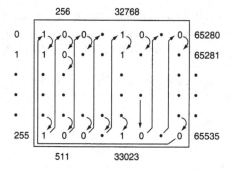

Read the entire RAM beginning with the cell following the second column of pairs, including the test columns, then write the complement of each cell into the cell location so that test columns are restored to their original background state.

Column #2

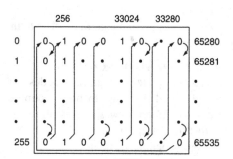

Repeat the sequence for every such pair of columns in the RAM, e.g., for column pairs:

0,128

1,129

...

127,255.

Step 3:

Write a background pattern to the RAM of all 1s and repeat Step 2.

4.3.2.12 CHEKCOL. A double checkerboard walking column pattern which is used to detect internal multiple-address selection failures and destruction of stored data due to noise coupling within a column. Test time is $2[(\frac{1}{4})n^{3/2} + (\frac{1}{4})n^2 + 2n]$.

Pattern generation scheme:

Step 1:

Write a background pattern to the RAM of alternating 1s and 0s.

The CHEKCOL pattern consists of a succession of 256-cell patterns, which move through the RAM as four cells at a time, for a total of 16,384 times. The procedure for first 256-cell test pattern is shown in the following steps.

0	0	0	•	•	•	0	65280	
1	0	0	•			0	•	
2	1	1	•			1	•	
3	1	1	•			1	•	
•	•	•				•	•	
•	•	•				•	•	
•	•	•				•	•	
124	0	0				0	•	
125	0	0				0	•	
126	1	1				1	•	
127	1	1				1	•	
128	0	0				0	65408	
129	0	0				0	•	
130	1	1				1	•	
131	1	1				1	•	
•	•	•				•	•	
•	•	•				•	•	
•	•	•				•	•	
252	0	0				0	•	
253	0	0				0	•	
254	1	1				1	•	
255	1	1				1	65535	

Step 2:

Write a test pattern consisting of 128 1s followed by 128 0s. Read each RAM cell sequentially, beginning in the next cell after the test pattern, and including the test pattern.

Test Pattern (Cells 0-255)
Reads Starting in Next Cell
(Cells 256-65535, 0-255)

0	1	0	• •	•	0	65280
1	1	0			0	•
2	1	1	•		1	•
3	1	1	•		1	•
•	•	•			•	•
•	•	•			•	•
•	•	•			•	•
124	1	0			0	•
125	1	0			0	•
126	1	1			1	•
127	1	1			1	•
128	0	0			0	65408
129	0	0			0	•
130	0	1			1	•
131	0	1			1	•
•	•	•			•	•
•	•	•			•	•
•	•	•			•	•
252	0	0			0	•
253	0	0			0	•
254	0	1			1	•
255	0	1			1	65535

Restore the first four cells of the test pattern to their original background state.

Step 3:

The fifth cell of the test pattern becomes the first cell of a new 256-cell test pattern. Repeat the procedure of Step 2 until the "first four cells" restored to their original background state occupy RAM locations 65532–65535.

Step 4:

Write a background pattern to the RAM which is complementary to the original, i.e., alternating pairs of 1s and 0s, and repeat Steps 2 and 3.

First Test Pattern in Cells 0-255
Second Test Pattern in Cells 4-259

0	0	0	0	•	•	0	65280
1	0	0	0			0	•
2	1	1	0			1	•
3	1	1	0			1	•
4	•	•				•	•
5	•	•				•	•
•	•	•				•	•
124	1	0				0	•
125	1	0				0	•
126	1	1				1	•
127	1	1				1	•
128	1	0				0	65408
129	1	0				0	•
130	1	1				1	•
131	1	1				1	•
•	0	•				•	•
•	0	•				•	•
•	0	•				•	•
252	0	0				0	•
253	0	0				0	•
254	0	1				1	•
255	0	1				1	65535

4.3.2.13 HAMPAT.

A sequential column test which is used to detect the destruction of stored data within a column due to multiple writes into a cell within that column. Test time is $2(n^{3/2} + 5n)$.

Pattern generation scheme:

Step 1:

Write a background pattern to the RAM of all 0s.

0	0	0	•	•	•	65280
1	0	0	•	•	•	65281
2	0	0	•	•	•	65282
•	•					•
•	•					•
•	•					•
255	•					65535

Step 2:

For a column of the RAM, perform the following sequence: write a 1, write a 0, write a 1 into a cell of the test column.

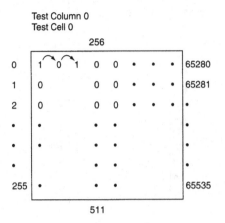

Test Column 0
Test Cell 0

Read the remaining cells of the test column, beginning in the cell after the one written into.

Read the test cell, then restore it to its original background state. Repeat the sequence for every cell of the test column. Repeat the sequence for every column of the RAM.

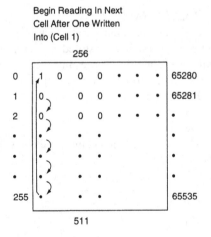

Begin Reading In Next
Cell After One Written
Into (Cell 1)

Step 3:

Write a background pattern to the RAM of all 1s and repeat Step 2.

4.3.2.14 MOVI. An alternating multiple-column, multiple-row pattern (moving inversion) test which detects faulty access or write

into a cell, noise coupling within a column, slow sense amplifier recovery, faulty address transitions between each cell and the cells rows, and refresh sensitivity in the DRAMs. Test time is $2[(n^{3/2} + (\frac{3}{2})n]$.

Pattern generation scheme:

Step 1:
Write a background of 0s.

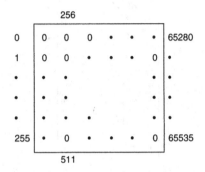

Step 2:

For alternating test columns starting at cell 0 (column 0), write the complement of the background. Read the pattern in the same sequence it was written.

	256	512	768				65024		
0	1	0	1	0	•	•	1	0	65280
1	1	0	1	0			1	0	65281
•	•	•	•	•	•		•	•	•
•	•	•	•	•	•		•	•	•
•	•	•	•	•	•		•	•	•
255	1	0	1	0			1	0	65535
	511	767	1024				65279		

0	1	1	•	•	•	•	•	1	65280
1	0	0	•	•	•	•	•	0	65281
2	1	1	•	•	•	•	•	1	65282
3	0	0	•	•	•	•	•	0	65283
•	•	•					•	•	
•	•	•					•	•	
254	1	1	•	•	•	•	•	1	65534
255	0	0	•	•			•	0	65535

	256	512	768	64512	64768/65024				
0	1	1	0	0	1	1	0	0	65280
1	1	1	0	0	1	1	0	0	65281
•	•	•	•	•	•	•	•	•	•
•	•	•	•	•	•	•	•	•	•
255	1	1	0	0	1	1	0	0	65535
	511	767	1023	64767	65023/65279				

Step 3:

Restore the background to the test columns. Now, write the complement of the background into alternating test rows starting at cell 0 (row 0). Next, the pattern is read in the same sequence in which it was written.

Step 4:

Restore the background to the test row, then write the complement of the background into alternating pairs of columns starting at cell 0, column 0. Next, the pattern is read in the same sequence in which it was written.

256

0	1	1	•	•	•	•	1	65280
1	1	1	•	•	•	•	1	65281
2	0	0	•	•	•	•	0	65282
3	0	0	•	•	•	•	0	65283
4	1	1	•	•	•	•	1	65284
5	1	1	•	•	•	•	1	65285
6	0	0	•	•	•	•	0	65586
7	0	0	•	•	•	•	0	65287
•	•	•				•	•	•
•	•	•				•	•	•
252	1	1	•	•	•	•	1	65532
253	1	1	•	•	•	•	1	65533
254	0	0	•	•	•	•	0	65534
255	0	0	•	•	•	•	0	65535

Step 5:

Next, restore the background to the test columns, then write the complement of the background into alternating pairs of columns starting at cell 0, column 0. Now, the pattern is read in the same sequence in which it was written.

Continue Steps 2–5, doubling the number of columns and rows for each step until the number of columns and number of rows written equals $(\frac{1}{2})n^{1/2}$.

Step 6:

Now, write a background of 1s into the RAM and repeat Steps 2–5.

$\dfrac{256}{2} = 128$ columns

$\dfrac{n^{1/2}}{2}$ rows 01

$\dfrac{256}{2} =$ 127

128

129

128 rows 255

	1	1	•	•	•	•	1	65280
	1	1	•	•	•	•	1	65281
	•	•				•	•	•
	1	1	•	•	•	•	1	65407
	1	1	•	•	•	•	0	65408
	1	1	•	•	•	•	0	65409
	•	•	•	•	•	•	•	
	0	0	•	•	•	•	0	65535

	256				32512/32768					
$\dfrac{n^{1/2}}{2}$ 0	1	1	•	•	1	0	0	•	0	65280
	•	•	•		•	•	•		•	
	•	•	•		•	•	•		•	
columns	•	•	•		•	•	•		•	
255	1	1	•	•	1	0	0	•	0	65535
	511				32767/33023					

4.3.2.15 BUTTERFLY. It is a combination of a galloping-row and galloping-column pattern developed to further reduce the number of operations required by the GALPAT test. It detects internal multiple-address selection and noise coupling within the columns and/or rows. Test time is $2(4n^{3/2} + 3n)$.

Pattern generation scheme:

Step 1:
Write a background pattern of 0s to the RAM.

```
  0 │ 0  •  •   •   •  •   0 │ 65280
  1 │ 0  •  •   •   •  •   0 │ 65281
  • │ •                    •
  • │ •                    •
  • │ •                    •
255 │ 0  •  •   •   •  •   0 │ 65535
```

Step 2:
Write the complement of the test cell into the test cell.

Step 3:
Read the location specified by the test cell $(T_c) + (n^{1/2}) \times I$, where $I = 1$ to $(\frac{1}{2})n^{1/2}$ iterations. Steps 3–6 equal I iterations and $I_{max} = (\frac{1}{2})n^{1/2}$.
Read the test cell.

Step 4:
Read the location specified by $T_c + (1)I$.
Read the test cell.

Step 5:
Read the location specified by $T_c - (n^{1/2})I$.
Read the test cell.

Step 6:
Read the location specified by $T_c - (1)I$.
Read the test cell.

Step 7:
Steps 3–6 are repeated until $I = (\frac{1}{2})n^{1/2}$.

Step 8:
Write a background of 1s and repeat Steps 2–7.

4.3.3 Functional Test Pattern Selection

Section 4.3.2 discussed some of the various RAM test algorithms developed and used by the manufacturers and the end users. The selection of a particular set of test patterns for a given RAM is influenced by the type of failure modes to be detected, memory bit density, and hence the test time, and the memory ATE availability [31]. Table 4-6 summarizes the various attributes of test algorithms discussed (except for Zero–One and Checkerboard) to aid in the selection of suitable test patterns for a given

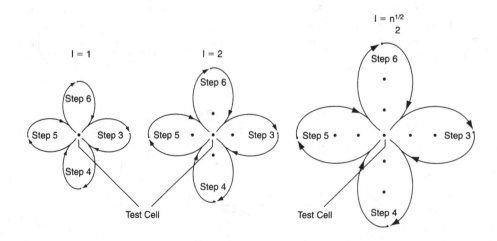

TABLE 4-6. RAM Functional Test Pattern Selection. (a) Test Time Factor Formulas and Relative Test Times. (b) Failure Detection Capabilities Matrix. (c) Failure Codes.

Pattern Name	Test Time Factor* Formula	Test Time Factor 64K RAM	Relative Test Time** 64K RAM
MASEST	$10n$	655,360	0.00004
MARCH	$10n$	655,360	0.00004
WAKPAT	$2(n^2 + 3n)$	8,590,196,544	0.50
GALPAT	$2(2n^2 + n)$	17,179,933,536	1.0
GALDIA	$2(2n^{3/2} + 5n)$	67,764,220	0.004
GALCOL	$2(2n^{3/2} + 1n)$	67,239,932	0.004
DIAPAT	$2(2n^{3/2} + 2n)$	67,371,004	0.004
WAKCOL	$2(2n^{3/2} + 3n)$	33,947,646	0.002
DUAL WAKCOL	$2[(1/2)n^{3/2} + 3n]$	17,170,431	0.001
CHEKCOL	$2[(1/4)n^{3/2} + (1/4)n^2 + 2n]$	2,156,134,350	0.1255
HAMPAT	$2(n^{3/2} + 5n)$	34,013,182	0.002
MOVI	$2(n^{3/2} + 1/2n)$	16,973,823	0.001
BFLY	$2(4n^{3/2} + 3n)$	134,217,720	0.008

(a)

Pattern Name	F1	F2	F3	F4	F5	F6	F7	F8	F9	F10	F11
MASEST	X										
MARCH		X									
WAKPAT			X	X	X						
GALPAT				X	X	X					
GALDIA				X	X		X				
GALCOL				X				X	X		
DIAPAT			X	X	X						X
WAKCOL				X	X						
DUAL WAKCOL				X	X						
CHEKCOL			X	X							
HAMPAT										X	
MOVI		X		X	X			X	X		
BFLY			X	X				X			

(b)

Failure Code	Failure
F1	Faulty device address decoders
F2	Faulty access or write into a cell
F3	Internal multiple address selection
F4	Noise coupling within a column
F5	Slow sense amplifier recovery
F6	Faulty address transitions between each cell and every other cell
F7	Faulty address transitions between each cell and the cell diagonal
F8	Faulty address transitions between each cell and the cell row
F9	Refresh sensitivity in dynamic RAMs
F10	Multiple writes into a fixed cell of a column
F11	Faulty sense amplifiers

(c)

RAM. Table 4-6(a) lists the pattern name, test time formula, test time factor for a 64K RAM, and relative test time for a 64K RAM. To obtain the actual test time using column 2, evaluate the formula for n, the number of device cells in the given memory, and then multiply this number by the device cycle time. The relative test time shown in column 4 is a quantity obtained by dividing the test time factor value for a desired pattern by the factor value for GALPAT, the pattern that takes the longest time. The relative test time is important for selecting the fastest test to reveal a specific failure type.

Table 4-6(b) and (c) list the pattern name, its utility in detecting certain types of failures identified by "X" in the appropriate column, and the description of various failure codes. For example, 64K RAM is known to exhibit occasional internal multiple-address selection failures. Then, according to Table 4-6(b), four patterns which have "X" in column F3 are candidates: WAKPAT, DIAPAT, CHEKCOL, and BFLY. Of these four patterns, the one with the fastest relative test time is the DIAPAT, which is shown as 0.004 in column 4 of Table 4-6(a).

4.4 RAM PSEUDORANDOM TESTING

Section 4.3 discussed memory testing using deterministic techniques which require well-defined algorithms and memory test input patterns with corresponding measurements of expected outputs. In deterministic testing for the memories, the control data for the RAM under test and reference data have some predetermined value. Then the output response data of a RAM are compared with the reference data to make a pass/fail decision. For the PSFs, various algorithms, such as the ones by Suk and Reddy [22] and by Hayes [23], were described. An alternative way of testing the memories has been investigated that uses pseudorandom test patterns as the input stimuli in a nondeterministic (probabalistic) way. The advantage of this approach is that it allows the detection of some complex faults such as k-coupling faults with fairly high probability and within reasonable test time.

Random (or pseudorandom) testing consists of applying a string of random patterns simultaneously to a memory device under test and to a reference memory (which may be either hardware- or software-configured) and comparing the outputs of those two memories. The reference RAM may be replaced by compressing the reference data (or the simulation model of that RAM) using a compressor such that only a limited number of bits remain which are called the "signature." This test, when repeated with the same input control data, should produce the same reference signature. This condition for application of the same control data as the ones used for generating the reference signature requires the need for pseudorandom control data which are repeatable instead of purely random data.

This pseudorandom testing (PRT) is widely acceptable since test pattern generation is independent of a given set of faults. The amount of fault coverage with PRT cannot be guaranteed very accurately as can be done for the deterministic testing with the MATS+ and March algorithms. The PRT is associated with an escape probability (e_p), i.e., for a particular fault given, an $e_p = 0.005$ represents fault coverage better than 99.5%. The probability of detection of a fault is an increasing function of test length (i.e., the number of random patterns) sequence and can be made close to 1. The PRT can be used both for SAFs and k-coupling faults.

Consider an n word $\times$ 1 b RAM where input lines are N address lines (i.e., $n = 2^N$), one data line, and one R/W line. Then the random input pattern is an $(N+2)$-bit vector, i.e., there are $4N$ possible random patterns. In the PRT, before a random test sequence is applied to the RAM, a known pattern must be written in it by a preset sequence, e.g., all 0s, all 1s, or a checkerboard. Figure 4-8(a) shows the basic scheme for a PRT which consists of the reference RAM (C_0), the RAM under test (C_1), the comparator (C_2), and the flip-flop (C_3) [32].

Let

f = fault under consideration

L = test length

Q_i = number of internal states of the machine C_i

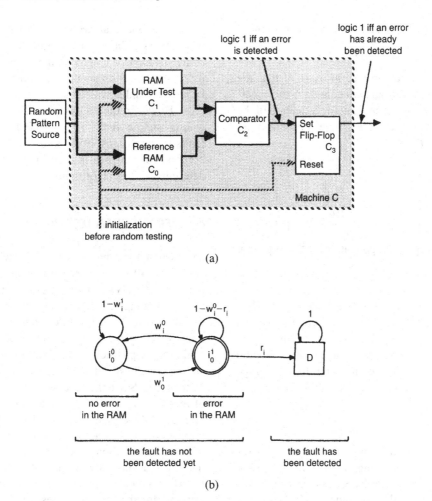

Figure 4-8. (a) The basic scheme for pseudorandom testing (PRT). (b) A typical Markov chain representation of a stuck-at-0 fault for cell "*i*." (From [32], with permission of IEEE.)

Q_D = detection uncertainty (i.e., probability that the fault has not been detected when L test patterns have been applied).

Then the number of internal states of the machine is given by the product

$$Q = Q_0 Q_1 Q_2 Q_3.$$

For an n word $\times$ 1 b memory, $Q_0 = 2^n$. For a faulty memory, Q_1 may be equal to Q_0 or different, depending on the fault. Since the comparator is combinational, $Q_2 = 1$. When the random patterns are applied to the machine C, its behavior can be modeled by a Markov chain [33]. Figure 4-8(b) shows a typical Markov chain representation of a stuck-at-0 fault for cell "*i*" [32]. Three states are diagramatically represented as follows: (a) the i_0^0 state, indicating that there is no error in the RAM and no error has been detected yet, (b) the i_0^1 state (shown by the double circle), indicating that there is an error in the RAM, but no error has been detected yet, and (c) the D state (shown by a square) in which the fault has already been detected.

Let

$$P_r[i_0^0] = \text{possibility of being in state } i_0^0 \text{ at some time } t.$$

The state probability vector is given by

$$P(t) = P_r[i_0^{\,0}], P_r[i_0^{\,1}], P_r[D].$$

The Markov matrix M is such that the entry at the jth row and kth column is the conditional probability of making the transition of the kth state, given the present state is jth, the order of vector $P(t)$.

Therefore, $P(t) = P(t-1)\,M$ where

$$M = \begin{bmatrix} 1 - w_i^{\,1} & w_i^{\,1} & 0 \\ w_i^{\,0} & 1 - w_i^{\,0} - r_i & r_i \\ 0 & 0 & 1 \end{bmatrix}.$$

The repeated application of this equation gives

$$P(t) = P(0)M^{\,t}.$$

Since D is an absorbing state, the $P_r[D]$ is an increasing function of time t, and the random test length is defined by $P(L)$ such that $P_r[D] \geq 0.999$. An important parameter is the length coefficient (h) which corresponds to the average number of times each cell is addressed:

$$h = \frac{L}{n}.$$

The Markov chain can also be used for modeling two-coupling inversion and two-coupling idempotence faults, as well as the passive and active PSFs. A comparison between random testing and deterministic testing has shown that random testing is particularly useful when detecting interactive faults. For example, in an interactive fault detection scheme, the random test length required was $219n$ for $Q_D = 10^{-3}$ compared to the deterministic algorithm test time requirement of $36n + 24n \log n$ [32]. Another useful application area for random testing is for three-coupling or PSFs involving influential cells anywhere. For the three-coupling faults, deterministic algorithms usually require $O(n \log n)$ time, whereas random testing requires a time which remains linear as a function of n: $447n$ for $Q_D = 10^{-3}$. The PRT has advantages over the deterministic algorithms which have excessive test lengths such as the GALPAT. It is usually considered better to use random testing with a reasonable test length (e.g., a few hundreds of n) instead

of an inefficient algorithm like the GALPAT which takes $2(2n^2 + n)$ test time. Therefore, in many situations, a long random test sequence offers better detection of well-defined faults and a larger set of unknown faults.

However, the PRT is less efficient for detecting simple faults (e.g., SAFs) and other faults such as the ANPSFs in a five-cell neighborhood for which it is easier to construct the deterministic test algorithms.

4.5 MEGABIT DRAM TESTING

As discussed in Chapter 2, the DRAM die size has grown 40–50% every generation, while the memory cell size has decreased by 60–65%, and this scaling of memory cell size is expected to continue to 1 Gb level densities. In the current megabit generation, the 16 Mb DRAMs are available as production devices, while 64 Mb and higher density devices are in the development stages. Since the process cost is a function of memory capacity, the production cost of future DRAMs will increase significantly. Since the memory test time increases proportionately with memory capacity, the test costs should also increase substantially. Figure 4-9 shows the DRAM test time ($t_{\text{cycle}} = 150$ ns) versus memory size for three different test modes [34]. It can be seen that the test cost for a 64 Mb DRAM may be 240 times that for a 1 Mb device. Further analysis shows that the test cost ratio to total cost (assembly process + cost) of a 64 Mb DRAM can amount to 39%, which is an unacceptably high figure.

The increase in density has required a corresponding increase in the address bits from 22 address lines for 4 Mb single I/O devices to 24 address lines for 16 Mb single I/O devices. This means that the memory tester data bus and pattern generator must also increase accordingly to provide the signal paths for an increase in the address and data bits. To increase yield, redundant elements (e.g., extra rows and columns of cells) are becoming standard in the megabit DRAM designs, and this redundancy analysis imposes additional demands on the memory test

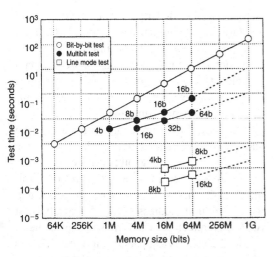

Figure 4-9. DRAM test time (t_{cycle} = 150 ns) versus memory size (bits) for three different test modes. (From [34], with permission of IEEE.)

equipment. To reduce test time, new test equipment has to use dedicated hardware to minimize redundancy analysis times, and eventually to reduce the device test time. Currently, the megabit DRAMs are available in a variety of I/O configurations, e.g., 16 Mb memories as 16 Mb × 1, 4 Mb × 4, and 2 Mb × 8 devices. Therefore, the memory automated test equipment (ATE) must be capable of handling the various I/O configurations.

In an effort to reduce the test times, wider I/O access is provided in many internal memory organizations, e.g., 1 Mb devices are organized as 256K × 4, 128K × 8, or 64K × 16, which reduces the test time. However, even single-bit-wide devices can also facilitate parallel testing if they are appropriately designed with internal partitioning and other suitable test features. The partitioned designs provide both test and application flexibility. A device structured as 16 internally addressable blocks may be used as 16, 8, 4, or single I/O devices [35]. Many newer DRAM designs offer special test modes that allow the tester to read a × 1 part in a wider configuration, either through the use of special pads at wafer sort or via special functional modes on the packaged parts. For example, a

16 Mb DRAM configured as 16Mb × 1, assuming 100 ns access time, would take 1.68 s (16M × 100 ns) to read the entire contents of the device. However, if the same device is configured as 1Mb × 16, the test time would 0.105 s (1M × 100 ns). Therefore, by providing parallel access, internal test modes, and/or self-test, the test times can be further reduced. Although the internal test mode is advantageous, it does have limitations in certain situations. For example, if a location in a 16 Mb device fails, then the test program must return to the full 16 Mb of addressing to identify the faulty location and failure mode.

In testing of redundancy-based reconfigurable RAMs, there are some algorithms which provide early detection of device unrepairability to guard against wasted test time. In principle, many memory testers continually accept and evaluate defect location information, apply redundancy repair algorithms, continue testing as long as the device appears repairable, and abort testing if it is not repairable.

Some of the methods currently proposed for reducing the test time of megabit DRAMs include using the multibit test, the line mode, and the built-in self-test (BIST) [34]. The line mode test can reduce test time significantly, but has poor error-detection capabilities when there is more than one error present on the single line. In another approach, the use of the testing acceleration chip (TAC) has been proposed, which acts as an interface between the device under test (DUT) and a simplified, low-cost memory tester. This TAC is fabricated using the same design rules as peripheral logic circuits of the memory DUT. It contains pin electronics, built-in digital timing comparators to verify the ac performance parameters of the DUT. Therefore, the simplified tester does not require I/O pin electronics. These TACs, along with megabit DRAM devices, can be distributed by the manufacturers to users and system developers to help reduce their test costs.

The high-performance memory testers (e.g., Hewlett Packard HP 83000) currently used for testing memories can be interfaced to an algorithmic pattern generator (APG) which allows easier and faster testing. The function of APG, which is available in both hardware and software configurations, is to automatically generate the address vectors for each of the memory address locations to be tested. In either the hardware or software version of the APG, the user does not have to manually generate the vectors needed to control the memory operations and for data transfer. The hardware APG version has the advantage that the address vectors are generated in real time. The APG uses an algorithm to produce a sequence of addresses which can be incremented, decremented, or even scrambled to reflect the memory topologies. The addresses are then synchronized with the repeating read or write sequences of the control and data vectors. The APG allows memory testing of very large devices without the impact of vector downloads.

A computer-aided method for rapid and accurate failure analysis of high-density CMOS memories has been developed which is based on the realistic fault modeling using the defect-to-signature "vocabulary" [54]. This vocabulary consists of multidimensional failure representations obtained through the functional and IDDQ testing of the memory devices. A high-resolution defect-to-signature vocabulary can be generated by an inductive fault analysis (IFA) using software such as VLASIC [55]. The IFA is a simple procedure consisting of the following basic steps: (1) generation of the defect characteristics, (2) defect placement, (3) extraction of the schematic and electrical parameters of the defective cell, and (4) evaluation of the results of testing the defective memory. This sequence of steps can be performed using the Monte Carlo approach (e.g., by using VLASIC) or by systematically choosing the defect characteristics (such as the location of defect, the defect size, and the affected layer of the memory layout).

In a study performed, this defect diagnosis strategy was implemented on the 64 kb SRAM fabricated in a 1.2 μm twin-tub CMOS process with p epitaxial layer on a p+ substrate using single-polysilicon and double-metal interconnections. The test results showed that realistic fault modeling, along with functional and IDDQ

testing, can significantly improve the diagnostic resolution of traditional bit mapping. The diagnostic information can be accurate if the Monte Carlo defect simulation strategy is used to get a completed defect-to-signature vocabulary. This defect diagnosis approach can lead to rapid failure analysis for providing detailed feedback on problems resulting from equipment malfunction and process variations during the production and testing of multimegabit memories.

Advanced memory architectures for ultra-high-density memories are being designed that use a column address-maskable parallel-test (CMT) architecture which makes it possible to handle the various test patterns and search failed addresses quickly during the parallel test operation [36]. These designs for testability (DFT) and BIST techniques will be discussed in Chapter 5.

4.6 NONVOLATILE MEMORY MODELING AND TESTING

Nonvolatile memory technology, including ROMs, PROMs, EPROMs, and EEPROMs, was discussed in Chapter 3. The functional model for the ROM (or PROM) is basically derived from the RAM chip functional model (see Section 4.2). Figure 4-10 shows the functional model of an EEPROM. In a PROM, EPROM, or EEPROM, the number of functional faults that can occur is limited in comparison to the faults that may occur in the RAMs. In nonvolatile memories, the functional faults that may usually occur are:

- Stuck-at faults (SAFs), e.g., stuck cell, stuck driver, read/write line stuck, data line stuck or open.
- Address decoder faults (AFs), e.g., address line stuck or open, shorts between the address lines, open decoder, wrong access, or multiple access.
- Bridging faults (BFs) or state coupling faults (SCFs), e.g., shorts between the data lines which can be detected with a test for the SAFs.

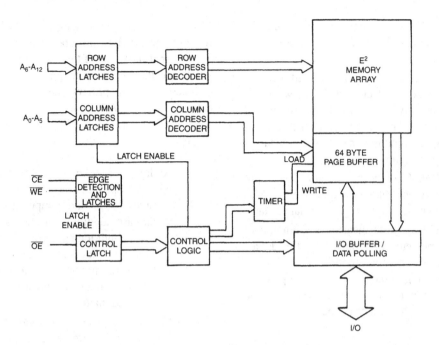

Figure 4-10. Functional model of an EEPROM.

The coupling faults (CFs) and pattern-sensitive faults (PSFs) are unlikely to occur in the ROMs (or PROMs) which have permanent bit patterns stored either through mask programming or fuse/antifuse mode programming. Also, in the EPROMs (UVEPROM or EEPROM), the coupling faults and PSFs are less likely to occur as compared to the RAMs since the cell charge is stored in the capacitor formed by the floating gate, which is a nonvolatile mode. However, the CFs and PSFs can occur in the EPROMs, and certain functional algorithms are performed as part of the reliability characterization to verify proper operation. Transition faults (TFs) can only be sensitized during the write operation, and can be detected during postprogramming functional verification testing.

The major failure modes in EPROMs, such as the SAFs, AFs, BFs (or SCFs), can be detected through the functional test algorithms. The simplest functional test is the postprogramming verification by addressing each memory location and comparing the stored data with the original memory bit map program (expected data). If the bit map program stored in a PROM (or EPROM) is not available, then the memory devices can be functionally tested using error-detection techniques such as parity checking, checksum verification testing, and cyclic redundancy. These will be discussed in Chapter 5.

The electrical testing for PROMs and EPROMs consists of the dc parametrics, dynamic and ac (switching) electrical measurements, and functional tests for program pattern verification. As an example, the following section describes electrical testing performed as a part of extensive reliability characterization on two nonvolatile memories: (1) a 256K UVE-PROM, and (2) a 64K EEPROM [37].

4.6.1 DC Electrical Measurements

The dc electrical measurements for both devices consisted of the following:

- VOH, VOL tests to measure the resultant logic high- and low-voltage levels on the outputs when driving a load. For this, the test currents are forced into the output and the resulting voltage is measured.

- IIH, IIL tests are performed to measure the amount of current drawn when the inputs have a specified voltage applied to them. This test is performed by driving an input high and low, and measuring the resultant current.

- IOZH, IOZL tests measure the output leakage currents of the device when its outputs are in the high-impedance state.

- ICC, ISB tests measure the amount of supply current required by the device for two operating modes: (1) ICC corresponding to the device typical operating mode when it is being accessed, and (2) ISB, the standby mode.

4.6.2 AC (Dynamic) and Functional Measurements

4.6.2.1 256K UVEPROM. To allow functional testing of the device, the memory array has to be programmed with a pattern. Three patterns were used for testing this device:

1. The first pattern was designed by the manufacturer for worst case programming with regard to internal address decoder delays, and consisted of programming 95% of the locations in the cell array with 0s and the remaining 5% with 1s. The 5% locations programmed with 1s were selected to provide the worst case address access time.

2. The second pattern, called a "TRUE pattern," was representative of an actual program consisting of a random sequence of 1s and 0s.

3. The third pattern, called a "FALSE pattern," was created by inverting the "TRUE pattern" to ensure that every cell can be programmed with a "1" and a "0."

The ac characterization not only determines the operating limits of the timing parameters, but also verifies the integrity of the internal functions under dynamic operating conditions. The internal functions of the memory can be divided into the following major blocks: address decoders, sense amplifiers, memory array, and data I/O circuitry. All of these functional blocks were verified during the following ac measurements:

- Address access time was evaluated for all three patterns at different supply voltages (V_{CC}) and temperatures. The worst case condition for address access time was high temperature (125°C) and low supply voltage. In all cases, the first pattern (programmed by the manufacturer) was found to be less stringent than the "FALSE pattern."

- Chip enable access time was evaluated for all three patterns. The worst case conditions for chip enable access time were high temperature (125°C), low supply voltage (4.5 V), and the "FALSE pattern."

- Output enable access time is a major concern in system design because of the bus contention issues. Therefore, the system timing must be such that each line is driven only by one output at a time. The output enable access time worst case conditions were high temperature (125°C), low supply voltage (4.5 V), and the first pattern. The data analysis showed that since the manufacturer pattern consisted of 95% zeros, the capacitors in the load circuit were discharged to a lower level, which took the outputs a longer time to rise to the VOH comparator level.

- Address hold time measures the amount of time the data are valid after the addresses change state, assuming that chip enable and output enable are still active.

- Chip enable/Output enable float time measures the amount of time required to tristate the data outputs from an active state. Two control pins (Output Enable and Chip Enable) have this same function. At turn-off, the voltage at the output pin largely becomes a function of the time constant of the capacitance and the load resistances.

The ac measurement test results showed one UVEPROM failing address access time only while it was programmed with the "FALSE pattern." This failure was caused by a faulty address decoder, and was not detected by the other two patterns because the locations which were selected by the faulty decoder had the same data for comparison as the tester reference data. Therefore, during the UVEPROM ac and functional characterization, the use of multiple program patterns at worst case temperatures and supply voltages is recommended.

4.6.2.2 64K EEPROM.
The functional testing of EEPROMs consisted of several different patterns which exercised the device to detect any address sequencing and data pattern sensitivities. The input timing relationships were set to verify the ac parameters during the write cycle, and the output strobe delay was set to verify the address access time during the read cycle. The EEPROM was tested like a static RAM by applying eight test patterns which contained write cycles and read cycles, briefly described as follows.

PATTERN #1 (WRITE ALL 0S, READ ALL 0S). The purpose of this pattern was to prove the operation of a "page mode," and most of the other patterns were developed from this pattern. This pattern remained in the program because it preconditioned the device for GALPAT0 functional test.

PATTERN #2 (GALPAT0, DATA BACKGROUND = ALL "00"). This pattern tested for the read disturbances. The byte under test is read alternately with other bytes in the same column and the same row.

PATTERN #3 (CHECKERBOARD). This pattern ensures that every cell can be written to a "1" and a "0," and the adjacent cells which are programmed to the opposite state do not affect the cell under test. It is desirable to have a

"bit-checkerboard," i.e., every bit in every byte is programmed to the opposite state of the bits above and below, and to the bits right and left of it. The page mode was used for this test pattern.

PATTERN #4 (WRITE ALTERNATING COLUMNS OF "00" AND "FF" AND VERIFY). This pattern loaded the data which were read during the ac parametric tests. The page mode was used for this test.

PATTERN #5 DRAD (DIAGONAL READ OF ALTERNATING DATA). This pattern reads the data loaded during the previous pattern. The DRAD pattern has no write cycles; it only reads the data along the diagonals. This pattern was used during the measurements of the read cycle ac parameters. It generates the worst case access times because the address decoders change every cycle and the data complement every cycle.

PATTERN #6 (BIT UNIQUE TEST). This pattern verifies that every bit within a byte is unique, i.e., the two bits in a byte are not shorted. The page mode was used for this test.

PATTERN #7 (GALPAT1, DATA BACKGROUND = ALL "FF"). This pattern tests for read disturbances. The byte under test is read alternately with other bytes in the same column and the same row. The page mode was used for this test.

PATTERN #8 (MARCH, DATA BACKGROUND = ALL "FF"). This pattern tests for address uniqueness and multiple selection. The byte write mode was used for this test.

The functional testing consumed the bulk of the total test time. Some special features of the EEPROM, such as the mass chip erase/program modes and DATA polling, can be used to reduce this time. However, it is necessary to distinguish between failures due to the pattern sensitivities and failures due to the mass programming modes or DATA polling. For example, if the mass erase chip mode failed to precondition a device prior to the GALPAT1 pattern, a read disturb problem could not be detected. Therefore, separate tests were developed to characterize the mass chip program/erase modes and the DATA polling.

The dynamic characterization consisted of measuring the read cycle and byte write cycle ac parameters. Most of these parameters were verified during the functional testing, but they were measured again to determine the margin between actual values and the manufacturer's specification limits. The testing was performed at supply voltages of 4.5 and 5.5 V and temperatures of −55, +25, and +125°C. The following measurements were performed:

READ CYCLE AC MEASUREMENTS.   The device under test was programmed with alternating columns of "FF" and "00" by using the page mode. A pattern was written that read the data along the diagonals of the memory array. Incrementing both row and address locations checks for the worst case address access time, and reading alternate data requires the sense amplifiers to respond with the complement data every cycle. Also, all eight outputs are switching every cycle. Three access times were measured: address, chip enable, and output enable. As expected, the address access time increased with increasing temperature. The Schmoo plots of V_{CC} versus address access time were generated at three test temperatures. The chip enable access time is very dependent on temperature and insensitive to V_{CC} change. The output enable access times (TOLQV1 and TOLQV2) increased with temperature increase.

The chip enable to output active and the output enable to output active times were also measured since they are important in the prevention of bus contention. For these tests, the time interval from the enable line low to the I/O pins active is measured since this interval is the time required for tristated outputs to become active. I/O disable times measured the amount of time required to tristate an output from an active state. The delays from chip enable to all I/O pins tristated and from output enable to all I/O pins tristated were measured.

BYTE WRITE CYCLE AC MEASUREMENTS. Byte write cycle parameters were measured because most of the functional tests used the page

mode. During the functional testing, the timing and waveshapes were assigned in such a manner that the page write cycle parameters were verified. Also, the following byte write cycle parameters were verified:

• Address setup and hold time
• Data setup and hold time
• Output enable setup time and hold time
• Chip enable setup time and hold time.

The test results showed that all measured access times were more sensitive to temperature than to V_{CC}. The Schmoo plots of address access time versus V_{CC} indicated little or no variation in the access time over a wide variation of supply voltage.

EEPROM SPECIAL FEATURES MEASUREMENT.   Many EEPROMs are often supplied with special features to overcome the inherent technology shortcomings such as long write cycles and inadvertent write cycles. The 64K EEPROMs tested offered three techniques to reduce the effective write cycle time, and a noise protection feature on the Write Enable input which ensured that glitches did not initiate a write cycle. The mass erase/program allows the user to erase or program the entire device in one write cycle. After a mass erase cycle, all memory locations contain "FF" (floating gates are stripped of electrons and are positively charged), whereas after the mass program cycle, all memory locations contain "00" (floating gates trap electrons and result in net negative charge). The mass program write cycle differs from the byte write or page write cycle in that a cell is not erased before it is programmed. This is desirable because the test results indicate that it may be necessary to perform several sequential mass program cycles to ensure that all bits are cleared, especially at low temperatures and low V_{CC}.

The 64K EEPROMs tested were designed with a V_{CC} sense circuit which did not allow a write cycle to be initiated for V_{CC} typically less than 3 V (no minimum value specified). The V_{CC} sense circuitry protects against inadvertent write cycles during the device power-up and power-down. The test results showed that the V_{CC} sense voltage decreased as the temperature increased, and it was generally above 3 V.

In the EEPROM tested, the DATA polling was a special feature which allowed the system software to monitor the status of a write cycle. During a read cycle, the most significant bit of the last byte written indicates when the write operation is complete. The completion of internal write cycle is detected when I/O7 reflects the true data. The DATA polling allows the user to take advantage of the typical byte write time or page write time without the expense of added hardware or extra outputs. Two tests were run to verify the operation of DATA polling and to measure the following two parameters: (1) a write recovery test to measure the delay required between completion of the DATA polling indicated by reading true data at I/O7 and the next write cycle, and (2) a test to measure the delay from the rising edge of the Write Enable signal to the start of the DATA polling indicated by reading complement data at I/O7. The use of DATA polling in test programs can reduce the test time by 50% since functional testing consumes most of the time.

An EEPROM must be able to distinguish between a valid write pulse and a glitch on the write enable pin which can cause an inadvertent write cycle to occur. The 64K EEPROM tested handled this problem by ignoring any pulse that was less than 20 ns wide. This is a noise protection feature which specifies that a write cycle will not be initiated if the pulse width of the write enable signal is less than 20 ns. The test was performed using the byte write, which is faster than page write, and writing to one location was sufficient since the Write Enable signal latches the address data and I/O data into the internal registers. The internal write cycle copies the data in the register into the memory array at the location specified by the address latch. Therefore, the minimum pulse width is not a function of the address. The test results showed that the minimum pulse width increased as the temperature increased, but none of the pulse widths was less than 20 ns.

A write cycle for the EEPROMs actually consists of an automatic erase cycle before the desired data are written. For example, if "0F" (00001111) is written at the address location 0, every bit at location 0 is automatically erased (floating gates are stripped of electrons) before "0F" is written. A write cycle time is defined as the time from the falling edge of Write Enable for the first valid write cycle to the falling edge of Write Enable for the second valid write cycle. The DATA polling was used to measure the write cycle time. The test results indicated that the write cycle times can be significantly reduced if DATA polling is implemented in the design and component test program.

4.7 IDDQ FAULT MODELING AND TESTING

The conventional approach to testing digital logic circuits has been to use single stuck-at fault (SSF) modeling, where each defect is modeled as a single gate input or output node held to a logic 0 or a logic 1. The RAM modeling techniques discussed in Section 4.2 included the stuck-at and stuck-open faults, bridging faults, coupling faults, and pattern-sensitive faults. The defect simulation experiments have shown that the majority of all local defects in MOS technology cause changes in circuit description that result in transistor stuck-ons, breaks, and bridges. Many of these circuit faults which are manifested as logical faults are detected by the SSFs, but the others that remain undetected may result in performance degradation and eventual failure. Several recent studies have shown that the failure mode for many of these faults includes elevated quiescent supply currents (IDDQ), and therefore monitoring of this IDDQ may provide another alternative for screening out the potential failures.

IDDQ for a normal functioning SRAM may be on the order of tens of nanoamperes, whereas the defects like bridging faults (or shorts) and coupling faults may cause a state-dependent elevated IDDQ which is several orders of magnitude greater than the quiescent current for a fault-free device [38]. In a testing conducted by Sandia National Laboratories on 5000 SRAMs, 433 failed functional testing [39]. When the same units were tested using an IDDQ screen, the identical 433 functional failures were identified with an additional 254 (5%) IDDQ-only failures. In general, the current testing is based upon the bridging fault model. The quiescent-current-based test generation detects a higher percentage of bridging faults than stuck-at fault testing because the exact conditions required to detect a bridging fault (i.e., the specific transistors that must be preconditioned to create a V_{DD} to ground path) can be determined directly from the model [40].

A study was performed to evaluate the effectiveness of IDDQ testing in memories by selecting Philips 8K × 8 SRAM, 1.2 μm technology parts as the test vehicle [41]. The objective of this evaluation was to compare the relative merits of IDDQ testing versus conventional memory functional testing performed, such as the March tests. The defect model assumed that all faults are caused by the spot defects which introduce either a hard open in a track or a hard short. In order to find all possible defects, the inductive fault analysis (IFA) technique was used, which is based on the layout of the circuit [42]. This IFA method generates all possible defects using a well-defined defect model and grouping the defects. All defects in one group have a similar electrical circuit, and each group had a different electrical circuit than the other group. By using the IFA method, a total of 33 distinct groups were found, which were then analyzed with respect to their electrical and functional behavior.

The analysis showed that 86% of the defects will show an increase of IDDQ level of at least 50 μA. A write cycle can force most of the modes to a specific voltage which will cause the IDDQ effects when shorted nodes are forced to different voltages. Therefore, most of the faults will only show elevated IDDQ during a write operation. A development batch of memories was tested that had been previously subjected to wafer level testing with E-sort program consisting of the March 13N test pattern for functional testing. The final test program consisted of the following:

- Continuity check
- Functional tests consisting of a set of 13N March patterns in the X and Y array directions (X_{fast} and Y_{fast}), and a k set of 6N March patterns X_{fast} and Y_{fast}.
- Parametric measurement unit (PMU) tests consisting of five current tests: standby, static read and static write, dynamic read and dynamic write.
- IDDQ tests using current monitor which can measure the IDDQ at speeds up to 200 kHz with the detector threshold set for 20 μA. A simple IDDQ test consists of writing and reading both a "0" and a "1" to each cell in the memory matrix.

A total of 7072 RAMs were tested, out of which 371 devices (5.3%) failed continuity testing and were excluded from further testing. The remaining 6701 devices were tested, out of which 2814 devices (39.8%) passed all tests and 3887 devices (54.9%) failed the functional and IDDQ testing. Table 4-7 summarizes the details of functional and IDDQ test failures [41].

The test results showed that March 6N functional test patterns produced the same results as 13N patterns in over 99.9% of the tested devices. It is clear that the functional tests were able to detect 97.7% (53 + 6.3 + 2.1 + 36.3) of the defective RAMs, whereas the IDDQ tests were able to detect 40.7% (0.1 + 2.1 + 2.2 + 36.3) of faulty RAMs. All IDDQ faults in the memory matrix will result in a functional error that will be detected by the March tests. However, there were IDDQ faults present in the peripheral circuits that did not show up as functional errors. A sequence of tests consisting of functional and IDDQ tests is capable of detecting 100% faulty devices.

IDDQ testing appears to be a good technique for detecting bridging failures. It can also be helpful in detecting faults that may cause a transistor gate to "float," i.e., the polysilicon path is broken and the electrical conductivity to gate is lost. The SAF model for logic devices such as memories does not model the class of faults due to openings in the transistor gate con-

TABLE 4-7. Summary of Functional and IDDQ Test Failures for 8K × 8
SRAM. (From [41], with Permission of IEEE.)

Test Failed	Number of Devices	Percentage
Functional only	2060	53%
Current only	0	0%
I_{DDQ} only	5	0.1%
Functional and current	245	6.3%
Functional and I_{DDQ}	81	2.1%
Current and I_{DDQ}	87	2.2%
Functional and current and I_{DDQ}	1409	36.3%
Total	3887	100%

ducting path. Figure 4-11(a) shows the layout of a transistor circuit with a large open (break) in the polysilicon path causing the gate to "float," and (b) the equivalent circuit model with coupling capacitances [43]. The break effect is modeled by the following two capacitances: (a) C_{pb} (poly–bulk) which depends on the polysilicon track from the gate to break, and (b) C_{mp} (metal–poly) which depends on the overlapping metal–polysilicon tracks. The capacitance values are calculated from the circuit layout measurements and associated technology parameters. The charge-oriented model [44] was used with the SPICE electrical simulator.

The test results for "large" opens which caused the gate to "float" showed that, although the widely used SAF model can be acceptable for failures that bring the faulty transistor to operate in the subthreshold region, it is inadequate for a wide range of realistic situations. The capacitive coupling induces enough voltage at the "floating" gate of a

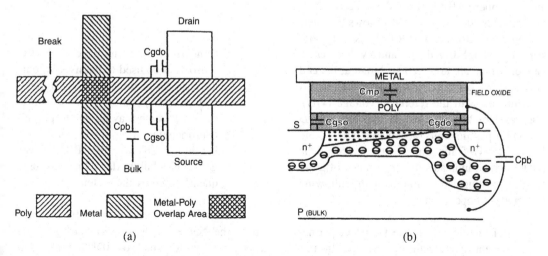

Figure 4-11. (a) Transistor circuit layout showing a break in the polysilicon path causing a gate to "float." (b) Equivalent circuit model with coupling capacitances. (From [43], with permission of IEEE.)

defective transistor to force operation above the threshold, which creates a conductive path for the quiescent current (IDDQ). The defect is characterized in two clearly identified regions represented by the metal–polysilicon and polysilicon–bulk capacitances. In one of the regions, the transistor operates at subthreshold mode, and testing for the defect is restricted to logic testing using two test vectors. In the remaining region, the logic testing becomes inefficient, and IDDQ testing can be performed with just one test vector.

There are several methods commonly used for performing IDDQ measurements, such as precision measurement unit (PMU) testing and built-in current (BIC) monitoring. Figure 4-12(a) shows the basic PMU used in conjunction with commercial automated test equipment (ATE) [45]. In normal test operation, Q_1 is turned on, connecting the device under test (DUT) V_{SS} pin to ground. For making an IDDQ measurement, Q_1 is turned off, which directs the current through the PMU. The capacitor C_1 is used to protect the device V_{SS} pin from the voltage spike from PMU while switching ranges. This voltage spike can cause logic upset at the DUT. The major disadvantage of this approach is the slow measurement speed.

Another technique currently under investigation is built-in current (BIC) monitoring, and a number of BIC sensors have been developed and tested. Figure 4-12(b) shows the basic structure of a BIC sensor [46]. It basically consists of a voltage drop device and a voltage comparator, which are arranged such that at the end of each clock cycle, the virtual ground (*VGND*) is compared with the reference voltage (*Vref*). The value of *Vref* is selected such that *VGND* < *Vref* for passing units (acceptable IDDQ), and *VGND* > *Vref* for units with unacceptable IDDQ, which indicates the occurrence of a defect. BIC testing can be applied in the following two modes of operation:

1. Die selection mode to detect permanent manufacturing defects. In this mode, the test vectors can be generated off-chip or on-chip.

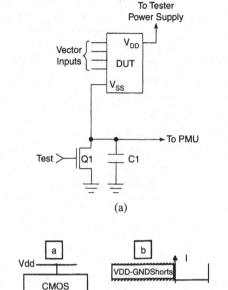

(a)

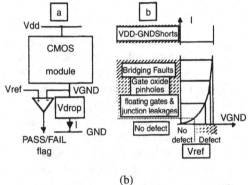

(b)

Figure 4-12. IDDQ measurement techniques. (a) PMU test method. (b) BIC sensor. (From [46] with permission of IEEE.)

2. Concurrent mode in which the BIC sensor can be used to passively monitor currents in the circuit during normal system operation. When a fault is detected which may be due to permanent manufacturing defects and/or reliability problems, the BIC sensor generates a failure flag which is subsequently used by the system.

In a study performed, an SSF simulator called the Nemesis was modified to generate tests for transistor stuck-on IDDQ faults and bridging IDDQ faults [47]. The fault simulator was modified so that the transistor stuck-on faults and bridging faults were flagged as detected when they were stimulated (with no requirement for error propagation), and the portion of the test pattern generator that extracts formulas was also modified to exclude the necessity of error propagation [48]. The test results showed that since many defects cause nonlogical faults, IDDQ monitoring during the application of test patterns to a DUT can provide significantly higher defect coverage than using only conventional testing. Also, the test generation for stuck-on faults is quicker and more complete than equivalent SSFs, and a vast majority of the IDDQ bridging faults are detected by the patterns generated for stuck-on faults. Instead of explicit IDDQ test generation, IDDQ monitoring can be added to existing SSF testing for increased defect coverage.

However, many SRAM designs use p-channel pull-up transistors which are not turned off during a write cycle and may typically draw 10–15 mA current. This large current will mask out any defect-related anomalous IDDQ current measurements which are typically several orders of magnitude smaller. Therefore, the application of IDDQ testing to memories requires a careful design analysis and selection of proper IDDQ threshold limit. An issue that needs to be investigated is the effectiveness of IDDQ testing in detecting defects which give rise to the delay faults [49]. In general, the available data indicate that there is much less correlation between the IDDQ failures and timing failures than between the IDDQ failures and voltage failures.

The goal of IDDQ testing for memories is to achieve 100% physical defect coverage, and the best approach is to develop a baseline test set that provides 100% node state coverage. Such a test set, when coupled with the IDDQ current measurements, appears to provide the most comprehensive coverage of physical defects. A tool (QUIETEST) has been used for identifying and selecting small subsets of vectors within large functional test sets which can provide the same nodal coverage as the entire functional test set [50]. For two VLSI circuits tested, the QUIETEST was able to select less than 1% of the functional test vectors from the full test set for covering as many weak (leakage) faults as would be covered if the IDDQ were measured upon application of 100% of the full vector set [51].

4.8 APPLICATION-SPECIFIC MEMORY TESTING

4.8.1 General Testing Requirements

Some of the application-specific memory device technology was discussed in Chapter 2. Many of these specialty memory devices, such as video RAMs, cache tag RAMs, FIFOs, registered memories, synchronous static, and dynamic RAMs, have a wide variety of modes under which they can be operated. The major objective of these application-specific memories is to increase performance. A characteristic which complicates testing of the devices such as video RAMs and FIFOs is that many of them have more than one data port, with each data port having different data rates and varying data widths. Also, these specialty devices have a wide variety of operating modes.

The major test issues for application-specific memories are speed and accuracy, rise and fall times, minimum pulse widths, and capacitive loading [44]. The emitter-coupled logic (ECL) memories currently available can operate at speeds of 100 MHz or more. The testing of these high-speed devices requires memory test system drivers to have very fast rise and fall times on the order of 1 ns/V. The capacitive loading of the tester, which plays a significant role in the high-speed memory testing environment, should be kept to a minimum. Many specialized memory devices have complex timing needs, and may require a test system capable of providing 20 or more independent clocks. These clock signals require formatting control to switch back and forth between a logic true and a logic false at full pattern speed.

For application-specific memories that have multiports which can be operated asynchronously, keeping track of timing restrictions for each data port becomes quite a difficult task that requires appropriate software tools. An example is a pattern generation tool (PGTOOL) which offers the programmer visual help for program writing, establishing timing edges, and keeping track of asynchronous operations [52].

The greater number of operational modes available on specialized memory devices requires more cycles dedicated to setting up various modes, and timing considerations for these modes introduce additional test complexity. An example is a typical video RAM which combines a 256K × 4 DRAM along with a 512 × 4 serial access memory (SAM). The test program for this requires a pattern in which random data are provided as input into the DRAM portion and serial data are read out of the serial port. Therefore, separate cycles, each having unique timing, are required for parallel input operations, transfer operations between the DRAM and the serial ports, and serial output operations. In a video RAM, the errors can occur during the following operations: (1) writing data to the DRAM, (2) data transfer from the DRAM to the serial port, and (3) reading of data from the serial port. A software tool which provides a visual display of the failing bits can help in program debugging and device characterization.

The cache DRAM (CDRAM) technology was discussed in Section 2.3.7. The CDRAM testing is a complex task which can be broken down into several subtasks. As an example, a 256 kword × 16 b CDRAM consists of a 256 kword × 16 b DRAM core, 8 w × 16 b read data buffer (RB), 8 w × 16 b write data buffer (WB), and 1 kw × 16 b SRAM [57]. A testing of this CDRAM includes a DRAM functional test, SRAM functional test, data transfer between the DRAM to RB or WB, and data transfer between the SRAM to RB or WB. One characteristic CDRAM function is the "concurrent operation of the DRAM and SRAM" to improve the system performance. This is enabled by separate address pin and control pin assignments for the DRAM and the SRAM. In general, the testing of a CDRAM includes the following subtasks: (1) DRAM core and data buffer test, (2) SRAM test, (3) high-speed operation test, and (4) concurrent operation test.

Application-specific memories often contain many different stages of latches and gates in their inputs and outputs. These latches and gates introduce propagation time delays in the data throughput which desynchronizes expected data from the actual data appearing at the device outputs. An example of this is the synchronous SRAM with a stage of latches in its output such that the data appearing at the outputs during a READ cycle actually correspond to the address latched by the synchronous SRAM in its previous cycle. A delay of one cycle or more requires tester hardware capable of being programmed to an arbitrary number of delay cycles corresponding to the DUT requirements. This variable expect data pipeline architecture is being provided in many new memory tester systems hardware.

The new memory test systems are equipped with dual-pattern generators which can provide the capability for interleaving the cycles and operating at twice the speed of single-pattern generators. The dual-pattern generators are better at creating independent expect and drive data. For cycle interleaving, the two pattern generators must be synchronized with each other in order to permit easy detection and reproduction of errors. This synchronization issue is quite important in devices such as video RAMs in which one port is controlled by one pattern generator performing all the input operations, while the second pattern generator controlling the output port is responsible for error monitoring and detection.

As an example, the high-bandwidth DRAMs (such as the RDRAM™) pose new challenges in testing and characterization for a production environment that has memory testers in the 60 MHz range capability. In this case, special hardware and software are required to overcome the speed limitations. Also, when the clock rates are in the 100 MHz range, the signal noise becomes a major concern, and it gets worse as the number of devices tested in parallel increases. In an RDRAM, the interface is connected to the Rambus channel, a 500 MHz, 9 b wide, bidirectional, multiplexed, and address data bus which is optimized for block data transfers [56]. To enable high-bandwidth data exchange, the Rambus channel consists of a set of microstrip transmission lines. The RDRAM package leads are very short to reduce stray inductance, and the length of the bus lines is limited to reduce the bus timing delays. The phase-locked loops (PLLs) are used to compensate for the delays of I/O circuits, allowing the RDRAM to sample inputs and drive outputs at very high timing accuracy, which is necessary when the data are valid for fewer than 2 ns.

The high-speed interface on the RDRAM leads to special test considerations to verify not only the logic, but also more stringent timing tolerances. This requires special fixturing for the tester run at high data rates (>500 MHz), the ability to measure setup and hold times of the inputs, and output timings with high accuracy. The PLLs used in the RDRAM to achieve high precision timing are sensitive to on-chip noise. This may result in jitter which affects I/O timings and maximum operating frequency. These special test issues and considerations for a high-speed interface RDRAM are addressed by an integrated approach that uses a high-speed tester in conjunction with some novel test equipment, on-chip test features, and custom designed fixturing. The test fixture, along with a microprober, allows characterization of the critical paths, PLL jitter, and current sources on the device under test (DUT).

Therefore, the application-specific memories with their complex timing requirements and multiple setup modes require a suitable mix of sophisticated test hardware and a design for testability (DFT) and built-in self test (BIST) approach.

4.8.2 Double-Buffered Memory (DBM) Testing

Computer memory design and applications have mainly concentrated on the use of conventional SRAMs and single-buffered memories (SBMs). The SBM differs from a SRAM since an SBM cell has separate read and write ports, although both cells contain only one

memory location. Therefore, the memories containing either single-buffered or SRAM memory cells are called SBMs. In digital signal processing and telecommunication applications, some examples of application-specific memory configurations are the double-buffered memories (DBMs) and pointer-addressed memories (PAMs). Figure 4-13 shows a comparison between the memory transistor layouts for a SRAM, an SBM cell, and a DBM cell [53]. A DBM cell consists of a master component and a slave component in a master–slave configuration with a buffering between the two portions. This conditional buffering is controlled by a global transfer signal which transfers the data from the master to the slave.

In a DBM, the writing of data into the master can occur completely independent of reading data from the slave. Before the beginning of a new time frame, the data in every master are transferred to the corresponding slave. A DBM consists of three basic structures: (1) the memory array, (2) read/write logic, and (3) address generation or decoder logic. The test algorithms discussed for the SRAMs and single-buffered memories are inadequate for testing of the DBMs because of the following reasons: (1) they do not account for the buffering effect, and (2) they do not take into consideration the data access or the address generation method such as combinational decoders or/and address pointers.

In the conventional testing approach, faults in the decoder and read/write logic are mapped onto functionally equivalent faults in the memory array [6]. However, in the new approach, a different model is adopted for each of the three blocks, and by considering the memory array and the address generation logic separately, the effect of the data access method on the resulting memory test algorithm can be investigated. An efficient test algorithm for the DBMs and PAMs was developed based on a realistic fault model. This involved using inductive fault analysis (IFA), which is a systematic method for detecting all faults that have a great likelihood of occurrence. The IFA of the DBM array assumes that only one defect per

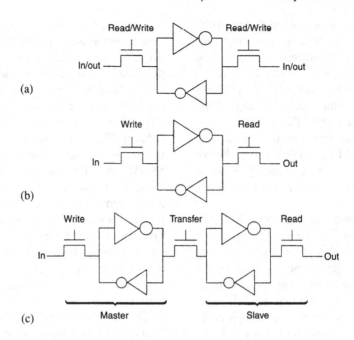

Figure 4-13. Memory cell transistor layouts. (a) SRAM. (b) Single-buffered memory (SBM). (c) Double-buffered memory (DBM). (From [53], with permission of IEEE.)

memory cell is present, and only the spot defects or fabrication defects are considered. Because of the regular structure of the DBM array, analysis used only one memory cell and its neighboring cells. For Read/Write logic, the fault model included the following three fault classes:

- One or more output lines of the sense amplifier logic or the write driver logic stuck-at-0 or stuck-at-1.
- One or more data input/output lines having shorts or hard coupling faults between them.

- One or more output cells of the sense amplifier logic or write driver logic are stuck open.

The address generation methods were divided into two separate categories of DBMs and PAMs in order to describe the effect of different data access methods on the test algorithm. The DBM test array algorithm developed had a length of 53 NW (number of words) and assumed that a word consists of 1 b only. The generalization of this algorithm to word-oriented DBMs involves reading or writing all cells in a word simultaneously using a process called data background (see Section 4.3).

REFERENCES

[1] A. J. van de Goor, *Testing Semiconductor Memories, Theory and Practice*. New York: Wiley, 1991.

[2] V. D. Agarwal and S. C. Seth, *Tutorial: Test Generation for VLSI Chips*. IEEE Computer Society Press.

[3] M. A. Breuer and A. D. Friedman, *Diagnosis and Reliable Design of Digital Systems*. Rockville, MD: Computer Science Press, 1976.

[4] M. S. Abadir and H. K. Reghbati, "Functional testing of semiconductor random access memories," *Computing Surveys,* vol. 15, pp. 175–198,

Sept. 1983; reprint: V. D. Agrawal and S. C. Seth, *Tutorial: Test Generation for VLSI Chips.* IEEE Computer Society Press.

[5] J. Knaizuk and C. Hartmann, "An optimal algorithm for testing stuck-at faults in random access memories," *IEEE Trans. Comput.*, vol. C-26, pp. 1141–1144, Nov. 1977.

[6] R. T. Nair *et al.*, "Efficient algorithms for testing semiconductor random access memories," *IEEE Trans. Comput.*, vol. C-27, pp. 572–576, June 1978.

[7] J. M. Acken and S. D. Millman, "Accurate modeling and simulation of bridging faults," in *Proc. IEEE Customs Integrated Circuits Conf.*, 1991, pp. 17.4.1–17.4.4.

[8] J. Rearick and J. H. Patel, "Fast and accurate CMOS bridging fault simulation," in *Proc. IEEE ITC 1993*, paper 3.1, pp. 54–62.

[9] R. E. Bryant, "COSMOS: A compiled simulator for MOS circuits," in *Proc. ACM/IEEE Design Automation Conf.*, June 1987, pp. 9–16.

[10] D. G. Saab *et al.*, "CHAMP: A concurrent hierarchical and multilevel program for simulation of VLSI circuits," in *IEEE Int. Conf. Proc., Comput.-Aided Design*, Nov. 1988, pp. 246–249.

[11] T. Lee *et al.*, "A switch-level matrix approach to transistor-level fault simulation," in *IEEE Int. Conf. Proc., Comput.-Aided Design*, Nov. 1991. pp. 554–557.

[12] W. Nagel, "SPICE2: A computer program to simulate semiconductor circuits," Ph.D. dissertation, Univ. California, Berkeley, 1975.

[13] A. J. van de Goor *et al.*, "Locating bridging faults in memory arrays," in *Proc. IEEE ITC 1991*, paper 25.3, pp. 685–694.

[14] A. J. van de Goor *et al.*, "Functional memory array testing," in *Proc. IEEE COMPEURO '90 Conf.*, 1990, pp. 408–415.

[15] S. D. Millman *et al.*, "Diagnosing CMOS bridging faults with stuck-at fault dictionaries," in *Proc. IEEE ITC 1990*, paper 37.3, pp. 860–870.

[16] C. Papachristou *et al.*, "An improved method for detecting functional faults in semiconductor random access memories," *IEEE Trans. Comput.*, vol. C-34, pp. 110–116, Feb. 1985.

[17] A. J. van de Goor, "Using March test to test SRAMs," *IEEE Design and Test of Comput.*, pp. 8–14, Mar. 1993.

[18] J. Savir *et al.*, "Testing for coupled cells in random access memories," in *Proc. IEEE ITC 1991*, paper 20.2, pp. 439–451.

[19] B. F. Cockburn *et al.*, "Near optimal tests for classes of write-triggered coupling faults in RAMs," in *Proc. IEEE ITC 1991*, poster paper 3.

[20] J. P. Hayes, "Detection of pattern sensitive faults in random access memories," *IEEE Trans. Comput.*, vol. C-24, pp. 150–157, Feb. 1975.

[21] M. Franklin and K. K. Saluja, "An algorithm to test RAMs for physical neighborhood sensitive faults," in *Proc. IEEE ITC 1991*, paper 25.2, pp. 675–684.

[22] D. S. Suk and S. M. Reddy, "Test procedures for a class of pattern-sensitive faults in semiconductor random access memories," *IEEE Trans. Comput.*, vol. C-29, pp. 419–429, June 1980.

[23] J. P. Hayes, "Testing memories for single-cell pattern sensitive faults," *IEEE Trans. Comput.*, vol. C-29, pp. 249–254, Mar. 1980.

[24] J. Sosnowski, "In-system transparent autodiagnostic of RAMs," in *Proc. IEEE ITC 1993*, paper 38.3, pp. 835–844.

[25] J. M. Dumas *et al.*, "Investigation of interelectrode metallic paths affecting the operation of IC MESFETs," in *Proc. 1987 GaAs IC Symp.*, pp. 15–18.

[26] S. Mohan and P. Mazumder, "Fault modeling and testing of GaAs static random access memories," in *Proc. IEEE ITC 1991*, paper 25.1, pp. 665–674.

[27] R. Dekker *et al.*, "Fault modeling and test algorithm development for SRAMs," in *Proc. IEEE ITC 1988*, paper 25.1.

[28] J. P. Shen *et al.*, "Inductive fault analysis of MOS integrated circuits," *IEEE Design and Test of Comput.*, vol. 2, no. 6, pp. 13–26, 1985.

[29] M. Sachdev and M. Verstraelen, "Development of a fault model and test algorithms for embedded DRAMs," in *Proc. IEEE ITC 1993*, paper 38.1, pp. 815–824.

[30] Local Preliminary Data Sheet Specifications for 1 Mb RH SRAM.

[31] Honeywell, Inc., Application Note, "Static RAM testing."

[32] R. David *et al.*, "Random pattern testing versus deterministic testing of RAMs," *IEEE Trans. Comput.*, vol. 38, pp. 637–650, May 1989.

[33] M. Issacson, *Markov Chains: Theory and Applications*. New York: Wiley, 1976.

[34] M. Inoue *et al.*, "A new testing acceleration chip for low-cost memory tests," *IEEE Design and Test of Comput.*, pp. 15–19, Mar. 1993.

[35] G. Jacob, "Rexamining parallelism in memory ATE," *EE Evaluation Eng.*, Dec. 1991.

[36] Y. Marooka *et al.*, "An address maskable parallel testing for ultra high density DRAMs," in *Proc. IEEE ITC 1991*, paper 21.3, pp. 556–563.

[37] "Electrical characterization of VLSI RAMs and PROMs," Rome Air Development Center Tech. Rep., RADC-TR-86-180, Oct. 1986.

[38] C. F. Hawkins and J. M. Soden, "Electrical characteristics and testing considerations for gate oxide shorts in CMOS ICs," in *Proc. IEEE ITC 1985*, paper 15.5, pp. 544–555.

[39] L. Horning *et al.*, "Measurement of quiescent power supply current for CMOS ICs in production testing," in *Proc. IEEE ITC 1987*, pp. 300–309.

[40] T. Storey *et al.*, "CMOS bridging fault detection," in *Proc. IEEE ITC 1990*, paper 37.1, pp. 842–851.

[41] R. Meershoek *et al.*, "Functional and I_{DDQ} testing on a static RAM," in *Proc. IEEE ITC 1990*, paper 41.1, pp. 929–937.

[42] J. Shen *et al.*, "Inductive fault analysis of MOS integrated circuits," *IEEE Design and Test of Comput.*, pp. 13–26, 1985.

[43] J. A. Segura *et al.*, "Current vs logic testing of gate oxide short, floating gate and bridging failures in CMOS," in *Proc. IEEE ITC 1991*, paper 19.3, pp. 510–519.

[44] K. Koo *et al.*, "A testing methodology for new generation specialty memory devices," in *Proc. IEEE ITC 1989*, paper 20.3, pp. 452–460.

[45] D. Romanchik, "I_{DDQ} testing makes a comeback," *Test and Measurement World*, Oct. 1993.

[46] M. Patyra and W. Maly, "Circuit design for built-in current testing," in *Proc. IEEE Custom Integrated Circuits Conf.*, 1991, pp. 13.4.1–13.4.5.

[47] T. Larrabee, "Efficient generation of test patterns using Boolean difference," in *Proc. IEEE ITC 1989*, pp. 795–801.

[48] F. J. Ferguson and T. Larrabee, "Test pattern generation for current testable faults in static CMOS circuits," in *Proc. IEEE VLSI Test Symp.*, 1991, paper 14.3, pp. 297–302.

[49] P. C. Maxwell *et al.*, "The effectiveness of I_{DDQ}, functional and scan tests: How many fault coverages do we need?," in *Proc. IEEE ITC 1992*, paper 6.3, pp. 168–177.

[50] W. Mao *et al.*, "QUIETEST: A quiescent current testing methodology for detecting leakage faults," in *Proc. 1990 IEEE Int. Conf. Comput.-Aided Design*.

[51] R. K. Gulati *et al.*, "Detection of undetectable faults using I_{DDQ} testing," in *Proc. IEEE ITC 1992*, paper 36.2, pp. 770–777.

[52] Y. Kawabata *et al.*, "PGTOOL: An automatic interactive program generation tool for testing new generation memory devices," in *Proc. IEEE ITC 1988*.

[53] J. Van Sas *et al.*, "Test algorithms for double-buffered random access and pointer-addressed memories," *IEEE Design and Test of Comput.*, pp. 34–44, June 1993.

[54] S. Naik *et al.*, "Failure analysis of high-density CMOS SRAMs," *IEEE Design and Test of Comput.*, pp. 13–22, June 1993.

[55] H. Walker and S. Director, "VLASIC: A catastrophic fault yield simulator for integrated circuits," *IEEE Trans. Comput.-Aided Design*, vol. CAD-5, pp. 541–556, Oct. 1986.

[56] J. A. Gasbarro *et al.*, "Techniques for characterizing DRAMs with a 500 MHz interface," in *Proc. IEEE ITC 1994*, pp. 516–525.

[57] Y. Konishi *et al.*, "Testing 256k word × 16 bit cache DRAM (CDRAM)," in *Proc. IEEE ITC 1994*, p. 360.

5

Memory Design for Testability and Fault Tolerance

5.1 GENERAL DESIGN FOR TESTABILITY TECHNIQUES

In general, the testability of an IC device such as a memory is defined as the property of a circuit that makes it easy to test. The testability is a function of variables such as the circuit complexity and design methodology. The general guidelines for a logic design based on practical experience that makes a circuit testable can be listed as follows:

- It should not contain any logic redundancy. A line in a circuit is redundant if the stuck value on the line is undetectable (i.e., the function performed by the circuit is unchanged by the presence of a stuck-at fault). Since the stuck fault is undetectable, no test can ever be found for it, and its presence can cause either a detectable fault to become undetectable, or another undetectable fault to become detectable.
- Avoid the use of asynchronous logic. The absence of clock synchronization in the asynchronous sequential circuits creates problems for conventional test generation and for random pattern built-in testing.
- The clocks should be isolated from the logic circuitry. If the clock signals are

combined with the logic signals and gated to the data input of a clocked latch, racing conditions will occur. It is important from the testability point of view since isolating the clock signals enables the tester to control the device under test (DUT) at the speed of the tester rather than the speed of the DUT.

- The circuits designed should be easily initializable.
- Partition large circuits into smaller subcircuits to reduce the test generation costs.
- The circuit should be easily diagnosable.

These rules of thumb are also applicable to built-in self-test (BIST) schemes used to facilitate testing of the memories. Two key concepts in the design for testability (DFT) of VLSI and ULSI devices are controllability and observability. The circuit design is analyzed to identify the critical nodes which are not observable (or not controllable), and then test points can be added to increase the testability. Then, techniques such as scan path and level-sensitive scan design (LSSD) can be used to initialize certain latches to reduce the controllability problems associated with sequential machines.

The two basic approaches to DFT commonly used in the semiconductor manufacturing industry to help solve VLSI test problems are: (1) ad hoc design, and (2) a structured design approach. Ad hoc techniques which are similar to good design practices are not directed toward solving general testability problems associated with sequential circuits such as flip-flops, latches, and memories. In comparison, the structured design approach addresses the testability problem with a special design methodology that will result in efficient fault simulation and reduced cost of test generation [1].

In the ad hoc design approach, testability analysis tools such as SCOAP [2] are often used to guide the designer in making testability enhancements which are unique to the circuit under consideration. The algorithms for testability analysis require logic gate level circuit models. The major objective of modeling memories for DFT analysis is not to analyze the memory itself, but to measure the testability of the logic which is peripheral to the memory. This is especially significant for large processor designs that contain embedded memories. A simple memory model can be constructed under the assumption that only one word is selected during any memory operation, and the memory peripheral logic can be tested through the easiest to select word in the memory. The easiest to select word has a select line with the lowest of the controllability values [3]. This allows a memory model which has a size proportional to the sum of the number of words in the memory and the number of bits per word. The model analysis includes word addressing and data bit analysis.

5.1.1 Ad Hoc Design Techniques

5.1.1.1 Logic Partitioning. It has been observed that the ATE time for test generation and fault simulation time (T) of a VLSI device is roughly proportional to the number of logic gates (N) to the power "r," i.e.,

$$T = K N^r$$

where K is a proportionality constant and $r = 2$–3 [4].

This relationship does not take into account the drop-off in ATE test generation capability due to the sequential circuit complexity. Therefore, as the number of gates (N) has risen exponentially with the device complexity, a significant amount of the DFT effort has been directed at partitioning the logic to facilitate a "divide and conquer" approach. Another strong motive for the partitioning is to implement a possible exhaustive testing strategy. Some of the commonly used techniques in partitioning are the use of modular block level design, and the use of degating logic for isolation. The capability of isolating high-speed internal clocks eliminates the need for synchronizing a high-performance tester to the DUT, and thus permits switching off the clock for stuck-at fault testing of the remaining circuit logic.

In testing of the RAMs or ROMs embedded within the combinational or sequential logic of a memory module or a memory card, the faults may exist within the input logic block driving the memory chip, the memory chip itself, or the output logic block being driven by the memory device. Therefore, depending upon the fault location, different test strategies are required. For example, let us consider the input logic block driving the memory chip. If the fault appears at the memory data input port, its propagation to the memory outputs requires only a write operation to an arbitrary address, followed by a read operation at the same address. However, when a fault occurs in the memory address input logic, it will result in the selection of an actual but incorrect address. A ROM address fault is detectable, provided the word stored at the valid address location is different from the word at the faulty location addressed. The detection of faults in the output logic circuitry driven by a memory chip requires specific binary values on the memory data output ports. The faults in the memory chip itself and various functional test algorithms were discussed in Chapter 4.

5.1.1.2 Input/Output Test Points. Another ad hoc DFT technique that has been used to increase the controllability and observability of a circuit such as an embedded RAM has been to add some test points for the inputs and observation points for the outputs. This can also help reduce the number of tests required for the fault detection and in improving the diagnostic resolution of the tests. The test points can be added by using I/O pins, access at the designated points through temporary probes, and a "bed of nails" test fixture. The major constraint with using the test points is the large demand on the I/O pins. To reduce the number of I/O pins, multiplexer circuits can be used. However, since they allow observation of one point at a time, the test time also increases. To reduce the pin count even further, a counter can be used to drive the address lines of a multiplexer. Another method to reduce the I/O overhead with a multiplexer and a demultiplexer is to use a shift register.

5.1.2 Structured Design Techniques

Structured design techniques are based upon the concept of providing a uniform design for the latches to enhance logic controllability. If these latches can be controlled to any specific value and if they can be observed easily, then the test generation problem is reduced to that for a combination logic network. Thus, a control signal can switch the memory elements from their normal mode of operation to an alternate test mode which is more controllable and observable. The structured design philosophy has the following major advantages: (1) it allows easy implementation of the BIST techniques, and (2) it results in synchronous logic which has a well-defined and reduced timing dependency. The most popular structured DFT technique used for external testing is referred to as the "scan design" since it employs a scan register. "Scan" refers to the ability to shift into or out of any state of the network. There are several major structured design approaches which differ primarily in the scan cell designs and their circuit level implementation. Some of these approaches are discussed in the following sections.

5.1.2.1 Level-Sensitive Scan Design.
Level-sensitive scan design (LSSD) is a full serial, integrated scan architecture originally developed by the IBM Corporation [5], [6]. A circuit is said to be level-sensitive if and only if the steady-state response to any of the allowed input changes is independent of the transistor and wire delay in that circuit [7]. A key element in the design is a "shift register latch" (SRL). The LSSD approach requires a clocked structure on all memory elements which are converted into the SRLs, making it possible to shift stored logic values into and out of the latches by way of the scan path. This technique enhances both the controllability and observability by allowing examination of the internal states during a scan mode. A disadvantage is the serialization of the test which takes longer to run. This circuit is quite tolerant to most anomalies in the ac characterization of the clock, requiring only that it remain high (sample mode) at least long enough to stabilize the feedback loop before being returned to the low (hold mode) state.

Figure 5-1 shows the LSSD (a) SRL symbolic representation, and (b) the NAND level implementation. The system data input (D) and clock (C) form the elements for normal mode memory function. The scan data input (I), shift A (clock A), shift B (clock B), and latch L2 are the elements of additional circuitry used for the shift register functions. These SRLs are "daisy-chained" together into one or more shift register strings. The test access to these SRL strings is through a scan-in primary input and scan-out primary output. Separate shift clocks are used to scan serial data into or out of the string. Each cycle of the shift A and shift B clocks moves the data one step down the shift register string. The A and B clocks must be nonoverlapping to ensure correct operation of the string. The implementation of an LSSD requires four additional package pins, of which two are used as the scan-in and scan-out ports of the shift register string, and two others for the nonoverlapping shifting clocks.

The two LSSD structures commonly utilized are the single-latch design and the double-latch design. Figure 5-1(c) shows the block diagram of a double-latch design in which both L1 and L2 are used as the system latches, and the system output is taken from L2 [1]. This

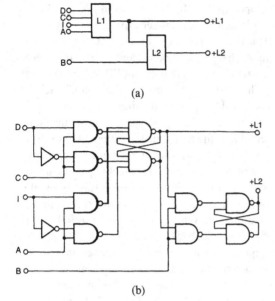

(a)

(b)

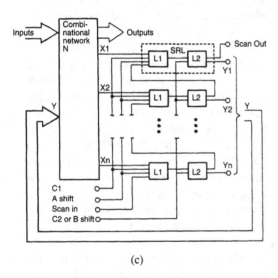

(c)

Figure 5-1. LSSD (a) SRL symbolic representation.
(b) SRL NAND level implementation.
(c) Block diagram of double-latch design. (From [1], with permission of IEEE.)

design requires the shift *B* clock to the *L2* latch also to be a system clock. The rules for an LSSD implementation can be summarized as follows [7], [8]:

• All internal storage should be in the clocked dc latches so that the data in the latches would not change when the clocks are off. This requires synchronous dc design.

• The latches should be controlled by two or more nonoverlapping clocks.

• The circuit must have a set of clock primary inputs from which the clock inputs to all the SRLs are controlled either through: (1) single-clock distribution trees, or (2) logic which is gated by the SRLs or/and nonclock primary inputs.

• Clock primary inputs cannot feed the data input to latches, either directly or through the combinational logic.

In considering the cost and performance tradeoffs for an LSSD implementation, the negative impacts are as follows:

• An SRL is more complex than the simple latches since it requires four extra gates for a scan design. Therefore, if *K* is the ratio of the combinational logic gates to latches in a nonscan design, then for an LSSD approach, the gate overhead is $(4/K+3) \times 100$ percent.

• In an LSSD, four additional I/O pins are required at the package level for control of shift registers.

• For a given set of test patterns, the test time is increased because of the need to shift the pattern serially into the scan path.

• A slower clock rate may be required because of the extra delay in the scan path flip-flops or latches that may cause performance degradation.

• For a memory array embedded within the combinational logic, the testing must access the memory through the embedded logic, which is always difficult. Therefore, to minimize this difficulty, a special memory embedding rule requires that under a particular set of primary input (or SRL values), a one-to-one correspondence must exist between each memory input and a primary input (or SRL), and between each memory output and a primary output (or SRL) [5].

5.1.2.2 Scan Path.

The scan path DFT, introduced by Nippon Electric Company (NEC) in 1975, is similar to an LSSD in its implementation of storage by using the shift register stages for scanning test data in and test response out [9]. A significant difference between the two techniques is that the scan path is edge-sensitive, whereas an LSSD is a level-sensitive operation. Figure 5-2 shows the memory elements used in the scan path approach, referred to as the raceless D-type flip-flops (Latch 1 and Latch 2). Clock 1 is the only clock in the system operation for this D-type flip-flop. When clock 1 is at 0 logic level, the system data input can be loaded into the Latch 1, providing sufficient time to latch up the data before turning off. As it turns off, it will make Latch 2 sensitive to the data output of Latch 1. As long as Clock 1 is at logic level 1, the data can be latched up into Latch 2, and reliable operation occurs.

The scan path configuration can also be used at the logic card level by selecting the modules in a serial scan path such that there is only one scan path.

5.1.2.3 Scan/Set Logic.

This DFT technique, originally proposed by Sperry-Univac, is a partial scan-structured design similar to an LSSD and the scan path, but not as rigorous in its design constraints [10]. The basic concept of this approach is to use shift registers (as in LSSD or scan path), but these shift registers are not in the system data path and are independent of all system latches. An auxiliary Scan/Set register is connected to the original combinational or sequential network for the purpose of either scanning the selected outputs or setting values in the storage registers, or both. Since the original circuit remains unchanged, the addition of control (or observation) points increases circuit complexity and the silicon area overhead.

Figure 5-3 shows an example of the Scan/Set logic, also referred to as the bit-serial logic, with a 64 b serial shift register. It allows capture of 64 b of observation data in parallel from the main circuit with a single clock. Once the 64 b are loaded, the following operations can be performed: (1) scan operation consisting of a shifting process for serial unload from the scan-out pin, and (2) the serially loaded bits in its scan-in port for parallel application to the test point or storage control points in the main circuit (during the set operation). In general, the Scan/Set logic is integrated onto the sequential

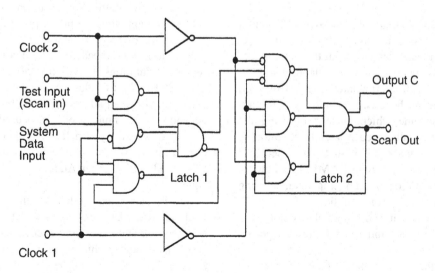

Figure 5-2. Scan path memory elements (raceless D-type flip-flops).

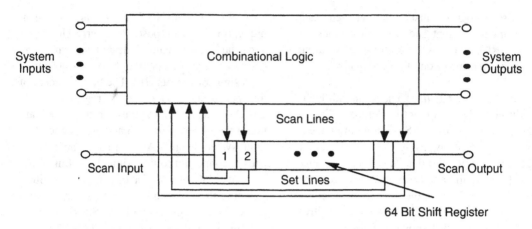

Figure 5-3. Example of Scan/Set logic (bit serial shift) with 64 b serial shift register.

system logic chip. However, there are applications in which the bit-serial Scan/Set logic is off-chip, and the bit-serial Scan/Set logic only sampled outputs or drove the inputs to facilitate in-circuit testing.

The Scan/Set logic implementation has more flexibility than an LSSD or the scan path since the set function does not require that all system latches be set, or that the scan function scan all system latches. Another advantage of the Scan/Set technique is that the scan function can occur during system operation such that a "snapshot" can be obtained and off-loaded without any system performance degradation.

5.1.2.4 Random Access Scan.   Random access scan is similar to an LSSD and the scan path DFT techniques, and with the same objectives of having complete controllability and observability of all internal latches [11]. However, it differs from those two techniques in the addressing scheme which, instead of using the shift registers, allows each latch to be uniquely selected in a manner similar to that used in a random access memory (RAM). The random access scan approach can be implemented using two basic latch configurations. Figure 5-4 shows the schematics of (a) a polarity-hold type addressable latch, and (b) a set/reset type addressable latch.

In the polarity-hold type addressable latch, a single latch has an extra scan data-in port (SDI) attached. The data are clocked into the latch by the SCK clock which can only affect this latch if both X and Y addresses are logic 1. Also, when X and Y addresses are logic 1 level, the scan data output (SDO) points can be observed. In the normal (system) mode, the system clock CK loads data into this latch.

The set/reset type addressable latch does not have a scan clock to load data into the system latch. It is first cleared by the CL line which is connected to other latches that are also set/reset type addressable latches. This sets the Q output to logic 0. The latches that are required to be set at the logic 1 level for a particular test are preset by addressing each one of these latches and applying the preset pulse label PR. This sets the Q output logic to 1.

The I/O overhead for the random access scan configurations is five pins shown as SDI, SDO, SCK, -CL, PR, plus two additional X-scan address and Y-scan address pins, both of which depend upon the number of latches used. Figure 5-4(c) shows the system level implementation of a random access technique by using the Clocks, SDI, SDK, Scan Address, and SDO test signals.

The SDI scans in data for the addressed latch, and the SDO scans out the data. The pin overhead for addresses can be reduced by using a serially loadable shift register as an address counter. The test hardware consists of an X and a Y decoder and the addressable storage elements

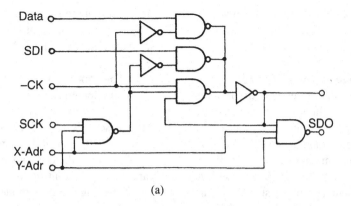

(a)

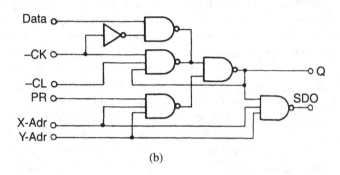

(b)

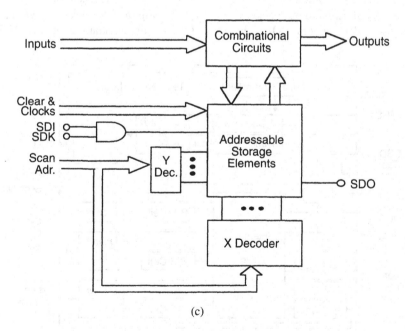

(c)

Figure 5-4. Random access scan. (a) Polarity-hold type addressable latch.
(b) Set/reset type addressable latch. (c) System level implementation.

(the random access scan cells). The random access scan technique allows observability and controllability of all system latches.

5.1.2.5 Boundary Scan Testing.   Boundary scan testing (BST) is essentially an extension of scan DFT methodology to the testing of complex, densely packed printed circuit boards (PCBs) or multichip modules (MCMs) containing the ASICs and memory chips with limited accessibility. IEEE/ANSI standard 1149.1 has been published which defines a standard boundary scan architecture. The standard defines a four-pin test access port (TAP) with the following I/O signal pins: test data-in (TDI), test data-out (TDO), test clock (TCK), and test mode select (TMS). A fifth pin is optional for an active-low test reset (TRST). The test is performed by shifting the test vectors and instructions serially into a device through the TDI pin for a serial readout through the TDO. The BST conforming device contains a state machine called the TAP controller. The standard defines a data sequence that must reset the controller, and also requires that the TAP controller power-up in its reset state. A conforming device also contains several registers dedicated to BST, such as

an instruction register (IR), and at least two test-data registers (TDRs), of which one is a boundary scan register (BSR) and the other is a bypass register.

This boundary scan architecture (BSA) can be implemented in complex memory PCBs and MCMs, provided all the ICs contain the additional scan logic. All the devices with scan logic have their TDI/TDO pins daisy-chained together, while the test signals are propagated in parallel to the control lines TCK, TMS, and TRST. During the BST mode, the initial step is to verify the scan chain integrity. The following step is the use of shifting patterns from one device and capturing them at the next to detect the shorts and opens. An automatic program generator can usually produce these tests. However, currently, the use of BST is rather limited since very few off-the-shelf ICs are available with built-in additional scan logic.

An application of BST is the structure that permits comprehensive testing of LSSD components with high signal input/output (I/O) pin counts while using relatively inexpensive reduced-pin count automated test equipment (ATE). Figure 5-5 shows a block diagram of an LSSD implementation of the IEEE 1149.1

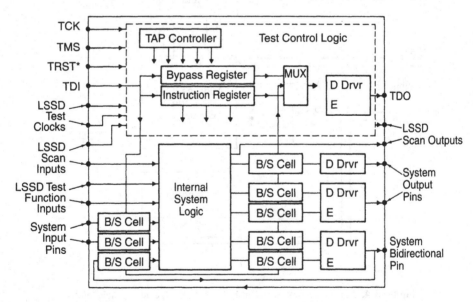

Figure 5-5. A block diagram of LSSD implementation of IEEE 1149.1 boundary scan structure. (From [12], with permission of IEEE.)

boundary scan structure [12]. This structure includes the test control logic and boundary scan cells, in addition to internal system logic. The TAP consists of five dedicated test-function pins providing access to the test control logic, which includes a TAP controller, an instruction register, and a bypass register. The TAP includes the five signals discussed earlier. All standard test operations are accessed via the TAP and synchronized either to the rising or the falling edge of TCK. The LSSD guidelines require level-sensitive operations using two (or more) nonoverlapping clocks. This structure provides race-free operation during a component test.

In normal operation, a device is unaffected by the presence of boundary scan circuits. The scan circuits have a silicon overhead which is a function of the chip's complexity. In simple chips, the penalty can be as high as 20 percent, whereas in more complex ICs, the penalty is often less than 3 percent [12]. In addition, the BSA implementation requires a four–five pin overhead which, in devices with a small number of pins, represents a significant percentage increase. In packages with hundreds of pins, the relative percentage of BSA pin increase is negligible.

5.2 RAM BUILT-IN SELF-TEST (BIST)

The built-in self-test is the merger of a built-in test (BIT) and self-test concepts. It is the capability of a chip (or a circuit) to test itself. The BIST techniques can be classified into two categories as follows:

1. On-line BIST, which includes the concurrent and nonconcurrent techniques, and occurs during normal operation of a chip, i.e., the device under test is not placed into a test mode where normal functional operation is blocked out. The concurrent on-line BIST is performed simultaneously during normal functional operation, whereas the non-concurrent on-line BIST is performed with the chip in the idle state.

2. Off-line BIST, which includes functional and structural approaches, refers to a test mode with the chip not in its operational mode. This can be performed using on-chip test pattern generators (TPGs) and output response analyzers (ORAs). Therefore, the off-line BIST does not detect errors in real time, as is possible with the on-line concurrent BIST technique. Functional off-line BIST refers to the test execution based on a functional description of the DUT, and uses a high-level functional fault model. Structural off-line BIST refers to test performance based on the device (or circuit) structure with a structural fault model.

The BIST is usually performed by applying certain test patterns, measuring the output response using linear feedback shift registers (LFSRs), and compressing it. Assuming that the circuit being tested is an n-input, m-output combinational circuit, the various test pattern generation (TPG) methodologies for a BIST implementation can be summarized as follows:

1. *Exhaustive Testing:* This approach requires the testing of an n-input combinational circuit where all 2^n inputs are applied. A binary counter can be used as the TPG. The exhaustive testing guarantees that all of the detectable faults that do not produce a sequential behavior will be detected. It is not generally applicable to the sequential circuits such as memories.

2. *Pseudorandom Testing:* Pseudorandom testing requires test patterns that have many characteristics of the random patterns, but are generated deterministically, and hence are repeatable. This technique, applicable to RAM testing, was briefly discussed in Chapter 4 (see Section 4.4). The pseudorandom test patterns without replacement can be generated by an autonomous LFSR. The test sequence length is

selected to achieve an acceptable level of fault coverage. Test pattern generation can be either: (1) weighted, for which the distribution of 0s and 1s produced on the output lines is not necessarily uniform, or (2) adaptive test generation, which also uses the weighted TPG, but the weights can be modified based on the results of fault simulation.

3. *Pseudoexhaustive Testing:* This TPG approach achieves many benefits of exhaustive testing, but requires many fewer test patterns. It depends on various forms of circuit segmentation (partitioning) and exhaustive testing of each segment. This segmentation can be logical or physical. In logical segmentation also, there are several techniques such as: (1) cone segmentation where an m-output circuit is logically segmented into m cones, each consisting of all logic associated with one output [13], and (2) sensitized-path segmentation [14], [15]. The TPG techniques based on logical segmentation methods are as follows:

- Syndrome-Drive Counter: This scheme uses a syndrome-drive counter (SDC) to generate the test patterns [16]. The SDC can either be a binary counter or an LFSR and contains only p-storage cells.
- Constant-Weight Counter: A constant-weight code counter (CWC), also known as an "N-out-of M code," consists of the set of codewords of M binary bits where each codeword has exactly N number of 1s. The complexity of a CWC goes up for large values of N and M.
- LFSR/SR Combination: This is another scheme to produce a pseudoexhaustive test by using a combination of an LFSR and a shift register (SR). The design implementation cost of this approach is less than that for a constant-weight counter, but the resulting circuit generates more test vectors [17].
- LFSR/XOR: A pseudoexhaustive test can also be generated using a combination of an LFSR and a XOR (linear) network based on the use of linear sums [18] or linear codes [19].

There are some other variations of pseudoexhaustive techniques based on logical segmentation such as the condensed LFSRs and cyclic LFSR designs. The pseudoexhaustive testing based on physical segmentation is implemented by partitioning into subcircuits [20]. It can also be achieved by inserting the bypass storage cells in various signal lines. A bypass storage cell in the normal mode acts as a wire, but in the test mode, it can be part of an LFSR circuit such as the ones used in the boundary scan designs.

Most of the BIST techniques for VLSI devices are based on pseudorandom pattern generation test methods, which are usually effective for combinational circuits, but not very efficient at testing high-density random access memories. Chapter 4 discussed various deterministic functional algorithms such as the ATS, March tests for detection of the SAFs, coupling, and pattern-sensitive faults (PSFs). The RAM BIST strategies depend upon the type of device being tested: standalone memories, or embedded memories. The main use of a self-testable, standalone memory is for system maintenance and support. The disadvantages of a self-test for the standalone memories is the need for additional pins, which is a performance loss because of additional hardware in both the data path and the control path, and a need for the clock signal for RAM during the test mode. The BIST approach is quite helpful for the embedded memories that have limited controllability and observability.

The BIST implementation strategies for the memories have to take into consideration the following requirements:

- Silicon overhead area should be used as minimally as possible. Therefore, the cells and interconnections are designed to optimize the use of silicon area.

- Since the memory access time is a main concern, the additional logic circuitry used for the BIST should not interfere with normal RAM operation.

- On-chip BIST circuitry should be targeted for the highest fault coverage.

- A BIST circuit should be able to operate in a normal as well as in the test mode. This may require a few additional pins. However, the number of additional pins should be kept to a minimum and the multiplexing of signals should be carefully weighed against performance loss.

- The BIST hardware should do all the tests that can be done without an external tester. In addition, the BIST hardware itself must be testable by making it either self-testable or by using an external tester.

In general, two BIST approaches have been proposed for the RAMs that utilize either random logic or microcoded ROM. Figure 5-6 shows block diagrams of BIST schemes using (a) random logic, and (b) microcoded ROM [21]. For an $m \times m$ cell memory array, the BIST implementation consists of the following three major components:

1. *Address Generation Logic (AGL):* This can be implemented as $\log_2 m$ bit counters for row and column address accessing.

2. *Test Generation Logic (TGL):* This should be capable of generating the required test patterns and Read/Write signals, which also control the AGL, as well as compaction and detection logic.

3. *Compaction and Detection Logic (CDL):* This is required for monitoring the output, compressing the readout sequences by using appropriate com-

paction functions, and comparing the compressed sequence with reference values at appropriate intervals.

In a microcoded ROM implementation of the BIST, a ROM chip replaces the TGL, while the AGL and CDL are still required as in the random logic BIST approach. It has been found that for RAMs larger than 64K, the use of microcoded ROM as the TGL hardware is a practical alternative. The major advantages associated with microcoded ROM over the use of random logic are a shorter design cycle, the ability to implement alternative test algorithms with minimal changes, and ease in testability of the microcode [21]. The following sections discuss examples of the various RAM BIST implementation strategies.

5.2.1 BIST Using Algorithmic Test Sequence

A RAM BIST design strategy was based upon the algorithmic test sequence which requires only four memory accesses per address to detect all the single and multiple SAFs in the memory address register (or MAR), the memory data register (MDR), the address decoder, and the memory array. Figure 5-7 shows a block diagram of a typical RAM structure with an N-bit MAR, an address decoder, an MDR, and 2^N words of memory storage [22]. Let the 2^N addresses be numbered from 0 to 2^{N-1}. Then, to apply the ATS, each address is divided modulo-3, and the result (0, 1, or 2) is used to assign each address to either G_0, G_1, or G_2. Then the ATS procedure consists of write and read operations as shown in Table 5-1 [22]. The element symbols used are as follows:

$$W_r = \text{write}$$
$$R = \text{read}$$
$$W_0 = \text{word consisting of all 0s}$$
$$W_1 = \text{word consisting of all 1s.}$$

The basic steps of an ATS procedure are: (1) Word W_0 is written to all the addresses in G_1 and G_2, (2) W_1 is written to all the addresses in group G_0, and (3) group G_1 is read, and the

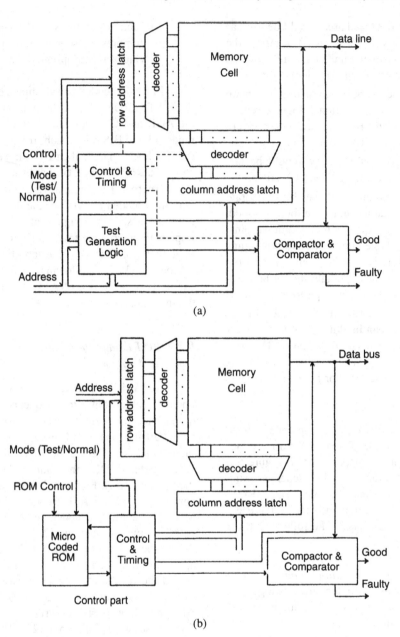

(a)

(b)

Figure 5-6. BIST scheme block diagrams using (a) random logic, and
(b) microcoded ROM. (From [21], with permission of IEEE.)

expected word is W_0. This sequence continues repeating, as shown in Table 5-1. The ATS can be modified to detect open-bit rail faults.

During a test mode, the MAR in Figure 5-7 is transformed into a count-by-3 circuit which generates the word addresses for G_0, G_1, and G_2 in real time. The count-by-3 circuit can be built by using the shift register latches (SRLs). In addition, the following hardware is needed for a BIST implementation of the ATS test algorithm:

TABLE 5-1. The ATS Algorithm Test Sequence.
(From [22], with Permission of IEEE.)

State	G_0	G_1	G_2
1		WrW_0	WrW_0
2	WrW_1		
3		RW_0	
4		WrW_1	
5			RW_0
6	RW_1	RW_1	
7	WrW_0, RW_0		
8			WrW_1, RW_1

(1) a sequence generator or test control circuit which is operational only during the test mode, (2) a data-compare circuit on the RAM output data bus, and (3) a means of setting the MDR to all 0s or all 1s word.

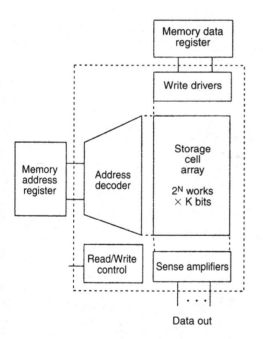

Figure 5-7. A typical RAM structure. (From [22], with permission of IEEE.)

5.2.2 BIST Using 13N March Algorithm

This RAM BIST approach used the 13N March algorithm plus a data-retention test that was discussed in Chapter 4. For a word-oriented SRAM, the words of data instead of the single bits are written to or read from the memory. To detect coupling faults between the cells at the same address, several data words called data backgrounds and their inversions are applied to the SRAM. The register cells are modified to make them scannable so that during the scan test, the SRAM is isolated from its environment. Figure 5-8(a) shows the modified register cell and its logic truth table [23]. A control signal $C1$ controls the register cell and enables the operation mode which is either normal or scan. A second control signal $C2$ enables the self-test mode for data, address, and control generation.

Figure 5-8(b) shows the global architecture for the BIST RAM which requires several logical blocks as follows [23]:

- An address generator (address counter) to generate the test sequence $0 \cdots (N - 1)$.
- A wait counter that can be combined with the address counter to count down the wait time for the data-retention test.
- A data generator to produce the data backgrounds and inverted data backgrounds.
- A data receptor that stores or compares the data received from the SRAM during a Read operation.
- A self-test controller to implement the test algorithm, including the control of the RAM and other blocks of self-test logic.

Figure 5-8(c) shows a state diagram of self-test controllers [23].

The self-test begins when all of the flip-flops are initialized and all register cells are in the self-test mode during which the self-controller has full control. Only four states are needed to generate the 13N test algorithm, and the remaining states are used for generating the data-retention test. The conditional jumps

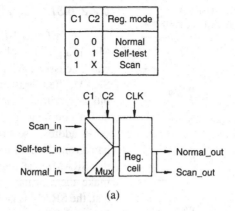

C1	C2	Reg. mode
0	0	Normal
0	1	Self-test
1	X	Scan

(a)

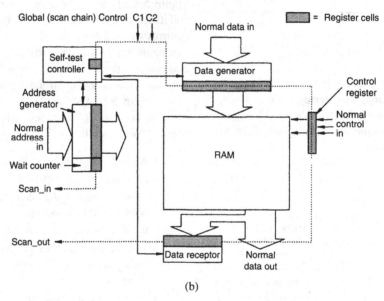

(b)

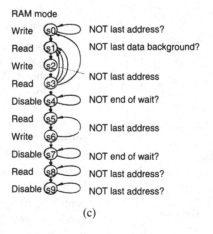

(c)

Figure 5-8. RAM BIST implementation. (a) Modified register cells. (b) Global architecture. (c) State diagram of the self-test controller. (From [23], with permission of IEEE.)

define the sequence of read and write actions, the additional order, and the data. Data receptor accepts the data generated by the RAM during read portions of the test algorithm. These data are compressed on the chip by polynomial division using the parallel signature analyzer (PSA) to produce a "signature." A hold mode prevents the loss of final signature in the PSA. After the completion of the self-test, the final signature is compared off-chip with a signature calculated by a dedicated software tool, and serially shifted out of the PSA. This BIST plan implemented for a 1K × 8 synchronous SRAM in a 1.5 μm double-metal process showed about 4% silicon overhead without the data-retention test.

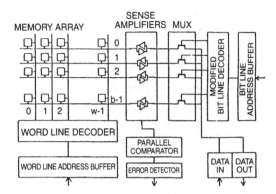

Figure 5-9. A BIST RAM organization with modified decoder and comparator. (From [24], with permission of IEEE.)

5.2.3 BIST for Pattern-Sensitive Faults

A BIST implementation has been proposed for high-density RAMs which can detect both the static and dynamic pattern-sensitive faults (PSFs) over a nine-cell neighborhood area [24]. This strategy for parallel testing, which can be considered as fault-syndrome-based, is several orders of magnitude faster than the conventional sequential algorithm for detecting the PSFs. This is achieved by a slight modification of the decoder circuitry such that in the test mode, the decoder makes multiple selection of the bit lines which allows the writing of some data simultaneously at many storage locations on the same word line. In the read mode, a multibit comparator concurrently detects errors, if any.

Figure 5-9 shows the RAM organization with the modified decoder and comparator [24]. The parallel comparator senses the output of sense amplifiers connected to bit lines which are selected in parallel. It determines whether the contents of all the multiple accessed cells are either all 0s or all 1s. If, due to some fault, a write operation on a cell fails or the contents of some cells change, the comparator triggers the error latch to indicate the occurrence of a fault. The proposed algorithm is generated by making an interlace scanning both along the bit lines and the word lines. The bit lines are scanned concurrently, and the word lines are scanned sequentially.

For a 256K RAM organized into four identical square subarrays, this technique re-

quired only 19 flip-flops and an overall silicon overhead of 0.4%. The algorithm test time for an n-bit RAM to detect both the static and dynamic PSFs over a nine-cell neighborhood for every memory cell was found to be $195\sqrt{n}$.

Another class of RAM PSFs is the row- and column-sensitive faults in which the contents of the cell are assumed to be sensitive to the contents of rows and columns containing the cell [25]. Considering a memory cell square organization of $n \times n$, each cell can have two possible values (0 or 1). The row and column weights of cells can vary between 0 and $n - 1$. Therefore, each cell can have $2n^2$ states. Figure 5-10(a) shows the row/column neighborhood of an arbitrary cell (i,j) [25]. The fault model is based upon the following definition of a row/column-weight-sensitive fault: A memory cell is said to be faulty if it can never be in one or more number of $2n^2$ possible states defined by the triplet (v, r, c) for a cell.

It is assumed that the read and write operations are fault-free, and only the presence of a single fault in the memory array is considered. It is desirable to find a test sequence of length $O(N^{3/2})$. This is done by adopting a memory test strategy of "divide and conquer" by recursive partitioning. The testing algorithm consisted of two procedures, $A1$ and $A2$. For each base cell, the two parameters p and q were calculated based on its row and column address. The cells with even row and column addresses were

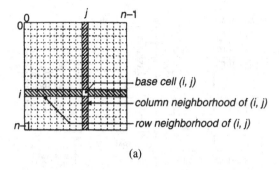

(a)

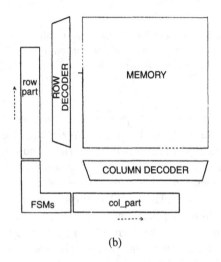

(b)

Figure 5-10. (a) The row/column neighborhood of an arbitrary cell (i, j). (b) RAM BIST chip floorplan. (From [25], with permission of IEEE.)

1. *Control Logic:* This was realized by the finite-state machines (FSMs), used as the main control elements to control the timing and start/stop testing. Six FSMs were hierarchically designed in such a way that they were closely related to the control flow in the test algorithm.

2. *Address Generation Logic:* For the 4 Mb RAM, address generation consisted of six 11-b (log n-bit) counters, three for the row address and three for the column address. The counters were designed using latches, and could be cleared or preset by the FSMs.

Figure 5-10(b) shows the chip floorplan for the BIST logic with: (1) FSM_1 to FSM_6, (2) row_part being the layout for the row address generation logic, and (3) col_part as the layout for the column address generation logic [25]. These three regions are arranged in an L-shape to minimize global routing. The BIST unit was designed using dynamic latches and two nonoverlapping clocks. Assuming a 4 Mb CMOS DRAM fabricated using a 0.9 μm triple-poly, single-metal process, the area occupied by the BIST logic will be only about 0.8% of the chip area.

This BIST logic needs only two additional pins, one for the clock and the other for test initiation. If the clock is available on-chip, then only one additional pin is required. This algorithm for testing cells in a sequential order to generate the cell addresses requires counters which occupy a significant portion of the BIST logic. Since the cells can be tested in any order, the algorithm can be modified to replace the counters by LFSRs which would require less chip area.

tested in any order by calling Procedure A1 with the appropriate parameters each time. The remaining cells get tested automatically. When Procedure A1 is applied to a cell, all column weights are applied while keeping the row weights fixed at each one of the required values. When Procedure A2 is applied to a cell, all row weights are applied while keeping the column weights fixed at each one of the required values. Procedure A1 is chosen for $p \geq q$; otherwise, Procedure A2 is applied [25].

During an implementation of this algorithm in a BIST for a 4 Mb memory, the BIST logic was divided into the following two blocks:

5.2.4 BIST Using Built-In Logic Block Observation (BILBO)

The BILBO technique includes test vector generation and response evaluation on the chip. It uses the scan path, LSSD, and signature-analysis concepts for self-testing of a circuit.

Signature analysis suits bus-oriented structures such as microprocessors and microcontrollers. The BILBO technique provides test generation and fault simulation for the embedded combinational logic. Figure 5-11 shows the basic operating mode of a BILBO register [26], [27]. The BILBO register consists of a set of latches, where the inverted output of each latch is connected to a data input of the following latch through the combination of a NOR and an XOR gate. A multiplexer (MUX) permits either a serial input to the latches in one operational mode ($B_1 = 0$), or a connection of the XOR feedback ($B_1 = 1$) in another mode of operation. The BILBO can be operated in four different modes as follows: (1) latch mode, (2) shift mode, (3) multiple-input shift register (MISR), and (4) reset mode. The actual mode of operation is determined by the value of control inputs B_1 and B_2.

The BILBO approach allows functionally converting the system of flip-flops to an LFSR. It saves both the off-line test generation time and the test application time. The BILBO circuit implementation requires two pins: one for setting up the chip in test mode, and the other for reading go/no-go information from the signature register. The BILBO technique silicon

overhead is about 5–15%, and there is some loss in performance from the delays in Exclusive-OR gates. Concurrent built-in logic block observers (CBILBO) are used both for test application and response analysis. A study performed showed that structures using the CBILBO or BILBO have the shortest test lengths as compared to other BIST structures that use scan registers [28], and BIST using checking experiment methods [29].

5.3 EMBEDDED MEMORY DFT AND BIST TECHNIQUES

In scan path testing of the embedded RAMs, long test vectors are required to shift in the test stimulii and shift out test results. A solution for reducing shift operations has been the implementation of JTAG IEEE 1149.1 Boundary Scan Architecture, which can provide a short scan path for each embedded RAM. BIST methods using LFSRs have been developed that help reduce the number of test vectors. A design methodology for testing the embedded RAMs based on the scan path method has been proposed that achieves significant reduction of the test vectors by using a flag-scan register (FLSR)

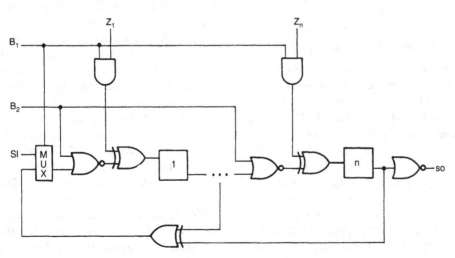

Figure 5-11. The basic operating mode of a BILBO register. (From [26], [27] with permission of IEEE.)

[30]. This FLSR is based upon the comparators, and not on the multiple-input signature registers (MISR), which helps avoid fault masking problems. The test algorithms of $O(n)$ complexity such as the conventional March test, unidirectional, or bidirectional random March test are considered practical for embedded DRAM testing. These can detect the SAFs, address decoder faults, and adjacent cell coupling faults.

Figure 5-12 shows the schematic of an FLSR structure based upon an LSSD [30]. These FLSRs are connected serially according to the bit configuration of a RAM. The FLSR consists of two-input master/slave latches and a comparator which is composed of an XOR gate and two NAND gates. These are the various I/O signals for FLSR:

$$SI = \text{shift-in signal}$$
$$SO = \text{shift-out signal}$$
$$SCK1M, SCK2M = \text{two phase clocks for a shift operation}$$
$$DIX = \text{write-data signal from the random logic}$$
$$DI = \text{write-data signal to the RAM}$$
$$DOX = \text{read-data signal to the random logic}$$
$$DO = \text{read-data signal from the RAM}$$
$$TM = \text{test mode enable signal for the comparator}$$
$$CMP = \text{comparison signal}$$
$$STBM = \text{strobe logic signal}$$

A RAM test circuit has been designed to take advantage of exhaustive random addressing and the FLSR. The test circuit used unidirectional random March testing (RMT). As the FLSR is based upon an LSSD, other scan registers in the RAM test circuit are also based on the LSSD. For the multiple embedded RAMs, setting all test circuits into a RAM test mode, the same address can be applied to all the RAMs. Therefore, a concurrent test can be executed for multiple RAMs. This DFT approach was used to experimentally develop the single-port and dual-port RAM generators. The same design methodology can also be extended to the BIST.

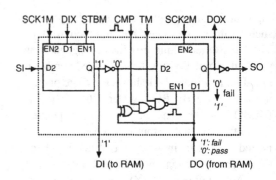

Figure 5-12. A schematic structure of flag scan register (FLSR). (From [30], with permission of IEEE.)

A self-test controller with an exhaustive random pattern generator may be embedded on a chip, e.g., a JTAG tap controller as a self-test controller. However, in this case, the BIST may not be a reasonable solution because of a higher silicon area overhead and testing problems in the self-test controller.

A deterministic RAM BIST scheme implements a test procedure that aims at a specific type of faults. An alternative design of the self-testable RAMs uses random or pseudorandom test patterns. A RAM BIST scheme based on the random patterns has the potential to provide high-quality testing of the embedded RAMs of up to several megabits size within a test time of a few seconds. It has been shown that a random BIST technique of circular self-test path (CSTP) can be efficiently used to test the embedded RAMs. The CSTP was originally developed for circuits consisting of arbitrarily connected combinational logic blocks and registers.

Figure 5-13(a) shows a typical embedded RAM module with n-address lines ($N = 2^n$ words) and m data lines [31]. This configuration has existing (or extra test-dedicated) registers such as the address input register (ADDR), data input register (DIN), data output register (DOUT), and the read/write control input (flip-flop R/W'). These registers are controlled by a common clock signal. The CSTP technique implementation requires these registers to be modified and included into a circular path, as shown

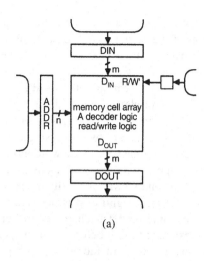

(a)

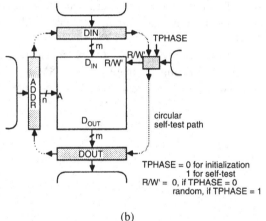

TPHASE = 0 for initialization
1 for self-test
R/W' = 0, if TPHASE = 0
random, if TPHASE = 1

(b)

Figure 5-13. (a) A typical embedded RAM module. (b) Self-testable embedded RAM using the CSTP technique. (From [31], with permission of IEEE.)

in Figure 5-13(b) [31]. It is not necessary to have the modified RAM registers as the consecutive segments of the circular paths. However, the modified registers must be provided with a means of initialization such as a global reset or serial scanning. During the self-test mode, these registers are used not only for testing the memory, but to also serve as the test pattern generators or test response compactors for the adjacent RAM chip modules.

This technique provides high coverage of various classes of RAM faults such as the SAFs,

dynamic and static two-coupling faults, unidirectional faults, and other unknown faults which represent defects identified by inductive fault analysis (or observed in the manufactured RAMs), but not represented by the fault model developed for this study [32]. To detect the data-retention faults, the test procedure should include some idle (wait) periods during which the control of memory is kept intact before it is read for comparison with the expected value. A comparison was made between the effectiveness of the CSTP-based RAMs and the deterministic RAM BIST schemes. The results of this study can be summarized as follows [31]:

- The CSTP-based RAM scheme has the potential to provide very high fault coverage for virtually any size of embedded RAM modules. This self-test procedure can provide very high yield with a defect level of only few ppm if supplemented by some other form of electrical testing, e.g., parametric and current testing.

- The test time for the CSTP RAMs for typical size embedded RAMs is in the range of a few hundred milliseconds. This test time delay is an acceptable compromise since it reduces the need for complex automatic test equipment.

- The CSTP-based schemes are less sensitive to RAM implementation details as compared to a RAM BIST approach, and are also less expensive because: (1) circular path registers are simpler than the modified registers such as the data generators, address counters, and signature analyzers used in a deterministic approach, and (2) the use of different BIST techniques on a single-chip embedded RAM, resulting from a dedicated deterministic scheme, require a more complex on-chip test controller.

A significant problem with the CSTP schemes is the long simulation cycle time for fault-free signature of the chip design and associated test development time requirements.

BIST implementation for a chip that has multiple embedded RAMs of varying sizes and port configurations is a complex task. One solution is a serial interfacing technique which allows the BIST circuit to control a single bit of RAM (or group of RAMs) input data path [33]. This means that while the test algorithms are executed, only one bit of the output data path is available to the BIST circuits for observation. The other bits are controlled and observed indirectly through the serial data path using the memory itself and a set of multiplexers.

Figure 5-14 shows (a) the block diagram of a SRAM (rectangular box enclosed by dotted outline) with serial-data path connections in the BIST mode, and (b) the general architecture of a BIST circuit using the serial shift technique [33]. The address bus is not latched, and is applied directly to X (row) and Y (column) decoders. When the Read strobe is high, a word is read and transferred to the memory output.

When the strobe goes back to low, the transparent latches at the output of the sense amplifier retain the data until the next operation. When the write strobe is high, a word is written into the memory at the location determined by the address. The multiplexers along the I/O data path are used to implement the shift operation which shifts the full word by one position. The memory cell stores the LSB of the word written with the data provided by the BIST circuit (serial input). The bit that was originally in the cell that stored the most significant is now only at the corresponding output latch. The BIST circuit must examine this bit before the read operation at any location, including the one just read. This output, called the serial output of the memory, is only one output available to the BIST circuit.

In this study, four algorithms considered for the BIST implementation were: SMarch, SMarchdec, SGalpat, and Swalk. These are

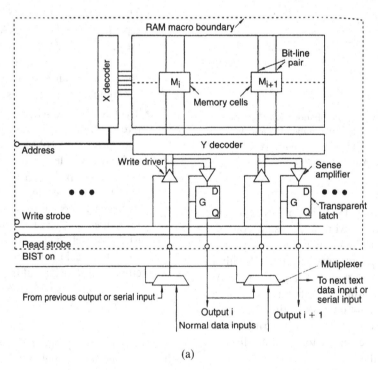

(a)

Figure 5-14. (a) Block diagram of a SRAM showing serial-data path connections in the BIST mode.

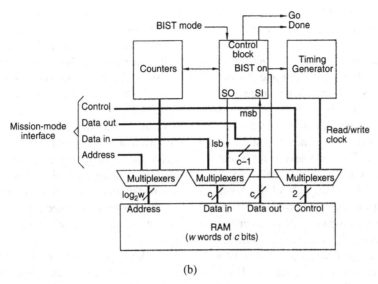

(b)

Figure 5-14 (cont.). (b) The general architecture of a BIST circuit using the serial shift technique. (From [33], with permission of IEEE.)

modifications of the well-known algorithms discussed in Chapter 4. The choice of algorithm depends upon factors such as the fault coverage required, chip area available, test or simulation time restrictions, and number of memories to be tested. The fault model used to evaluate the test algorithms was based upon the SAF model for the memory array, the address decoder, and the read/write circuitry, and the static and dynamic coupling faults in the memory cells.

In addition, the fault model also takes into consideration the following:

- Two types of sequential faults, with $S1$ being the stuck-open faults observed in the static address decoders, and $S2$ the faults from the inaccessible cells such as the open access transistors or stuck word lines.
- Two additional faults for the dual-port memories, with $D1$ consisting of shorts between the bit lines belonging to different ports in the same column, and faults $D2$ consisting of shorts between the word lines between different ports in the same row.

The general architecture of a BIST circuit implementation using the serial interface technique, as shown in Figure 5-14(b) has the following characteristics:

- Data path multiplexers to set up the serial shift mode.
- Address and control lines multiplexers to switch between the mission-mode and test-mode access to the RAM.
- A set of counters whose length depends on the RAM's dimensions and the test algorithm being used.
- An FSM that emulates the actual test algorithm, controls the counters, and generates the serial data stream of expected data for comparison with the RAM output.
- A memory control timing generator.

All BIST circuitry is synchronous and scan-testable. A useful feature of the serial BIST scheme is the ease with which several memories can share the test circuit. The most commonly used schemes for sharing the BIST circuit are daisy-chaining and multiplexing. For dual-port memories, the tests are performed twice, once for each port.

In system applications which require periodic component testing, it is desirable that the test algorithms must not destroy the contents of the RAMs. This means that before testing each RAM, the chip contents should be saved somewhere, and later at the completion of testing, restored in the RAM. This is a severe constraint, especially if the RAM is an embedded block of some VLSI circuit. A technique has been developed which allows transformation of any RAM test algorithm to a transparent BIST algorithm [34]. This transparent BIST does not reduce the fault coverage of initial algorithms, and uses slightly greater overhead than the standard BIST. As an example, for a 32 kbyte RAM, the BIST area overhead using conventional techniques is 1%, while with the transparent BIST, it is 1.2%.

5.4 ADVANCED BIST AND BUILT-IN SELF-REPAIR ARCHITECTURES

5.4.1 Multibit and Line Mode Tests

Chapter 4 discussed memory fault modeling and testing using various algorithms. In general, N^2 patterns can detect all the coupling and pattern-sensitive faults, but due to long test times, N size patterns are preferred for multimegabit memories. As the density of DRAMs increases, higher costs and long test times justify advanced DFT and BIST concepts to further reduce the test time. The functional test time of the memories depends on the number of cells tested simultaneously. Several kinds of parallel-testing methods have been developed, such as the multibit test (MBT) and the line mode test (LMT). In one of the parallel test method approach, on-chip test circuits are introduced, and all the memory cells connected to a word line are tested simultaneously [35].

Figure 5-15 shows the block diagram of a memory device with an on-chip test circuit that uses the word-line test mode [35]. The on-chip test circuit consists of parallel write and parallel compare circuits. The parallel write circuits are

used to write data W_1 into the memory cells connected to a word line. The parallel compare circuits are used to compare all data from the cells with their expected data, and to generate a flag F if the cell data do not coincide with their expected data. This mode allows all memory cells to be tested in cycles equal to the number of word lines. Therefore, the test pattern length is proportional to the square root of memory capacity, i.e., $N^{1/2}$. However, during the memory functional testing using the word-line mode, circuit blocks such as the address decoders and I/O circuits (shown as hatched areas in Figure 5-15) are not tested, and have to be tested by some other means.

The complete line mode test for memories has three modes to reduce the pattern length, with each mode testing different circuit blocks in the following sequence:

1. In this first test, not only the peripheral circuits, but also the cells connected to a word line are verified for correct functionality.

2. Then by using the verified word line, the on-chip test circuits are tested.

3. Finally, these circuits are used to test memory cells by the word-line test mode.

All three modes of the line mode test are executed in cycles proportional to $N^{1/2}$. This test methodology applied to a 2 Mb memory cell array resulted in test time reduction by two orders of magnitude. However, there are certain limitations to this approach. The memory cell data are restricted to certain patterns due to the parallel write operation in one cycle. Therefore, these test modes may not be able to adequately detect some pattern-sensitive faults. Another problem with this method is the possibility of lower yield due to defects in the test circuits. In some 4 Mb DRAMs, the multibit test mode is provided as a standard function.

In a new architecture for parallel testing, the use of the line mode test (LMT), which enables random pattern tests along the word line, has been implemented on a 16 Mb DRAM [36].

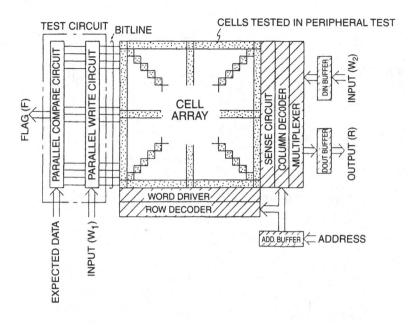

Figure 5-15. Block diagram of a memory device with on-chip test circuit that utilizes the word-line mode. (From [35], with permission of IEEE.)

In the conventional DRAM array architecture consisting of the cell blocks divided by column decoders, the data register and the comparator must be provided for every bit-line pair of each block which increases the circuit complexity and chip size. In the new architecture proposed, the LMT is realized by adopting a hierarchical I/O bus line structure consisting of main-I/O and sub-I/O bus lines, together with a multidivided bit-line structure. Figure 5-16(a) shows the memory array with a twisted bit-line structure which is effective in the reduction of inter-bit-line coupling noise [36]. The memory array is divided into several basic subarray blocks of 256 kb by alternate shared (ALS) sense amplifier bands. One of the four bit-line pairs is selectively connected to a sub-I/O bus-line pair by a sense amplifier connecting signal (SACi). The signal is decoded by the column address inputs $CA0$–$CA1$. Each pair of sub-I/O bus lines has a set of multipurpose registers (MPRs) and a comparator.

Figure 5-16(b) shows the circuit diagram of the MPR, which consists of a full CMOS latch, and functions as a buffer register in a copy write and an expected test data latch utilized for parallel comparison in the LMT [36]. The comparator is an Exclusive-OR circuit like a content-addressable memory (CAM) cell, and forms a coincidence detector with a wired-OR match line. The MPR performs several other functions, such as:

- an intermediate amplifier in normal access mode
- a high-speed data register in the fast page and static column mode
- a cache register independent of the DRAM array operation

This combination of MPR and comparator can achieve the LMT with random test patterns along a word line. Figure 5-16(c) shows the flowchart of the LMT sequence used in the 16 Mb DRAM array architecture [36]. The LMT can be used with various patterns such as the Column Stripe and Checkerboard. It can be activated by the $\overline{WE}$ and $\overline{CAS}$-before-$\overline{RAS}$ signal sequence, together with an external test signal. The 16 Mb

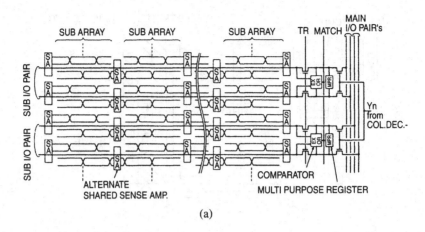

(a)

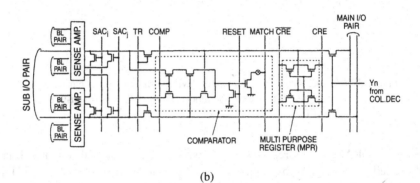

(b)

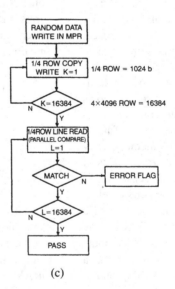

(c)

Figure 5-16. A 16 Mb DRAM array. (a) Twisted bit-line architecture. (b) Multipurpose register (MPR) and comparator. (c) Flowchart of line mode test. (From [36], with permission of IEEE.)

memory has 16 spare columns for array reconfiguration. However, the LMT mode is applicable even for the reconfigured array. The 1024 b concurrent testing achieves a test time reduction of three orders of magnitude as compared to conventional bit-to-bit testing. The LMT circuit implementation for a 16 Mb DRAM results in 0.5% excess area penalty.

5.4.2 Column Address-Maskable Parallel Test (CMT) Architecture

In the MBT technique used for 1 and 4 Mb DRAMs, test bits at each cycle are limited at 4–16 b, which reduces the test time by about one-fourth to one-sixteenth. In the LMT discussed for 16 Mb DRAMs, the test time is reduced to 1/1K, but an area penalty has to be paid for an additional comparator or register at each bit-line pair. To distinguish the failed addresses on the line, the conventional bit-by-bit test is required in the LMT. To improve fault coverage, other parallel test methods have been reported that require decoder modifications, and thereby impose a large area penalty [37], [38]. A column address-maskable parallel test (CMT) architecture has been developed which is suitable for ultrahigh-density DRAMs [39]. It provides the capability to check test pattern sensitivity and to search failed addresses quickly by masking the column address. This architecture reduces the test time to $\frac{1}{16}$K with a small area penalty.

Figure 5-17(a) shows the block diagram of a memory with CMT architecture [39]. A separate read and write data line is used, which also works as a match line during the test mode operation so that no additional circuit and routing lines are required in the memory array to perform the parallel test. The test circuit is placed in the peripheral area, which minimizes the overhead and does not degrade the memory array access in the normal read/write operation. Figure 5-17(b) shows the details of the CMT architecture [39]. The switch elements SW0 and SW1 are controlled according to the operational mode. In normal mode operation with the SW0 in the on state and SW1 in the off state, only one column decoder output (CDi) is activated. This transfers the write data to the selected bit-line

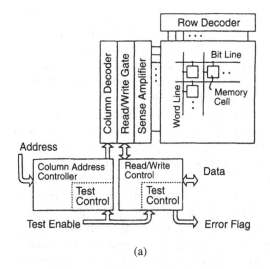

(a)

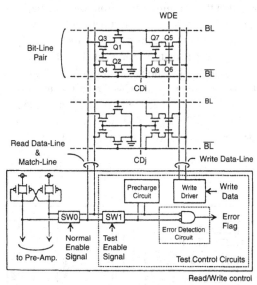

(b)

Figure 5-17. (a) Block diagram of a memory with CMT architecture. (b) The details of the CMT architecture. (From [39], with permission of IEEE.)

pair through Q5–Q8 in the write cycle, and the selected bit-line data are read out to the read data line through Q1–Q4 in the read cycle. In the test mode with the SW0 in the off state and SW1 in the on state, multiple column decoder outputs are activated. The same data are written into selected plural bit-line pairs simultaneously during the test mode write operation. In the test mode read operation, the read data lines are used as match lines, which are initially precharged at a high level. All selected bit-line pairs are compared through $Q1$–$Q4$, respectively.

An error-detection scheme is used in the test mode operation. For no error, all bit-line pairs hold the same data. When an error occurs, the error bit-line pair holds the opposite data to the other bit-line pairs, and the error flag is generated. The test pattern capability and searching of failed addresses are accomplished by the column masking technique (CMT). To apply CMT, the test bits which are tested simultaneously must hold the same data. The masking operation is performed in the column address controller. A major feature of the CMT is high test pattern sensitivity to reject marginal memory cells of the DRAMs by using various patterns for a cell disturb test.

Figure 5-18 shows the chip layout of a 64 Mb DRAM with the CMT architecture implementation [39]. The memory array is divided into 16 4 Mb blocks, each of which is further divided into 16 256 kb subarrays. A column de-

coder is located at the center of the chip. In each 4 Mb array, the R/W control blocks, which include the test control circuits, are located across the row decoder block. Column address controllers are located in the peripheral circuit area. One-half of the 16 256 kb subarrays in two 4 Mb arrays (see shadowed tubes in Figure 5-18) are activated at the same cycle, and the error detection in the CMT module is performed in each 256 kb block. The error flag output from each 256 kb block, which comes from 1024 b of data (512×2), is combined by 16 b OR to represent a total of 16 kb of data (1024×16). This CMT architecture reduces the test time to $\frac{1}{16}K$, and the additional circuit area penalty is less than 0.1% of the entire chip area.

5.4.3 BIST Scheme Using Microprogram ROM

An alternative to the multibit test (MBT) and line mode test (LMT) to reduce test time has been to implement the ROM-based BIST function on the random access memory chip itself, which enables testing many more chips on the

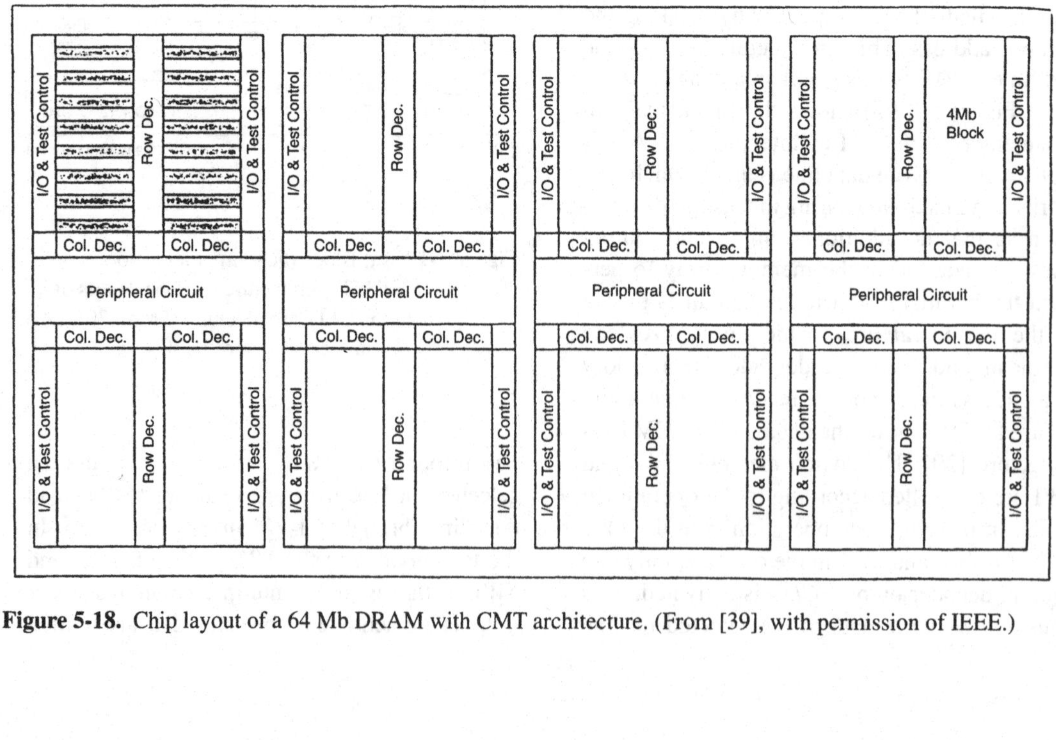

Figure 5-18. Chip layout of a 64 Mb DRAM with CMT architecture. (From [39], with permission of IEEE.)

board simultaneously. An improved micropro-gram ROM BIST scheme has been developed which has the following features [40]:

- The ROM stores test procedures for generating test patterns, writing the test patterns into memory cells, and compar-ing data read out from the memory cells with previously written data (called "ex-pected data").
- Self-test is performed by using BIST circuits controlled by the microprogram ROM.
- A wide range of test capabilities due to ROM programming flexibility.

There are several functional test algo-rithms in sizes ranging from the N to N^2 patterns for microprogram ROM BIST implementation. Many of these algorithms were discussed in Chapter 4. A representative N-size pattern such as a March test can detect all the SAFs. Figure 5-19(a) shows a block diagram of a memory with microprogram ROM BIST for the N pattern [40]. The BIST circuits consist of the following func-tional blocks: (1) microprogram ROM to store the test procedure, (2) program counter which controls the microprogram ROM, (3) address counter, (4) test data generator and comparator, and (5) clock generator which controls the oper-ational timing for the BIST circuitry.

A typical N-pattern, 9N March test con-sists of the following sequence of steps:

1. Initialize the BIST circuits.
2. Write data "D" into all of the memory cells.
3. Read the data from the address "0" cell, and compare it with "D."
4. Write the inverted data "$\overline{D}$" into the address "0" cell.

   Then steps (1)–(4) are repeated from the address "0" cell to the ad-dress "N-1" cell.
5. Read the data from the address "0" cell, and compare it with "$\overline{D}$."
6. Write "D" into the address "0" cell.

Then steps (5)–(6) are repeated from the address "0" cell to the ad-dress "N-1" cell.

7. Set the address counter to the maxi-mum value of "N-1."

8–11. Repeat steps (3)–(6) from the address "N-1" cell to the address "0" cell.

12. Test ends.

The 12-step March test procedure is di-vided into ten unit processes which are assigned 10 b microcodes, as shown in Figure 5-19(b) [40].

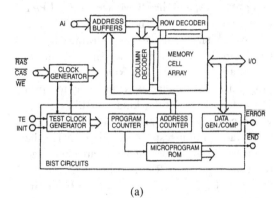

(a)

STEP		MICROCODE
1	CLEAR.	0000000010
2	WRITE(D), INC AC, IF AC=MAX THEN INC PC.	1010000100
3	READ(D).	0000001000
4	WRITE($\overline{D}$), INC AC, IF AC=MAX THEN INC PC ELSE DEC PC.	1110010100
5	READ($\overline{D}$).	0000011000
6	WRITE(D), INC AC, IF AC=MAX THEN INC PC ELSE DEC PC.	1110000100
7	DEC AC.	0001000000
8	READ(D).	0000001000
9	WRITE($\overline{D}$), DEC AC, IF AC=0 THEN INC PC ELSE DEC PC.	1101010100
10	READ($\overline{D}$).	0000011000
11	WRITE($\overline{D}$), DEC AC, IF AC=0 THEN INC PC ELSE DEC PC.	1101000100
12	STOP.	0000000001

PROGRAM COUNTER (PC) CONTROL	IF AC=MAX/0 ELSE DEC PC
ADDRESS COUNTER (AC) CONTROL	INCREMENT AC DECREMENT AC
DATA CONTROL	EXCLUSIVE OR INVERT/NORMAL
WRITE CONTROL	COMPARE/MASK WRITE/READ
FLAG CONTROL	CLEAR TEST END

(b)

Figure 5-19. (a) Block diagram of a microprogram ROM BIST for N pattern. (b) March test procedure and microcodes. (From [40], with permission of IEEE.)

These microcodes are stored in a 12 word $\times$ 10 b ROM. The N^2 patterns such as the GALPAT are capable of detecting most of the DRAM faults such as SAFs, cell-to-cell coupling, and pattern-sensitive faults. The ROM BIST circuit implementation for N^2 patterns compared to that for the N-pattern BIST contains three additional circuits as follows: (1) base register which stores a base cell address, (2) loop counter which changes the base cell, and (3) switch which exchanges the base cell address and surrounding cell addresses.

The GALPAT test can be divided into 16 unit processes, and the ROM size for GALPAT is 16 word $\times$ 16 b. The data-retention test, which is N size, can also be performed by the Checkerboard scan write/read pattern, and requires an 8 word $\times$ 11 b ROM. This microprogram ROM scheme was implemented in a 16 Mb DRAM with three kinds of BIST circuits which consisted of N pattern, N^2 patterns, and the data-retention tests. A comparison for the BIST overhead showed that the area required for a microprogram ROM is less than 1/10 of the whole BIST circuit area. The total BIST circuit area overhead for the N pattern was only 0.8%, whereas for the N^2 patterns and data-retention tests, the area overhead was less than 2%.

An important requirement is that the BIST circuits themselves should be testable. Figure 5-20 shows the flowchart for BIST circuitry testing which includes: (a) data write function test, and (b) compare function test [40]. This testing is performed by interrupting the BIST operating at any cycle by setting the TE clock at the "Low" state. While TE = "Low," the BIST circuits hold the condition when they were interrupted. Then the self-test can be resumed by returning the TE to the "High" state.

A single 5 V supply 16 Mb DRAM has been fabricated using a high-speed latched sensing scheme and the BIST function with a microprogram ROM in double-level 0.55 μm CMOS technology [41]. This DRAM has a 55 ns typical access time and 130 mm^2 chip area, out of which the BIST circuitry occupies about 1 mm^2 area.

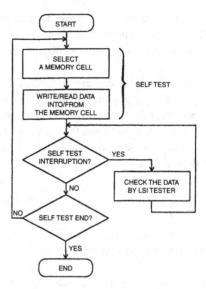

Figure 5-20. Flowchart for 16 Mb DRAM BIST circuitry testing. (From [40], with permission of IEEE.)

5.4.4 BIST and Built-In Self-Repair (BISR) Techniques

The current megabit memory chips use the laser and fuse/antifuse techniques to compensate for the faulty memory cells. A common approach to fault tolerance includes introducing the spare rows and columns (redundancies) into the memory array. However, these techniques are suitable for a production environment, and are difficult to implement in any field-operation-related error-correction and fault-tolerance scheme. These disadvantages can be overcome by an effective self-repairing technique that uses on-chip logic structure, which can be activated in an operational environment.

Figure 5-21 shows the basic configuration of a self-testing and self-repairing scheme in the lowest level memory blocks [42]. Self-testing is performed by the 13N algorithm. Test enable is a global signal which controls whether the memory blocks are in the self-testing/self-repairing mode or in the normal operation mode. The address generator provides incremental or decremental addresses during the March tests. During self-testing, the address generated by the

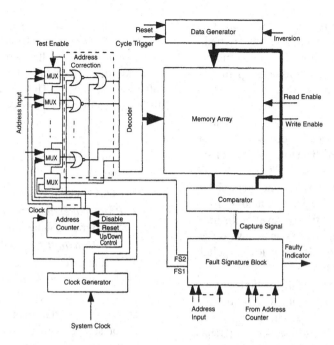

Figure 5-21. The basic organization of a self-testing and self-repairing scheme in the lowest level memory blocks. (From [42], with permission of IEEE.)

address generator is sent to the address control block which passes the address directly to the address decoder. The address generator also generates the addresses for redundant memory cells. If a fault is found in a redundant location, its addresses will be latched into the fault signature block. The LSB and MSB of the faulty addresses stored in the fault signature block are used in the address correction block to make sure that a faulty incoming address is correctly diverted to a functional readout location.

Test patterns are generated by the data generator. When the self-test starts, the "Reset" signal sets the outputs of the data generator to "0." When the entire 13N algorithm is completed for one word pattern, a pulse on the "Cycle Trigger" signal generated by the clock generator activates the next word pattern. The "Inversion" signal switches the data generator output between the data pattern and its inverted pattern, as required by the test algorithm. The redundant memory cells can be introduced at different levels of hierarchy. At the lowest level,

redundant words are introduced. If the local self-repairing logic can repair all the faults at the local level, the whole memory system can be restored to its fullest capacity. However, if a memory block has an excessive number of faults which the local self-repairing logic is unable to restore to its intended capacity, this memory block must be excluded from being accessed during normal operations.

This approach was used to build a prototype 512×8 b memory chip organized in three hierarchical levels with the following major features:

- Each lowest level memory block was organized in a 16×8 array and contained 16 redundant memory cells organized in a 2×8 array.
- Four redundant memory blocks (16×8) at the top level.
- Four registry blocks in the fault assembler.

• For self-testing and self-repairing functions of the chip, five sets of "fault-inject" pins to allow for any combinations of fault locations and up to five faulty blocks in the chip.

The chip was designed in a 2 μm, double-metal, single-poly CMOS process with a die size of 9.4×8.58 mm². The total amount of silicon area used for the self-testing and self-repairing at the lowest level of the hierarchy is about 0.25 mm², including the redundant rows in the memory array. At the top level of the hierarchy, about 10.73 mm² of silicon area was used by the self-testing and self-repair logic, including four fault registry blocks.

To improve the yield of high-density megabit RAMs, in addition to the built-in self-test, techniques such as the built-in self-diagnosis (BISD) and built-in self-repair (BISR) have been investigated. A BISD which can detect and locate fault in the embedded RAMs has been developed that uses a low-silicon overhead area [43]. It is based upon a diagnosis-and-repair (D&R) unit which includes a small reduced-instruction-set processor (RISP) that executes instructions stored in a small ROM. For the embedded RAM which is externally repairable, i.e., the RAM has spare rows and columns that can be programmed by blowing the fuses with laser beam pulses, the D&R unit first locates the faults, and then implements a repair procedure to control the laser beam. However, for the internally repairable RAM (e.g., nonvolatile RAM with EEPROM cells), the D&R unit can test, diagnose, and repair more than one embedded RAM on the same chip.

The test and diagnosis algorithms include various combinations of the March tests. For the standalone memories, it is assumed that the circuitry performs read and write operations directly on all memory cells. However, for the embedded memories, it may be necessary to perform read and write operations indirectly during the testing process using the following memory codes: (1) serial access, (2) parallel access, (3) combined serial/parallel access, and (4) modular memory access. Figure 5-22 shows (a) the block diagram of a repairable embedded RAM with a self-

diagnosis and self-repair capabilities, and (b) the details of *D&Runit* [43]. The D&R unit serves as a programmable controller which tests the memory array, locates the faults, and then programs the soft fuses for repairs. This *D&Runit* consists of the following functional blocks:

• *InstrregisterFSM* as a finite-state machine configured around an instruction register.
• *ROM* containing stored software.
• *Progcounterstack* consisting of a stack of program counters for the *ROM*.
• *InstrdecodingPLAs* and *Clkgen2* as decoding circuits which convert the current instructions into various control signals.
• A data path that creates the test vectors for storage into memory, examines the test vectors retrieved from the memory, and programs the soft fuses for memory reconfiguration.
• *ALU* as the arithmetic and logic unit.
• *Readonlyregs* as a bank of read only registers containing numerical constants and data patterns.
• *Readwriteregs* to serve as a bank of scratch-pad registers.
• *Busport* as an interface between the data path bus and the D&R bus.

In addition to the *D&Runit*, the self-test, repairable, embedded RAM, using local redundancy, consists of the following other major functional components, as shown in Figure 5-22(a):

• *Addgenerate:* This BIST address generator can be implemented using either a counter or a pseudorandom pattern generator (PRPG). For the March test algorithms, the address generators based on a PRPG are quite suitable and can be built by using the LFSRs.
• *SRAM:* This consists of *M* modules (or subarrays) along with the peripheral circuitry made up of the cells *Addrdecoder, Addrbuffer, Databuffer,* and part

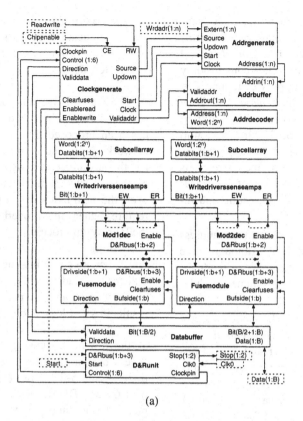

(a)

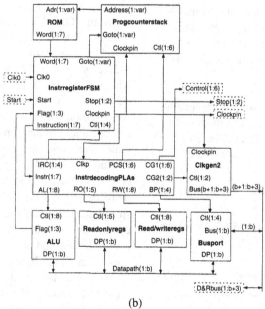

(b)

Figure 5-22. (a) Block diagram of an embedded RAM with self-diagnosis and self-repair capabilities. (b) The details of *D&Runit*. (From [43], with permission of IEEE.)

of *Clockgenerate*. Each module (or subarray) contains the cells *Subcellarray, writedriverssenseamps, Mod#dec,* and *Fusemodule.* Each module contains only one spare bit line to provide redundancy for fault repair. In general, the large memories may contain an arbitrary number of such modules. In addition to *D&Runit,* the SRAM is supported by *Addrgenerate* to control the address bits during testing and by part of *Clockgenerate* to synchronize all testing and repair functions.

- *Mod#dec:* This cell contains an address decoder for module #. Although there can be an arbitrary number of modules, each one requires a unique module-address decoder.

- *Fusemodule:* This cell provides a programmable interface between the "*b*" bits of the data buffer and "*b*+1" bits (including one spare) of the memory module. The data-carrying signals on both sides of this cell are bidirectional. The "Clearfuses" signal forces the fuses into a predetermined default setting. The "Enable" signal from one of the module address decoders specifies which module is being tested or being reconfigured by means of fuses.

This study showed that for the implementation of a built-in self testing, self-diagnosing, and self-repairing embedded RAM, some of the major design considerations are as follows:

- Minimal area overhead requirements by the additional circuitry.

- Minimal test and diagnosis time while maximizing the fraction of faults located by the BISD.

- Optimization of the repair algorithm's allocation of spare resources to repair faults.

- Maximizing self-testability of additional circuitry which performs the self-diagnosis and self-repair.

- Maximizing the design flexibility for application to various other memory configurations such as the dual-port RAMs, etc.

An alternative approach for the on-chip self-repair algorithms and yield improvement has been proposed that uses electronic neural networks [44]. The neural networks technique has the advantage of a significantly higher percentage of successful repair allocations. However, the neural network implementation may require higher silicon overhead and changes to the fabrication process.

In megabit density memories, the scaling down increases interbit-line coupling noise and reduces the readout sense signal level from the memory cells, and the presence of faulty cells results in yield loss. A 64 Mb DRAM has been developed that uses advanced techniques such as the low-noise sensing method, high-speed access design, and the BIST and BISR functions. Figure 5-23 shows (a) the block diagram of a 64 Mb DRAM with the BIST/BISR implementation, and (b) the associated timing diagram [45]. The BIST circuits include microprogram ROM that stores microcoded test procedures for a Marching test and a scan WRITE/READ test with the checkerboard test. Figure 5-23(c) shows the BISR circuit block diagram, which consists of an address pointer, a fail address memory (FAM), word drivers, and a spare memory [45]. The FAM is composed of content-addressable memory (CAM) cells which can store and compare input addresses. During a test and repair cycle, when a faulty cell is detected, the address pointer selects one word line of the FAM and the error address is stored there. During the normal READ/WRITE cycles, all comparators in the FAM compare externally applied addresses with the stored addresses in a simultaneous and parallel manner. If a stored address and an input address match, the FAM sends a MATCH signal to the word drivers. These word drivers generate an address match signal ADM and activate one word line of the spare memory, corresponding to the match signal.

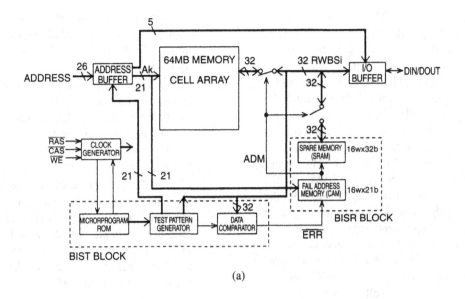

(a)

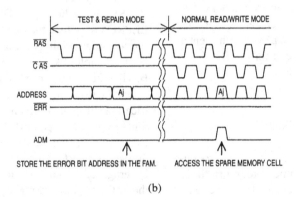

(b)

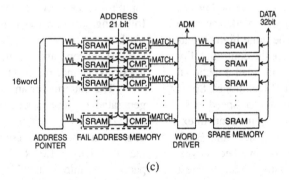

(c)

Figure 5-23. (a) Block diagram of 64 Mb DRAM with BIST/BISR. (b) The associated timing diagram. (c) BISR circuit block diagram. (From [45], with permission of IEEE.)

The 64 Mb DRAM has a 32 b I/O bus width for a 32 b parallel test scheme. Therefore, the faulty cells are replaced by 32 b units. For the BISR operations, there is no access time overhead. The TE and REPAIR pins perform the BIST and BISR operations. When both the TE and REPAIR signals are at a high level, the DRAM executes memory test and faulty cell address storing. When the TE signal is high and the REPAIR signal is low, the DRAM executes only a memory test. Since the FAM and the spare memory consist of volatile SRAMs, the BIST and BISR functions must be executed every time the chip is powered on.

A 64 Mb chip was fabricated in 0.4 μm CMOS design using a double-level metal, triple-well, and hemispherical grain (HSG) storage node, stacked capacitor cell structure. The cell arrays were divided into two 32 Mb subarrays, each consisting of 16 2 Mb cell array blocks. The Y decoders, control clock generators, and the BIST and BISR circuits were placed at the center of the die. The area overhead for the BIST and BISR circuits was about 0.8%.

5.5 DFT AND BIST FOR ROMs

ROMs are often used in processor and microcontroller designs for storing microprograms. Since the ROMs are usually embedded functional blocks in these critical applications, their controllability and observability are a design for testability concern that needs to be addressed. While the scan techniques can provide access to test data, the test time can be quite high. An attractive alternative to enhance the controllability and observability of the embedded functional block I/Os is through the BIST techniques. Some of the BIST advantages include "at speed" testing capabilities and reduced requirements in expensive ATE for test pattern generation. BIST schemes for RAMs and PLAs have been developed that are based on exhaustive testing and test response compaction [46], [47].

Conventional exhaustive test schemes for ROMs have used the following compaction techniques:

- Parity-based compaction, where an extra parity bit is appended to each column in the ROM in an odd or even parity scheme.

- Count-based compaction, also known as checksum checking, which assumes that the ROM contains n data words of m bits and uses a checksum formed by the modulo-k arithmetic sum of the n data words, where k is arbitrary.

- Polynomial-division-based compaction (signature analysis) in which the bit stream from a DUT is fed to the polynomial divider implemented by an LFSR to get a remainder (characteristic signature) that can be compared with the known good signature [48].

A key requirement in the BIST approach for ROMs including PROMs, EPROMs, and EEPROMs is that the adopted scheme should provide high fault coverage and result in the least possible likelihood of error escapes. A BIST scheme has been proposed for ROMs, called the exhaustive enhanced output data modification (EEODM), which has quite a low possibility of error escape [49]. The EEODM uses an exhaustive testing technique in which all addresses of the ROM are read to provide complete coverage of all combinational faults arising in the cell array, address decoder, and I/O logic. The compaction scheme is based on a modified polynomial-based division which uses a multiple-input feedback shift register (MISR) and has the following advantages: (1) it eliminates error masking by using the output data modification (ODM) concept, and (2) the use of bidirectional MISR to enhance the BIST scheme.

The patterns for an exhaustive test can be generated by any process that cycles exhaustively through the address space, e.g., a binary counter, a Gray code counter, or a modified maximal length LFSR. The ODM is a compaction technique in which the detection volume can be made extremely small compared to other schemes such as the signature analysis. This is done by perfect modification of the n-bit test output stream q from the DUT by performing an Exclusive-OR (EXOR) operation between q and a modifying bit stream s. Ideally, if q is the same as the expected bit stream originating from the DUT (i.e., $s = q$), then the EXOR of the two yields a stream of only zeros. Then a simple "ones detector" can be used to detect any single or multiple bit error in the modified stream q' induced by the DUT.

In EEODM, since the data to be compacted are an $n \times m$ matrix and not a bit serial stream, the ODM concept is modified using the MISR as a space compactor (parallel-to-serial compactor). Then the MISR equivalent serial bit (quotient) stream is modified using a divisor polynomial [50]. For this, an extra column of bits is stored in the ROM, and the column is programmed such that it corresponds to the quotient bit stream that is shifted out of the MISR during the read operation of words stored in the ROM. Therefore, only a "ones detector" is required to monitor the modified bit stream which should be all 0s in the case of a fault-free ROM.

The complete EEODM scheme implementation requires a fairly reasonable hardware overhead consisting of: (1) a modified LFSR to address all the words of the ROM which includes an extra column of bits to perform the ODM, and (2) a bidirectional MISR and the "ones detector." The EEODM test procedure sequence of steps can be summarized as follows:

1. Load a predetermined seed for the MISR initialization.
2. Perform a sequential read operation of the contents of the augmented ROM (including the ODM column) into an MISR of $m + 1$ stages, shifting left.
3. Verify the MISR contents.
4. Concurrently, perform the following operations:
   - Sequentially read all the words of the ROM into an m-stage MISR, shifting right.
   - Modify the quotient bit stream using the ODM column.
   - Feed the modified stream to the "ones detector."

5. Verify the final contents of the MISR that should have "all 0s" for a fault-free ROM.

The EEODM hardware and test time overhead are comparable to those of other known BIST schemes, while the fault coverage is much higher.

Another approach has been presented for the design of built-in self-diagnostics, which extends the concept of BIST to provide fault masking in the ROM normal operating mode. This technique is based upon generalized Hamming codes. Figure 5-24(a) shows the ROM

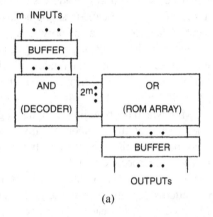

(a)

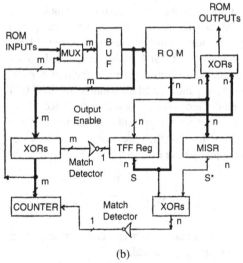

(b)

Figure 5-24. (a) ROM functional block diagram. (b) Block diagram of built-in self-diagnostic (BISD) ROM. (From [51], with permission of IEEE.)

functional block diagram [51]. In general, the ROM is a two-level AND–OR structure which is commonly implemented by a NOR–NOR array. The AND array is implemented through a decoder with m inputs and 2^m outputs called the word lines, and the OR array implements the n-bit word memory cells. The I/O buffers usually consist of static or dynamic D flip-flops.

The analysis of the ROM transistor circuit switch level model shows that the single stuck-open and stuck-on faults can occur in the decoder circuitry, ROM arrays, and the buffers. Most of the single stuck-open faults and the single stuck-on faults (95% or greater) result in erroneous outputs when the content of only one address is read. These types of errors are called "single-address errors" [51]. Therefore, for a memory array of (m, n) size, a built-in self-diagnostic (BISD) scheme which identifies the magnitude and addresses of the single-address errors is desirable. Figure 5-24(b) shows a block diagram of the BISD ROM in which the bold lines and the "regular" lines represent the components involved in the normal operating mode and the test/diagnostic mode, respectively [51].

In the test mode, the counter generates all 2^m input patterns beginning from the address 0 up to address $2^m - 1$. For each input pattern applied, the contents of the corresponding cell are loaded in parallel into the T flip-flops (TFF) registers and simultaneously into the MISR (parallel-input LFSR) which is shifted to its next state. This procedure is repeated for the next input pattern until all contents of the ROM are read. Once all the contents of the ROM are read, signature S represents the contents of the TFF register. For the case where error only occurs at an address i, $0 \le i \le 2^m - 1$, i.e.,

$$e(t) = 0$$

where $e(t) \in GF(2^n)$ for all $t \ne i$ (single-address error).

The signature $S = e(i)$ is the error pattern in the content of the address i.

Let S^* be the content of the MISR after $T = 2^m - 1$ state transitions. Then the theorem which summarizes the proposed signature scheme for identifying a single-address fault can be stated as follows: The contents of address i contain error pattern $e(i) \ne 0$ iff

$$S^* = \alpha^{T-i} S, \qquad 0 \le i \le T$$

where $S = e(i)$ is the signature appearing at the TFF register.

Thus, the diagnostic process provides the magnitude of an error as the content of the TFF register, and the corresponding error address as the content of the counter. In the normal operating mode, the ROM input is monitored for a possible match with the error address. When the address appears as the input, the errors are corrected by Exclusive-OR operation between the contents of the TFF register and the ROM output. This action is controlled by the signal from the match detector to the output enable of the TFF register. This scheme allows the ROM to perform its functions correctly, even in the presence of faults, and the silicon area overhead is on the order of 15% or less.

5.6 MEMORY ERROR-DETECTION AND CORRECTION TECHNIQUES

The errors in semiconductor memories can be broadly categorized into: (1) hard errors caused by the stuck faults or permanent physical damage to the memory devices, and (2) soft errors caused by alpha particles, or ionizing dose and transient radiation typically found in the space environment (see Chapter 7). The error-correcting codes (ECC) can be used to correct the hard as well as soft errors. The errors for which locations but not values are known are called erasures, which are easier to correct than random errors. In memory applications, the hard errors can be considered erasures if their locations can be identified. The hard failures include single-cell failures as well as failures of the row and column selection circuitry. Advanced megabit DRAM redundancy architectures include the BISD and BISR techniques discussed in Section 5.6, which are useful for dealing with the row and column failures. As the memory capacity per chip increases, the net effect of hard and soft errors becomes too great to handle at high-density memory cards and the board level systems. Therefore, for higher reliability, it is desirable to have some on-chip error-correction coding capabilities.

The simplest approach for memory error detection in a bit stream (0s and 1s) is called a parity check (odd or even) scheme. For example, an even parity binary coded decimal (BCD) code is obtained directly from the BCD code, and then an extra bit "p," called a parity bit, is added to make the number of ones in each codeword even. If a single error occurs, it changes the valid codeword into an invalid one which can be detected. Another simple error-detection technique is called the checksum method, in which the checksum of a block of "q" words is formed by adding together all the words in a block module "n," where n is arbitrary. Then the block of "q" words together with its checksum constitutes a codeword. Before the message is transmitted, a checksum of the data is prepared and stored. This checksum is compared with the checksum calculated after receiving the data to detect any discrepancy which would indicate the occurrence of a fault.

However, the error-detection codes commonly used for semiconductor memories are binary linear block codes. Consider the code shown below, in which each of the four coded characters differs from the other three coded characters in at least three of the five bits.

$$A = 0\,0\,0\,0\,0$$
$$B = 1\,1\,1\,0\,0$$
$$C = 0\,0\,1\,1\,1$$
$$D = 1\,1\,0\,1\,1$$

Then the minimum distance (d_{min}) of this code is three. In general, the minimum distance of a code is defined as the minimum number of bits in which any two characters of a code differ. The weight of a codeword is the number of nonzero components in the codeword. For example, if the character (byte) D in a memory serial output has the last two bit errors which make it 11000, it still would not be confused with the other coded characters A, B, and C. This observation can be generalized to the following statement: the errors in two or fewer bits

may be detected in any code with d_{min} of 3. If, in some applications, the probability of an error in two bits is sufficiently small to be negligible, the minimum distance-3 code can also be used as a single error-correcting code. Suppose a two-bit error actually occurs in a character of d_{min} of 3 code set up to correct the single-bit errors. The correction process is still carried out on the erroneous character, which will be coded to the wrong bit value. This means that the two-bit error will go undetected. Therefore, it is necessary to choose between the following two modes of operation: (1) reject all erroneous characters and require retransmission, or (2) correct all single-bit errors on the spot and pass two-bit errors.

The most convenient minimum distance-3 code is referred to as the Hamming code. In this code, the number of bit positions is in sequence from left to right, and those which are powers of 2 are reserved for the parity check bits, whereas the remaining bits are information bits. An example is a seven-bit code shown below, for which P_1, P_2, and P_4 indicate the parity bits; X_3, X_5, X_6, and X_7 are the information bits.

$$\begin{array}{ccccccc} 1 & 2 & 3 & 4 & 5 & 6 & 7 \\ P_1 & P_2 & X_3 & P_4 & X_5 & X_6 & X_7 \end{array}$$

In general, a binary (n, k) linear block code is a k-dimensional subspace of binary n-dimensional vector space [52], [53]. An n-bit codeword of the code contains k data bits and $r = n - k$ check bits. An $r \times n$ parity check matrix $\mathbf{H}$ is used to describe the code.

Let $\mathbf{V} = (v_1, v_2, \cdots, v_n)$ be an n-bit vector word.

Then $\mathbf{V}$ is a codeword iff [54]

$$\mathbf{H} \cdot \mathbf{V}' = 0 \qquad (1)$$

where $\mathbf{V}'$ is the transpose of $\mathbf{V}$, and all the additions are modulo-2. The encoding process of a code consists of generating r check bits for a set of k data bits. The $\mathbf{H}$ matrix is expressed as

$$\mathbf{H} = [\mathbf{P}, \mathbf{I}_r] \qquad (2)$$

where $\mathbf{P}$ is an $r \times k$ matrix, and $\mathbf{I}_r$ is the $r \times r$ identity matrix. A code specified by this $\mathbf{H}$ matrix is called a "systematic code."

A word read from the memory may not be the same as the original codeword written in the same location.

Let the word read from the memory be $U = (u_1, u_2, \cdots, u_n)$.

Then the difference between U and the original codeword V is defined as the error vector $E = (e_1, e_2, \cdots, e_n)$, i.e., $U = V + E$. The ith position of U is in error iff $e_i \neq 0$.

The decoding process consists of determining whether U contains errors and finding the error vector. To do that, an r-bit syndrome S is calculated as follows:

$$S = H \cdot U' = H \cdot (V' + E')$$
$$= H \cdot E'.$$

If S is an all-zero vector, the word U is assumed to be error-free. If S is a nonzero vector, then it is used to find the error vector.

The error-correcting capability of a code is closely related to the minimum distance of the code. For a linear code, d_{min} is equal to the minimum of the weights of all nonzero codewords. A code is capable of correcting l error and detecting $l + 1$ errors iff $d > 2l + 1$. A typical RAM memory of $N \times 1$ size consists of a two-dimensional array of N memory cells with word lines along the rows and bit lines along the column. These $N \times 1$ chips are organized on boards to create byte-wide memory systems such that the bits from the same address on each chip make up a byte (i.e., each chip provides one bit of the byte). In a larger memory system, the board may consist of many rows of chips to form a multipage memory board. In a board-level correction scheme, the rows are encoded with an (n, k) error-correcting code. These are some of the codes commonly used in the error detection and correction of the memory chips and memory cards, as well as board-level systems:

1. *SEC–DED Codes:* These are single-error-correction and double-error-detection (SEC–DED) codes for which the minimum distance is equal to or greater than 4. Thus, the H matrix of the SEC–DED code must satisfy the following conditions:

- The column vectors of the H matrix should be nonzero and distinct.
- Any set of three columns of H matrix are linearly independent, i.e., the sum of the two columns is nonzero, and is not equal to a third column of the H matrix.

In the semiconductor memory ECC implementation, SEC–DED codes are often shortened so that the code length is less than the maximum for a given number of check bits to meet certain objectives for a specific application. The maximum code length of an SEC–DED odd-weight column code with r check bits is 2^{r-1}. The maximum number of data bits k must satisfy $k \leq 2^{r-1} - r$. A few examples of the number of check bits required for a set of data bits is shown below:

Data Bits (k)	Check Bits (r)
8	5
16	6
32	7
64	8
128	9

The memory array chips on the board (or memory card) are usually organized such that the errors generated in a chip failure can be corrected by the ECC scheme. In the SEC–DED codes, the most effective design implementation is such that each bit of a codeword is stored in a different chip, which ensures that any type of failure in a chip can corrupt, at the most, one bit of a codeword. Therefore, as long as the errors do not occur in the same codeword, the multiple-bit errors in the memory are correctable.

2. *SEC–DED–SBD Codes:* These are a class of the SEC–DED codes that are capable of detecting all single-byte errors in memory array applications where it is required that all chips be

packaged in b bits per chip organization. In this case, a chip failure or a word-line failure in the memory array would result in a byte-oriented error that contains from 1 to b erroneous bits. For $b \leq 4$, in practical applications, most of the SEC–DED codes can be reconfigured to detect single-byte errors.

3. *SBC–DBD Codes:* These are the byte-oriented error-correcting codes used for higher reliability than that provided by the SEC–DED codes for the memory arrays with a b bit per chip configuration. A codeword of SBC–DBD consists of N b-bit bytes. Considering a binary b-tuple as an element of the finite field $GF(2^b)$ of 2^b elements, then the SBC–DBD is a linear code over $GF(2^b)$ with a minimum distance of $d \geq 4$ [52]. The code can also be defined by the parity check matrix **H** of equations (1) and (2), with the component of the matrices and vectors that are elements of $GF(2^b)$. An example of this code is the Reed–Solomon code with applications in terrestrial data communication networks.

4. *DEC–TED Codes:* A large memory array system with high chip failure rates (both soft and hard errors) may use a double-error-correction and triple-error-detection (DEC–TED) code to meet its reliability requirements. It is capable of detecting any combination of hard and soft double errors that can be corrected automatically without system interruption. A minimum distance of the DEC–TED code is $d \geq 6$. The parity check matrix **H** of a DEC–TED code has the property that any linear combination of five or fewer columns of **H** is not an all-zero vector.

A general class of codes suitable for on-chip ECC are the linear sum codes (LSCs) which are generalizations of the bidirectional parity check code to allow for multiple error correction. Linear sum codewords are similar to product codewords, except for some constraints on decoding which limit the error-correcting capability of the codes to a function of sum of the minimum distances of the constituent codes rather than the product [55]. The memory chips are arranged in rectangular blocks, and the addressed bit is specified as a word line/bit line intersection. Then the LSC can be implemented on the word lines of a RAM, i.e., the bit-line address could be split into the smaller addresses to form a two-dimensional sum codeword in each line. The decoding constraint implies that to decode any given bit, only those bits sharing one of the smaller addresses with the desired bit are ever considered.

As an example, consider a $k_2 \times k_1$ information array of q-ary memory cells containing an arbitrary pattern of symbols. A simple case of the binary LSC is being considered for which $q = 2$, i.e., the codes can be constructed from the parity checks, Hamming codes, and extended Hamming codes. To each of the k_2 rows, add $r_1 = n_1 - k_1$ additional symbols such that each row constitutes a codeword from the systematic (n_1, k_1) linear block over F_q (i.e., F_q is a finite field containing q elements). Similarly, for each of k_1 columns, add $r_2 = n_2 - k_2$ parity symbols such that each column forms a codeword from a (possibly different) systematic (n_2, k_2) linear block. Thus, all possible configurations constructed for the given constituent codes are defined as an (n_1, k_1, n_2, k_2) linear sum code over F_q. Figure 5-25(a) shows the general layout of this linear sum codeword [55].

The linear sum codes have a minimum distance property, which means that the minimum distance of a product code is equal to the product of the minimum distances of the constituent codes. It can be shown that the parity checks, Hamming codes, and extended Hamming codes can be used to implement the single-, double-, and triple-error-tolerating LSCs. Figure 5-25(b) shows the typical organization of a one-bit wide DRAM chip encoded with a LSC [55]. The chip consists of a rectangular array of open bit memory cells, sense amplifiers, refresh

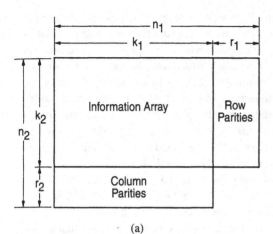

(a)

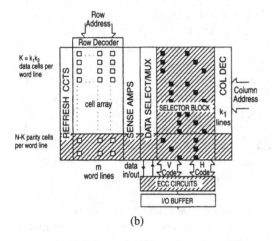

(b)

Figure 5-25. (a) The general layout of an (n_1, k_1, n_2, k_2) linear sum codeword. (b) Typical organization of a one-bit wide DRAM chip encoded with the LSC. (From [55], with permission of IEEE.)

circuitry, selecting and addressing logic, and the ECC circuitry. The memory cells are organized as m N-bit word lines, each of which uses $K = k_1 k_2$ cells for the data and $N - K$ cells for parity.

Therefore, total user's capacity of the RAMs
$$= m \, k \text{ bits.}$$
Number of address bits $= \log_2 m + \log_2 K$.

These address bits are decoded by a row encoder and a column encoder. A major functional unit in the coded RAM is the V–H code

selector block which outputs the vertical and horizontal component codewords associated with the bit being addressed. These codewords, together with the addressed data bits, are fed to the ECC unit which performs any necessary error correction. In Figure 5-25(b), the shaded area represents the ECC overhead [55]. No extra column or row decoders are required, as compared to an uncoded memory.

The simplest LSC is the one in which the two constituent codes are single-error-detecting (SED) parity checks which have a minimum distance of two. These may be called SED/SED codes. An LSC constructed from two SED codes is a single-error-tolerating code, and for the binary case, it is the bidirectional parity check code. A single-error correction to a codeword containing $2L$ bits in the information array can be provided with a $(2^{l1} + 1, 2^{l1}, 2^{l2} + 1, 2^{l2})$ SED/SED code, where $l1 + l2 = L$. The number of parity bits in such a codeword is $2^{l1} + 2^{l2}$.

The advantage of this scheme is its high rate and simple decoding algorithm. However, this is only capable of correcting a single error in each codeword. To design a two-error-tolerating LSC, the following condition must be satisfied:

$$d^1{}_{min} + d^2{}_{min} \geq 6$$

where $d^1{}_{min}$ and $d^2{}_{min}$ are the minimum distances of the constituent codes. The two alternatives that satisfy this equation are:

1. An SEC/SEC linear sum code consisting of single-error-correcting (SEC) Hamming codes ($d_{min} = 3$).

2. An SEC–DED/SED linear sum code consisting of a combination of single-error-correcting/double-error-detecting (SEC–DED) extended Hamming code ($d_{min} = 4$) and a single-error-detecting (SED) parity check ($d_{min} = 2$).

In comparing the two double-error-tolerating coding schemes, the SEC–DED/SED is preferable because of its optimal configuration range of 10–20% better than those of the SEC/SEC codes over the most reasonable values of L (i.e., $8 \leq L \leq 12$). The complexity of implementation of these codes in terms of the number of gates required is essentially the same. A penalty has to be paid in terms of the code rate for increasing the on-chip EDAC capabilities. However, relatively high rates can be achieved with the multiple-error-tolerating schemes for reasonable values of L. For example, by placing 256 data bits on each word line, a two-error-tolerating SEC–DED/SED scheme can be implemented at a rate $R*(L)$ of 0.744. Table 5-2 shows the rates, complexity data for the optimal (i.e., best rate) configurations for various linear sum codes with several values of L, and the hardware implementation requirements [55].

A new memory on-chip soft-error-detection method called the interlaced vertical parity (IVP) scheme has been proposed [56]. This is based upon the two-dimensional parity and multiple-cell accessing technique. The memory block is partitioned into a set of small check spaces, and a built-in parity checking scheme performs concurrent vertical parity on a check space in a single memory cycle during normal memory operation. IVP checking can be used with any word-wide error-detection or correction scheme for implementation of the interlaced two-dimensional parity (ITP) checking. The number of check bits required for ITP is less than that of a Hamming code. The error-detection and correction capability is higher than that of a two-dimensional parity scheme. The IVP scheme used with a memory parallel testing technique, such as the multiple access with read-compaction (MARC), can reduce high memory test cost. For a RAM chip, the IVP checking scheme can be added to the MARC function to provide efficient on-line and off-line test and error-correction methods.

Figure 5-26(a) shows a memory block organization with the MARC implementation [56]. A modification is made to the column decoder and I/O data path such that the test data applied to a data input pin can be written into a set of cells simultaneously. In multiple-access function implementation, the column decoder is divided into two stages such that the first stage column decoder selects k data lines out of n data lines connected to the column sense amplifiers,

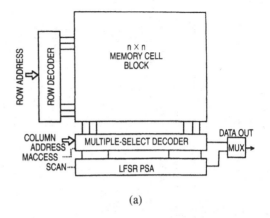

(a)

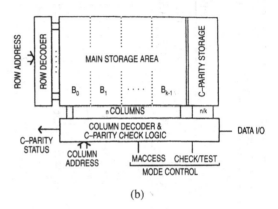

(b)

Figure 5-26. (a) Memory block organization with multiple-access read compaction (MARC) implementation. (b) Block diagram of memory block modified for IVP checking. (From [56], with permission of IEEE.)

and the second stage column decoder selects one data line out of these k data lines. In this parallel test scheme, k is a design parameter equal to 2^i, where "i" is an integer less than or equal to $\log_2 n$. The modification needs an additional $\log_2 k$ gates for each column selection logic. The MARC function can also be utilized with test algorithms of $O(N^2)$ complexity for the RAMs by using several possible test strategies.

In a conventional two-dimensional parity scheme, two kinds of parities are generated as follows: (1) a horizontal parity whose check space is each word or byte, and (2) a vertical parity whose check space is each bit plane. In

TABLE 5-2. Rates, Complexity Data for Optimal Configurations for Various Linear Sum Codes with Several Values of *L*, and Hardware Implementation requirements. (From [55], with Permission of IEEE.)

Code	Error Toler.	L	R*(L)	n₁	k₁	n₂	k₂	Select. ANDs	Syndr. XORs	Correc'n ANDs	Encoder ANDs	Encoder XORs	Encoder ORs	Overall
SED/SED	1	4	0.667	5	4	5	4	40	8	0	2	2	6	252
		6	0.800	9	8	9	8	144	16	0	2	2	14	748
		8	0.889	17	16	17	16	544	32	0	2	2	30	2508
		10	0.941	33	32	33	32	2112	64	0	2	2	62	9100
		12	0.970	65	64	65	64	8320	128	0	2	2	126	34572
		14	0.985	129	128	129	128	33024	256	0	2	2	254	134668
SEC/SEC	2	4	0.400	7	4	7	4	56	18	30	6	6	12	560
		6	0.500	12	8	12	8	192	36	72	8	8	28	1464
		8	0.615	21	16	21	16	672	76	170	10	10	66	4188
		10	0.727	38	32	38	32	2432	162	396	12	12	150	13004
		12	0.821	71	64	71	64	9088	358	910	14	14	344	43656
		14	0.889	136	128	136	128	34816	800	2064	16	16	784	155616
SEC-DED/ SED	2	4	0.471	13	8	3	2	50	26	50	6	6	20	696
		6	0.615	22	16	5	4	168	52	108	7	7	45	1666
		8	0.744	39	32	9	8	600	104	238	8	8	96	4440
		10	0.842	72	64	17	16	2240	224	528	9	9	215	13366
		12*	0.908	137	128	33	32	8608	504	1170	10	10	494	44212
		14	0.950	523	512	33	32	33632	2262	5654	12	12	2250	179836
SEC-DED/ SEC-DED	3	4	0.333	8	4	8	4	64	24	48	8	8	16	736
		6	0.444	13	8	13	8	208	48	100	10	10	38	1772
		8	0.571	22	16	22	16	704	96	216	12	12	84	4712
		10	0.696	39	32	39	32	2496	192	476	14	14	178	13892
		12	0.800	72	64	72	64	9216	416	1056	16	16	400	45344
		14	0.877	137	128	137	128	35072	944	2340	18	18	926	159196

*Note: Since $L = 12$ is of the form $2^m - m$, there are two SEC–DED/SED configurations which achieve $R*(L)$.

comparison, the IVP checking scheme is a partitioned vertical parity checking in that each parity check space is much smaller than that of the conventional vertical parity, and the parity is checked or generated within a single memory cycle in parallel with the normal memory operation. For an IVP checking scheme in an N size block, an additional memory array of size N/k for c-parity bits and c-parity checking and generation logic is required. Figure 5-26(b) shows a block diagram of a memory block modified for IVP checking [56]. Error detection and correction using IVP and word parity can also be implemented on a multichip organized memory system.

Most of the current generation megabit DRAMs use three-dimensional cell topographies where the charge is stored on vertically integrated trench-type structures. These storage capacitors are vulnerable to the double-bit soft errors induced by alpha particles and other transient effects such as the power supply voltage spikes. In a DRAM chip, more than 98% of the failures occurring during normal operation are the radiation-induced soft errors [57]. The conventional on-chip error-correcting SEC/DED codes are inadequate for correcting such double-bit/word-line soft errors. Also, other conventional double-bit error-correcting codes such as the Bose–Chaudhury–Hocquenghem (BCH) [58], Reed–Solomon [59], and Golay [58] are not easily applicable to correct double-bit errors in a DRAM chip. The coding/decoding circuits for these codes utilize multibit LFSRs which, if used in a DRAM chip, can introduce unacceptably high access time delays.

A functional knowledge of the memory cycle is needed to understand the bit-line mode soft error. The cell topography is a significant factor in analyzing the soft error rate (SER) for a memory-cell mode upset. The Monte Carlo simulations performed in a study have shown that as the feature width and critical charge in storage capacitors decrease, the double-bit soft errors dominate over single-bit errors [60]. A new double-error-correcting (DEC) coding circuit has been developed that can be integrated in a DRAM chip to reduce the SER and improve the reliability of the memory system. This new code is an augmented product code (APC) which uses diagonal parities in addition to horizontal and vertical parities. A typical APC implementation on a DRAM chip with m-bit wide bit lines will contain $3\sqrt{m} + 1$ redundant bits per bit line.

Figure 5-27(a) shows the APC implementation for a DRAM where each word line has nine data bits (organized into a 3×3 matrix) and ten parity bits [61]. The distribution of horizontal, vertical, and diagonal parities on the word line is also shown. The vertical parities can be easily computed by the parity trees (shown by PARITY blocks). The horizontal and diagonal parities are computed by the external Exclusive-OR (EXOR) gates. The suitable bit is selected by the MUXs. If there is no error in the selected bit line, then all inputs $X(i)$, $Y(j)$, and $D(t)$ are 0s.

Figure 5-27(b) shows the error-correcting circuit using the APC gate level implementation which can automatically correct all single-bit errors [61]. If a single-bit or a double-bit error occurs and the $b'(i, j)$ Erroneous signal is high, the bit $b(i, j)$ is automatically corrected and written back. The "Fatal Error" flag indicates that a double-bit error has occurred, and in such a double-error event, all the bits in the vertical column of the APC are read to determine whether the selected bit $b(i, j)$ is faulty. If the selected bit is faulty, it is complemented and rewritten by separate hardware. The ECC circuit consists of multiplexers, parity trees, additional sense amplifiers, and encoders. The area overhead for chips with the proposed ECC scheme has been compared with the no-ECC and with the SEC-type ECC schemes. It was found that for 4 Mb or higher memory chip sizes, the DEC-type APC requires twice the area for the SEC-type product code schemes. However, this higher area overhead can be justified by an order of magnitude improvement in the storage reliability over an SEC code. For a 16 Mb DRAM chip with 16 partitions, the chip area overhead due to the APC is about 8% of the overall chip area.

A high-speed 16 Mb DRAM chip with on-chip ECC has been developed by the IBM Corporation for a 3.3 or 5 V supply operation which

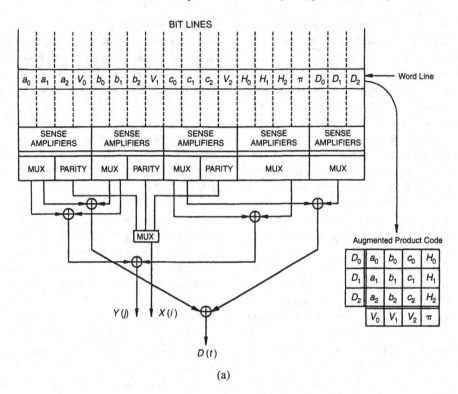

(a)

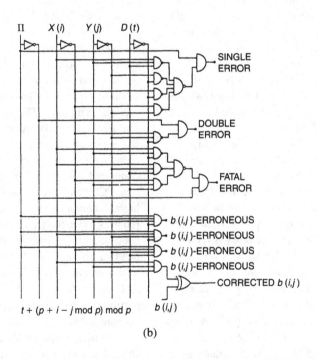

(b)

Figure 5-27. (a) Augmented product code (APC) implementation in a DRAM. (b) Error-correcting circuit using APC. (From [61], with permission of IEEE.)

is capable of operating in a fast page mode, static column mode, or toggle mode [62]. This design uses redundant word and bit lines in conjunction with the ECC to produce an optimized fault tolerance effect. The 16 Mb chip is divided into four quadrants, each of which is an independent memory that can operate in either a 4 Mb × 1 or 2 Mb × 1 mode. Also, each of the quadrants has its own individual bit redundancy and bit steering, word redundancy system, error-correction circuitry, SRAM, and off-chip drivers (OCDs). These circuits are arranged in a "pipeline" order which provides the most direct channel for data flow. The quadrants are further divided into four array blocks, with each block containing 1024 word lines. Each word line contains 1112 b, of which 16 b belong to the redundant bit lines and the remainder to eight ECC words. An ECC word is 137 b long and consists of 128 data bits and 9 check bits. These ECC words are interwoven along the word lines so that eight adjacent bits belong to eight different words. This allows clustered faults affecting the adjacent memory cells to be separated as individual cells of different ECC words. Figure 5-28(a) shows the quadrant layout, and (b) the block diagram of a functional array block of the 16 Mb DRAM chip [62].

The use of ECC circuits on-chip require a tradeoff among the performance, chip size, and yield (or reliability) enhancement. This 16 Mb DRAM employs an odd-weight DED/SEC Hamming code scheme that uses 128 data bits and 9 check bits. The check bits indicate the correct logic state of the data bits. The ECC logic tests the data bits by using the check bits to generate syndrome bits that indicate which bits in the ECC word are faulty. The ECC logic uses this information to correct the faulty bits. The ECC circuits are designed using a differential cascade voltage switch (DCVS) logic tree, and they can be very effective in correcting single-cell failures. However, any additional faults in an ECC word must be fixed with the chip level

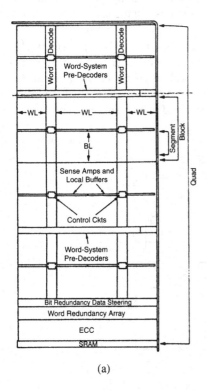

(a)

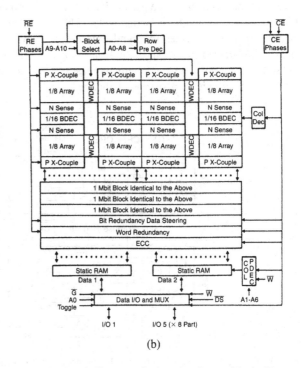

(b)

Figure 5-28. A 16 Mb DRAM chip. (a) Quadrant layout. (b) Block diagram of a functional array block. (From [62], with permission of IEEE.)

redundancy. To improve chip fault tolerance, it is necessary to replace the ECC words that contain more than one faulty cell. This can be accomplished by use of bit-line redundancy and chip layout optimization.

Another example of memory on-chip architecture with the embedded ECC technique is the Mitsubishi Electric Corporation's 16 Mb DRAM with a combination of a page mode hierarchical data bus configuration and a multipurpose register (MPR) [63]. The ECC unit consists of the MPR, Hamming matrix, data latch, syndrome decoder, and rewriting driver. These elements are embedded at the end of the memory array, while the ECC control circuits are placed in the peripheral area. The MPR has a full-CMOS latch, two pairs of transfer gates, and can operate as a bit inversion register. Figure 5-29 shows a detailed circuit diagram of the memory cell array [63].

The background ECC operation starts in parallel with the dual-stage amplification. The data bits and parity bits sent from the MPRs are checked and corrected in the read and $\overline{CAS}$-before-$\overline{RAS}$ refresh cycles. The parity bits on the MPRs are also rewritten in the write cycles. The Hamming matrix generates the syndrome signals to assign an error position. In the Hamming code scheme, as the data bit length increases, the additional circuit area overhead increases, while the ratio of the number of parity bits to data bits decreases. The 16 Mb DRAM with an array embedded ECC scheme using a data bit length of 128 b resulted in an estimated 15% chip area penalty which was smaller than the bidirectional parity code ECC approach.

A new self-checking RAM architecture has been developed in which on-line testing is performed during normal operations without destroying the stored data, and a delayed write capability allows data to be "forgotten" if discovered to be erroneous [64], [65]. This approach is different from the schemes that involve the use of Hamming codes to detect and correct errors after the data have been read out of a memory. Figure 5-30(a) shows a functional block diagram of this self-checking, on-line testable static RAM [64]. This design includes a self-checking memory cell which detects transient errors at the site of each memory cell instead of conventional checking across the rows or word lines.

Figure 5-30(b) shows a structural schematic of the specially designed memory cell which contains a static subcell (two inverters) and a dynamic subcell that is part of a built-in comparator [64]. The CMOS pass transistors (G and $\overline{G}$) store the true and complement values of a static subcell as charge on gates A and B of the comparator. When a register cell is written, the data are stored in both subcells, and then the

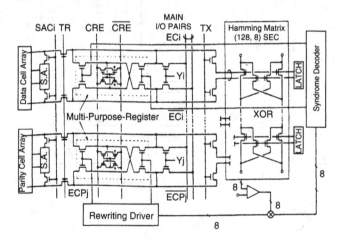

Figure 5-29. Detailed circuit diagram of the memory cell array for Mitsubishi Electric Corporation's 16 Mb DRAM. (From [63], with permission of IEEE.)

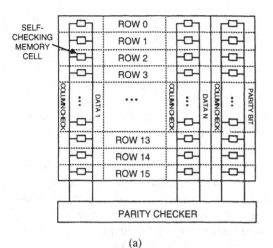

(a)

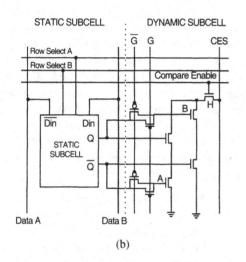

(b)

Figure 5-30. Self-checking, on-line testable static RAM. (a) Functional block diagram. (b) Schematic of memory cell consisting of static subcell and dynamic subcell [64].

pass transistors are turned off so that both subcells are isolated. The static and dynamic subcells are continuously compared to each other by the built-in comparator. The output of the comparator is connected to transistor H which is normally turned on. The output of transistor H is connected to a column error signal (CES) line shared by all memory cells in the same column. A transient fault that flipped (soft-error/upset) either subcell of a memory cell in a column would be instantly detected by the CES.

A recovery from transient error is possible by using a parity bit. A built-in binary search procedure allows detection of errors by cross-checking the column error signals. The faulty row is then read out for parity checking to determine if the CES is caused by the static or dynamic subcell. If a parity error is found, the error must be in a static subcell; otherwise, the CES is caused by a flipped dynamic subcell. The error can be corrected by either reversing the faulty static subcell or just refreshing the faulty dynamic subcell. The search of faulty rows can be performed during normal memory read/write operations since the CES lines are independent of the data lines. When a faulty row is isolated, it will be read out, and the faulty bit indicated by CES will be reversed. There is no need to decode the read data before the faulty bit can be located, as is done by conventional coding techniques.

This self-checking, on-line testable static RAM design can correct single-bit errors in a number of words in memory. The worst case time needed to identify a faulty row is $\log_2 (2N)$, where N is the number of rows in the memory. However, using the method of backtracking, the average identification time decreases as the number of faulty rows increases. It should be noted that the self-checking feature is not capable of detecting latent stuck-at faults because, if the static subcell has a stuck-at fault, the dynamic subcell would also have an incorrect value. The memory cell design can be further modified to make the RAM fully testable in fewer test cycles, without affecting the stored data.

5.7 MEMORY FAULT-TOLERANCE DESIGNS

An approach to increased reliability is a fault-tolerant design that uses redundancy in the form of hardware, software, information, or time. The most common form of redundancy is hardware implementation, such as the extra rows and columns provided in megabit DRAMs to substitute for hard cell failures and row/column failures. Dynamic redundancy techniques reconfigure the system components in response to failures to prevent the faults from affecting the system operation. Therefore, hardware "duplication" is the most commonly used technique in fault-tolerant computer designs. The error-detection and correction approaches discussed in Section 5.6 use various ECC techniques to provide redundancy. A great advantage of ECC schemes is that they are capable of identifying and correcting fault conditions immediately, and no interruption is required in the normal memory system operation.

In most fault-tolerant memory designs, SEC–DEC codes are used because of the higher silicon area overhead and circuit complexity for codes which detect multibit error patterns. Fault-masking methods such as triple-modular redundancy (TMR) are not practical for memories because of large hardware redundancy requirements, and the resulting reliability decrease for times longer than the component mean life [66]. Fault-tolerance synergism for memory chips can be obtained by the combined use of redundancy and the ECC. In a 16 Mb DRAM, the ECC can be very effective in correcting single-cell failures. The ECC schemes commonly used correct only one faulty bit in an error-correcting codeword. Therefore, the occurrence of two or more faulty cells in such a word can disrupt the data integrity of the chip. This event of finding more than one fault in an ECC word is referred to as "multiple fault alignment" [67].

As an example, consider a 16 Mb DRAM chip which contains 131,072 pages and 131,072 ECC words with 137 b per ECC word. Then, starting with a fault-free chip, the chance that first cell failure will not be coincident with another failing cell is absolute certainty. For the chance that the second faulty cell will not occur in the same word as the first one, the probability is $131,071 \div 131,072$, and for third random failing cell not to occur in the same word as the previous ones, the probability is $131,070 \div 131,072$. In general, for N single-cell faults on a chip, the conditional probability of nonalignment can be expressed as

$$P \text{ (no alignment} \setminus N\text{)} = \frac{W!}{W^N(W - N)!}$$

where $W = 131,072$ represents the ECC word.

Current megabit-density, high-quality DRAM manufacturing typically averages less than one fault in a million bit chip, i.e., fewer than 16 faults for a 16 Mb chip. If these were all single-cell failures, the ECC can bring the yield up to 99.9%. However, some of these faults affect the chip circuits, and the word and bit lines. Therefore, the fault-tolerance approach for 16 Mb chips includes redundant word and bit lines to substitute for the defective ones. A single-cell failure can be repaired or fixed with a single redundant word or bit line, while a short circuit between the word lines requires two redundant word lines for repair. Similarly, a dielectric defect causing a short circuit between a word line and bit line usually requires replacement by a redundant word line and a redundant bit line. Computer modeling of an early pilot line for 16 Mb DRAMs and eventual fabrication showed that the synergistic combination of the ECC with redundancy produced superior yields, which outweighed the increase in chip complexity.

A RAM fault-tolerance approach uses dynamic redundancy which allows the treatment of memory chip faults using the following two schemes [68]:

1. *Standby Reconfiguration Method:* The basic idea of the memory standby system is to provide a number of s spare bit slices to tolerate faults within the memory array. Therefore, for each word cell of length w bits, there are s spare bit cells; the function of up to s defective bit cells in each word cell can be replaced. In case access to a word cell in a faulty memory block is required, the I/O lines of the memory can be dynamically switched to spare bit slices. This is done by the switching network implemented at the memory interface. During each memory access, the network is controlled by a fault-status table (FST) which memorizes the fault conditions of each memory block. This FST can be implemented outside the main memory by a VLSI chip that can be added modularly to the memory interface. As

the FST access is performed concurrently with the main memory access, there is no performance loss.

2. *Memory Reconfiguration by the Graceful Degradation Scheme:* In some cases, for an existing memory system, it may be impractical and too complicated to include additional bit slices to the memory structure already implemented. Therefore, a modification of the dynamic standby reconfiguration method may be required. One scheme is to use entire memory blocks as replaceable units, i.e., if a word cell is defective, the entire faulty memory block and its functions are replaced by a good memory block. In this case, the wastage of memory space is considerable. An alternate scheme similar to the dynamic standby system logically deactivates only defective bit-slice modules. The basic concept is to store the fault-coincident bits of a defective memory block at some other fault-free location in the memory in correction-bit fields (CBFs). Therefore, a data word that is fetched from a logical memory block containing defective bit-slice modules must be completed outside the memory by its fault-coincident bits (FCBs). These FCBs have to be fetched from the associated CBFs during a second memory cycle. This flexibility of the degradation method is achieved at the expense of performance losses.

The problem of transient upsets and recovery in fault-tolerant memory designs has been studied using the memory-scrubbing techniques [69]. Memory scrubbing can be an effective method to recover from the transient effects resulting from environmental disturbances such as space radiation effects, as well as intermittent faults due to circuit transients. Single-error-correction and double-error-detection (SEC/DED) codes can be used by in-serting extra parity bits in each memory word. Then the memory scrubbing is performed by the memory word readout, single-error detection, and rewriting the corrected data in their original location. The scrubbing interval can be probabilistic or deterministic. There are two different models used for memory scrubbing, as follows:

a. *Probabilistic Model Scrubbing:* In the probabilistic model scrubbing, the memory words are accessed by the operating system based on an exponentially distributed time interval with a rate μ. This may not be a realistic assumption for cases where the location of the programs causes some parts of the memory to be accessed more often than the others. However, the exponential distribution for the scrubbing interval can be implemented by the operating system. Also, the memory allocation scheme can be modified to make the location access random and uniform over the memory system. An advantage of this method is that it allows memory scrubbing without interrupting normal program execution. The memory-scrubbing recovery phase can be modeled using a Markov chain, with the number of states depending upon the type of ECC scheme used (e.g., three states for single-error-correcting codes).

b. *Deterministic Model Scrubbing:* The deterministic scrubbing model depends on a mechanism which consists of cycling through the memory system, reading every word, checking for their correctness, and scrubbing them periodically. The interval between two accesses of the same word is deterministic, and intervals of 100 s or greater can be easily implemented in software with very low time overhead. However, the software overhead gets higher as the interval is reduced (<10 s), and a dedicated hardware approach is required. The hardware may include scrubbing processors appended to every bank of an interleaved memory system, which can be quite complex and should only be considered for a high error-rate environment. If a word contains soft (transient) errors, these errors are corrected, and the word is rewritten in its original location. If SEC/DED codes are used, an accumulation of two errors in the same word cannot be corrected, thus leading to the failure of the memory system.

These memory-scrubbing models can be used to calculate the reliability rate $R(t)$ and the MTTF of unprotected memory systems, SEC memory systems, and SEC systems with probabilistic or deterministic scrubbing techniques. A comparison of the MTTF between an unprotected memory system and a memory system using the scrubbing technique showed an improvement on the order of 10^8 for 1 Mbyte memory size. For the same average scrubbing interval, the deterministic scrubbing approach provides an MTTF about twice that for the probabilistic scrubbing technique.

A redundancy technique of binary-tree-dynamic RAM (TRAM) has been proposed to overcome the performance and test time limitations of traditional memory chip architectures [70]. The TRAM architecture partitions the RAM into modules, with each of the modules appearing as a leaf node of the binary interconnection network. Figure 5-31 shows the area model of the TRAM approach in which the leaf nodes are organized in groups of four, forming an *H*-tree, which are further connected hierarchically [71]. Each nonleaf node in the TRAM is a switch node that is a simple 1-out-of-2 decoder with buffers. The memory nodes are memory modules which have traditional four-

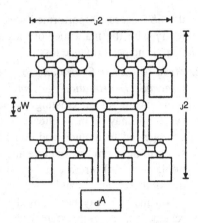

Figure 5-31. Area model of binary-tree-dynamic RAM (TRAM) architecture. (From [71], with permission of IEEE.)

quadrant layout with independent control units. Therefore, a TRAM chip with a "depth = 1" is equivalent to a conventional memory chip. For chips with "depth >1," the address/data/control bus is connected to the root node which decodes the most important part of the address and generates a select for a left or a right subtree. The other signals are buffered and propagated down the tree. This action is repeated at each level until a single-memory mode is selected. Then the remaining address bits are used to select a cell within the node.

The TRAM device has the following two modes of operation: (1) the normal mode in which it functions as a traditional RAM, and (2) the test mode for verification of the presence of faults. The advantages of the TRAM architecture over the conventional approach can be summarized as follows:

- It is partially self-testing with small test times for very large RAM sizes (e.g., 64 Mb).
- Low chip area overhead.
- Improved performance.
- It is partitionable, which can improve the effective yield.
- It includes both the static and operational fault-tolerant strategies that can

reconfigure the chip without loss of memory capacity.

Advanced redundancy configurations for fault-tolerant designs of wafer scale integration (WSI) static RAM modules have been developed. Two examples of these are: (1) word duplication and selection by horizontal parity checking (WDSH), and (2) error correction by horizontal and vertical parity checking (ECHV) [72]. Both of these techniques have single-bit error-correction capability in every word of read data, and are effective both for soft errors (logic upsets) and small defects in RAM cell arrays. However, this single-bit error correction is not adequate for burst errors caused by the defects in bit lines, word lines, and memory peripheral circuitry. Therefore, additional modifications such as automatic testing and replacement are proposed for these two techniques to enhance their defect and fault-tolerance capabilities.

Figure 5-32(a) shows the basic configuration of (a) a WDSH scheme implemented on a single RAM unit, called WDSH-B, and (b) an ECHV scheme implemented on a single RAM unit, called ECHV-B [72]. In the case of WDSH-B, with a 16 b I/O port, during the write mode operation, a 1 b horizontal parity is generated and attached to an input 16 b word. Then the 17 b word is duplicated, and stored in a 34 b word storage area of a single RAM unit. In the read mode operation, horizontal parity checking to both 17 b output words is achieved individually, and a normal one is selected as the final output. For ECHV-B, during the write mode operation, an input 16 b word is arranged into a 4 × 4 matrix. Then four horizontal parity bits and four vertical parity bits are generated, and attached to the matrix. These data are rearranged into a 24 b word consisting of four 5 b subwords and a 4 b subword, and stored in a 24 b word storage area of a single RAM unit. Similarly, in the read mode operation, an output 24 b word is arranged, and then single-bit error correction is performed based on the horizontal and vertical parity checking.

The two techniques, WDSH and ECHV, have been evaluated in a 1.5 μm CMOS tech-

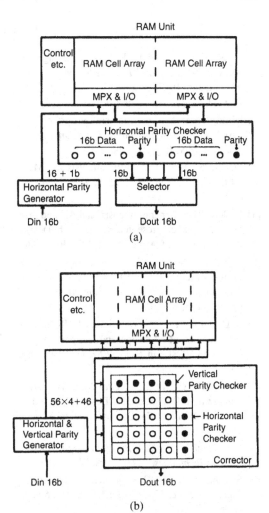

Figure 5-32. (a) Word duplication and selection by horizontal parity checking on a single RAM unit module (WDSH-B). (b) Error correction by horizontal and vertical parity checking on a single RAM unit module (ECHV-B). (From [72], with permission of IEEE.)

nology, 128 kb static RAM module. The test results showed that the ECHV-B had a slightly larger defect and fault-tolerance capability than the WDSH-B. However, a very high degree of defect and fault-tolerance capability can be obtained with advanced WDSH-based configurations employing both the automatic and external replacement procedures.

REFERENCES

[1] T. W. Williams and K. P. Parker, "Design for testability—A survey," *Proc. IEEE*, vol. 71, pp. 98–112, Jan. 1983.

[2] L. A. Goldstein *et al.*, "SCOAP: Sandia controllability/observability analysis program," in *Proc. 17th Design Automation Conf.*, Minneapolis, MN, July 1980, pp. 190–196.

[3] D. M. Singer, "Testability analysis of MOS VLSI circuits," in *Proc. IEEE ITC 1984*, pp. 690–696.

[4] P. Goel, "Test generation costs analysis and projections," paper presented at the 17th Design Automation Conf. Minneapolis, MN, July 1980.

[5] E. B. Eichelberger and T. W. Williams, "A logic design structure for LSI testability," *J. Design Automation & Fault Tolerant Computing*, vol. 2, pp. 165–178, May 1978.

[6] S. Dasgupta *et al.*, "A variation of LSSD and its implementation on design and test pattern generation in VLSI," in *Proc. IEEE ITC 1982*, pp. 63–66.

[7] M. Abramovici *et al.*, *Digital Systems Testing and Testable Design*, ch. 9: "Design for testability." Rockville, MD: Computer Science Press, 1990.

[8] P. H. Bardell *et al.*, *Built-In Test for VLSI: Pseudorandom Techniques.* New York: Wiley, 1987.

[9] S. Funatsu *et al.*, "Test generation systems in Japan," in *Proc. 12th Design Automation Symp.*, June 1975, pp. 114–122.

[10] J. H. Stewart, "Future testing of large LSI circuit cards," in *Dig. Papers, 1977 Semiconductor Test Symp.*, IEEE Publ. 77CH1261-C, Oct. 1977, pp. 6–17.

[11] H. Ando, "Testing VLSI with random access scan," in *Dig. Papers, COMPCON '80*, IEEE Publ. 80CH1491-0C, Feb. 1980, pp. 50–52.

[12] S. F. Oakland, "Combining IEEE Standard 1149.1 with reduced-pin-count component test," in *Proc. IEEE ITC 1991*, pp. 78–84.

[13] E. J. McCluskey, "Verification testing—A pseudoexhaustive test technique," *IEEE Trans. Comput.*, vol. C-33, pp. 541–546, June 1984.

[14] O. Patashnik, "Circuit segmentation for pseudoexhaustive testing," Center for Reliable Computing, Stanford Univ., Tech. Rep. 83-14, 1983.

[15] J. G. Udell *et al.*, "Pseudoexhaustive test and segmentation: Formal definitions and extended fault coverage results," in *Dig. Papers, 19th Int. Symp. Fault Tolerant Computing*, June 1989, pp. 292–298.

[16] Z. Barzilai *et al.*, "The weighted syndrome sums approach to VLSI testing," *IEEE Trans. Comput.*, vol. C-30, pp. 996–1000, Dec. 1981.

[17] D. T. Tang *et al.*, "Logic pattern generation using linear codes," *IEEE Trans. Comput.*, vol. C-33, pp. 845–850, Sept. 1984.

[18] S. B. Akers, "On the use of linear sums in exhaustive testing," in *Dig. Papers, 15th Annu. Int. Symp. Fault Tolerant Comput.*, June 1985, pp. 148–153.

[19] N. Vasanthavada *et al.*, "An operationally efficient scheme for exhaustive test-pattern generation using linear codes," in *Proc. IEEE ITC 1985*, pp. 476–482.

[20] E. C. Archambeau, "Network segmentation for pseudoexhaustive testing," Center for Reliable Computing, Stanford Univ., Tech. Rep. 85-10, 1985.

[21] K. K. Saluja *et al.*, "Built-in self-testing RAM: A practical alternative," *IEEE Design and Test of Comput.* pp. 42–51, Feb. 1987.

[22] P. H. Bardell *et al.*, "Built-in test for RAMs," *IEEE Design and Test of Comput.*, pp. 29–37, Aug. 1988.

[23] R. Dekker *et al.*, "Realistic built-in self test for static RAMs," *IEEE Design and Test of Comput.*, pp. 26–34, Feb. 1989.

[24] P. Mazumder and J. H. Patel, "An efficient built-in self testing for random access memory," in *Proc. IEEE ITC 1987*, paper 45.2, pp. 1072–1077.

[25] M. Franklin *et al.*, "Design of a BIST RAM with row/column pattern sensitive fault detection capability," in *Proc. IEEE ITC 1989*, Paper 16.3, pp. 327–336.

[26] B. Koenemann *et al.*, "Built-in logic block observation techniques," in *Proc. IEEE ITC 1979*, pp. 37–41.

[27] B. Koenemann *et al.*, "Built-in test for complex digital integrated circuits," *IEEE J. Solid-State Circuits*, vol. SC-15, pp. 315–318, June 1980.

[28] E. B. Eichelberger, *et al.*, "Random-pattern coverage enhancement and diagnosis for LSSD logic self-test," *IBM J. Res. Develop.*, vol. 27, pp. 265–272, Mar. 1983.

[29] S. Z. Hassan, "An efficient self-test structure for sequential machines," in *Proc. IEEE ITC 1986,* pp. 12–17.

[30] H. Maeno *et al.,* "LSSD compatible and concurrently testable RAM," in *Proc. IEEE ITC 1992,* paper 31.2, pp. 608–614.

[31] A. Krasniewski *et al.,* "High quality testing of embedded RAMs using circular self-test path," in *Proc. IEEE ITC 1992,* paper 32.3, pp. 652–661.

[32] R. Dekker *et al.,* "Realistic built-in self-test for static RAMs," in *Proc. IEEE ITC 1988,* pp. 343–352.

[33] B. Nadeau-Dostie *et al.,* "Serial interfacing for embedded-memory testing," *IEEE Design and Test of Comput.,* pp. 52–63, Apr. 1990.

[34] M. Nicolaidis, "Transparent BIST for RAMs," in *Proc. IEEE ITC* 1992, pp. 598–607.

[35] J. Inoue *et al.,* "Parallel testing technology for VLSI memories," in *Proc. IEEE ITC 1987,* paper 45.1, pp. 1066–1071.

[36] Y. Matsuda *et al.,* "A new array architecture for parallel testing in VLSI memories," in *Proc. IEEE ITC 1989,* paper 16.2, pp. 322–326.

[37] T. Sridhar, "A new parallel test approach for large memories," in *Proc. IEEE ITC 1985,* pp. 462–470.

[38] P. Mazumder and J. H. Patel, "Parallel testing for pattern-sensitive faults in semiconductor random-access memories," *IEEE Trans. Comput.,* vol. 38, pp. 394–407, Mar. 1989.

[39] Y. Morooka *et al.,* "An address maskable parallel testing for ultra high density DRAMs," in *Proc. IEEE ITC 1991,* paper 21.3, pp. 556–563.

[40] H. Koike *et al.,* "A BIST scheme using microprogram ROM for large capacity memories," in *Proc. IEEE ITC 1990,* pp. 815–822.

[41] T. Takeshima *et al.,* "A 55-ns 16-Mb DRAM with built-in self-test function using microprogram ROM," *IEEE J. Solid-State Circuits,* vol. 25, pp. 903–911, Aug. 1990.

[42] T. Chen *et al.,* "A self-testing and self-repairing structure for ultra-large capacity memories," in *Proc. IEEE ITC 1992,* pp. 623–631.

[43] R. Treuer *et al.,* "Built-in self-diagnosis for repairable embedded RAMs," *IEEE Design and Test of Comput.,* pp. 24–33, June 1993.

[44] P. Mazumder and J. S. Yih, "A novel built-in self-repair approach to VLSI memory yield enhancement," in *Proc. IEEE ITC 1990,* pp. 833–841.

[45] A. Tanabe *et al.,* "A 30ns 64-Mb DRAM with built-in self-test and self-repair function," *IEEE J. Solid-State Circuits,* vol. 27, pp. 1525–1531, Nov. 1992.

[46] R. Truer *et al.,* "Implementing a built-in self-test PLA design," *IEEE Design and Test of Comput.,* vol. 2, pp. 37–48, Apr. 1985.

[47] S. M. Reddy *et al.,* "A data compression technique for built-in self-test," *IEEE Trans. Comput.,* vol. 37, pp. 1151–1156, Sept. 1988.

[48] H. Thaler, "Pattern verification and address sequence sensitivity of ROMs by signature testing," in *Proc. IEEE Semiconductor Test Symp.,* Oct. 1978, pp. 84–85.

[49] Y. Zorian and A. Ivanov, "EEODM: An efficient BIST scheme for ROMs," in *Proc. IEEE ITC 1990,* pp. 871–879.

[50] Y. Zorian, "An effective BIST scheme for ROMs," *IEEE Trans. Comput.,* vol. 41, pp. 646–653, May 1992.

[51] P. Nagvajara *et al.,* "Built-in self-diagnostic read-only memories," in *Proc. IEEE ITC 1991,* pp. 695–703.

[52] W. W. Peterson and E. J. Weldon, Jr., *Error Correcting Codes,* 2nd ed. Cambridge, MA: M.I.T. Press, 1972.

[53] C. E. W. Sundberg, "Erasure and error decoding for semiconductor memories," *IEEE Trans. Comput.,* vol. C-27, pp. 696–705, Aug. 1978.

[54] C. L. Chen and M. Y. Hsiao, "Error-correcting codes for semiconductor memories applications: A state of the art review," *IBM J. Res. Develop.,* vol. 28, pp. 124–134, Mar. 1984.

[55] T. Fuja, "Linear sum codes for random access memories," *IEEE Trans. Comput.,* vol. 37, Sept. 1988.

[56] S. H. Han and M. Malek, "A new technique for error detection and correction in semiconductor memories," in *Proc. IEEE ITC 1987,* pp. 864–870.

[57] R. J. McPartland, "Circuit simulations of alpha-particle induced soft errors in MOS dynamic RAMs," *IEEE J. Solid State Circuits,* vol. SC-16, pp. 31–34, Feb. 1981.

[58] W. W. Peterson and E. J. Weldon, *Error Correcting Codes.* Cambridge, MA: M.I.T. Press, 1972.

[59] S. Lin, *An Introduction to Error-Correcting Codes.* Englewood Cliffs, NJ: Prentice-Hall, 1980.

[60] G. A. Sai-Halasz *et al.,* "Alpha-particle induced soft error rate in VLSI circuits," *IEEE J. Solid-State Circuits,* vol. SC-17, pp. 355–362, Apr. 1982.

[61] P. Mazumder, "Design of a fault-tolerant three-dimensional dynamic random access memory with on-chip error correcting circuit," *IEEE Trans. Comput.,* vol. 42, Dec. 1993.

[62] H. W. Kalter *et al.,* "A 50-ns 16-Mb DRAM with a 10-ns data rate and on-chip ECC," *IEEE. J. Solid-State Circuits,* vol. 25, pp. 1118–1128, Oct. 1990.

[63] K. Arimoto *et al.,* "A speed-enhanced DRAM array architecture with embedded ECC," *IEEE J. Solid-State Circuits,* vol. 25, pp. 11–17, Feb. 1990.

[64] S. Chau and D. Rennels, "Self-testing static random access memory," *NASA Tech. Brief,* vol. 15, no. 2, item # 140 [JPL Invention Rep. NPO-17939 (7433)].

[65] S. Chau *et al.,* "Self-checking on-line testable static RAM," U.S. Patent 5,200,963.

[66] D. P. Siewiorek and R. F. Swartz, *The Theory and Practice of Reliable System Design.* Digital Press, 1982.

[67] C. H. Stapper, "Synergistic fault-tolerance for memory chips," *IEEE Trans. Comput.,* vol. 41, pp. 1078–1087, Sept. 1992.

[68] K. E. Grosspietsch, "Schemes of dynamic redundancy for fault tolerance in random access memories," *IEEE Trans. Rel.,* vol. 37, pp. 331–339, Aug. 1988.

[69] A. M. Saleh *et al.,* "Reliability of scrubbing recovery-techniques for memory systems," *IEEE Trans. Rel.,* vol. 39, pp. 114–122, Mar. 1990.

[70] N. T. Jarwala and D. K. Pradhan, "TRAM: A design methodology for high-performance, easily testable, multi-megabit RAMs," *IEEE Trans. Comput.,* vol. 37, pp. 1235–1250, Oct. 1988.

[71] B. Ciciani, "Fault-tolerance considerations for redundant binary-tree-dynamic random-access-memory (RAM) chip," *IEEE Trans. Rel.,* vol. 41, pp. 139–148, Mar. 1992.

[72] N. Tsuda, "A defect and fault tolerant design of WSI static RAM modules," in *Proc. IEEE 1990 Conf. Wafer Scale Integration,* pp. 213–219.

6

Semiconductor Memory Reliability

6.1 GENERAL RELIABILITY ISSUES

Memory yield definition may vary from one manufacturer to another, although it is commonly expressed as a the ratio of the passing devices during an initial test such as wafer sort or wafer level acceptance testing versus the total number manufactured. Quality is measured by the proportion of chips (or packaged parts) that fall out during a screening test flow (e.g., screening per MIL-STD-883, Method 5004). Reliability is measured by the device failures that occur during the qualification and quality conformance inspection (QCI). Field failures are devices which fail in the actual operating environment (subsequent to all processing steps). A memory failure is defined as the loss of the ability to perform functionally or parametrically as per the device specification. The failures are typically grouped into one of three categories, depending upon the product's operating life cycle stage where the failures occur. These categories are as follows: infant mortality, useful life, and wearout failures. The infant mortality failures for a device type occur during the early product life cycle, and by definition, the rate of failure decreases with age. The major causes for early failures include design-related defects, flawed materials and contamination, poor manufacturing assembly and workmanship, improper process monitoring and lack of quality controls,

and inadequate screening (or burn-in). The second type of failures occur during the useful life of the device for which the rate of failure is assumed to be constant by a widely used model. During this stage, the causes of failures include inadequate derating, improper startups and voltage transients, and higher than expected thermal and electrical stresses. The third type are the wearout failures which occur late in the device operating life cycle. The cause of the wearout failures include aging phenomenon, degradation, and material fatigue, and by definition, the rate of failure during this cycle increases with age.

In accordance with this failure model, if the failure rate is plotted against the operating life test data taken from a statistically large sample size placed in operation at $t = 0$, then the resulting curve for medium- to low-level stress takes the shape of a "bathtub curve." For this bathtub curve, the high initial rate of failure represents infant mortality failures, the flat central portion of the curve indicating constant failure rate corresponds to the useful life of the product, and the increasing rate of failure with advanced age is due to the wearout failures. Figure 6-1 shows a series of curves corresponding to three different levels of stresses: high, medium, and low.

The reliability of a semiconductor device such as a memory is the probability that the device will perform satisfactorily for a given time at a desired confidence level under the specified

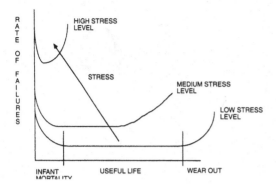

Figure 6-1. A typical bathtub curve.

operating and environmental conditions. The mathematical theory for reliability includes fundamental concepts such as the reliability function, the probability density function, the hazard rate, the conditional reliability function, and mean time to failure (MTTF). Several types of statistical distributions are used to model the failure characteristics, the hazard rate, and the MTTF. The most commonly used statistical distributions include normal, exponential, Weibull, and lognormal. The methodology used for calculation of the failure rates of microelectronic devices such as the memories is based upon device models and procedures outlined in MIL-HDBK-217 (Reliability Prediction of Electronic Equipment) and other commercial reliability prediction models. These will be discussed in more detail in Section 6.4.

The memory device failures are a function of the circuit design techniques, and materials and processes used in fabrication, beginning from wafer level probing to the assembly, packaging, and testing. Memory device fabrication is a complex combination of many materials—semiconductors, metals and their alloys, insulators, chemical compounds, ceramic, polymers—and its successful operation (or failure) depends on the interaction of these various materials and their interfaces. This section discusses general reliability issues for the memories applicable to bipolar and MOS technologies, such as: dielectric-related failures from the gate-oxide breakdown, time-dependent dielectric breakdown (TDDB), and ESD failures; dielectric-interface failures such as those caused by ionic contamination and hot-carrier effects; conductor

and metallization failures, e.g., electromigration and corrosion effects; and assembly- and packaging-related failures. However, there are some memory-specific failure modes and mechanisms applicable to the random access memories (RAMs) and nonvolatile memories, which will be discussed in Sections 6.2 and 6.3, respectively.

Table 6-1 summarizes the most common failure-related causes, mechanisms, and modes for the microelectronic devices such as memories (including soft errors) [1].

Hitachi has compiled memory users' data on failure modes such as pinholes, oxide film failures, defective photolithography, foreign materials contamination, etc. Figure 6-2 shows examples of

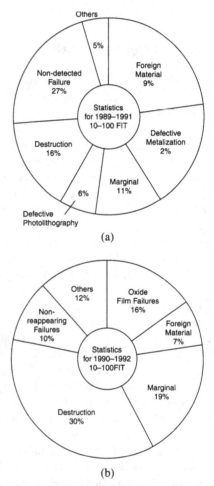

Figure 6-2. Relative percentage of common failure mode rates for memories. (a) BiCMOS. (b) MOS. (From [1], with permission of Hitachi America Ltd.)

TABLE 6-1. General Failure-Related Causes, Mechanisms, and Modes for Microelectronic Devices such as Memories. (From [1], with Permission of Hitachi America Ltd.)

Failure Modes	Failure Mechanisms	Failure-Related Causes	
Withstanding voltage reduced, Short, Leak current increased, hFE degraded, Threshold voltage variation, Noise	Pin hole, Crack, Uneven thickness, Contamination, Surface inversion, Hot carrier injected	Passivation	Surface oxide film, Insulating film between wires
Open, Short, Resistance increased	Flaw, Void, mechanical damage, Break due to uneven surface, Non-ohmic contact, Insufficient adhesion strength, Improper thickness, Electromigration, Corrosion	Metallization	Interconnection, Contact
Open, Short, Resistance increased,	Bonding runout, Compounds between metals, Bonding position mismatch, Bonding damaged	Connection	Wire bonding, Ball bonding
Open, Short	Disconnection, Sagging, Short	Wire lead	Internal connection
Withstanding voltage reduced, Short	Crystal defect, Crystallized impurity, Photoresist mismatching	Diffusion, Junction	Junction diffusion, Isolation
Open, Short, Unstable operation, Thermal resistance increased	Peeling, chip, Crack	Die bonding	Connection between die and package
Short, Leak current increased, Open, Corrosion disconnection, Soldering failure	Integrity, moisture ingress, Impurity gas, High temperature, Surface contamination, Lead rust, Lead bend, break	Package sealing	Packaging, Hermetic seal, Lead plating, Hermetic package & plastic package, Filler gas
Short, Leak current increased	Dirt, Conducting foreign matter	Foreign matter	Foreign matter in package
Short, Open, Fusing	Electron destroyed	Input/output pin	Electrostatics, Excessive voltage surge
Soft error	Electron hole generated	Disturbance	Alpha particle
Leak current increased	Surface inversion	Disturbance	High electric field

the relative percentage of these failure mode rates for both BiCMOS and MOS devices [1].

In general, for microelectronic devices such as memories, the failure mechanisms can be broadly categorized as: (1) semiconductor bulk failures, (2) dielectric failures, (3) semiconductor–dielectric interface failures, (4) conductor and metallization failures, and (5) assembly- and packaging-related failures. These failure mechanisms, as applicable to random access memories and nonvolatile memories, are discussed briefly in the following sections.

6.1.1 Semiconductor Bulk Failures

The bulk failure mechanisms are those occurring within the memory semiconductor material itself, and include the chip fracture, cracks or pits in the chip or passivation layers, secondary breakdown, latchup, displacement damage due to the ionizing radiation, and single-event upset (SEU). The secondary breakdown which was first observed in early bipolar transistors results in reduced voltage, and an increase in localized current density that may lead to permanent device degradation or failure [2]. Latchup in the bipolar or CMOS circuits occurs when a parasitic p-n-p-n structure is turned on due to a silicon-controlled rectifier (SCR) action. In the latchup condition, the gain of interconnected bipolar n-p-n and p-n-p transistors exceeds unity, which provides a low-impedance current path from the power supply to ground. The phenomenon of latchup is inherent in CMOS technology, and may be caused by the stray voltage transients [3–5]. It can also be caused by transient ionizing radiation and heavy particle space environment, which is discussed in Chapter 7 [6–8]. Techniques such as dielectric isolation have been used for hardening against the radiation-induced latchup for both the bipolar and CMOS memories. For bulk CMOS memories, another popular latchup prevention technique uses an epitaxial layer of appropriate thickness. Insulator technologies such as the silicon-on-sapphire (SOS) and silicon-on-insulator (SOI), which are latchup-immune because of their nonparasitic p-n-p-n structure, are particularly attractive for CMOS memories for use in high-radiation space and military environments.

In bipolar transistors, ionizing radiation can produce semiconductor bulk displacement damage which reduces the lifetime of minority carriers, thereby reducing the current gain. Total dose and transient radiation effects for both the bipolar and MOS memories are discussed in Chapter 7.

6.1.2 Dielectric-Related Failures

Memory dielectric defects in the deposited dielectric (e.g., pinhole)-related failure mechanisms can be attributed to causes such as: time-dependent dielectric breakdown (TDDB) at the defect sites in thermally grown oxides, dielectric wearout mechanisms due to the high field stresses, and dielectric rupture due to ESD or electrical overstress (EOS) [9–11]. The dielectrics used in memories include the oxides over the base and collector regions of the bipolar devices, thermally grown gate oxides and field oxides for the CMOS devices, tunneling oxides, and intermetal dielectrics in dual (or triple) levels of metallic conductors. The passivation layers include low-temperature deposited passivation materials over patterned metal, and high-temperature materials such as phosphosilicate glass (PSG) or boron phosphosilicate glass (BPSG). Thermally grown silicon dioxide has a low thermal activation energy of typically 0.3 eV, but a high-voltage acceleration factor (γ) of 10^7/MV/cm. As the MOS memories are scaled for higher densities, the margin between the intrinsic oxide breakdown voltage and the device operating voltage is considerably reduced [12]. The TDDB has been a significant failure mechanism in metal-gate as well as silicon-gate devices, and DRAMs with trench capacitors [13–17].

The dielectric wearout modes for very thin (<200 Å) thermally grown silicon films are caused by very high electric field stresses [18–20]. The wearout failures occur due to trapped charge in the oxide that increases the localized electric field value above the avalanche breakdown. This dielectric failure phenomenon is a significant reliability concern for the floating-gate technology nonvolatile EPROM (or UVPROM) and the EEPROMs that typically use 100 Å thin thermally grown oxide films. These dielectric failure modes and mechanisms for the nonvolatile memories will be discussed in more detail in Section 6.3.

The dielectric fractures can be caused by the thermal coefficient of expansion mismatch between the dielectric film and the underlying materials such as bulk semiconductor, polysilicon, or conductor metal. Dielectric film rupture over an aluminum metal line may be caused by the encapsulant plastic material, or the growth of hillocks from electromigration in the metal lines. The passivation layers are often deposited over the metallized pattern to provide mechanical protection and serve as a contamination barrier [21], [22]. However, cracks, pinholes, and other defects in the passivation layers can accelerate the rate at which corrosion can occur and cause electrical opens [23], [24]. A widely used passivation material is silicon nitride (SiN_xH_x) through the plasma-enhanced chemical vapor deposition (PECVD) system. Silicon nitride films prove to be an effective alkaline ion barrier, although they tend to reduce the stability of surface-sensitive device processes such as the MOS gate oxides and tunnel oxides in the EEPROMs.

The dielectric breakdown due to electrostatic discharge (ESD) or electrical overstress (EOS) is a major failure mechanism in the microelectronic devices. The memories are usually designed with input protection circuits to protect against the ESD hazards. The military specification for microcircuits MIL-M-38510 designates three levels of ESD failure threshold classification as follows: Class 1 (0–1999 V); Class 2 (2000–3999 V), and Class 3 (4000 V and above). The ESD characterization testing is performed per MIL-STD-883, Method 3015 for Electrostatic Discharge Sensitivity Classification.

6.1.3 Semiconductor–Dielectric Interface Failures

Memory semiconductor–dielectric interface failures are caused by the motion of charges in a dielectric or across the interface between the dielectric and underlying semiconductor material. The most significant failure mechanisms across the Si–SiO$_2$ interface are caused by alkali ion migration from contamination, hot carrier effects, slow trapping instabilities, and surface charge spreading [25–27]. Ionic contamination results from the sodium (and to lesser extent potassium) ions which are present in various materials and packaging processes in a typical memory IC manufacturing cycle. The mobility of sodium ions in silicon dioxide is quite high, and the presence of a high electric field at temperatures typically above 100°C causes these ions to drift toward the Si–SiO$_2$ interface. It is the surface potential which determines whether or not a conducting channel exists. Therefore, any drift or variation in the potential influences the gate threshold voltage, the carrier regeneration–recombination rate, and the depletion width of p-n structures [28]. The net effect of this ionic drift is to reduce the threshold of n-channel transistors and increase the threshold voltage of p-channel transistors. The positive ionic contamination can degrade information which is stored in the EPROMs as negative charge trapped within the floating gate.

The buildup of charge in the interface traps can result in the TDDB which presents a problem in the static MOS RAMs. These failures differ from field-dependent oxide breakdowns in that the breakdowns occur with a lognormal distribution of times after the application of field. The level of alkali ions in the MOS memories is monitored at the wafer level by determining the stability of the capacitance–voltage (C–V) curves and the flat-band voltage shift (V_{FB}) on MOS capacitor test structures in accordance with MIL-STD-883, Method 5007 Wafer Lot Acceptance Testing. The passivation layers typically used in memory device fabrication provide protection against the alkali ion contamination. However, because of imperfections in passivation layers, the contaminants can migrate to the defect sites and cause device degradation.

The wafer level mobile charge contamination is given by the equation [29]

$$Q_m = C_{ox} \cdot \Delta V_{FB}$$

where C_{ox} = oxide capacitance
Q_m = mobile charge per unit area.

The voltage variations due to mobile charge contamination can be expressed as

$$\Delta V = A t^{1/2} E^{1/2} \exp\left(\frac{-E_a}{kT}\right)$$

where A = normalizing constant
       E = electric field
       E_a = activation energy (typically
              1.0–2.0 eV).

In submicron memory designs, a good understanding of the hot-carrier effects in an actual circuit environment is essential to ensure product reliability in early stages of process optimization. Hot-carrier effects are generated in short-channel (<2 μm) devices such as DRAMs when operated at a voltage near the breakdown voltage (BV_{DS}) by the presence of a high drain–source electric field which may cause injection of some carriers across the Si–SiO$_2$ interface in the region of channel in the drain [30–33]. Hot carriers are electrons and holes in the transistor channel and pinch-off regions that have gained enough energy either from the drain–source field or due to thermal currents that would allow them to surmount the energy barrier at the silicon channel and tunnel into the dielectric.

There are three types of hot carriers depending upon the source origin: (1) channel hot carriers which, in crossing the channel, experience very few lattice collisions and can thereby accelerate to high energies, (2) substrate hot carriers which are thermally generated below the inversion layer in the substrate and drift toward the interface, and (3) avalanche hot carriers which are created in the avalanche plasma and result from multiple-impact ionizations in the presence of strong lateral electric fields [34]. The hot-electron injection can increase both the oxide-trapped charge (N_{ot}) and the interface-trapped charge (N_{it}) in the channel, which can affect both the threshold voltage (V_{TH}) and transconductance (gm) of the MOS transistors. As these hot carriers gain energy (≈1.5 eV in silicon) and penetrate into the dielectric, they have the capability of impact ionization, i.e., the creation of e–h pairs which increases the source–drain current and produces a substrate current. The hot carriers that continue to gain energy (≈3 eV) are injected over the energy barrier into the SiO$_2$ where they become trapped. Figure 6-3 shows the cross-sectional view of an MOS transistor operating in the saturation mode, showing impact ionization [32].

The hot-electron injection into the gate oxide has been modeled using various approaches [35], [36]. An example is Berkeley Reliability Tool (BERT), which is a public domain simulator for hot-carrier as well as oxide and electromigration reliability simulation [37]. It includes both the NMOSFET and PMOSFET hot-carrier models. The hot-carrier effects are a

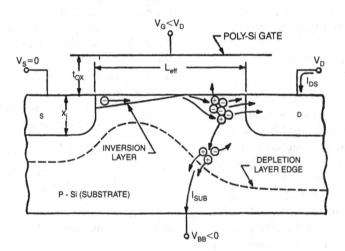

Figure 6-3. Cross-sectional view of an MOS transistor operating in the saturation mode, showing impact ionization. (From [32], with permission of IEEE.)

significant reliability concern for high-density MOS RAMs, and will be discussed in more detail in Section 6.2.

The semiconductor–dielectric slow-trapping instability effects refer to immobile charge density and surface charge density when a negative bias is applied to the gate of an MOS device [38]. This can increase the threshold voltage of p-channel transistors, reduce circuit speed, and eventually cause functional failures [39].

6.1.4 Conductor and Metallization Failures

Conductor and metallization major failure mechanisms in microelectronic devices such as memories include electromigration, metallization cracks and voids, corrosion faults and effects, and contact spiking [40–42]. Electromigration is a phenomenon which involves the transport of atoms of a metallic conductor such as aluminum because of high current density. It occurs at elevated temperatures under the influence of a temperature gradient or current density in excess of 10^6 A/cm^2 [43–46]. When the ionized atoms collide with the current flow of scattering electrons, an "electron wind" is produced [1]. This wind moves the metal atoms in the opposite direction from the current flow, which generates voids at a negative electrode, modules or hillocks, and whiskers at the opposite end electrode. The generated voids increase wiring resistance and cause excessive current flow in some areas, which may lead to open metallization. The whiskers generated at the electrode can cause a short circuit in multimetal lines. The electromigration-related failure modes can be summarized as follows [47]:

- Catastrophic open failures due to voiding for single-layered Al-alloy metallization.
- Resistance increases due to the multi-layered Al-alloy metallization systems.
- Interlevel and intralevel shorting due to the electromigration-induced extrusions.

- Resistive contact formation and junction spiking (leakage).

Stress-induced void formation can reduce the cross-sectional area in Al–Si metal lines which makes them more susceptible to electromigration or microcracking [48]. The microcracks usually occur at oxide steps and at the contact windows. Advances in memory processing techniques can minimize these effects. Techniques such as reactive ion etching have been used to achieve photolithographic patterns with near vertical sidewalls (e.g., in vias for double-level metal systems) and bias sputtering to improve step coverage.

Numerous electromigration studies have been performed on Al and Al-alloy metallization systems used in the memories. The mean time to failure (t_F) has been shown to vary as a function of the temperature and current density. In an early model proposed by Black and based on the classical assumption that the electromigration failures have lognormal distribution, the median-time-to-failure (t_{50}) is given by [49]

$$t_{50} = AJ^{-N}\exp\left[\frac{Q_A}{kT}\right]$$

where J = continuous dc current density in A/cm^2

Q_A = activation energy (typically 0.5–0.6 eV for pure Al and Al–Si alloys, 0.7–0.8 eV for Al–Cu alloys)

A = process-related parameter (depends upon factors such as the sample geometry, film grain size, temperature, and current density gradients)

N = constant (reported values from 1 to 7, with 2 being typical)

k = Boltzmann's constant

T = absolute temperature.

The electromigration is a significant reliability concern in memories with multilayer metallization systems such as Ti–W/Al–Si, etc. [50]. The techniques commonly used to reduce the susceptibility of Al metallization to electromigration failures includes the use of copper alloyed metal systems (e.g., Al with 4% Cu) and titanium alloyed metal composition systems. However, the Cu containing Al alloys is more susceptible to corrosion effects than the Al or Al–Si films.

Another electromigration-related phenomenon, contact spiking, is the localized penetration of metal into the semiconductor ohmic contact region at elevated temperatures. The high temperature may be caused by failure of the chip-to-substrate bond, thermal runaway of the device, or by the electrical overstress conditions (EOS). Electromigration-induced contact failure issues include the resistive contact formation and junction spiking (leakage), and impact of the silicides and barrier layers due to material transport. The contact time-to-failure (t_F) is given by Black's equation, with Q_A typically 0.8–0.9 eV for Al–Si to the nonsilicided junctions, and 1.1 eV for the Al–Si to silicided junctions and/or barrier layers.

The standard wafer electromigration accelerated test (SWEAT) technique for the EM or metal integrity testing was introduced to study metal reliability characteristics. The American Society for Testing and Materials (ASTM) has developed a standard test method entitled: "Estimating Electromigration Median-Time-to-Failure and Sigma of Integrated Circuit Metallization." The X-ray stress measurements can be used to evaluate the dependence of the thermal stress distribution of the passivated aluminum conductors on line width thickness. Aluminum yield stress increases with decreasing line width and thickness, and has a major impact on the stress states in aluminum [51]. A scanning electron microscope technique called charge-induced voltage alteration (CIVA) has been developed to localize open conductor voids on both the passivated and nonpassivated chips [52]. The CIVA images are produced by monitoring the voltage fluctuations of a constant power supply as an e-beam is scanned over the chip surface.

6.1.5 Metallization Corrosion-Related Failures

The corrosion of aluminum (or alloyed Al) metallization systems in semiconductor devices such as memories in plastic-encapsulated packages is a major failure mechanism [53–55]. This corrosion can be caused by the presence of halide ion contaminants which may originate from following sources: (1) plastic-encapsulation-related incomplete curing process, resulting in the ionic impurities, presence of halogens, and other corrosive elements; (2) atmospheric pollutants and moisture; (3) fluxes used for lead finishing; (4) impurities in die attach epoxies and molding compounds; (5) byproducts from reactive ion etching of metallization; and (6) human operators' fingerprints, spittle, etc. The hermetically sealed packages including metal, glass, ceramic, and other types can develop leaks under severe environmental operating conditions and collect above normal moisture (H_2O) content. The corrosion of aluminum metallization systems can occur at the anode as well as the cathode, and can be galvanic due to two different metals in contact, or electrolytic under the biased conditions. In plastic-encapsulated fusible link PROMs, the corrosion of nichrome elements has been reported to be a failure mechanism [56].

In plastic packages, a mismatch of thermal coefficients of expansion between the silicon die and encapsulant materials, as well as chemical shrinkage, can cause mechanical stresses. These stresses can lead to cracks in the plastic package and/or die, including the passivation layers and metallization. In thin plastic packages, cracks can also develop if the die overheats due to an electrical overstress, or latchup in CMOS devices. The cracks may be caused by high-pressure thermomechanical stresses inside the package created by the outgassing of plastic encapsulant material surrounding an overheated die [57]. A polyimide passivation is often used in the memory plastic packages to improve reliability and provide mechanical protection of the die surface (e.g., to reduce film cracking) [58]. However, the polyimide etch residues or hydrolysis products generated in a humid environment may promote intermetallic growth which can produce gold ball bond degradation.

Advances in memory plastic packaging technologies have been made to reduce the corrosion susceptibility by monitoring and controlling contamination levels, improvement of chip passivation layers, and reduced thermal coefficient of expansion mismatch. C-mode scanning acoustic microscopy (C-SAM), in conjunction with the accelerated aging conditions (e.g., 85°C in 30–90% relative humidity), can be used to characterize the moisture sensitivity of the plastic-packaged memory surface mount devices [59].

For high-reliability applications, the memory ceramic packages are screened and quality conformance inspection is performed for hermeticity, including fine and gross leak tests, residual gas analysis (RGA), etc., in accordance with MIL-STD-883 Test Methods. For example, a dew point test is performed per MIL-STD-883, Test Method 1013, to detect the presence of moisture trapped inside the package in sufficient quantity to adversely affect the device parameters. A thin-film corrosion test per MIL-STD-883, Method 1031 is particularly suitable for devices containing thin-film conductors such as resistors or fuses (e.g., nichrome in PROMs) which are susceptible to corrosion as a result of excessive water vapor content inside the package. A highly accelerated stress (HAST) test was developed for high temperature/high humidity (85°C/85% RH) as part of the package acceptance and qualification testing. These tests will be discussed in more detail in Section 6.7.

6.1.6 Assembly- and Packaging-Related Failures

The major reliability issues for memory chip assembly processes and packaging operations are related to failure mechanisms pertaining to: (1) die bonding, (2) wire bonds, (3) contaminant effects, and (4) corrosion of aluminum metallization systems. The die bonding to a metallic (header/lead frame) or a ceramic substrate provides a mechanical support for the chip and a low-resistance path for heat conduction from the chip. There are several die bonding materials used, such as gold–silicon (Au–Si) or Au–germanium eutectic alloy, silver-filled epoxy, and polyimide adhesives. Some of

the major causes for die bonding failures include: voids in the die bonding materials, contamination, formation of brittle intermetallics at the bonding surface, higher than normal thermal resistance, and die cracking from the die-bond-induced stresses [60–62]. Some of these die bonding process-related problems may cause reduced heat conduction, higher chip temperature, and eventual disbonding.

To avoid die-bond-induced chip cracking, advanced assembly techniques include making distortion measurements using sensitive gauges, the use of finite element analysis, and thermal cycling test results. Screening for the die bond quality assurance inspection include tests such as the centrifuge (acceleration), temperature cycling, thermal shock, vibration, X-ray analysis, ultrasonic scan, and die shear (as part of the destructive physical analysis) in accordance with MIL-STD-883 Test Methods.

Wire bonding is a critical assembly process, and is automated to avoid manual assembly-induced defects. Aluminum and gold are the two materials used for wire bonds, and the most common methods of wire bonding are the thermocompression and ultrasonic techniques. Thermocompression bonding, commonly employed for gold wire bonds, uses a combination of heat and force to cause the metal atoms of the wire to diffuse into the metallization on the silicon or substrate to create a bond. Al wire bonding uses an ultrasonic technique in which the wire is placed in direct contact with the metallization, and then the bonding pad and wire are vibrated to scrub off the oxide and create sufficient heat to form a bond. Some of the common wire bonding failure mechanisms are: poor bonding machine control causing underbonding or overbonding, contamination effects that may produce poor lead adhesion, improper lead dressing resulting in wire loops or sags, and bondwire-pad intermetallic formation (also called purple plague) [63–65]. The use of adequate process controls and optimum wire bonding parameters has produced very high-quality gold ball bonds [66].

For high-reliability military and space applications, a nondestructive wire bond pull test is performed in accordance with MIL-STD-883, Test Method 2023. Other screens and quality conformance inspection tests for assuring high wire bond integrity include: centrifuge (acceleration), temperature cycling, thermal shock, vibration, and X-ray analysis (see Section 6.7).

The development of very high-density monolithic memory chips has been accompanied by a typical decrease in package outline and volume, and a switch from through-the-hole to surface mount technologies. In addition, high chip integration densities have been accompanied at the cost of the following: (1) reduced feature size, (2) the use of new materials, e.g., complex silicides and oxides, (3) multilevel layers of metallization and polysilicon, and (4) increased 3-D processing such as trenches and pillars [67].

Thermomechanical stresses inside the plastic packages include shear stresses on the chip surface in the corners and compressive stresses throughout the bulk silicon. The stress-related failures for chips include mount-induced silicon and passivation layer cracks, shifted and smeared metal, interlevel oxide cracks, bond wire chipouts, lifted ball bonds, and electrical parameters shifts. The stress-related package failures can be listed as follows: cracks between the leads, crevices between the encapsulant plastic and leads, cracks around the mount pad and chip, delamination from the chip/lead frame, etc. Military specifications for memories (MIL-M-38510 and MIL-I-38535) require package qualification and QCI testing, which will be discussed in Section 6.7.

6.2 RAM FAILURE MODES AND MECHANISMS

The memory manufacturing process and field failure data analysis for certain failure modes and mechanisms have shown a strong relationship among the yield, quality, and reliability [68]. Section 6.1 discussed general reliability issues, including the failure modes which are applicable to microelectronic devices in general, including the memories. This section discusses reliability issues and failure modes which are of special concern for random access memories, both SRAM and DRAM. These issues include the gate oxide reliability defects, hot-carrier degradation, DRAM capacitor charge-storage and data-retention properties, and the DRAM soft error phenomenon.

6.2.1 RAM Gate Oxide Reliability

A gate oxide short is a transistor defect that may produce a low-impedance path between a MOSFET gate and the underlying silicon substrate (or p- or n-well, source, or drain) [69]. The gate oxide shorts in memories are serious reliability concerns, and may result in an increased quiescent current (I_{DDQ}), degraded logic voltage levels, and propagation delay times. The defective gate oxides in low-density ICs (e.g., 1K SRAMs) may be detected parametrically by the I_{DDQ} tests which are performed by writing and reading "0s" and "1s" at all memory locations. For high-density memory gate oxide defect detection, more sophisticated fault modeling and special functional I_{DDQ} test vector sets may be required (see Chapter 4).

A study was performed to evaluate the gate dielectric reliability issues for a CMOS 3.6 V submicron 144K SRAM with 140 Å gate oxide thickness, and wafer channel lengths of 0.5 ± 0.15 μm [70]. The fabrication process used NFET with a double-implanted, lightly doped drain source to minimize the hot-carrier and short-channel effects [71]. The PFET used a buried channel design with an n-well doping. To obtain matched channel lengths, several levels of spacers were used to define the NFET and PFET diffusions in separate and joint process steps. Figure 6-4 shows the NFET and PFET cross-sections [70].

The test structures used were fabricated by designing a unique metal personalization mask identical to the one used on first-level metal for static RAM wafers. These test structures included: (1) thick-oxide bounded thin-oxide epitaxial capacitors, (2) finger-like n-channel device configurations with diffusion-bounded structures going up and down thick oxide islands, and (3) a personalized SRAM

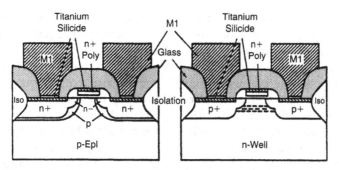

Figure 6-4. 144K SRAM NFET and PFET cross-sections. (From [70], with permission of IBM Corporation.)

gate array structure. The gate oxide reliability evaluation testing was performed at room temperature by using double-voltage ramp stress in excess of 10 MV/cm for the oxide breakdown. The stress test results showed a high fallout due to gate oxide shorts. Analysis of the failed units revealed adverse effects of exposing the gate dielectric to an ion implantation process and pre-polysilicon cleans that introduced foreign material contamination. Therefore, process optimization is typically performed during the photoresist removal and reactive ion etching (poly, gates, contacts/vias, and metal). The device reliability improved by a factor of $5\times$ when these gate oxide processes were eliminated and an in-line wafer inspection/defect monitoring system was introduced.

The memory gate dielectric integrity and reliability are affected by all processes involved in the gate oxide growth. Another study was performed to identify the critical gate oxide process parameters in IBM 0.5 μm 16 Mb DRAM CMOS technology that utilizes shallow trench isolation between the deep-trench storage capacitors [72]. The gate dielectric used is a 13 nm planar oxide. These DRAMs use p-channel transfer devices which are connected to the deep-trench capacitors via a doped polysilicon surface trap. Gate dielectric yield is measured on segments of 16 Mb arrays, and additionally on long serpentine antenna structures consisting of 128 kb cells. Two types of antenna test structures were used: one with the deep-trench capacitors, and the other without them.

The step stress testing was performed on wafers from various process options to identify the process that produced the most reliable gate dielectric. The voltage step stress testing ranged from 5.5 to 8 V on 16 Mb array segments which had the gates shorted to each other by the first layer of metal. A double-ramp breakdown stressing from 0 to 12 V, at 5 V/s, was done at room temperature on a 128 kb antenna structure. Gate oxide failures were analyzed using the scanning electron microscope (SEM) and electron-beam-induced current (EBIC) techniques. Test results of this study are summarized as follows:

- A lower temperature furnace anneal (e.g., 950°C for 20 min) is preferable to a higher temperature rapid thermal anneal (RTA) at 1090°C, for an optimum dielectric reliability. However, the adverse effects of high temperature may be mitigated if other process parameters such as gate oxide preclean and passivation are optimized.

- The use of pregate oxidation megasonics preclean improved the dielectric reliability even for the process that used high-temperature RTA. The gate oxide pinhole defects were also reduced to a minimal level by the use of megasonics.

- The charging effects occurring during the deposition of the passivation layer (PECVD PSG) above the gate conductor also degraded the gate dielectric integrity and reliability.

• The transient response of the RF matching network used in PSG deposition also affected the gate dielectric yields. Test structures that used deep-trench capacitors required additional processing and had higher defect levels than simpler structures. The use of an upgraded RF matching network in reliability modeling improved yield and reliability.

6.2.2 RAM Hot-Carrier Degradation

The reduced MOS transistor geometries from scaling of the memory devices for higher densities has made them more susceptible to hot-carrier degradation effects. Several studies have been performed to correlate the experimental data obtained from transistor stress test results to derive circuit device lifetime. These mean time to failure calculations often include some assumptions regarding the transistor's lifetime criteria for parametric variations (e.g., threshold voltage and transconductance shifts), duty cycle effects, ac-enhanced degradation effects, annealing effects, and usually unknown sensitivity of the circuit performance to transistor degradation [73], [74].

A study was performed to investigate the relationship between the transistor's degradation and hot-carrier lifetime for 64K (8K × 8) full CMOS SRAMs [75]. These SRAMs featured a conventional six-transistor memory cell fabricated with a single-poly, double-metal 1.2 μm twin-tub CMOS process with p-epi on a p^+ substrate. The 1.2 μm LDD n-channel and 1.4 μm conventional p-channel transistors have $L_{eff(min)}$ of 0.75 μm and have n^+ and p^+ doped polysilicon gates in the matrix, respectively [76]. This SRAM is organized in 16 sections, with each section consisting of a cell array matrix of 128 rows × 32 columns.

In this study, the SRAM stress conditions consisted of dynamically operating the devices at high speed, low ambient temperature ($-30°C$), at different voltages (V_{dd}) ranging from 7.5 to 10 V, using a worst case pattern with all input voltage levels raised to V_{dd} to limit power dissipation in the input/output buffers. Static transistor degradation experiments were performed at 20°C ambient temperature. During a stress test, this circuit parameter monitoring using a Teradyne J386 memory tester showed an increase of access time (T_{aa}) as a function of stress time and degradation of $V_{dd(min)}$. The location of physical damage was determined by an electrical analysis (and confirmed by circuit simulation) voltage microprobing of the memory cell transistors and photoemission microscopy. The detailed analysis showed that the hot-carrier degradation of CMOS SRAMs results in an increase of the minimum operating voltage (V_{dd}), the write time parameter (T_{dw}), and the access time. The circuit simulations showed that the write time degradation is caused by the presence of a metastable state in the degraded cell during the write operation. The degradation of $V_{dd(min)}$ and T_{dw} depends only on the access transistor of the memory cell. The relationship between the hot-carrier lifetime of statically stressed transistors and SRAM devices showed that the SRAM (product) lifetime is significantly larger (roughly by a factor of 50) than the transistor lifetime. This discrepancy may be the result of the relatively low sensitivity of SRAMs to the transistor degradation, and by duty cycle effects. Therefore, the conclusion of this study was that the SRAM lifetime can be severely underestimated if it is derived solely from the static transistor lifetime data.

The impact of hot-carrier degradation on DRAM circuit functionality has also been investigated. The DRAMs are particularly susceptible to hot-carrier damage due to the bootstrapping techniques used to boost the internal voltage levels to twice V_{cc}. Hot-carrier effects may cause a change in the MOSFET threshold voltage, transconductance, and subthreshold leakage. The loss of bootstrapped levels during a pause allows precharged high-balanced sense nodes to drift prior to sensing. For hot-electron effects evaluation, a low-temperature and high-voltage accelerated life test (dynamic) was performed on several 64K and 256K DRAMs [77].

In this study, the DRAM data-retention time, which is the maximum interval that the devices can remain inactive and still function properly, was characterized. Schmoo plots of the data-retention time versus V_{cc} were generated to accurately model the degradation rate for the devices that had been subjected to hot-electron stress tests. Analysis showed multiple pass and fail regions as the pause time was extended. The DRAM parametric shifts observed were due to an interaction of degraded precharge clocks and sense circuits. The bit map results of stressed devices show that the refresh time degradation is due to improper sensing of the stored data after an extended pause. Therefore, the data loss is due to peripheral circuit malfunctioning rather than cell retention failure. Also, an increased degradation rate was observed for the DRAMs passivated with a plasma-deposited silicon nitride layer, which is known to contain significant quantities of hydrogen.

Dual-dielectric SiO_2/Si_3N_4 nanometer thick films have been reported to be promising material for DRAM fabrication because of their high integrity compared with thermal silicon dioxide [78]. These dual-dielectric films can trap charge in the film. Their trap/detrap properties have been investigated to explain I–V characteristics or a dielectric breakdown mechanism. A study was performed to investigate the relationship between trapped charge and signal voltage shift in a DRAM storage cell and to discuss the limitations of SiO_2/Si_3N_4 films [79]. An evaluation was performed on dual-dielectric films fabricated by deposition of an 8 nm thick LPCVD Si_3N_4 layer followed by thermal oxidation on a storage electrode of n-type polysilicon. The top oxide is about 1 nm thick, and the equivalent SiO_2 thickness of this SiO_2/Si_3N_4 film is 6.5 nm. The trap/detrap characteristics of this film were examined on the wafer by C–V measurements using a capacitor test structure. To investigate the shift of a DRAM cell signal voltage, measurements of retention characteristics of 4 Mb DRAMs with stacked capacitor cells were made [80].

Figure 6-5(a) shows charge trap characteristics of SiO_2/Si_3N_4 film as a function of stress time (T) and stress bias applied to the plate electrode [79]. Q_{trap} is the net negative charge trapped in the film independently of the polarity of stress bias and increases with an increasing stress bias. It also increases logarithmically with stress time (T_{stress}) and saturates after

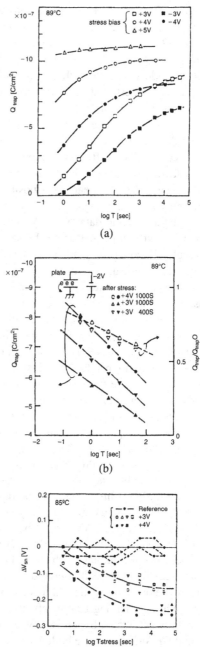

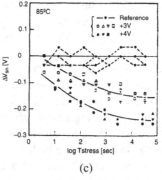

Figure 6-5. (a) Charge trap characteristics of SiO_2/Si_3N_4 film as a function of stress time T. (b) Detrap characteristics of SiO_2/Si_3N_4 film under plate bias of -2 V just after positive stress test. (c) Shift of DRAM cell signal voltage, ΔV_{sn} versus stress time. (From [79], with permission of IEEE.)

a certain time. Figure 6-5(b) shows the detrapping characteristics of trapped charge under the plate bias of -2 V versus time as a parameter of trapped charge (Q_{trap0}) just after the positive stress test. Q_{trap} decreases logarithmically with time T, after a stress test, and the detrap rate of Q_{trap}/Q_{trap0} is independent of Q_{trap0} [79].

The test results showed that the detrapping causes deterioration of the signal voltage of a DRAM cell, and also affects the maximum time (also called pause time). This reduction in cell signal voltage (ΔV_{sn}) appears after positive peak bias stress, i.e., when logic data "1" (4 V) are written just after logic data "0" (0 V) have been repeatedly written or read from the same cell. This logic "1" voltage degradation is a reliability concern since leakage current through a p-n junction between the storage node and p-type substrate reduces the storage node potential. Figure 6-5(c) shows a ΔV_{sn} degradation versus stress time (T_{stress}) plot [79]. In the case of negative plate bias stress, i.e., when logic data "0" (0 V) are written into a cell just after logic data "1" (4 V) have been repeatedly written or read for the same cell, ΔV_{sn} increases.

The conclusion of this study was that trap/detrap characteristics are strongly dependent upon the stress voltage. Therefore, the thickness and plate bias should be optimized by considering not only the leakage current through the SiO_2/Si_3N_4 film, but also the detrapping effects. These findings are especially significant for low-voltage DRAM operation and the use of thinner dual-dielectric films. Advanced processing techniques such as lightly doped drain (LDD) and graded drain transistors have been developed to reduce high electric fields and the hot-carrier degradation effects.

In a study performed by TI, two types of experimental DRAM chips were built using two different processing technologies: (1) a 2 μm NMOS process with abrupt junction transistors, and (2) a 1 μm CMOS process with graded junction devices [81]. These DRAMs were hot-carrier-stressed while monitoring the performance degradation of access time, precharge time, and refresh time. The impact of hot-carrier stress on DRAM access time can be assessed by

monitoring the drain current degradation in both the linear and saturation regions. In a DRAM, the precharge time defines the rate at which the bit/sense lines are equalized to their preset value before the next cycle of operation. For performance degradation due to hot-carrier stressing, it would take longer to precharge both the bit lines. It was observed that the precharge time degradation for NMOS DRAMs tested was due to linear drive current degradation in the precharge device. It could be minimized by slightly increasing the channel length of the precharge transistor. In the CMOS DRAMs (e.g., 1 and 4 Mb devices), the bit lines are usually precharged to $V_{DD/2}$, and hence the channel hot-electron stress on the precharge transistor would be much less. The CMOS DRAMs (1 μm design) showed very little shift in precharge time with hot-carrier stress.

In a DRAM, the leakage of a stored logic "1" level determines the time required to refresh the data. The burst refresh (or pause refresh) time, the TREF degradation due to hot-carrier stress was found to be a problem for the NMOS DRAMs rather than the CMOS devices. The TREF degradation observed in NMOS DRAMs was attributed to the loss of charge (due to increased saturation region subthreshold current) in the booted node of a critical clock circuit. The transistor degradation due to hot-carrier stress can also adversely impact the overall functionality of DRAMs by affecting the on-chip circuitry such as the sense amplifiers, decoders, and output buffers.

6.2.3 DRAM Capacitor Reliability

6.2.3.1 Trench Capacitors. High-density DRAMs (e.g., 1–4 Mb) use trench capacitor technology, which was discussed in Chapter 2. The amount of charge needed to be stored by each cell has basically remained constant at roughly 250–300 fC. This implies that scaling of the cell size may have to be achieved by reducing the oxide thickness of the planar capacitor. However, very thin oxides are susceptible to failure mechanisms such as the TDDB. Another solution has been the use of a 3-D trench, which requires only a small surface area for the capacitor, and the capacitance can be controlled by the depth of the trench as well as the insulator thickness. The major reliability issues for trench capacitors are electrical properties of the oxides grown in trenches, conduction mechanisms, breakdown characteristics, wearout phenomenon, and the soft-error rate (SER) relative to the planar structures.

Figure 6-6(a) shows the schematic of a trench capacitor cell [82]. The maximum electric field that this capacitor can handle is determined by the trench profile. A fabrication process which ideally produces rounded trench corners can result in maximum electric field strength similar to that for the planar capacitors. However, any sharp corners (or rough edges) at the top or bottom of the trench can lead to poor voltage breakdown distribution. Also, the oxide thinning and electric field enhancement in trench corners can result in a severe increase in leakage currents, even at low voltages. Accelerated testing by application of high electric fields and elevated temperatures has shown that the electric field acceleration is dependent upon temperature and the activation energy is field-dependent. Figure 6-6(b) shows typical data for the wearout of oxides in trenches (T_{ox} = 150Å) which fits the standard lognormal distribution [82]. The wearout phenomenon is similar to that for planar oxides, except for a slightly more accelerated wearout for the trench oxides.

The trench capacitor's reliability can be significantly improved by using an oxide film thicker than the ones used for planar capacitor DRAMs. A comparison of DRAM's soft-error rates (SER) of the trench capacitor versus the planar capacitor showed no significant increase in the SER for trench structures. In fact, the SER for trench cell DRAMs can be reduced considerably by selecting adequate cell capacitance, proper design of cell to sense amplifier charge transfer, and optimized bit-line structure. The use of silicon nitride films in conjunction with silicon dioxide (also called ONO) insulators in trench capacitors has been investigated for use in high-density DRAMs.

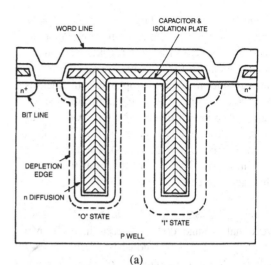

(a)

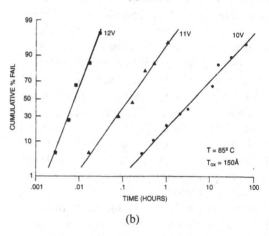

(b)

Figure 6-6. (a) Schematic of a trench capacitor cell. (b) Data for wearout of oxides in trenches, T_{ox} = 150 Å. (From [82], with permission of IEEE.)

A study was performed on thin-layer $SiO_2/SiN/SiO_2$ insulators for 9 μm deep trench capacitors fabricated on high boron concentration substrate using various combinations of thicknesses for the bottom oxide, middle nitride, and top oxide layers [83]. These thicknesses were measured by TEM analysis as well as capacitance measurements. The TDDB test was performed for both positive and negative stress to the polysilicon gate electrode. There were no early failures due to insulator pinholes. Test results showed that the leakage current and TDDB characteristics for ONO insulator trench capacitors were comparable to those for the planar capacitors. For ONO film, the breakdown voltage is dependent on the gate bias polarity and the ratio of bottom oxide to top oxide thicknesses for electron current (i.e., the current for positive gate bias is controlled by the top oxide thickness, and for negative gate bias, it is controlled by the bottom oxide thickness). The mean time to failure (MTTF) becomes longer as the nitride thickness increases.

6.2.3.2 Stacked Capacitors.

The conventional storage dielectric used in 4–16 Mb stacked DRAM chips consists of thin oxide/nitride (ON) or oxide/nitride/oxide (ONO) composite films. For 64 and 256 Mb DRAMs in development, stacked capacitor cells (STCs) are being used that require ultrathin dielectric materials which can provide high capacitance values. These STCs depend heavily on the quality and storage capacity of the dielectric film sandwiched between two heavily doped polysilicon electrodes. Silicon nitride (Si_3N_4) films are considered attractive for use in the DRAM capacitors because of their high dielectric constant. However, the application requires films with low leakage currents, high breakdown voltage, and stable flatband voltage (V_{FB}). Also, the nitride films are known to contain many trap levels which cause leakage currents and V_{FB} shifts. A study was performed to investigate the mechanism of leakage current and V_{FB} shift, and the dielectric breakdown behavior of stacked film capacitors $SiO_2/Si_3N_4/SiO_2$ with an oxide-equivalent thickness ranging from 100 to 200 Å [84].

The leakage current of stacked capacitors, time-zero dielectric breakdown, and the TDDB characteristics were measured. Test results showed that the leakage current of stacked films is not that significant. The V_{FB} shift under the operating voltage of a DRAM was less than 1 V. The capacitors with stacked film exhibited a lower failure rate for time-zero dielectric breakdown and TDDB than a planar MOS capacitor. The electric field acceleration factor of the stacked capacitor was less than that for a MOS capacitor with the same oxide equivalent thickness. The activation energy of the stacked capacitor was comparable to that reported for a MOS capacitor. The conclusion of this study was that the stacked film dielectric layer tested was highly reliable for DRAM applications.

In another study performed to evaluate ultrathin dielectric material for the DRAMs, it was found that capacitance values of up to 12.3 fF/μm^2 can also be obtained with ultrathin nitride-based layers deposited on rugged polysilicon storage electrodes [85]. Two-step annealed low-pressure chemical vapor deposition (LPCVD) tantalum pentoxide (Ta_2O_5) films have been used for capacitance values as high as 12.3 fF/μm^2 for 1.5 V operated DRAMs. Thinned silicon nitride layers yield large capacitance values. However, because of the oxidation punchthrough mechanism, there is an upper limit to the maximum capacitance value that can be reached with the ON dielectrics. The capacitance values obtained by using the ON films on rugged polysilicon were comparable to the maximum reported for Ta_2O_5 layers. Test results showed that the ON-based rugged capacitors presented lower leakage current levels for positive bias stress conditions. However, because of reliability considerations, the nitride film thickness cannot be scaled below 6 nm.

It has been shown that the existence of the bottom oxide layer degrades the TDDB characteristics of stacked ONO dielectrics. Also, the local electric field intensification and current density at the tips of asperities on the rugged poly-Si surface can enhance the electron trapping, which can trigger a fast breakdown of the bottom oxide layer. A study was performed that used rapid thermal nitridation (RTN) of the rugged poly-Si surface prior to Si_3N_4 deposition [76]. It was found that the use of RTN treatment of rugged poly-Si prior to Si_3N_4 deposition resulted in ON films with significantly reduced defect density, improved leakage current and TDDB characteristics, suppressed electron and hole trappings, and improved reliability against capacitance loss during stress.

6.2.4 DRAM Soft-Error Failures

The DRAM alpha-particle-induced soft-error phenomenon was briefly discussed in Chapter 2. The alpha particles from radioactive trace impurities present in the packaging materials can cause soft errors (bit upsets) in memory storage cells. One approach to realize a DRAM cell with a smaller area, larger capacitance, and with reduced soft-error rate (SER) is to minimize the surface area of the storage node in the substrate. This has been done by investigating the use of 3-D storage cell structures such as trench capacitors, stacked capacitor cells (STCs), buried storage electrodes (BSEs), etc. Another approach uses a combination of trench capacitor cell and a p-well on p-substrate to protect against minority carrier collection and help reduce the SER. An evaluation was performed for measuring the SER in experimental 256K NMOS DRAM (2 μm feature size) with a grounded cell plate and fabricated with several structures on p-type substrate [86]. The trench capacitor cells were formed either directly on a uniform p-type substrate or on a p-well in a p-type substrate. Some planar capacitor DRAMs were also fabricated for comparison purposes. These DRAMs were subjected to accelerated testing using an alpha-particle source. The SER was measured at different supply voltages and various cycle times. Two types of soft-error modes were observed: (1) the cell mode for which the SER is independent of the cycle time, and (2) the bit-line mode whose SER is inversely proportional to the cycle time.

The test results of this evaluation showed that both the cell mode and bit-line mode soft-error rate can be effectively reduced by about 1/200 or lower by utilizing the trench capacitor with a p-well structure on p-type substrate. Soft-error phenomenon modeling has shown that two kinds of charge mechanisms are involved: funneling and diffusion [87]. The experimental data showed that SER is primarily dependent on acceptor concentration (N_A), which implied that the dominant mechanism for the SER was funneling rather than diffusion. A simple model based on the funneling mechanism has been proposed. In the funneling mechanism, the amount of charge collected (Q_{coll}) is proportional to the depletion width (or acceptor concentration N_A). This relationship can be expressed by

$$Q_{coll} \propto N_A^{-1/2}.$$

Also, the critical charge (Q_{crit}) can be written as

$$Q_{crit} = C_s (V_{cc} - V_o)$$

where C_s = cell capacitance
V_{cc} = supply voltage
V_o = constant representing the sense amplifier imbalance and sensitivity.

Since the cell capacitance (C_s) is almost constant,

$$Q_{crit} \propto V_{cc} - V_o.$$

Soft errors occur when $Q_{coll} > Q_{crit}$. The SER is invariant if the unit of electric charge is changed, and is a function of Q_{coll}/Q_{crit}:

$$SER = K [N_A^{1/2} (V_{cc} - V_o)]$$

where K is a universal function which is determined as the physical cell dimension is fixed.

Another funneling length model proposed shows that a "scaling low" for soft errors exists which determines a limitation for the planar and trench cells [88]. Effective funneling length can be scaled down to some extent because of the p-n junction size effect and the proximity effect of adjacent cells. Funneling length does not depend on the alpha-particle track length in the depletion layer. The barrier effect of n^+-p structures is influenced by the power supply (V_{cc}), and reduced V_{cc} has a significant effect on critical charge collection. To reduce the soft-error rate in megabit DRAMs, advanced design techniques such as an n^+-p barrier for a switching MOS transistor and large storage capacitor cells are necessary.

A significant improvement in the SER for a BiCMOS process was achieved by implementing a triple-well structure. Figure 6-7 shows a schematic cross-section of: (a) the standard 0.5 μm BiCMOS process, and (b) the modified triple-well BiCMOS structure which incorporates the n-well under the p-well by extending the n^+ buried layer [89]. For the triple-well process, additional doping is needed in the p-well to provide adequate isolation between the n^+ source/drain and n^+ buried layer. Accelerated soft-error rate (ASER) testing was performed on

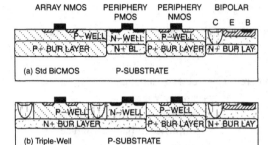

Figure 6-7. Schematic cross-section of (a) standard 0.5 μm BiCMOS process. (b) Modified triple-well BiCMOS structure. (From [89], with permission of IEEE.)

4 Mb SRAMs made from three different processes: 0.5 μm standard BiCMOS, an elevated p⁺ buried layer process ("*P + BL*"), and a triple-well ("*TW, RW*") process that used a high-energy boron implant to form a retrograde p-well. The charge collection measurements were performed using a thorium film. Test data showed the SER to be the highest for the standard BiCMOS process and lowest for the triple-well process. The conclusion of this evaluation was that an optimized triple-well process can improve the SER by over two orders of magnitude without compromising the device performance.

The SER of a DRAM is typically measured using the following two methods [90]:

- Accelerated test using a highly radioactive alpha source such as thorium to bombard the exposed die for a fixed time at a given V_{cc} and loaded patterns (e.g., "0s," "1s," checkerboard) and a refresh period.

- Real-time SER test method that uses fully packaged devices which are placed in a real-time soft-error test system, and tested for extended periods under nominal operating conditions while monitoring for any nonrecurring errors classified as "soft errors" to differentiate them from permanent bad bits (or hard errors).

The accelerated alpha test method is commonly used by most manufacturers as a process monitor for critical fabrication steps. A study was performed to test the concept of built-in reliability (BIR) for the soft errors and its implementation on a 256K DRAM family [90]. Accelerated testing was conducted at various levels such as final wafer probe on several lots with single-level and multilevel films and other processing variables, sample wafer batches that had used hydrofluoric acid to remove any native oxide, and at device packaging and lead frames. A direct correlation was observed between the high SER on the DRAM belonging to the wafer lot that had been processed with a contaminated batch of phosphoric acid. This study illustrates a practical example of building in the reliability concept through the implementation of process monitors that can identify the soft errors' critical material and process variability factors.

A soft-error evaluation method for the DRAMs has been developed to investigate local sensitive structures by using a proton microprobe as the radiation source [91]. In this technique, the nuclear microprobe flux and energy, the ion's angle of incidence, and spot position can be controlled to obtain two images: (1) a soft-error mapping consisting of bit state and microprobe coordinates, and (2) a secondary electron image consisting of probe-induced secondary electron yields and probe coordinates. The soft errors for both a cell mode and/or a bit-line mode are induced by the incident protons within 4 μm around the monitored cell. The locally sensitive positions against soft errors in the DRAM can be identified by overlapping these images and the pattern of Al wires around the monitored cell.

The test devices used were 64 kb DRAMs with 0.5 μm minimum feature size and stacked memory cell capacitors with both megaelectron-volt ion-implanted retrograde wells and conventional diffused wells. No soft-error events (cell mode or bit-line mode) were observed for the DRAMs with retrograde wells when monitored with a spot size of 1×1 μm² and 25 pA current. One of the conclusions was that the retrograde well structure reduces soft errors induced by the incident particles. Thus, the soft-error susceptibility of a DRAM caused by incident ions can be directly monitored by this method using a nuclear microprobe.

6.2.5 DRAM Data-Retention Properties

It has been reported that for ULSI devices such as megabit DRAMs, some of the defects are process-induced, such as stacking faults due to thermal oxidation and other defects due to As implantation or reactive ion etching (RIE). In submicron DRAMs, the failure due to charge leakage from bit failures often causes yield losses. It is often difficult to identify failure modes for devices that contain only a few failure bits. By observing failed bits in the DRAMs, it has been found that the dislocation lines originating at the sidewall edge of the cell plate cause data-retention errors [92]. A method has been proposed for direct observation of failed bits by transverse electron microscopy (TEM) analysis and improvement of charge-retention properties.

The samples used in this evaluation were conventional n-channel 256K DRAMs fabricated on p-type silicon, using conventional LOCOS isolation techniques. Then n^+ substrate leakage was evaluated by test devices using 1 Mb memory cells. The failed bit was defined as the bit which leaked charge accumulated at the capacitor during a short time below 300 ms. The samples were prepared for TEM observation of the failed bits. The TEM analysis of the failed bits showed that the dislocation loops were formed during the annealing process after As implantation, and the failure of n^+ substrate leakage in the cell increased in proportion to the dislocation density. These leakage failures were significantly reduced for the DRAMs that were fabricated by eliminating As implantation in the regions between the cell plate and the transfer gate.

A new failure mechanism related to the cell plate leakage phenomenon has been reported in 1 Mb DRAMs, with the planar cells consisting of thin capacitor gate SiO_2 and phosphorous-doped polycrystalline silicon as

the cell plate electrode [93]. This leakage caused significant degradation of the operating margin of the device for very small changes in the internal cell plate electrode voltage. Figure 6-8(a) shows a top view of the 1 Mb DRAM with planar-type storage cells, each consisting of one transistor and one capacitor, and (b) shows a cross-sectional view of the memory cells between A and A', as identified in part (a) [93]. The first polycrystalline silicon acts as a cell plate electrode, the second polycrystalline silicon is used for a conductor word-line signal and gate electrode of a transistor, and the third polycrystalline silicon acts as a bit line. The LOCOS structure is used for isolation.

Analytical techniques used include the focused ion beam (FIB) for circuit modification, plate emission microscopy to identify the location of the leakage in degraded cells, high-resolution scanning electron microscopy (SEM) and cross-sectional transmission electron microscopy (XTEM), and surface analysis by auger electron microscopy.

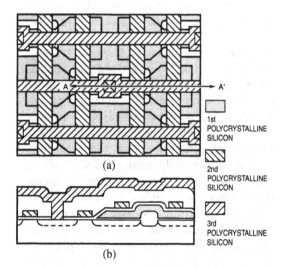

Figure 6-8. (a) A top view of memory cells of 1 Mb DRAM with planar-type storage capacitor cells. (b) Cross-sectional view of memory cells between A and A', as identified in part (a). (From [93], with permission of IEEE.)

The electrical measurements were performed to characterize fluctuations of the retention time. Schmoo plots were generated of retention time versus the cell plate electrode voltage which was supplied from an external voltage source in a normal cell and in a failed cell, respectively. Test results showed that the leakage failures represented a new failure mode related to the grain growth of polycrystalline silicon under the oxidation-induced compressive stress. The localized stress causes irregular grain growth and the development of silicon protuberance which results in a buildup of phosphorous-rich SiO_2 at the open grain boundary. This phosphorous diffuses into capacitor gate SiO_2 where it causes leakage through the phosphorous trapping centers. As the phosphorous concentration of the cell plate polycrystalline silicon became higher, the failure ratio increased.

6.3 NONVOLATILE MEMORY RELIABILITY

Section 6.1 reviewed general reliability issues such as the gate oxide breakdown, electromigration, hot-carrier degradation, metallization corrosion, etc., which are generic among various semiconductor technologies. However, there are a number of failure modes and mechanisms which are specific to memories, and those for the random access memories were discussed in Section 6.2. This section will discuss reliability issues related to the nonvolatile memories. For PROMs with fusible links (e.g., nichrome, polysilicon), the physical integrity and reliability of fusible links are a major concern. Nichrome (NiCr) fuses have exhibited a "growback" phenomenon which was first used to describe the apparent reversal of the programming process. This can result in excessive input/output leakage currents, changes in the bit logic states of the storage elements, and shorts.

In EPROMs and EEPROMs with floating gate (such as MNOS and SONOS) technologies, data-retention characteristics and the number of write/erase cycles without degradation (endurance) are the most critical reliability concerns. The stored charge (electrons) can leak away from the floating gate through various failure mechanisms such as thermoionic emission and charge diffusion through defective interpoly dielectrics, ionic contamination, and program disturb stresses in EPROMs. The opposite effect of charge gain can occur when the floating gate slowly gains electrons with the control gate held at V_{cc}, thus causing an increase in threshold voltage. The repeated ERASE/PROGRAM cycles can subject the memory transistor oxides to stresses that can cause failures, e.g., tunnel oxide breakdown for the EEPROMs. For flash EEPROMs, WRITE/ERASE endurance is a major concern since, during these operations, the charge can be trapped in the tunnel dielectric which can impact either the threshold voltage (V_{th}) window of the device, or the ERASE/WRITE time for subsequent cycles. Typical endurance characteristics for EPROMs and EEPROMs are specified as 10,000 ERASE/WRITE cycles.

The ferroelectric memory reliability concerns include the aging effects of temperature, electric field and number of polarization reversal cycles on the ferroelectric films used such as PZT, and the limited endurance cycling. The various reliability issues for nonvolatile memories are discussed in the following sections. Nonvolatile memories are also susceptible to radiation effects, which are discussed in Chapter 7.

6.3.1 Programmable Read-Only Memory (PROM) Fusible Links

The PROMs have traditionally used fusible links such as nichrome, polysilicon, and other materials as "programming elements" in series with each memory cell. The programming operation involves biasing the device to a high-voltage/high-current state while sequentially accessing each memory cell address which has to be programmed and blowing fuses at each of those locations [94]. There are some potential reliability issues associated with the programming techniques used and the proper implementation of the manufacturer-specified programming algorithm, and postprogramming electrical testing and functional verification. The infant mortality failures during the screening and early hours of operation can be grouped into three categories as follows: (1) 80–90% of the devices that exhibit anomalous input/output leakage currents and shorts due to process-related defects and/or stress during programming, (2) 5–15% failures from typical assembly/workmanship-related defects, and (3) 1–5% improperly programmed fuses that change resistance due to dendritic relinking (growback).

The normally blown fuses display high resistance (typically > 2 MΩ for some nichrome resistors) compared to 200–300 Ω for unprogrammed fuses and 1000–2000 Ω for the failing (improperly blown) fuses. A study performed to evaluate the cause of failing nichrome fuses (growback failures) in PROMs made several observations which can be summarized as follows [94]:

- The device-oriented mechanisms which limit current to the fuses are: (1) low breakdown on reverse-biased junctions which normally prevent programming current from flowing through the unwanted paths, and (2) failure of PROM circuit design that would assure sufficient programming current under the worst case conditions.

- The "growback failures" exhibited a current-limiting condition either within the PROM circuitry, or were current-limited by the programming equipment. A correlation was observed between the failures and limited programming (fusing) energy.

- The higher the breakdown voltage of the gap, the lower the probability that the "growback" can ever take place, no matter what voltage potential is applied across the gap. For $V_{BR} > 20$ V, the growback effect virtually disappears. In the blown fuses that demonstrate V_{BR} in the gap region of 0.5–2.0 V, growback occurs when a field is applied.

- The PROM devices should characteristically program on the rise time of the pulse if there is no current-limiting factor. If the fuse does not program, then the higher voltages should be tried to overcome the current-limiting condition.

- Most failures occurred during the first 100 h of room temperature operation. Therefore, the best screen for potential growback failures was a dynamic burn-in at maximum recommended data sheet specifications for V_{cc} and operating temperature (e.g., 85 or 125°C).

For high-reliability military and space applications, a postprogramming (96 h minimum) dynamic burn-in is recommended per MIL-STD-883, Method 5004. Military specifications for bipolar fusible link PROMs also require a "freeze-out test" to verify the structural integrity of the nichrome fuses. This test will be discussed in Section 6.7.

Failure analysis techniques used for the failed fuses include chemical analysis, fuse and contact resistance measurements, and X-ray diffraction (XRD). A SEM technique called "dot mapping" scans a predetermined area for a specified element (e.g., chromium) and produces a white-on-black background photographic image wherever the element is present. If the element is localized to a surface feature having a unique shape or contour (e.g., nichrome fuse in a PROM), then the white image will replicate this same shape or contour. This dot map can be generated for a blown nichrome fuse after all the glass passivation has been removed by a hydrofluoric (HF) etch.

Instead of using nichrome fuses, other materials such as polysilicon have been tried as fusing elements for the PROMs. For example, the fabrication of a 1024 b array Intel PROM follows normal bipolar processing through the emitter step, following which the polycrystalline silicon is deposited by standard silicon gate MOS technology [95]. After the deposition process, the fusible links are delineated with normal photographic techniques. A study was performed to understand the physics of the programming mechanism of phosphorous-doped polysilicon fuses and to develop a model for the fusing time and power relationship [96]. This was done initially by examining the I–V characteristics of polysilicon which, because of its grain boundaries, has a greater dependence on temperature and voltage than is observed in single-crystal silicon. Polysilicon resistivity and process simulation models were developed, and the one-dimensional heat transfer equation was solved.

Two groups of test fuse devices were fabricated for this study, with each group further subdivided into various size combinations ranging in thickness from 40 to 100 nm, width from 0.5 to 1.5 μm, and length from 2 to 3 μm. The fusing element, the voltage across the fuse, and the current through the fuse were monitored using various test equipment. It was found that the current–voltage (I–V) relationship during the fusing, up to and including the formation of molten filament in the fuse neck, can be well explained by the change of resistivity with temperature (TCR) and resistivity change with voltage (VCR) effects of polysilicon material. It was found that the time required to blow a fuse could be estimated from the power density, fuse thickness, fuse resistance, and fuse width. A 2-D heat flow analysis was used to predict the fusing time as a function of power density. In general, a high-power density fuse blow results in a well-programmed fuse; the gap is located near the center, and mass transport does not take place during such a short fusing interval. For the marginally programmed fuse, the input power is much less than that for a well-programmed fuse, and the current decreases with time mainly because of the change in resistivity as a function of mass transport, either by electromigration or field-activated migration.

SEM, transverse electron microscopy (TEM), selected area diffraction (SAED), and scanning auger microscopy were used to study the crystallographic and physical effects of the fusing event. Test results showed that the polysilicon has a preferred orientation (110) with roughly 0.3–0.4 μm grain size. After fusing, the grains adjacent to the fuse gap became elongated to typically 0.6–0.7 μm. Analysis of the physical remnants of the fusing event indicate the formation of a meniscus during fusing. It is theorized that the surface tension and fluid dynamics are the key influences for mass transport after a molten meniscus has formed.

6.3.2 EPROM Data Retention and Charge Loss

The floating-gate technology for EPROMs (or UVEPROMs) was discussed in Chapter 3. As explained earlier, in EPROMs, a hot-electron programming mechanism is used to inject electrons to the floating gate where they become trapped by the potential barrier (typically 3.1 eV) at the polysilicon–oxide interface. The erasure operation involves exposure to ultraviolet light of energy greater than 3.1 eV that discharges (or erases) a cell by photoemitting electrons over the potential barrier. A charge that has been stored on the floating gate through hot-electron injection tunneling must be retained in EPROMs for more than ten years, as per their typical specification requirements. According to the accelerated life testing of EPROMs using an Arrhenius model and temperature extrapolations for mean time to failure predictions, a typical nondefective cell will retain data for over 1000 years at 125°C. However, in practice, any defect in the oxide or oxide–nitride–oxide (ONO) stacked dielectric, mobile contamination, or any other mechanism that changes a cell's threshold voltage (V_{th}) can cause premature data loss in EPROMs. When the threshold instability is caused by a change in the stored charge Q, then the V_{th} shift relationship to the actual charge loss (or gain) can be expressed by the following equation:

$$\Delta V_{th} = \frac{\Delta Q}{C_{pp}}$$

where C_{pp} = capacitance between the floating gate and the access gate.

In a study performed on early generation Intel 2716 EPROMs, three significant data loss mechanisms for the charge loss or (charge gain), each with their own activation energies, were reported as follows: intrinsic wearout (1.4 eV), oxide defect (0.6 eV), and contamination compensation (1.2 eV) [97]. Some other Intel 2764 and 27128 EPROMs which failed reliability screens for data loss were used to characterize various failure mechanisms [98]. Analysis was performed on the defective devices after a data-retention bake, and after a bake with positive access-gate bias. A major conclusion of this analysis was that the charge loss in these EPROMs was primarily due to interpoly oxide defects. The rate of charge loss in a cell with an interpoly oxide defect increases exponentially with access-gate bias. This implies that interpoly oxide defect's *I–V* dependence is exponential as compared to the gate oxide defect that had a linear *I–V* dependence. Under a constant access-gate bias, the voltage across the interpoly oxide will decrease with time as the floating gate loses charge. The defect conductivity decreases with successive retention bakes, typically by 30% for every program/bake cycle, and is considered evidence of electron trapping along the path. Another EPROM charge loss mechanism investigated was from the contamination which occurs when the positive ions neutralize the negative charge stored on the floating gate. The contamination is a field-driven effect and depends upon the applied gate bias.

Another study was performed on the EPROM cell that uses an ONO stack as interpoly dielectric. Figure 6-9 shows a schematic cross-section of this cell which has a thermal oxide grown from crystalline silicon below the floating gate and an interpoly dielectric between the floating gate and access (or control) gate [99]. This interpoly dielectric, which historically consisted of thermal polysilicon oxide, has been substituted by many manufacturers with triple-layer ONO film for improvement of the data-retention characteristics by varying the thickness of three layers. In this evaluation, the EPROM charge loss mechanism was studied by performing the retention bake experiments on several hundred programmed cells in the EPROM arrays. The programming voltage was adjusted so that each cell had the same initial

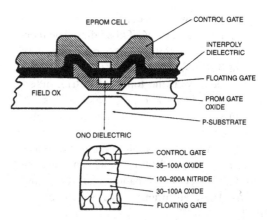

Figure 6-9. Schematic cross-section of an EPROM cell with an ONO stack as interpoly dielectric. (From [99], with permission of IEEE.)

threshold voltage. These programmed cells were baked at different temperatures with different external bias voltages for fixed intervals of time. At the completion of each bake cycle, V_{th} was measured bit by bit by using a special EPROM test mode.

The test results showed that there were three dominant charge loss mechanisms manifested in three distinct phases. The initial phase is a fast V_{th} shift that increases with nitride thickness. The second phase is caused by the electron's movement within the nitride which follows a linear ohmic conduction. The charge losses in the first and second phases are reversible by the electric field. The third (final) phase is caused be the electrons leaking through the top oxide, and is an irreversible process. This EPROM charge retention study was used to develop an ONO scaling model. According to this model, the middle nitride layer should be kept as thin as possible while making the top and bottom layers as relatively thick as possible.

The long-term charge loss characteristics of an EPROM cell with a stacked ONO film has been modeled by some other experiments also [100]. According to this charge loss model, the trapped electrons at the nitride–oxide interface can directly tunnel through a thin (≈ 30 Å) top oxide. The estimated tunneling barrier height is roughly 2.6 eV, which is consistent with the previous results based on MNOS studies. The thermal activation energy of long-term charge loss is about 0.37 eV. It is believed that the trapped electron density at the top oxide/nitride interface becomes larger at higher temperatures due to the thermal-activated electron movement during the second phase.

In general, there are four main charge loss mechanisms for EPROMs which can also be used for device qualification and as quality screens. These are briefly described as follows [101]:

- *DC Erase (or Word-Line Stress):* In this case, when a high voltage is applied to the polysilicon word line (poly 2), it causes electrons to be emitted from the poly 1 floating gate through the interpoly dielectric to the poly 2, resulting in a loss of charge and a reduction in the programmed threshold voltage of the cell. This may occur during the device programming, or under extended read cycling. This stress can be used to examine the quality of the interpoly dielectric.

- *Program Disturb (or Bit-Line Stressing):* This may occur when a high voltage is applied to the bit line (n^+ diffusion) of the cell and charge can be transferred from the floating gate to the drain, causing a reduction in the programmed threshold voltage. This stress examines the quality of the interfaces between the gate and the drain.

- *DC Programming:* In this charge loss mechanism, an erased cell may have its threshold voltage raised by an inadvertent charge from the transistor channel on to the poly 1, which makes an unprogrammed cell become programmed. This stress provides information on the quality of the gate oxide between the cell substrate and the floating gate.

- *Charge Loss Due to Bake Stressing:* EPROMs tend to lose floating-gate charge over a period of time, and an increase in temperature tends to accelerate the rate of charge loss. This high-temperature bake stress can be used to evaluate the overall quality of all oxides surrounding the polysilicon floating gate.

These four stresses can be used as in-line process monitors for an overall quality assurance plan that includes the statistical process control (SPC) and quality conformance inspection (QCI) requirements. The various fabrication processes also have a critical impact on the oxide quality. For example, the beam current level used in high-current implanters, such as those used for the source–drain implants, can degrade the oxide quality. The improper use of floodguns during a high-dose implant can adversely affect the EPROM cells by making them more susceptible to charge loss (and charge gain) mechanisms. In this case, a 250°C bake fallout versus defect density level monitoring can be used as a process monitor to maintain a high level of reliability in a production environment.

Figure 6-10 illustrates the relationship between the ONO layer thickness and dc erase fallout [101]. It is clear that for this particular process, the dc erase susceptibility increases significantly for ONO thicknesses greater than 400 Å. In a study performed, charge loss failure in the EPROMs was linked to the presence of carbon in oxidizing ambients [102]. The oxides grown in hydrogen chloride (HCL) and TriChloroethAne (TCA) were compared to the polysilicon and single-crystal substrates for several product runs of 32K EPROMs. Several charge loss failures were associated with the TCA and CO_2 grown floating-gate oxides, and amounted to a 90% yield drop when compared to the control group. The material analysis of silicon substrate oxidized in a carbon-bearing atmosphere revealed an accumulation of carbon at the oxide–silicon interface.

The charge loss due to the program disturb (or threshold voltage) shift of a programmed cell due to bit-line stress is a significant reliability issue. Program verification may cause failures during the programming

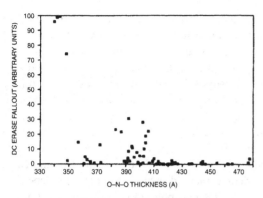

Figure 6-10. Curve showing relationship between ONO layer thickness and dc erase fall-out. (From [101], with permission of IEEE.)

verification or during extended read cycling [103]. The program disturb failures appear to be pattern-sensitive, and thus might pass screening, but later fail in the users' application environment. The study of program disturb failure mechanism has identified two types of memory devices as being: (1) those with defective cells that show V_{th} degradation, and (2) others with good cells that show no V_{th} shift due to the program disturb stress. It was found that the program disturb could be initiated as a result of a number of factors such as: high local E-field concentration around the floating gate that may be caused by the topology (asperities) of the polysilicon surface or particles in the gate oxide, and poor oxide quality that may introduce trapping sites and defects.

The objective of reliability testing for the program disturb in EPROMs is to determine voltage and temperature acceleration factors that would make it possible to predict defect levels, and to develop screens to identify the potential failures (marginal devices). The use of this testing can accelerate program disturb defects by high-voltage stress on the drain and grounding the gate through internal device modes. The programming stress selected is a worst case pattern which assumes that every bit on the column gets programmed.

The intrinsic UV-erasability performance of the nonvolatile memories depends upon the composition of passivation layers. Intrinsic charge loss which has not been well understood may affect all cells, but causes only a fraction of the stored charge to be lost. This charge loss of 0.5 V or less of threshold voltage drop in the first few days of 250°C bakes slows with time and eventually stops. There are three possible mechanisms for intrinsic charge loss: (1) contaminants' motion within the oxide, (2) conduction of electrons through the material, and (3) detrapping of electrons trapped within the oxides. The first possibility appears unlikely since ultraviolet light may affect electrons through photoemission from the floating gate or from oxide traps, but would not be expected to affect the ionic contamination. Test results on some Intel 2764 EPROMs suggested that the intrinsic charge loss is due to the detrapping of electrons trapped in the oxide during UV erase. The contaminated cells which exhibit either charge gain or charge loss can often be mistaken for cells with oxide defects.

An evaluation was performed on the nonvolatile memories with 1.0 μm design rules that utilized double-polysilicon technology with tungsten silicided poly control gates [104]. Analysis was performed on the variation in the refractive index of PECVD oxynitride layer and its effect on the UV erasability. The test results showed that an optimal erase performance was achieved with a planarized passivation system that utilized an ON layer of minimal thickness and low refractive index. No data-retention degradation was observed when the ON layer thickness was reduced. A humidity resistance study showed that the highest margin in moisture resistance and humidity-driven charge loss was achieved through a planarized passivation process which used a significantly thicker ON layer with a high refractive index.

In plastic-packaged EPROMs, the use of phosphosilicate glass (PSG) passivation can lead to failures due to moisture penetration and/or sodium concentration [105]. The most common failure mode is due to a threshold voltage shift causing a charge loss (or charge gain). Also, the data-retention characteristics at the wafer level are worse than for hermetically packaged devices. The manufacturers of one-time programmable (OTP) EPROMs commonly use oxynitride or UV-transparent nitride passivation, which makes the subsequent processing flow more complex. A new EPROM structure, which uses PSG passivation, but has superior data-retention capabilities compared to the conventional EPROM cell, has been developed [105]. This nitrided self-aligned MOS (NIT-SAMOS) uses a thin layer of LPCVD nitride between the double-poly-gate structure and the poly-metal isolation dielectric to reduce the possibility of contamination of the floating-gate area.

Figure 6-11 shows schematic cross-section of an NIT-SAMOS EPROM cell [105]. This is fabricated in 1.2 μm technology with poly 1 for the floating gate, and poly 2 for the control gate, enhancement transistor gates, and poly interconnects. A double-diffused drain (DDD) structure is used in the n-channel enhancement transistors. The programming for nitrided cells takes a little longer than the conventional cells, and the ultraviolet erasure properties are equivalent. However, the bakeout testing at 250°C demonstrated superior data-retention characteristics for this NIT-SAMOS EPROM compared to that of the conventional EPROM.

The scaling of interpoly dielectric is a significant reliability concern for the EPROMs. The polyoxide which was relatively thicker (~50 nm), used as an interpoly dielectric in earlier EPROMs, had some intrinsic limitations. The leakage current for polyoxide thermally grown on phosphorous doped poly-Si depends on polyoxide quality and surface texture [106]. Both of these factors are strongly correlated to the phosphorous concentration in oxidized first poly-Si (poly 1). Also, the thicker polyoxide defect density is strongly affected by the contamination level in poly-Si, and these defect levels are higher than for the oxide grown on a single-Si substrate (especially in the thinner film region). Thin polyoxide (12 nm) can be formed with high critical field strength by the diluted oxidation method which reduces the dependence on the phosphorous concentration and oxidation temperature used. However, the TDDB study for these devices has shown that the increased phosphorous concentration leads not only to oxide quality degradation, but also to an increase in defect density. Another limitation observed has been the "edge effect" resulting in high leakage current polyoxide on a patterned poly-Si of low phosphorous concentration. These limitations have been partially overcome by process improvements such as superclean technology, modified cell structure, and an enhanced doping or oxidation process to reduce the incorporation of phosphorous into the oxide.

Currently, the ONO stacked films (e.g., CVD Si_3N_4) are preferred over polyoxide film as an interpoly dielectric for the EPROMs because of their superior characteristics such as high critical field strength and lower defect densities. The ONO stacked films show small edge effects and very little leakage current dependence on the phosphorous concentration. However, the UV erase speed with ONO interpoly dielectric usually becomes slower compared

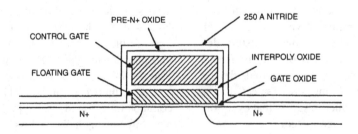

Figure 6-11. Schematic cross-section of an NIT-SAMOS EPROM structure. (From [105], with permission of IEEE.)

with that for the polyoxide dielectric [107]. This degradation in UV erase speed for the ONO cells is caused by a decrease in photocurrent through the ONO interpoly dielectric. This phenomenon is related to the electron trapping and carrier transport mechanism in the ONO film under UV irradiation. The ONO film composition must be carefully selected since it affects both the charge-retention capabilities and the dielectric reliability.

The EPROM ONO scaling methodology can be summarized as follows:

- The top-oxide thickness selected should be greater than the critical value (~3 nm), which can effectively block the hole injection from the top electrode.
- SiN scaling (e.g., 5 nm typically) can improve the data retention, but too much thinning can lead to degradation in the TDDB characteristics, and charge loss in a long time bake test.
- The thinning of the bottom oxide down to 10 nm does not lead to degradation if the oxide quality is controlled.
- Dielectric reliability considerations for high charge-retention characteristics must include improvements to reduce the process-induced damage.

6.3.3 Electrically Erasable Programmable Read-Only Memories (EEPROMs)

The EEPROM floating-gate technologies, including the FLOTOX and textured polysilicon EEPROMs, were discussed in Chapter 2. The general reliability issues for microelectronic devices reviewed in Section 6.3.2, as well as the EPROM data-retention characteristics and charge loss phenomenon, are also applicable to the EEPROMs. In addition to these, the endurance or the number of program/erase cycle specifications for EEPROM devices are a significant reliability concern. The EEPROMs, both FLOTOX and textured polysilicon, just like EPROMs, store data as charge on an electrically insulated polysilicon layer called the float-

ing gate. Both EEPROM types use Fowler–Nordheim (F–N) tunneling in moving charge to and away from the floating gate instead of hot-electron programming and UV erase techniques for EPROMs [108]. To achieve the high electric fields required for tunneling (~10 MV/cm), the FLOTOX cell uses thin oxide or ONO layers (<150 Å), whereas the textured poly cell uses a thick oxide with field enhancement at the textured polysilicon surface.

Figure 6-12 shows a schematic of superimposed energy-band diagrams of the FLOTOX and textured polysilicon tunnel oxides during the tunneling process [108]. Despite their different thicknesses, the two oxides can achieve the same electric field strength at the injecting surface. However, the FLOTOX tunnel oxide is symmetric in both directions, whereas textured polysilicon thick oxide has good tunneling characteristics only from the lower, textured poly layer. Therefore, the textured polysilicon cell requires two tunnel oxides (one for program operation, and the other for erase), while the FLOTOX requires only one. The EEPROM data-retention characteristics are limited by the oxide defects which allow the electrons to leak

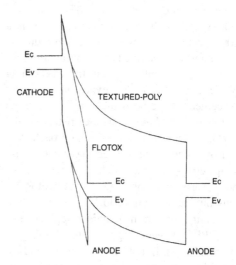

Figure 6-12. Schematic of the superimposed energy-band diagrams of FLOTOX and textured polysilicon tunnel oxides during the tunneling process. (From [108], with permission of IEEE.)

off the floating gate, or by a source of contamination which causes threshold voltage (V_{th}) instability. A major difference between the FLOTOX and textured polysilicon EEPROMs is their endurance (program/erase characteristics). In an extended cycling evaluation performed on an Intel 16–64 kb FLOTOX and textured poly EEPROMs, the test results showed the following: (1) the FLOTOX device endurance is more strongly dependent upon memory size, being better than textured poly at the lowest densities, but worse at the highest densities; and (2) the textured poly endurance is intrinsically limited even for low-density devices, although it is capable of meeting a 10,000 cycle endurance specification limit at both low and high densities.

The evaluation test results also showed that both the FLOTOX and textured poly cells can be cycled 1×10^6 times without oxide failure or degradation in retention characteristics. However, for devices with significant defect densities, there may be a few cells that have defective tunnel oxide which can break down under a high electric field during the tunneling. This tunnel oxide breakdown, mostly responsible for EEPROM endurance failures, is a gradual process, with oxide showing increasing leakage until the failure point. Therefore, the screening that includes program/erase cycling (10,000 cycles and above) is useful in reducing early life failures, but the fairly flat random failure rate may prevent screening from becoming fully effective. Since this evaluation, process enhancements have been made to improve the tunnel oxide breakdown characteristics of the FLOTOX memories. Another approach followed by some manufacturers has been to use on-chip error-correction schemes which add some processing complexity to the EEPROM design, but increase the yield.

Another failure mechanism applicable to both EEPROM technologies is the MOS gate oxide breakdown in both memory cells and associated logic circuitry. This may occur because of high field stress and the presence of defective oxide since both technologies use high voltage for the program and erase operations. The typical symptoms are failure of a byte, a row, or a column, or the whole device. Improved design

and processing can help reduce the gate oxide breakdown failures.

During the program/erase cycling, some tunneling electrons may become trapped, and the resulting negative oxide charge may inhibit further tunneling, causing the cell to require higher voltage and thus a longer time to write. This trap-up mechanism of failure to write (window closing) is responsible for the textured poly wearout, but is not a significant issue with the FLOTOX devices. A feature of the textured poly devices that contributes to the trap-up charge effects is the difficulty in controlling the number, size, and shape of polysilicon bumps, which cause wide variations in the tunneling characteristics. In comparison, the FLOTOX tunneling depends mostly on the tunnel oxide thickness, which may typically vary less than 5% across a memory array.

For a given application, the EEPROM reliability evaluation criteria should be based upon a comparison of overall failure rate, and not just the endurance failure rate. This implies that both the life test as well as endurance testing are important reliability screening tools. Table 6-2 shows a recommended endurance evaluation flow for two dominant failure mechanisms of the window closing and oxide breakdown [108].

The programming window degradation for the FLOTOX EEPROM structures is frequently attributed to a buildup of bulk oxide

TABLE 6-2. A Recommended Endurance Evaluation Flow for EEPROMs. (From [108], with Permission of IEEE.)

Window Closing	Oxide Breakdown
Small sample of devices (20)	Large sample of devices (>100)
Cycle 10K times	Cycle 10K times
Test for program/erase	Test for program/erase
Repeat to > 100K cycles	Program to non-equilibrium state
Extrapolate wearout curve to desired endurance limit (10K) to determine the failure rate	Retention bake
	Verify data

charge in the tunnel area during program/erase cycling. A theoretical model has been developed to explain the programming window degradation and the corresponding high and low threshold voltage shifts as a function of number of cycling operations [109]. Figure 6-13 shows the capacitive equivalent circuit of (a) a FLOTOX EEPROM cell, and (b) the floating-gate-to-drain electrode capacitor representing the presence of built-up oxide charge. According to this model, the programming window degradation, which is the difference between high and low threshold voltage shifts, can be expressed by the following equation:

$$\Delta W_p = \Delta Vth^h - \Delta Vth^l = \frac{\Delta Q_{ox}}{[C_{fd}/C_t]C_{pp}}$$

$$= \frac{Q_{ox}}{A_d C_{pp}}$$

where ΔQ_{ox} = incremental oxide charge buildup
C_t = total capacitance $(C_{pp} + C_b + C_{fd})$
C_{pp} = coupling capacitance
C_{fd} = tunnel oxide capacitance
C_b = floating-gate-to-substrate capacitance
A_d = drain coupling ratio which equals C_{fd}/C_t.

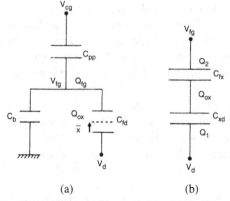

(a)   (b)

Figure 6-13. Capacitive equivalent circuit of (a) a FLOTOX EEPROM. (b) Floating-gate-to-drain electrode capacitor representing the presence of built-up oxide charge. (From [109], with permission of IEEE.)

The choice of doping concentration for the floating-gate electrode is a key parameter in the optimization of an EEPROM process. An evaluation was performed to study the effect of floating-gate polysilicon doping on electron injection barrier height, and therefore the program/erase window of an EEPROM cell [110]. The EEPROM cell used in this study was a part of merged memory module technology incorporating single-polysilicon EPROM cells, double-polysilicon EEPROM cells, and polysilicon capacitors. The test results showed that the introduction of dopant and the concentration of electrically active sites at the floating-gate polysilicon/tunnel oxide interface influences the electron injection barrier height during cell erase operation. The electron injection barrier increased up to 250 meV upon degenerate doping of the floating-gate polysilicon electrode as measured by the dark current voltage (*I–V*) characteristics.

To analyze the effects of processing variables on reliability, a study was performed on TI 16K EEPROMs [111]. These were fabricated in a standard twin-well CMOS process with the addition of a tunnel oxidation process module prior to deposition of first polysilicon gate material. The tunnel oxide was grown using dry oxidation technique at 900°C and pyrogenic (steam) oxidation at 850°C. The two process techniques evaluated for write/erase endurance characteristics showed that the EEPROMs with dry oxygen tunnel oxides performed better than those with pyrogenic oxides. Silicon nitride protective coating has been preferred as a barrier to moisture penetration when the devices are placed in nonhermetic plastic packages. The reliability effects of protective overcoats (e.g., oxide/nitride layers sandwich) on the EEPROMs bare die for use in plastic packages were also evaluated. According to this study, very little difference in performance was found between various overcoats for up to 1000 h bake at 300°C. The effect of overcoatings on the endurance characteristics, although not quite obvious, is related to trap formation in the tunnel oxide. Hydrogen, which may be inadvertently introduced during the fabrication process, can lead to accelerated device degradation.

A new technology has been developed to improve the program/erase cycling endurance characteristics for the EEPROM memory cells by using self-aligned, double-polycrystalline stacked structures [112]. The program/erase cycle limitation is caused by electron traps in the tunnel oxide which are generated by hot-hole injection during the erase operation. Therefore, the improvement of endurance characteristics is realized by reducing the lateral electric field between the drain and surface channel region during the erase operation. Figure 6-14(a) shows the improved structure whose source n^+ region is located within the depletion region of the surface channel area when high voltage is applied to the drain with the source left floating [112]. The source n^+ region voltage is charged up to almost the same level as that of the n^+ drain region during the erase operation. As the source region is charged up, the voltage differential between the drain and the source decreases, which reduces the lateral electric field between the drain and surface channel area to about 50% of that for conventional EEPROM structures. Figure 6-14(b) shows a schematic cross-section of a conventional EEPROM structure for comparison [112].

In this new memory cell design, the narrowing of the threshold voltage window, as well as degradation of the tunnel oxide, is suppressed, which results in good data-retention characteristics after a 10K write/erase cycle. It was experimentally confirmed that the endurance of these cells in 0.5 μm design rules, which are erased by Fowler–Nordheim tunneling, can be more than 10^7 write/erase cycles. They have good potential for scaling up to 4 Mb full-featured EEPROM or 16 Mb flash technology.

In a study performed on floating-gate electron tunneling MOS (FETMOS) devices, it was found that dramatic improvements in data-retention characteristics could be achieved by using a uniform write and uniform erase technology, performed by uniform injection and uniform emission over the whole channel area of the cell [113]. The EEPROM memory cell used in this evaluation was a self-aligned, double-polysilicon structure with a 1.0 μm gate length, 10 nm gate oxide thickness, and 25 nm thick interpoly dielectric (ONO). Figure 6-15 shows (a) the uniform erase operation by application of a positive voltage at the substrate, and (b) nonuniform erase by applying a positive voltage at the drain [113]. In uniform write and uniform erase technology, V_{th} of the erased cell (although not the written cell) depends on the number of write/erase (W/E) cycles. The data-retention characteristics of the FETMOS cell programmed by two erase and write technologies were measured at 300°C bake after a number of W/E cycles ranging from 10 to 10^6.

The test results of this FETMOS EEPROM evaluation showed that for uniform write and nonuniform erase technology, the stored positive charge slowly decayed as the baking time increased, resulting in threshold window

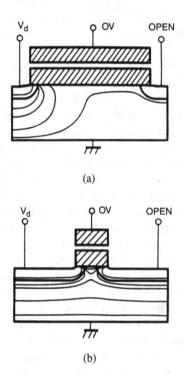

(a)

(b)

Figure 6-14. (a) Improved EEPROM structure whose source n^+ region is located within the depletion region of the surface channel area during the erase operation. (b) Conventional EEPROM structure. (From [112], with permission of IEEE.)

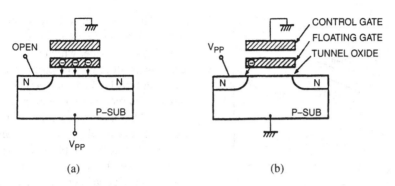

Figure 6-15. FETMOS EEPROM cell. (a) Uniform erase operation.
(b) Nonuniform erasing. (From [113], with permission of IEEE.)

narrowing. In comparison, for uniform write and uniform erase technology, the stored charge effectively increased during the retention bake due to detrapping of electrons from the gate oxide to the substrate. This significantly increased the data-retention characteristics for the FETMOS cell beyond 10^6 write/erase cycles.

A FETMOS EEPROM cell fabricated using a furnace N_2O oxynitridation process resulted in devices that exhibited about eight times high-endurance performance and good data-retention characteristics while maintaining defect densities comparable to those of the control thermal oxide devices [114]. The FETMOS EEPROM cell tunnel area, which is merged into the floating-gate transistor to reduce the cell size, requires a robust tunnel dielectric because of a higher tunnel oxide electric field at the gate corner during the program cycle as compared to the FLOTOX EEPROMs that have a separate tunnel area on the top of the drain. Techniques to improve the gate dielectrics have been developed that use nitrogen (oxynitridation) at the Si–SiO$_2$ interface to enhance charge-to-breakdown characteristics, lower electron trapping, and reduce interface state generation. The FETMOS EEPROMs used in this study were integrated into a multiple-well, double-level metal, and double-level polysilicon CMOS process. The tunnel oxide thickness and electron trap characteristics were measured with large-area MOS capacitors, the oxide thickness was evaluated using CV measurements, and single-

cell endurance was characterized using an automated test system.

Figure 6-16(a) shows the charge trapping characteristics of the N_2O oxynitride and the control oxide, under $100\ mA/cm^2$ constant current stress [114]. The curves show that both the gate oxide voltage shift and the rate of gate voltage change are reduced for N_2O oxynitride, which indicates suppression of the electron trap generation. Figure 6-16(b) shows endurance characteristics of a FETMOS EEPROM single-cell N_2O oxynitride which are about eight times better than those of control oxide devices [114]. At the end, both device types fail due to the tunnel oxide breakdown rather than the window closing, which is a typical characteristic of FETMOS EEPROMs. For time-zero defect density evaluation, the circuit yield showed that the N_2O oxynitridation process had a very low defect density compared to the control oxide process.

The scaling down of FLOTOX EEPROM tunnel oxide offers advantages such as reduction of memory cell or peripheral circuitry related to voltage operation, and reduced programming voltage (V_{pp}). However, the use of thin tunnel oxides can also result in oxide breakdown, charge generation, and trapping under electric field stressing. A study was performed to evaluate the reliability implications of thinning tunnel oxides by using capacitors and cell structures with oxide thicknesses ranging from 47 to 100 Å [115]. The polysilicon capacitors and EEPROM cells used were fabricated on n-type silicon and

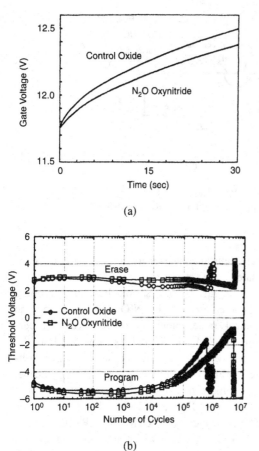

(a)

(b)

Figure 6-16. (a) Charge trapping characteristics of control oxide and N_2O oxynitride capacitors under a constant current stress. (b) Single-cell endurance characteristics of control oxide and N_2O oxynitride FETMOS EEPROMs. (From [114], with permission of IEEE.)

The test results showed that the stress-induced leakage is the major degradation mode in thin tunnel oxide, which is different from the positive feedback mechanism due to the hole trapping involved in dielectric breakdown behavior. Charge loss in retention tests of stressed EEPROM cells can be explained by this stress-induced current of tunnel oxide and annealing effect. Therefore, for the scaling of EEPROM tunnel oxide, the most serious limiting factor is the oxide leakage current caused by W/E cycling stress.

In order to scale down the effective thickness of the oxide–nitride–oxide (ONO) interpoly dielectric used in the EEPROM cells, several factors have to be taken into consideration, such as the electric field leakage current, the charge-retention characteristics, and the mean time to failure (MTTF). To investigate the leakage current in ONO films, extensive testing was conducted on double poly-Si planar capacitors and on the EPROM cells as test vehicles [116]. The EEPROM ONO scaling methodology based on the charge-retention characteristics can be summarized as follows:

- The top oxide thickness should be more than a critical value ($\sim$30 Å) such that it can effectively block hole injection from the top electrode.
- Although SiN thickness may be scaled down to 50 Å, it should be carefully carried out since too much thinning can lead to degradation in the TDDB characteristics.
- The bottom oxide layer can be scaled to 100 Å, provided the oxide quality is carefully controlled.

6.3.4 Flash Memories

Flash memories combine the advantages of EPROM density with EEPROM electrical erasibility. The reliability performance of the flash memories before program/erase cycling is comparable to that of the UVEPROMs. The cycling can induce instabilities (failure mechanisms) found in the UVEPROMs such as intrinsic charge loss and dc program disturb

p-type silicon substrate, respectively. They both had an n-region beneath the thin tunnel oxide. A write/erase (W/E) cycle stressing was performed by applying a 0.2 ms programming pulse alternately with a 0.2 ms interval to the control gate in the charging operation, and applying them to both the drain and substrate in the discharging operation. For endurance testing, the programming pulse was applied to the control gate and drain alternately with the same pulse width and interval (as for W/E stress).

mechanisms. An evaluation was performed on Intel 64K array flash memories based on a single EPROM-type floating-gate transistor without a select transistor [117]. These flash memory arrays were byte electrically programmable and byte electrically erasable. The flash memory cell uses a thinner tunnel gate dielectric and was manufactured using ETOX technology. Figure 6-17(a) shows a schematic cross-section of this cell [117]. The cell's important characteristics are a thin first-gate dielectric ($\sim$100 Å), abrupt drain and graded source junctions, and a scaled interpoly dielectric. A graded source junction permits high voltage (typically 12.5 V) for the erasure. Figure 6-17(b) schematically illustrates the flash memory programming by hot-channel electron (CHE) injection [117].

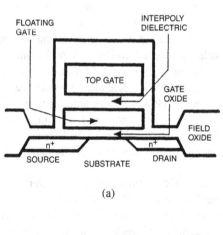

(a)

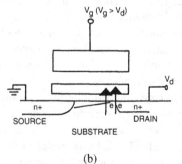

(b)

Figure 6-17. (a) Schematic cross-section of Intel 64K array flash memory cell.
(b) Schematic illustration of flash memory cell programming by CHE injection. (From [117], with permission of IEEE.)

The three principal memory disturbs for the flash memories that occur during programming are the same as for the UVPROMs, i.e., dc erase, dc program, and program disturb. In addition to these memory disturbs, other failure mechanisms in flash memories include charge loss due to the defective bits, oxide breakdown during program/erase cycling, and trap-up degradation during cycling. The cycling subjects the memory cell's dielectrics to high electric fields ($>$10 MV/cm for gate oxide), and also produces significant charge transfer through the gate oxide. The trap-up phenomenon refers to the negative charge trapping which can inhibit further tunneling and CHE injection, causing degradation of the erase and program margins with cycling. During the evaluation, 64K flash memories were subjected to greater than 10,000 program/erase cycles, a data-retention bake test at 250°C for pre- and post-100 cycle devices (measurements at 48, 168, and 500 h), and a high-temperature life test at 125°C for the devices programmed with checkerboard patterns.

The test results showed that for extensive cycling ($>$10,000 cycles), no failures due to dielectric breakdown were observed. However, changes in program and erase times corresponding to window widening and narrowing effects due to charge trapping (as in EEPROMs) were observed. No measurable shift was found in the dc erase and program disturb characteristics until several thousand cycles. The cycling did impact dc program behavior measured by allowable margin shift, e.g., 0.5 V shift for the erased state is considered to be the onset of parametric degradation. Also, charge loss was observed for devices with 100 cycle operation after 168 h bake at 250°C that resulted in enhanced margin shifts of up to 1 V. The presence of holes in gate oxide was considered to be the most likely failure mechanism for these degradation mechanisms. During the erase operation, the high source-to-substrate voltages are responsible for this hole injection. The CHE during programming does not play any significant role in the observed degradation. The cycling effects can be minimized by incorporating an additional flash memory cell threshold margin (guardband).

Hot-hole injection during operation has been reported to cause variations in the erased threshold voltage of the cells, and trapped holes in the oxide were shown to degrade the charge-retention characteristics of the flash memory cells. Since hole trapping in oxide may alter the tunneling characteristics of the oxide, a study was performed to compare the hole trapping produced by F–N tunneling with that produced by hot-hole injection [118]. The flash cell used was a structure similar to a conventional stacked-gate EPROM cell with a channel length of about 1 μm and gate oxide thickness less than 120 Å. A cell was electrically erased from the source side and another one from the drain side. A significant degradation was observed in the programmability of the cell that had been erased from the drain side. The trapped holes in the oxide near the drain junction appeared responsible for this degradation effect. The hole trapping in the oxide causes another problem called "gate disturb," which is the undesired increase in the V_{th} of an erased cell during programming of the other cells on the same word line. V_{th} shifts due to gate disturbs are used to monitor the amount of trapped holes in the oxide after cell erasure. The conclusion of this study was that the trapped holes are primarily injected from the junction depletion region rather than direct generation in the oxide by the F–N tunneling process. Therefore, for good flash memory design, hot-hole injection should be minimized.

The trapped oxide charge is often difficult to measure because of its location over the junction transistor region. A technique has been developed to determine the sign and effective density of the trapped oxide charge near the junction transistor region which is based on the measurement of the gate-induced drain leakage (GIDL) current [119]. This method was used to investigate the hot-carrier effects resulting from the erase operation and bit-line stress in flash EPROM devices. While the trapped oxide charge depends on the stress conditions, the test results indicated that a significant amount of hole trapping is likely to occur either during the erase or the bit-line stress, as long as sufficient potential difference existed between the floating gate and the junction for either an abrupt or graded junction device.

As program/erase cycling requirements increase and the tunnel oxide thickness is scaled down, the flash EPROM cell becomes more susceptible to tunnel oxide leakage. A mechanism for the read disturb failures caused by flash EEPROM cycling has been studied which causes unselected erased bits residing on selected word lines to gain charge under low-field conditions, causing them to appear programmed [120]. The read disturb stress condition occurs on erased bits which share a word line with a bit being read. Normally, the low-field stress associated with the read disturb is not significant enough to cause charge conduction onto or off the floating gate. However, a defect can occur which, in the low-field read disturb condition, may permit an erased bit to gain charge.

The two major criteria for evaluating the failure rate for read disturb are: (1) the number of cycles which determines the probability of creating a leaking bit, and (2) the stress time that determines whether a leaking bit would reach a failure point. However, the actual failure rate (FIT) varies significantly, and is a function of the distribution of cycles over the lifetime of the part and the total read stress time applied to each row.

The read disturb failure characterization was performed through accelerated voltage and temperature failure tests. The test results indicated that the read disturb failures occurred in randomly distributed cells. The failure rate with respect to cycling, temperature, and stress fields was characterized, yielding a temperature-dependent voltage acceleration factor. It appeared that the failures which occurred from the positive charge trapped during the program/erase cycling lowered the effective tunneling barrier by values that ranged from 0.39 to 0.88 eV.

For flash EPROM arrays that utilize hot-electron injection for programming, high-temperature programming requirements may further aggravate the program disturb effects by a strong increase in thermally generated electrons in the substrate [121]. This program disturb via substrate injection of thermal electrons

(PDSITE) is a significant concern for: (1) high-density memories with a large number of cells on a bit line, and (2) sector erase (word-line-oriented sectors) architectures where the disturb time is coupled to the memory chip endurance specification. For example, consider the case of a 1 Mb flash EPROM with 1000 word lines, a maximum programming time of 100 μs/byte at high temperature, and an endurance specification of 10K cycles. The program disturb time for this device with flash erase can be 100 ms, and with unrestricted sector erase, it can be 1000 s. Therefore, in designing high-density flash EPROMs that are susceptible to PDSITE, it is necessary to ensure program disturb immunity at the highest programming temperature.

The ETOX™ flash memory cell is a stack-gate EPROM cell with the bottom gate oxide replaced by a thin tunnel oxide. The cell is programmed with channel hot-electron injection from the drain side as in a conventional stack-gate EPROM, and is erased with F–N tunneling from the source side (graded junction) by grounding the gate (or putting a negative voltage on it) and floating the drain. The tunneling for erase takes place from a small overlapping region (typically 10^{-9} cm^2) between the source and floating gate [122]. This tunneling area, because of its small size, is considered immune to oxide or other defects, and as such: (1) identical erase behavior is expected from every bit in the flash memory array, and (2) erase characteristics should remain stable with cycling. However, in a study performed on an ETOX flash memory array, erratic erase behavior was observed which varied randomly from cycle to cycle. This erratic erase behavior was believed to be caused by hole trapping/detrapping in the tunnel oxide. Therefore, for high-density ETOX flash memory designs, proper techniques have to be developed for minimizing this erratic erase behavior.

In flash EEPROMs, the quality of the tunnel oxide affects the yield and reliability. A commonly used test to evaluate the tunnel oxide is the charge-to-breakdown measurement (Q_{bd}) with a constant or exponential ramped current stress on large-area test capacitors, which may not be a very reliable test for monitoring single-bit failures. To overcome the limitations of the Q_{bd} test, a cell array test (CAST) has been developed that can be implemented using automatic routines, and that provides a valuable tool as a process monitor to improve the tunnel oxide defect density [177].

6.3.5 Ferroelectric Memories

An area of high interest in advanced nonvolatile memories has been the development of thin-film ferroelectric (FE) technology to build ferroelectric random access memories (FRAMs). The FRAM cell basic theory and memory operations are discussed in Chapter 8. Several hundred ferroelectric materials have been investigated, out of which only a few (e.g., lead zirconate titanate, $PbZr_4Ti_{1-x}O_3$, commonly known as PZT) have been found useful and compatible with standard CMOS processing for integration with the nonvolatile memory devices. These ferroelectric memory devices have shown write endurance and speed which exceed that possible with existing nonvolatile memory technologies [123]. These ferroelectric memories are scalable to quite high densities and are inherently radiation-hard. Another promising ferroelectric material under investigation is phase-III potassium nitrate (KNO_3-III) [124]. There are a number of important chemical and structural properties of ferroelectric materials that need to be well understood and characterized, such as [125]

- various material bulk phases and their crystalline structures
- film orientation during film growth
- size (symmetry of unit cell and interatomic distances)
- thickness and chemical composition of interfacial layers (metal/ferroelectric)
- diffusion and segregation effects at the metal/ferroelectric interface.

A number of techniques exist to characterize the bulk, surface, and interface properties of ferroelectric materials, such as X-ray and neutron diffraction, *e*-beam microprobing, laser Raman spectroscopy (LSR), Auger electron spectroscopy (AES), secondary ion microscopy (SIM), scanning electron microscopy (SEM), transmission electron microscopy (TEM), and scanning tunnel microscopy (STM). Two of these techniques, X-ray diffraction and LSR, are used extensively to characterize the bulk phases, bulk crystal structures, and lattice vibration modes.

Ferroelectric films used as memory storage elements have some significant reliability concerns, such as the aging/fatigue effects from a large number of polarization reversal cycles, thermal stability, effects of electric field, and time-dependent dielectric breakdown (TDDB). The aging effect in PZT films is the predominant failure mechanism. Fatigue can exhibit itself as a degradation of the signal-to-noise ratios (loss of nonlinear charge) in which the "1s" and "0s" become indistinguishable after 10^7–10^8 read/write cycles. The experimental results on the fatigue and aging of ferroelectric are somewhat difficult to interpret because of a lack of clear understanding regarding physical mechanisms responsible for the fatigue. Several theories and models have been proposed to explain the aging effects on ferroelectric materials. The phenomenon of fatigue in ceramics that use silver-plate electrodes has been attributed to the electrode artifacts [126]. A second fatigue mechanism has been reported to be microcracking, particularly in high-voltage $BaTiO_3$ capacitors [127]. This is induced by large strains introduced into the ferroelectric material during the polarization-reversal process. A third likely mechanism is the trapping of mobile ions at the metal/ferroelectric interface. During the switching cycles, the polarization reversals may cause the mobile ions' movement back and forth in the ferroelectric material, ultimately getting trapped at the interfaces. This trapping of mobile charge increases the interfacial resistance, but reduces interfacial capacitance.

The polarization decay based on the trapping of the mobile space charges inside the ferroelectric has been expressed by the following equation [128]:

$$P = P_i [1 - e^{-t_{ON} R_1}(1 - e^{-to/\tau})]^n$$

where P = polarization value after n polarization reversals
  P_i = initial polarization value
  t_{ON} = applied voltage pulse ON time
  to = OFF time between pulses of opposite polarity
  τ = mobile space-charge constant
  R_1 = a constant related to the rate of nucleation of new domains.

A study performed to characterize the ferroelectric films through a series of tests over the range of extended read/write cycling ($<10^{11}$ cycles) showed that the primary loss mechanism for the ferroelectric signal was "aging" [129]. Aging is defined as a mechanism which causes signal loss during a data-retention period that does not recover after a rewrite and immediate read. "Aging time" makes reference to the time between the first write cycle and the current read point. Aging can be considered as a gradual stabilization of the domain structure. The physical model developed for possible failure mechanisms primarily involves electrical effects which include internal space charge within the ferroelectric ceramic or to stress relaxation [130], [131]. In either case, the ferroelectric film becomes less responsive to the applied fields and harder to switch.

This reliability evaluation was performed on capacitor test structures processed from the same ferroelectric and CMOS fabrication lots, and packaged for long-term aging study [129]. Various size capacitors were used to characterize the read/write cycling voltage and temperature. To ensure proper correlation between test capacitor data and the actual product, picoprobing measurement techniques were used on the nonencapsulated parts. An automated stress and measurement system (SMS) was developed for read and write operations in the ferroelectric capacitor cycling environment. The objective was to obtain sufficient data to accurately predict the endurance and retention capabilities of the ferroelectric memories, and to identify factors which could accelerate aging. The evaluation parameters and variables consisted of the polarization reversals (cycles), aging time, write and read voltages, and temperature effects.

The test results for cycling up to 10^8 cycles showed that the aging occurred at a constant rate with the log of time for all stress and test conditions. A remnant polarization of 0.5 μC/cm^2 was defined as the failure threshold. The aging rate was found to increase with temperature and the number of read/write cycles, as well as with an increase in cycling voltage. Figure 6-18 shows data for a write voltage of 4.5 V and the following: (a) aging rate versus number of write cycles at various temperatures, (b) aging rate versus temperature and number of cycles, (c) lifetime versus number of cycles and temperatures, and (d) log lifetime versus $1/kT$ and the number of cycles [129].

The experimental data showed that the ferroelectric film aging is a complex process, and the aging rate depends significantly on the temperature and cycling history. The cycling induces two competing effects: (1) higher initial signal levels for short aging times (increases lifetime), and (2) a higher aging rate (reduces lifetime). Since the aging rate is not constant, the standard Arrhenius plots are not valid for this mechanism. However, temperature is one of the degradation acceleration factors, and is a useful tool for accelerated lifetime testing. The slope of the curve for the plot of log lifetime versus $1/kT$ shows a strong dependence of lifetime on temperature. The log lifetime dependence suggests that aging has a range of activation energies instead of a single value.

Figure 6-18. FE capacitor test structure data for write voltage of 4.5 V and the following. (a) Aging rate versus number of write cycles and various temperatures. (b) Aging rate versus temperature and number of cycles. (c) Lifetime versus number of cycles and temperatures. (d) Log lifetime versus $1/kT$ and number of cycles. (From [129], with permission of IEEE.)

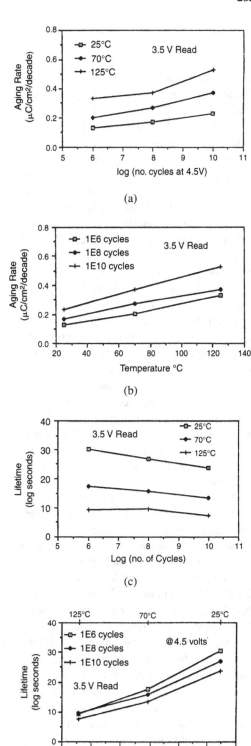

Thin-film ferroelectric materials such as the phase-III potassium nitrate (KNO₃-III) devices have been considered as attractive candidates for nonvolatile memory devices. An evaluation was performed to characterize the terminal fatigue characteristics of a vacuum-deposited KNO_3-III capacitor type memory test structures [132]. A number of electrical tests were conducted on these memory devices, such as: (1) pulse testing for output current versus time (I–t), (2) polarization hysterisis versus electric field (Q–V), (3) capacitance versus voltage (C–V), (4) conductance versus voltage (G–V), and (5) current versus voltage (I–V). A special memory tester was used for the fatigue testing. The fatigue characteristics were determined by observing the polarization decay after roughly 5×10^7 read/write cycles. The C–V plots taken before and after cycling showed quite different characteristics, indicating the buildup of mobile charges at the interface. The Auger depth profiles confirmed migration of the potassium ions into gold layers, leaving behind an oxygen-rich interface. A conclusion was that the change in electrical properties of thin-film ferroelectric memory due to the fatigue phenomenon was strongly dependent on the interface (between electrodes and ferroelectric film) characteristics.

The thermal stability of poled ferroelectric films such as the PZT integrated as nonvolatile memories has been studied extensively. A study was performed to evaluate the effects of thermal excursions, as well as fatigue and aging on the remnant polarization levels in polycrystalline film [133]. For this evaluation, a standard CMOS base process was used for the first-level dielectric. A stack consisting of conductive electrodes and ferroelectric capacitors of ⁵⁰⁄₅₀ PZT film deposited by sol-gel processing is formed above the dielectric and connected to the CMOS transistor circuitry by Al metallization. The integrated nonvolatile memory cell is configured like a DRAM with a select transistor and capacitor, and one additional control line (drive line).

To characterize fatigue, high-current cycling boxes were used to cycle $20 \times 20 \ \mu m^2$ capacitors at 6 MHz. The fatigue plot for ferroelectric films showed a steep degradation of switchable polarization (Pnv) with log (cycling) at about 10^{12} cycles. Aging is the loss of remnant polarization with time. The aging test results showed that both the fatigue (up to 10^{10} cycles) and temperature (125°C) have very little effect on the aging rate. In the experiment designed to model the charge loss in retained polarization as a function of temperature, polarization measurements were performed on two capacitors, before and after bakes of 100 s at four read/write temperatures. Table 6-3 shows the thermal stability test matrix [133]. The test capacitors were preconditioned with 10^5 cycles, and the polarization measurements were performed at selected read/write temperatures (−25, 25, 75, and 125°C). The capacitors were then baked as shown in the test matrix, and returned to read/write temperature for the final polarization measurement.

The test results showed that the negative temperature excursions do not affect the retained polarization. However, the positive thermal excursions cause a reduction in the polarization value when the device is reread, and the higher thermal excursions cause large polarization losses. Figure 6-19 shows a plot of polarization (Pnv) as a function of ΔT with Pnv normalized to its initial value at each write temperature [133]. Another effect observed was that the capacitors reaching higher peak temperature (relative to those subjected to lower peak temperature) generally exhibited a greater loss in polarization, although not as significant as the loss attributed to ΔT. A conclusion of this evalu-

TABLE 6-3. Capacitor Thermal Stability Test Matrix. (From [133], with Permission of IEEE.)

Read Write Temperature (°C)	Bake Duration (sec)	Bake Temperature (°C)
−25	100	−25,25,75,125
25	100	−25,25,75,125
75	100	−25,25,75,125
125	100	−25,25,75,125,175

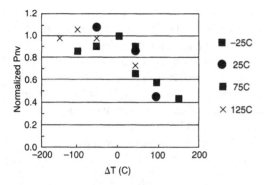

Figure 6-19. Capacitor polarization (*Pnv*) as a function of ΔT with *Pnv* normalized to its initial value at each write temperature. (From [133], with permission of IEEE.)

ation was that because of high ΔT degradation effects on the remnant polarization, significantly high margins for signal strengths should be designed in for the ferroelectric memory devices.

The use of PZT films as storage elements for DRAMs has also been investigated since, for this application, the ferroelectric film needs to be biased only in one polarity and long-term data retention (nonvolatility) is not a requirement. Also, the dielectric constant for PZT films is several times higher than that for thermal oxide. For this potential DRAM application, a study was performed on some PZT films to evaluate important characteristics such as the polarization, electrical conduction, and TDDB behavior [134]. The PZT films used were 4000 Å ($x = 0.5$) with effective SiO_2 thickness less than 17 Å that were prepared by sol-gel deposition and subsequent annealing above the crystallization temperature. Platinum was used for both the top and bottom electrodes.

The test results showed that the PZT films exhibited ohmic behavior at low fields and exponential field dependence at high fields. The conduction characteristics can be modeled accurately using expressions derived for ionic conductivity. During the TDDB evaluation at room temperature, the current density during constant voltage stress decreased by several orders of magnitude before a destructive breakdown, which appears to follow a power-law relationship. However, at high temperatures, a turnaround was observed before the breakdown. The lifetime extrapolations to worst-case operating conditions showed that the TDDB may be a serious limitation for use of PZT films in DRAM applications.

6.4 RELIABILITY MODELING AND FAILURE RATE PREDICTION

6.4.1 Reliability Definitions and Statistical Distributions

The reliability of a semiconductor device such as a memory is defined as its characteristics expressed by the probability that it will perform its required function under the specified conditions for a stated period of time. This reliability concept involves probability and the required function that includes the definition of a failure. A device is considered a failure if it shows an inability to perform within its guaranteed functional and parametric (e.g., voltage, current, temperature) specification limits. MIL-STD-721 (Definition of Terms for Reliability and Maintainability) defines failure rate as the total number of failures within an item population, divided by the total number of life units (total operating time) expended by that population, during a particular measurement interval under stated conditions [135]. The quantitative definition of reliability is based upon the characterization of three important parameters: hazard rate, failure rate, and mean time to failure (MTTF). These are related to some other reliability functions such as the cumulative distribution function (cdf) and failure probability density function (pdf). The basic definitions and relationships among various reliability functions are given as follows [136].

The cumulative distribution function $F(t)$ is defined as the probability in a random trial that the random variable is not greater than "t":

$$F(t) = \int_{-\infty}^{t} f(t)\, dt \qquad (1)$$

where $f(t)$ is the density function of the random variable, time to failure.

$F(t)$ may be considered as representing the probability of failure prior to some time "t." If the random variable is discrete, the integral is replaced by a summation sign. Reliability function $R(t)$ is defined as the probability of a device not failing prior to time "t," and is given by

$$R(t) = 1 - F(t) = \int_t^\infty f(t)\, dt. \qquad (2)$$

Differentiating equation (2) gives

$$\frac{-d\,R(t)}{d(t)} = f(t). \qquad (3)$$

The probability of failure in a given time interval t_1 and t_2 can be expressed by the reliability function as

$$\int_{t_1}^\infty f(t)\, dt - \int_{t_2}^\infty f(t)\, dt = R(t_1) - R(t_2). \qquad (4)$$

The failure rate $\lambda(t)$ between time intervals t_1 and t_2 is defined as the ratio of probability that the failure occurs in this interval (given that it has not occurred prior to t_1) divided by the interval length (l):

$$\lambda(t) = \frac{R(t_1) - R(t_2)}{(t_2 - t_1)R(t_1)} \qquad (5)$$

$$= \frac{R(t) - R(t + l)}{l\,R(t)} \qquad (6)$$

where $t = t_1$
$t_2 = l + t$

The hazard rate $h(t)$ or instantaneous failure rate is defined as the limit of failure rate as interval length $l \to 0$. Therefore,

$$h(t) = \lim_{l \to 0} \left| \frac{R(t) - R(t + l)}{l\,R(t)} \right|$$
$$\qquad (7)$$
$$= \frac{-1}{R(t)} \left| \frac{dR(t)}{dt} \right| = \frac{1}{R(t)} \left| \frac{-dR(t)}{dt} \right|$$

Using equation (3),

$$h(t) = \frac{f(t)}{R(t)}. \qquad (8)$$

This is one of the fundamental relationships in reliability analysis since it is indepen-

dent of the statistical distribution used. It implies that if the density function of the time to failure $f(t)$ and reliability function $R(t)$ are known, the hazard rate function $h(t)$ at any time t_1 can be found. Although the failure rate $\lambda(t)$ and hazard rate $h(t)$ are mathematically different, they are often used interchangeably in conventional reliability engineering.

A general expression for reliability function $R(t)$ can also be derived from equation (7) as

$$h(t) = \frac{-1}{R(t)} \left[\frac{dR(t)}{d(t)} \right]$$

$$\frac{dR(t)}{R(t)} = -h(t) \cdot d(t).$$

Integrating both sides of this equation,

$$\int_0^t \frac{dR(t)}{R(t)} = -\int_0^t h(t)\, dt$$

$$\ln R(t) - \ln R(0) = -\int_0^t h(t)\, dt.$$

But

$$R(0) = 0 \text{ and } \ln R(0) = 0$$

Therefore,

$$R(t) = \exp\left[-\int_0^t h(t)\, dt \right]. \qquad (9)$$

If $h(t)$ can be considered a constant failure rate (special case of exponential distribution), then

$$R(t) = e^{-\lambda t}.$$

The mean time to failure (MTTF) is defined as the average time to failure (or expected value of time to failure) for a population of devices when operated under a specific set of conditions for a given period of time:

$$\text{MTTF} = \int_0^\infty t\, f(t)\, dt$$

$$= \int_0^\infty t \left[\frac{-dR(t)}{d(t)} \right] d(t).$$

Integrating both sides and simplifying,

$$\text{MTTF} = \int_0^\infty R(t)\, dt. \qquad (10)$$

The MTTF can also be expressed in terms of failure rate $\lambda(t)$. If $\lambda(t)$ represents the number of failures expected over a specific time interval (say T_1–T_2), then the mean or average time to failure over T_2–T_1 ($=T$) is equal to the reciprocal of the failure rate over T:

$$\text{MTTF} = \frac{1}{\lambda(T_2 - T_1)} \tag{11}$$

for $T_i = 0$, $T_2 = T$, and $\text{MTTF}(T) = \frac{1}{\lambda(t)}$.

For a special case of exponential distribution,

$$\text{MTTF} = \frac{1}{\text{hazard rate}} = \frac{1}{\text{failure rate}}.$$

Therefore, the hazard rate equals the failure rate for any and all intervals of operating time T, i.e.,

$$\lambda = h(t) = \lambda(T).$$

It should be noted that these relationships will not be valid for distributions other than the exponential [137].

There are many standard statistical distributions, discrete and continuous, which are used to model the various reliability parameters. However, only those applicable to microelectronic devices (memories) will be reviewed in the following sections. Binomial and Poisson distribution are examples of discrete distributions. The continuous distributions commonly used are: Normal (or Gaussian), Exponential, Gamma, Weibull, and Lognormal. Table 6-4 summarizes the shape of density function $f(t)$, reliability function $R(t)$, and hazard rate function $h(t)$ for these five continuous distributions [136].

6.4.1.1 Binomial Distribution.

This distribution is used for situations where there are only two outcomes (such as success or failure), and the probability remains the same for all trials. The probability density function (pdf) of a binomial distribution is given by

$$f(x) = \binom{n}{x} p^x q^{(n-x)}$$

where

$$\binom{n}{x} = \frac{n!}{(n-x)! \, x!} \quad \text{and} \quad q = 1 - p.$$

$f(x)$ represents the probability of obtaining exactly x good devices and $(n - x)$ bad devices in a sample size of n; p is the probability of obtaining a good device (success), and q (or $1 - p$) is the probability of obtaining a bad device (failure).

6.4.1.2 Poisson Distribution.

This discrete distribution used quite frequently may be considered an extension of the binomial distribution when $n \to \infty$. If events (e.g., memory device failures) have a Poisson distribution, they occur at constant average rate, and the number of events occurring in any time interval is independent of the number of events occurring in any other time interval. For example, the number of failures (x) in a certain time interval would be given by

$$f(x) = \frac{a^x - e^{-a}}{x!}$$

where

a = expected number of failures.

For reliability analysis,

$$f(x; \lambda, t) = \frac{(\lambda t)^x e^{-\lambda t}}{x!}$$

where λ = failure rate

t = time interval under consideration.

The reliability function $R(t)$, or the probability of zero failures in time t, is given by

$$R(t) = \frac{(\lambda t)^o e^{-\lambda t}}{0!}$$

$= e^{-\lambda t}$ which represents the exponential distribution.

6.4.1.3 Normal (or Gaussian) Distribution.

One application of this distribution deals with the analysis of manufactured items, their variability, and their ability to meet specifications. The basis for the use of the Normal distribution in this application is the Central Limit Theorem, according to which the sum

TABLE 6-4. Density Function, Reliability Function, and Hazard Rate for the Normal, Exponential, Gamma, Weibull, and Lognormal Distributions. [136]

Name	Density Function $f(t)$	Reliability Function $R(t)$	Hazard Rate $h(t) = \frac{f(t)}{R(t)}$
Normal or Gaussian	$f(t)=\frac{1}{\sigma\sqrt{2\pi}}e^{-\frac{(t-\theta)^2}{2\sigma^2}}$	$R(t)=\int_t^{\alpha}\frac{1}{\sigma\sqrt{2\pi}}e^{-\frac{(t-\theta)^2}{2\sigma^2}}dt$	$h(t)=\frac{f(t)}{R(t)}$
Exponential	$f(t)=\frac{1}{\theta}e^{-t/\theta}$	$R(t)=e^{-t/\theta}$	$h(t)=\frac{1}{\theta}$
Gamma	$f(t)=\frac{1}{\alpha/\beta^{\alpha+1}}t^{\alpha}e^{-t/\beta}$, $\beta=1$ $\alpha=0,1$	$R(t)=\int_t^{\alpha}\frac{1}{\alpha/\beta^{\alpha+1}}t^{\alpha}e^{-t/\beta}$, $\beta=1$ $\alpha=0,1$	$h(t)=\frac{f(t)}{R(t)}$, $\beta=1$ $\alpha=0,1$
Weibull	$f(t)=\alpha\beta t^{\beta-1}e^{-\alpha t^{\beta}}$, $\alpha=1$ $\beta=1,2$	$R(t)=e^{-\alpha t^{\beta}}$, $\alpha=1$ $\beta=1,2$	$h(t)=\alpha\beta\, t^{\beta-1}$, $\alpha=1$ $\beta=1,2$
Lognormal	$f(t)=\frac{1}{\alpha t(2\pi)^{1/2}}e^{\left[-\frac{\ln(t-\theta)}{2\alpha}\right]^2}$	$R(t)=\int_t^{\alpha}f(t)\,dt$	$h(t)=\frac{f(t)}{R(t)}$

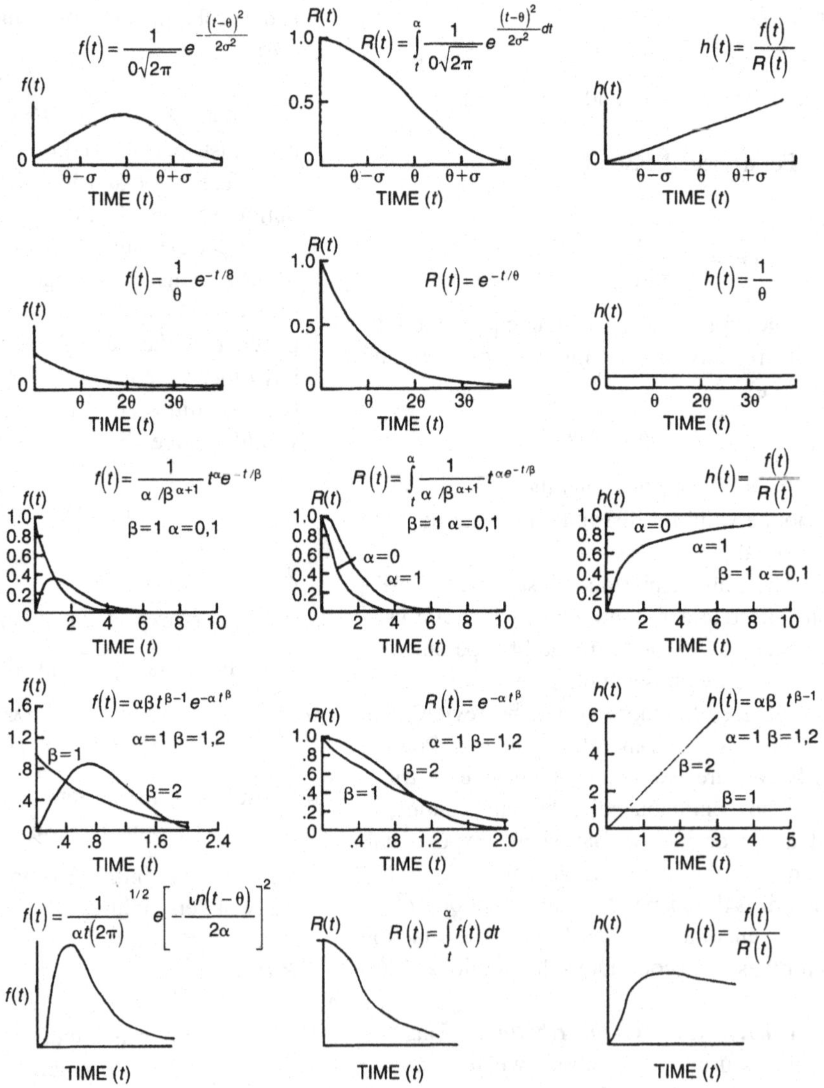

Note: For the functions shown above, θ and σ are the mean and standard deviations; α and β are the shaping parameters.

of a large number of identically distributed random variables, each with a finite mean and variance, is normally distributed. For example, in using this distribution for memory devices, the variations in their performance would be considered as normally distributed. The failure density function for the Normal distribution is given by

$$f(t) = \frac{1}{\sigma(2\pi)^{1/2}} \exp\left[\frac{-1}{2}\left(\frac{t-\mu}{\sigma}\right)^2\right]$$

where μ = population mean
  σ = population standard deviation, which is the square root of variance.

The standard Normal distribution density function is given by

$$f(z) = \frac{1}{(2\pi)^{1/2}} \exp\left(\frac{-z^2}{2}\right)$$

where $\mu = 0$
  $\sigma^2 = 1$.

For most practical applications, the probability tables for the standard Normal distribution are used.

6.4.1.4 Exponential Distribution. This is the most widely used distribution for reliability prediction of the electronic components and equipment (Ref.: MIL-HDBK-217). It is representative of the situation where the hazard rate (or failure rate) is constant and can be shown as being generated by a Poisson process. The Exponential distribution has the advantages of having a single, easily estimated parameter (λ). It is particularly suitable for devices for which the early failures (or infant mortalities) have been eliminated during a screening "burn-in." The failure density function for this distribution is given by

$$f(t) = \lambda\, e^{-\lambda(t)} \qquad \text{for } t > 0.$$

The reliability function is given by

$$R(t) = e^{-\lambda(t)}.$$

6.4.1.5 Gamma Distribution. This is used in reliability analysis for cases where partial failures can exist. An example is redundant systems where a given number of partial failures must occur before a device fails. It can also be considered as the time to second failure when the time to failure is exponentially distributed. The failure density function for this distribution is given by

$$f(t) = \frac{\lambda}{\Gamma(\alpha)}(\lambda t)^{\alpha-1}e^{-\lambda t} \qquad \text{for } t > 0$$

where λ = failure rate (complete failure, or the number of partial failures for complete failure)

$$\text{mean } \mu = \frac{\alpha}{\lambda}$$

$$\text{standard deviation (SD)} = \frac{\alpha^{1/2}}{\lambda}$$

$$\text{Gamma function } \Gamma(\alpha) = \int_0^\infty x^{\alpha-1}e^{-x}dx.$$

The Gamma function can be evaluated by using the standard tables.

When α-1 is a positive integer (which is usually the case, e.g., partial failure situation), then

$$\Gamma(\alpha) = (\alpha - 1)!$$

$$f(t) = \frac{\lambda}{(\alpha - 1)!}(\lambda t)^{\alpha-1}e^{-\lambda t}$$

which for $\alpha = 1$ (special case) becomes an Exponential distribution.

6.4.1.6 Weibull Distribution. This is a general distribution which is particularly useful since, by the adjustment of distribution parameters, it can be made to model a wide range of life distribution characteristics of different classes of the manufactured devices. A general form of failure density function is given by

$$f(t) = \frac{\beta}{\eta}\left(\frac{t-\gamma}{\eta}\right)^{\beta-1} \exp - \left(\frac{t-\gamma}{\eta}\right)^{\beta}$$

where β = shape parameter
  η = scaling parameter or characteristic life (i.e., life at which 63.2% of the population will have failed)
  γ = minumun life.

For most practical situations, $\gamma = 0$ (failures are assumed to start at $t = 0$), and $f(t)$ simplifies to

$$f(t) = \frac{\beta}{\eta}\left(\frac{t}{\eta}\right)^{\beta-1} \exp - \left(\frac{t}{\eta}\right)^{\beta}.$$

Depending on the value of β, the Weibull function can take the form of the following distributions:

$\beta < 1$ Gamma
$\beta = 1$ Exponential
$\beta = 2$ Lognormal
$\beta = 3.5$ Normal (approximately).

6.4.1.7 Lognormal Distribution. This is quite often used to model the failures during accelerated life testing of semiconductor devices such as memories, where failures occur after a certain damage threshold is reached. It is the distribution of random variables whose natural logarithm is distributed normally.

The Lognormal distribution density function is given by

$$f(t) = \frac{1}{\sigma\, t\, (2\pi)^{1/2}} \exp\left[\frac{-1}{2}\left(\frac{\ln t - \mu}{\sigma}\right)^2\right]$$

$$\text{mean} = \exp\left(\mu + \frac{\sigma^2}{2}\right) \qquad \text{for } t \geq 0$$

$$\text{SD} = [\exp(2\mu + 2\sigma^2) - \exp(2\mu + \sigma^2)]^{1/2}$$

where μ and σ are the mean and standard deviation of $\ln t$.

6.4.2 Reliability Modeling and Failure Rate Prediction

Reliability modeling is a key to failure rate prediction, and validated failure models are essential to the development of good prediction techniques. Inputs to the failure rate models are operational field data, test data, physical failure information, and engineering judgment [137]. There are six widely used reliability prediction procedures for predicting the electronic components reliability as listed below:

- MIL-HDBK-217E, Reliability Prediction of Electronic Equipment [138]
- Bellcore Reliability Prediction Procedure for Electronic Equipment (Bellcore RPP) [139]
- Nippon Telegraph and Telephone Corporation Standard Reliability Table for Semiconductor Devices (NTT Procedure) [140]
- British Telecom Handbook of Reliability Data for Components Used in Telecommunications Systems (British Telecom Handbook HRD4) [141]
- French National Center for Telecommunications Studies (CNET Procedure) [142]
- Siemens Reliability and Quality Specification Failure Rates of Components (Siemens Procedure) [143].

All of these procedures use the constant failure rate reliability model for which the reliability at time t is given by

$$R(t) = e^{-\lambda t}.$$

The failure rate (λ) is given by the device model, which is a function of parameters describing the device physical and operating characteristics and the device operating environment. Table 6-5 summarizes the failure rate model equations in each of these reliability prediction models [144]. MIL-HDBK-217 and CNET procedures provide models for both a "parts count analysis" and "stress analysis." The parts count model (called "simplified" in CNET) assumes typical operating parameters

TABLE 6-5. A Summary of Failure Rate Equations in Six Widely Used Reliability Prediction Procedures. (From [144], with Permission of IEEE.)

Procedure	Microelectronic Device Model
MIL-HDBK-217 (stress model)	$\lambda = \Pi_Q (C_1 \Pi_T \Pi_V + C_2 \Pi_E) \Pi_L$
MIL-HDBK-217 (parts count)	$\lambda = \lambda_G \Pi_Q \Pi_L$
Bellcore RPP	$\lambda = \lambda_G \Pi_Q \Pi_S \Pi_T$
British Telecom HRD4	$\lambda = \lambda_b \Pi_T \Pi_Q \Pi_E$
NTT Procedure	$\lambda = \lambda_b \Pi_Q (\Pi_E + \Pi_T \Pi_V)$
CNET Procedure (stress model)	$\lambda = (C_1 \Pi_t \Pi_t \Pi_V + C_2 \Pi_B \Pi_E \Pi\sigma) \Pi_L \Pi_Q$
CNET Procedure (simplified)	$\lambda = \Pi_Q \lambda_a$
Siemens Procedure	$\lambda = \lambda_b \Pi_U \Pi_T$

for a component (device) and is used when most of its operating parameters are not known. The stress model requires a detailed analysis of all the parameters on which the component failure rate depends.

- Π_Q = Quality factor which depends on the quality of the device as determined by inspection level and test procedures during the manufacturing process.
- C_1, C_2 = Failure rate constants. C_1 depends on the circuit complexity and technology. C_2 depends on the packaging type and package pin count. $\Pi_\tau, \Pi_B, \Pi_\sigma$ in the CNET model also depend on the circuit technology and function, the packaging technology, and the package pin count, respectively.
- Π_T, Π_t = Temperature acceleration factors which depend on the steady-state operating temperature of the device.
- Π_v, Π_u, Π_S = Voltage-stress factors which depend on the ratio of the applied voltage to the rated voltage of the component.
- Π_E = Environmental factor which depends on the device operating environment.
- Π_L = Device or process learning factor. This is related to the device technology maturity (length of time in production).
- λ_b = Base failure rate which depends on the device complexity and technology. λ_G and λ_a are generic or average failure rates, assuming average operating conditions.

The failure rate equations given in Table 6-5 consist of a base failure rate modified by the several pi (Π) factors. MIL-HDBK-217 and CNET stress models consist of two separate failure rates C_1 and C_2 for the parts technology and its packaging, respectively, and with each of these modified by appropriate pi factors. MIL-HDBK-217 tables for use with a parts count analysis give the generic factor rate (λ_G) for various microelectronic devices. The values depend upon the device usage environment, and are based on the stress model, assuming normal operating conditions and temperature for that environment. The CNET procedure also provides λ_G for use with a simplified device model. Failure rates for T_j = 40 and 70°C are given for each operating environment. Bellcore, RPP, and Siemens Procedures include tables of generic failure rates for many devices of varying size, type, technology, and complexity. In British Telecom HRD4, the base failure rate (λ_b) includes the time factor, and λ_b expressions for various bipolar and MOS device technologies including memories are given in Table 6-6 [144].

For memories, the device complexity is determined by the number of bits. The base failure rate (λ_b) in the NTT procedure is of the form

$$\lambda_b = k\,x^{0.25}(\text{or } k\,x^{0.50})$$

where k = constant depending on device type and technology

x = measure of a device complexity (e.g., number of memory bits).

TABLE 6-6. British Telecom HRD4 Formulas for Calculating the Base Failure Rate (λ_b) for Various Microelectronic Devices. (From [144], with Permission of IEEE.)

Component Category	Base Failure Rate
Bipolar digital logic	$\lambda_b = \dfrac{94}{t}\,G^{\,5/t}$
Bipolar and MOS linear	$\lambda_b = \dfrac{72}{t}\,N^{\,5/t}$
MOS digital logic	$\lambda_b = \dfrac{1500}{t^{1.5}}\,G^{\,5/t\;1.5}$
MOS DRAMs, EPROMs, EEPROMs, and CCDs	$\lambda_b = \dfrac{22}{t}\,B^{\,5/t}$
MOS SRAMs	$\lambda_b = \dfrac{43}{t}\,B^{\,5/t}$
Bipolar SRAMs and fusible link PROMs	$\lambda_b = \dfrac{94}{t}\,B^{\,5/t}$

G = number of gates; N = number of transistors; B = number of bits.

Π_Q is the quality factor. For high-reliability applications, MIL-M-38510 (General Specification for Microcircuits) and MIL-I-38535 (General Specification for IC Manufacturing) establish general requirements including manufacturing, screening, quality conformance inspection, and qualification. Each reliability prediction procedure given in Table 6-5 defines several levels of quality, and assigns a value of Π_Q to each level (e.g., 0.25–20). The British Telecom procedure specifies three levels of procurement quality. The CNET procedure identifies seven quality classes and several subclasses.

The environmental factor Π_E accounts for the effects of environmental stresses on device reliability. Most of the reliability prediction procedures list typical environments within their range of applicability and provide a corresponding value for Π_E. However, the Siemens procedure does not include an environmental factor, nor does it identify any operational environments. Therefore, the failure rates for a device such as a 64 kb DRAM are expected to vary widely, not only from one reliability prediction methodology to another, but also within each of the procedures, depending upon the type of operating environment selected.

Temperature acceleration factor (Π_T) is based upon device technology, and the Arrhenius model is commonly used to describe the effects of steady-state temperature on the component failure rates. Various failure mechanisms can have different activation energies, and in some cases, an average value corresponding to several failure mechanisms which might be used in failure rate prediction. The Arrhenius model includes the effect of temperature and activation energies of the failure mechanisms, and is expressed by the following equation:

$$A = \exp\left[\frac{-E_A(T_{J1} - T_{J2})}{k(T_{J1})(T_{J2})}\right]$$

where A = acceleration factor
E_A = activation energy
k = Boltzmann's constant
$(8.62 \times 10^{-5}\,\text{eV/K})$

T_{J1} = device junction temperature under ambient conditions (e.g., room temperature, 25°C or 298 K)

T_{J2} = device junction temperature under accelerated conditions (e.g., 125°C or 398 K).

Π_T is calculated from the reference temperature (T_r) and the activation energy. Table 6-7 shows the temperature acceleration factors for various prediction methodologies [144]. The reference temperature used is in the range of 25–70°C. A, A_1, A_2, E_{a1}, E_{a2} are constants defined within the corresponding procedure; and T_r, T_{J1}, T_{J2} are the device p-n junction absolute temperatures. These constants depend on the device technology and whether the component is in a hermetic or nonhermetic package. The range of Π_T values in the prediction procedure strongly depends on the device power dissipation (P_d). For devices with high P_d, differences in the estimates of θ_{JC} in the procedures cause much greater differences in the values of Π_T.

The voltage-stress factor (Π_v, Π_u, Π_S) is 1 in all procedures for all IC technologies except the CMOS. MIL-HDBK-217 also has a voltage stress derating factor table. The process maturity or learning factor (or Π_L) of the device manufacturing technology appears explicitly in the failure rate models for MIL-HDBK-217 and CNET procedures, and indirectly in the others. In MIL-HDBK-217, Π_L is usually 1, but can have other values also, which are listed in a table in the handbook.

The MIL-HDBK-217 procedure assumes that the device had sufficient screening burn-in to have eliminated infant mortality failures and have achieved constant failure rate. The Bellcore and CNET procedures provide a methodology for developing a failure rate adjustment factor for the devices that have not been completely burned in.

Failure Rate Calculations for a 64K DRAM (Example 1): Failure rate calculations were made on a 64K DRAM using all six failure prediction methodologies, and the results are listed in Table 6-8 [144]. All failure rates are expressed as the number of device failures per 10^9

TABLE 6-7. Temperature Acceleration Factors for Various Prediction Methodologies. (From [144], with Permission of IEEE.)

Procedure	Temperature Acceleration Factor
MIL-HDBK-217E	$\Pi_T = 0.1 \exp\left[-A\left(\dfrac{1}{T_j} - \dfrac{1}{298}\right)\right]$
HRD4	$\Pi_T = 1$, for $T_j \leq 70°C$; $2.6 \times 10^4 \exp\left[\dfrac{-3500}{T_j}\right] + 1.8 \times 10^{13} \exp\left[\dfrac{-11600}{T_j}\right]$ for $T_j > 70°C$
NTT Procedure	$\Pi_T = \exp\left[3480\left(\dfrac{1}{339} - \dfrac{1}{T_j}\right)\right] + \exp\left[8120\left(\dfrac{1}{356} - \dfrac{1}{T_j}\right)\right]$
	$= 2.9 \times 10^4 \exp\left[\dfrac{-3480}{T_j}\right] + 8.0 \times 10^9 \exp\left[\dfrac{-8120}{T_j}\right]$
CNET Procedure	$\Pi_T = A_1 \exp\left[\dfrac{-3500}{T_j}\right] + A_2 \exp\left[\dfrac{11600}{T_j}\right]$
Siemens Procedure	$\Pi_T = A \exp\left[E_{a1} \cdot 11605 \left(\dfrac{1}{T_{j1}} - \dfrac{1}{T_{j2}}\right)\right] + (1 - A) \exp\left[E_{a2} \cdot 11605 \left(\dfrac{1}{T_{j1}} - \dfrac{1}{T_{j2}}\right)\right]$

hours of operation (FIT). These calculations were based on the following assumptions: ambient temperature of 40°C (stress model), ground benign environment (ideal situation having controlled temperature, humidity, and nearly zero environmental stress), devices procured to good specifications with proper qualification program and manufacturing process controls, encapsulated (ceramic package) devices, screened de-

TABLE 6-8. Failure Rates (FIT) for a 64K DRAM in Various Prediction Methodologies. (From [144], with Permission of IEEE.)

Procedure	λ
MIL-HDBK-217 (stress model)	216 FIT
MIL-HDBK-217 (parts count)	219 FIT
Bellcore RPP	140 FIT
NTT Procedure	138 FIT
CNET Procedure (stress model)	631 FIT
CNET Procedure (simplified)	1950 FIT
British Telecom Procedure	8 FIT
Siemens Procedure	96 FIT

vices that have passed infant mortality period, and power dissipation (P_d) of 250 mW.

The calculated FIT rate agreement between the MIL-HDBK-217 stress and parts count models is coincidental. Assumptions of $T_A = 40°C$ and P_d of 250 mW give $T_J = 47.5°C$ in the stress model. The parts count model assumes $T_A = 30°C$, which leads to $T_J = 45°C$. The very low failure rate for the British Telecom procedure is due to the choice of the "base year," which is related to the technology maturity (Π_L) of the device family. Table 6-9 shows the failure rates in different methodologies for this 64K DRAM corresponding to various operating temperatures (20, 40, 60, and 80°C) [144].

Failure Rate Calculations Using Only MIL-HDBK-217E (Example 2):

- *Device Description:* An 8192 b n-channel MOS UVEPROM in a ground, fixed application, junction temperature of 55°C, procured to vendor equivalent B-2 quality level. The production line for this device has been in continuous production. The

TABLE 6-9. Failure Rates (FIT) for a 64K DRAM in Different Prediction Methodologies for Various Operating Temperatures. (From [144], with Permission of IEEE.)

Procedure	20°C	40°C	60°C	80°C
MIL-HDBK-217 (stress)	77	216	582	1495
MIL-HDBK-217 (parts count)	219	219	219	219
Bellcore RPP	45	140	378	910
NTT Procedure	84	139	259	541
CNET Procedure (stress)	353	631	1156	2140
CNET (simplified)	790	1950	—	—
British Telecom Procedure	8	8	10	19
Siemens Procedure	48	96	208	504
	(a) Hermetic packaging			
MIL-HDBK-217 (stress)	86	362	1523	5648
MIL-HDBK-217 (parts count)	380	380	380	380
Bellcore RPP	54	168	454	1092
NTT Procedure	216	410	827	1844
CNET Procedure (stress)	446	835	1687	4088
CNET (simplified)	790	1950	—	—
British Telecom Procedure	8	8	16	33
Siemens Procedure	69	149	341	845
	(b) Nonhermetic packaging			

device is a solder seal, ceramic/metal dual-in-line package (DIP), hermetic with 24 pins.

- From MIL-HDBK-217E (Section 5.1.2.5), the operating failure rate model is

$$\lambda_p = \pi_Q (C_1 \, \pi_T \pi_V + C_2 \pi_E) \, \pi_L$$

	MIL-HDBK-217E
C_1 for 8192 b = 0.06	Section 5.1.2.5
Quality level B-2: $\pi_Q = 5.0$	Table 5.1.2.7-1
Ground, fixed	
environment: $\pi_E = 2.5$	Table 5.1.2.7-3
NMOS, hermetic package,	
corresponding to π_T	Table 5.1.2.7-4
Table 5.1.2.7-7: $\pi_T = 0.59$	
$\pi_V = 1.0$	Table 5.1.2.7-14
24-pin hermetic DIP,	
solder seal: $C_2 = 0.009$	Table 5.1.2.7-16
$\pi_L = 1$	Table 5.1.2.7-2

Substituting these values in the failure rate equation,

$$\lambda_p = 5.0 \, [(0.06 \times 0.59 \times 1.0) + (0.009 \times 2.5)] \, 1.0$$
$$= 0.29 \text{ failures}/10^6 \text{ h}$$
$$= 290 \text{ FIT}.$$

6.5 DESIGN FOR RELIABILITY

The design for the memory reliability concept is very important, and adds only a small effort to the overall routine process of a "design for performance." The relevant reliability parameters of a given technology are obtained from the dc accelerated tests on test structures. Failure mechanism models in a simulator can then predict the reliability of complex circuits. Currently, the reliability assurance techniques and methodologies are mainly based upon failure detection at the end of a lengthy product design and development cycle. However, a circuit simulation and analysis at the design stage is an important tool for reliability assurance. A circuit reliability simulator may typically contain several models to simulate the major reliability failure mechanisms. For example, HOTRON is a proprietary simulator of the hot-electron effects developed at TI [145]. The RELY simulator is being developed at the University of Southern California, and its goal is to simulate hot-electron effects, oxide reliability, and electromigration [146]. The simulator BERT, under development at the University of California, Berkeley, contains the model for hot-electron effects, oxide reliability, electromigration, and bipolar transistor beta degradation [147].

The BERT (Berkeley Reliability Tool) contains four models for four areas of reliability concerns, as follows [148]:

- CAS for circuit aging simulation (or hot-electron effects reliability)
- CORS for circuit oxide reliability simulator
- EM for electromigration
- Bipolar transistor degradation.

The BERT is linked to the Simulation Program with Integrated Circuits Emphasis (SPICE) externally in a pre- and postprocessor configuration to form an independent simulator. For hot-carrier effects simulations, the fresh circuit is first simulated to determine the terminal voltage waveforms at all transistors over one cycle at any user-specified supply voltage. The postprocessor calculates the transient substrate current and the "age" of each device after any user-specified stress time. Then, according to the calculated "age," the simulator generates the corresponding new device parameters for each device by interpolating user-supplied process files of dc stressed transistors. Finally, the aged circuit is resimulated by SPICE. Another BERT module, CORS, predicts the probability of circuit failure as a function of the operating time, temperature, power supply voltage, and input waveforms. The simulation model uses "defect density" $D(\Delta X_{OX})$ extracted from the oxides grown in the University laboratory. The effect of burn-in on the oxide reliability also can be sim-

ulated. The TDDB characteristics can be predicted quite accurately using this model as well. Electromigration lifetime predictions from the vacancy recombination model can be implemented to calculate the current dependence of the electromigration time to failure [149]. This model is valid for interconnects as well as intermetallic contacts.

An end-of-life (EOL) model has been developed that uses semiconductor device aging to predict the end of mission-life device parameters by extrapolating from early environmental and electrical test measurements [150]. This model assumes that simple physical and chemical reactions between the impurities in the bulk material, on the semiconductor device surfaces within their package, and the host materials within the device cause its electrical parameters to shift from their typical specification values to their EOL values. For example, in a given semiconductor device such as a memory, the output voltage low (V_{OL}) is measured initially for a given lot of components. These devices are subjected to a 240 h burn-in screen test, followed by an elevated 1000 h accelerated life test, both at the same temperature (T_L). Then the devices whose pre–post life test parameter value shifts (ΔV_{OL}) exceed predetermined limits are rejected, while those that pass are accepted. The EOL model uses the preceding information to extrapolate a given parameter of interest for the acceptable lot to its EOL value. The principal mechanism used to extract EOL values is thermal stress, and the major parameter used to characterize this phenomenon is the activation energy (E_A) of the physical degradation processes as specified in the Arrhenius relationship. This model depicts the device damage as due to the long-term collective efforts of a number of amount-limited contaminant reactant sources.

The method of accelerated stress aging for semiconductor devices is commonly used to ensure long-term reliability. However, to correctly interpret the results of accelerated stress testing, the accelerating effects of the applied stresses must be well understood. A physical model for degradation of the DRAMs during accelerated stress aging was developed to understand and characterize the charge trapping and interface-state generation at the Si/SiO_2 interface [151]. In this evaluation, the pulse aging and radiation damage (Co^{60} gamma ray source) were used as tools for conducting experiments on the n-channel Si-gate MOSFETs. The test results showed that the injection of hot holes is significantly more damaging to the Si/SiO_2 interface compared to the injection of hot electrons. The combined result of the rate of damage and the rate of annealing determines the net degradation of DRAMs. Therefore, the failure rates of DRAMs would most likely be overestimated if the process of annealing is not appropriately accelerated.

The long-term integrity and stability of thin oxides used to fabricate gate and cell dielectrics are an area of major concern in memory reliability. The stress test involved accelerated voltage conditions combined with data acquisition during the manufacturing wafer-sort step. The use of test capacitors in a 64K DRAM memory array matrix is useful in determining the distribution in time of oxide failures [152]. The advantages of this approach of using an actual 64K DRAM array (versus test capacitors) for failure rate prediction are the following: (1) there are no processing or structural differences, and thus the latent defect densities are measured directly; (2) the TDDB data are obtained automatically at the wafer-sort test step on all RAMs; and (3) the DRAM cell structure provides very low current-detection capability. Standard functional testing is used in conjunction with an applied rectangular waveform to provide time distribution during a 1 s test. These data yield the TDDB lognormal cumulative distribution, the TDDB-induced shorts incidence, and the slope of the curve represents the failure rate under accelerated conditions. Extrapolations and derating of these curves can predict failure rates at in-system and under operational life test conditions. The TDDB data are obtained on the dice intended to be the end product. Since DRAM is inherently a sensitive electrometer capable of roughly 12 pA current resolution, even oxide ruptures and low-level leakages are detectable. This measurement ability is augmented

with the use of single-bit refresh tests. This method has the potential of automatic lot-by-lot failure predictions and screening ability.

AT&T Microelectronics has implemented a test methodology called the Operational Life Test (OLT) to monitor and quantify the early life reliability of selected semiconductor technologies and identify early life failure mechanisms [153]. This early life reliability monitor measures the effectiveness of screens and tests employed in order to identify and eliminate the infant mortality failure modes. In addition, this early life reliability monitor complements the data derived from highly accelerated long-term reliability tests. The major objective of OLT is to provide rapid feedback regarding infant mortality failure modes and mechanisms to early design and processing stages.

A typical OLT sequence is shown in Figure 6-20 [153]. A sample is selected from approved shipment products for pre- and post-testing, consisting of a standard set of device electrical test requirements. The OLT test attempts to simulate field conditions by operating the devices for an extended period of time under mildly accelerated conditions (nondestructive). The devices are dynamically operated at elevated temperature while being subjected to periodic power and temperature cycles. The devices are electrically configured to

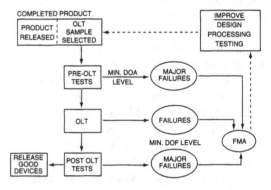

Figure 6-20. AT&T typical operational life test (OLT) sequence. (From [153], with permission of IEEE.)

represent the primary or most significant field application, and their in-test functional performance is monitored to detect nonfunctional and intermittent failures. The failure rate in the infant mortality region is modeled by the Weibull distribution, whereas in the steady-state region, it is modeled by an Exponential distribution (constant failure rate). An important feature of the Weibull distribution is that the failure rate plot becomes a straight line when the log of the failure rate is plotted versus the log of the operating time. To estimate the Weibull parameters for a product family, the failure times of OLT defects are analyzed using a computer software package called STAR (statistical reliability analysis). Then the resulting Weibull parameters can be used to make reliability predictions for the entire device population.

The data collected during the OLT monitor program have been quite helpful for AT&T in making continuous improvements in a product's design and processing cycle to detect and eliminate infant mortality failure mechanisms. The functional operation of the devices in the OLT (with continuous monitoring) at near-field use temperatures, combined with temperature and power cycling, detects different failure modes from those which are seen during long-term reliability testing.

For nonvolatile memories, the endurance characteristics are a major concern, and therefore endurance modeling is necessary in the design for reliability approach. Memory endurance describes the number of program/erase cycles that the device can withstand before failure, and the failure mode may involve dielectric breakdown or loss of data retention. In textured polysilicon floating-gate (TPFG) memories, the endurance failures are caused by trap-up, which is the result of the accumulation of trapped negative charge in the dielectric from the repeated passage of current. The endurance limit is reached when the potential due to this trapped charge grows large enough to suppress the F–N tunneling so that insufficient charge is transferred to change the state of the floating gate. Therefore, endurance modeling for the TPFG memories simply involves the modeling of

buildup of the negative charge as a function of the number of pulses of the F–N current through the tunnel dielectric [154].

The TPFG memories have a textured surface whose curved features generate a field enhancement effect that permits F–N emission at reasonable voltages. These features on the poly-surface have the shape of bumps, and modeling of emission from these surfaces is the basis of endurance predictions. The bumps can be approximated by spherical caps on the tip of truncated cones, allowing the fields to be found using Laplace's equation in the spherical coordinates. From the SEM and TEM inspections, the bumps were found to have a range of sizes. It was found that a simple two-parameter extreme value distribution of the cap radii was a sufficient description to get a good fit of the model to actual experimental data. The test pattern used to characterize endurance is the nonvolatile part of a Xicor NOVRAM cell whose dimensions are identical to all those in the NOVRAM memory array. The advantage of this cell is that all voltages can be applied directly to this cell. However, the experimental determination of endurance is a very time-consuming process, and single-cell data collection typically requires tens to hundreds of millions cycles. For product endurance specification, about 20 cells are needed, requiring simultaneous testing for any semblance of monitoring efficiency. The relationship between single-cell endurance testing and the product endurance specification follows from basic statistical considerations.

In general, for EPROMs that use hot electrons for cell writing obtained by charging the floating gate (FG) with energetic carriers injected over the energy barrier at the Si/SiO_2 interface, the modeling for the write process has been largely empirical or based on oversimplified models. Two approaches have been used for modeling the hot-electron currents injected into the gate oxide of MOS transistors: (1) the lucky electron model [155], and (2) Richard's thermoionic emission formula [156]. In general, the problem of hot-electron modeling is ideally solved by means of Monte Carlo device simulation, which is particularly suitable for dealing with the nonlocal effects involving energy thresholds. However, this approach is quite expensive and computation-intensive. An alternative approach for simple and efficient modeling of the EPROM writing has been proposed that uses a 2-D device simulator coupled with a postprocessor to compute the change into the FG by means of a thermoionic emission formula based on a non-Maxwellian hot-electron energy distribution (EED) [157]. This tool is validated by means of comparison with both experimental data obtained by directly measuring the hot-electron-induced gate current in normal transistors and the writing characteristics of EPROM cells. The test results were found to be in good agreement with the experiments performed on a group of devices with a gate effective channel length (L_{eff}) ranging from 1.4 to 0.8 μm for the transistors, and from 0.8 to 0.5 μm for the EPROM cells.

6.6 RELIABILITY TEST STRUCTURES

Accelerated life testing has been used historically to verify that a chip manufacturing line is meeting its reliability failure rate goals. Intrinsic wearout characteristics on failure mechanisms such as the electromigration and hot-electron effects have been measured on discrete test structures and extrapolated to product failure rates through known acceleration factors [158]. Failure rates due to the random defects are estimated based on accelerated product life test data which can be translated to nominal operating conditions once the precise failure mechanisms are understood. Failure analysis techniques have been developed for the memories which allow determination of the acceleration factors through identification of specific failure mechanisms.

However, the accelerated life testing and end-of-line product life test monitoring methodology is becoming less effective as complex technologies with submicron feature size and lower failure rate (FIT) goals are being developed. Also, the technology scaling is outpacing the development of advanced failure analysis

techniques, e.g., *e*-beam probing also has limitations due to multilayer metallization structures and small feature sizes which inhibit the isolation of defects in random logic. Numerous studies have shown that even in a well-controlled manufacturing facility, process variables can affect the quality of individual product lots or wafers. Statistical process control (SPC) is a tool which is often used to identify and control variations beyond the specified limits. One of the major goals for quality improvement during processing is to identify and eliminate those wafer lots ("mavericks") which have significantly higher defect densities than the rest of the population.

One of the approaches used in conjunction with end-of-line product testing has been the fast, highly accelerated wafer level reliability (WLR) tests on discrete structures which are designed and fabricated to monitor and evaluate failure mechanisms such as electromigration (e.g., through BEM [159], SWEAT [160]), TDDB [161], oxide breakdown, hot-electron effects, etc. Some of these techniques have been effective as process control monitors and predictors of outgoing product failure rates. However, if a process control monitor is defined as a monitor which detects when a process variable drifts out of specification and provides a quick feedback for corrective action, then the WLR may not be an effective process control monitor. The major reasons are: (1) a WLR test is generally performed at the end of the manufacturing line, which can result in a significant number of devices being affected by the process variable drifting out of specification before detection, and (2) a particular WLR test may not be sensitive enough to detect a process variable drifting out of specification even though there may be an increase in the product failure rate. However, the wafer level test structures are considered integral components of an overall reliability assurance plan that includes statistical process control (SPC) limits on in-line monitors, adequate screening, and a continuous process improvement philosophy.

For increasingly high-density megabit DRAMs operating at low supply voltages, it be-

comes difficult to realize large cell capacitances. The usage of a greater number of cells
per bit-line segment in a cell array reduces the
number of sense amplifiers and results in a
smaller chip size. However, the ratio of cell capacitance to bit-line capacitance is reduced, affecting the initial sense signal which is further
deteriorated from noise coupling between the
bit lines, and the bit lines to word lines. The
sense amplifiers which are very sensitive to
the imbalances have to be designed as highly
symmetrical. For example, in a typical 16 Mb
DRAM, 32,000 sense amplifiers are distributed
over the entire chip area. Consequently, they
are very susceptible to process tolerances that
affect the circuit symmetry such as differences
in transistor gate lengths, bit-line capacitances,
and threshold voltage variations. All of these
factors influence the minimum cell signal that
can be correctly sensed. A test structure has
been implemented which allows the determination of the minimum sense signal required for
correct sensing of the stored cell charge in a
16 Mb DRAM [162]. The analysis of the measured data provides a monitor for DRAM
process control.

Figure 6-21 shows the test structure
schematic which is based on a typical 16 Mb
DRAM sense amplifier mainly consisting of an
n-channel latch and a p-channel latch [162]. It
uses a folded bit-line (*BL*) concept in which the
voltage difference between the *BL* and the reference bit line (*BLN*) is amplified by the sense
amplifier. In this test structure, the bit lines are
monitored by 300 fF capacitors (*C*). In addition,
a capacitive mismatch of about 50 fF can be applied to either the upper or lower bit line supplying *VC1* or *VC2* with positive voltage. This
allows simulation of connection of the active
memory cell to one of the bit lines, as well as
tolerances in the bit-line capacitances, both of
which result in a capacitive asymmetry at the
sense amplifier input. Three different n-channel
parts of the sense amplifier can be separately activated.

This test structure determines the differences in the gate lengths of neighboring transistors in a cross-coupled pair due to the process
tolerances. The local variations in the threshold
are measured with an independent test structure
on the same wafer. Therefore, contributions of
the minimum sense signal due to variations in

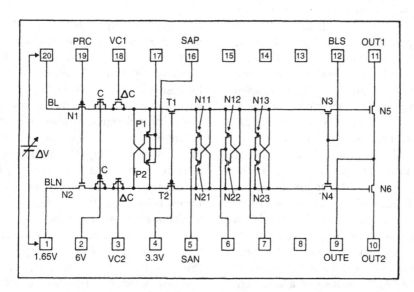

Figure 6-21. A 16 Mb DRAM test structure schematic based on a typical sense
amplifier consisting of an n-channel latch and a p-channel latch.
(From [162], with permission of IEEE.)

current gain and parasitic capacitance asymmetries can be separated from the threshold voltage mismatch. The main conclusion of this test structure evaluation is that a minimum cell capacitance of 35 fF is required for correct sensing, taking into account contributions of the required cell charge due to alpha particles, leakage currents, and coupling noise. Only a small amount of charge from the total available charge is required to compensate for the effects of mismatched device parameters. The alpha-particle contribution is greater, and therefore reducing its effect in future DRAMs may allow for reduced cell capacitance.

Test structures may be yield (or process monitors) typically incorporated at wafer level in kerf test sites and "drop-in" test sites on the chip. The test structures go through the same fabrication processing flow as the production chip, and typically consist of interdigitated metal serpentines (for metal opens and metal shorts), metal studs or via chains (for interlevel connections), and interlevel crossover patterns (for interlevel shorts). An example is an "on-chip" hot-electron test/stress structure developed which provides insight into the device degradation under real circuit operation, and thus makes it possible to accurately predict device lifetime in a specific circuit application [163]. This approach was used to analyze the hot-electron susceptibility of a 1 Mb ECL I/O SRAM. The circuit simulation was performed using the degraded transistor models extracted from experimental data which represented ac or dc stress results at operating voltage over a specific time interval for various subcircuits.

Figure 6-22(a) shows a block diagram of this circuit and the contribution of each subcircuit to the total SRAM access time, and (b) shows the increase of access time due to individual device degradation [163]. The correlation between device and circuit degradation was obtained through subcircuit analysis and simulation. Analysis showed that the CMOS inverters in the 1 Mb SRAM operating at a minimum cycle time of 12 ns were the primary cause of access time increase under the hot-electron stress. The experimental data showed that circuit relia-

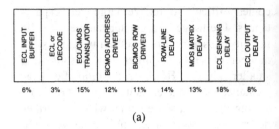

ECL INPUT BUFFER	ECL or DECODE	ECL/CMOS TRANSLATOR	BiCMOS ADDRESS DRIVER	BiCMOS ROW DRIVER	ROW-LINE DELAY	MOS MATRIX DELAY	ECL SENSING DELAY	ECL OUTPUT DELAY
6%	3%	15%	12%	11%	14%	13%	18%	8%

(a)

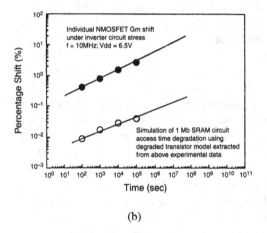

(b)

Figure 6-22. 1 Mb ECL I/O SRAM circuit. (a) Block diagram. (b) Access time as a function of individual device degradation. (From [163], with permission of IEEE.)

bility can be significantly improved by replacing the CMOS inverters with BiCMOS inverters. This is due to the domination of bipolar transistor drive capability. Therefore, by establishing the design guidelines in the subcircuits, the overall circuit reliability can be greatly improved. Hot-electron design guidelines are tradeoff considerations amidst chip area, circuit performance, and reliability. Design parameters such as channel length and drive capability are determined based on the circuit configuration, experimental hot-electron lifetime data, as well as the ac/dc hot-electron correlation.

The use of test chips for monitoring the total ionizing dose (TID) and single-event upset (SEU) is discussed in Chapter 7, under "Radiation Test Structures" (see Section 7.4.4.2). The SEU/TD Radiation Monitor Chip is a part of a comprehensive set of diagnostic test structures developed for high-reliability NASA/JPL space applications [164]. These chips were fabricated in a 1.6 μm n-well, double-level metal CMOS process fabricated by the MOS Implementation Services (MOSIS) foundry. The purpose of these chips was to analyze the quality of the fabrication run. The diagnostic test chips consist of process monitors PM_1 and PM_2, a reliability chip, and a fault chip. In order to reduce the overall fabrication cost, this wafer run (N06J) was shared with an ASIC and a Standard Cell (STD CELL) chip. Figure 6-23 shows a block diagram of a chip (die surface) including the comprehensive set of diagnostic test structures [165].

The fault chip consists of a set of test structures to characterize and provide statistical distribution on the VLSI defect densities for: (1) gate oxide pinholes, (2) poly–metal pinholes, (3) poly–poly shorts, (4) intermetal (metal 1–metal 1 or metal 2–metal 2) shorts, (5) poly wire opens, and (6) metal 1 or metal 2 wire opens. Defect density test structures used were comb resistor, serpentine resistor, p-pinhole capacitor, and n-pinhole capacitor. The gate oxide pinholes were dominating defects. These were analyzed using the Poisson yield expression

$$Y = \exp(-DA)$$

where   yield $Y = 1 - F$ (F is the failure rate)
$$D = \text{defect density}$$
$$A = \text{gate oxide area}.$$

For very small failure rates,

$$D = F/A.$$

For this MOSIS N06J run for which the gate oxide area had dimensions of $1.6 \times 2.4 \ \mu m^2$,

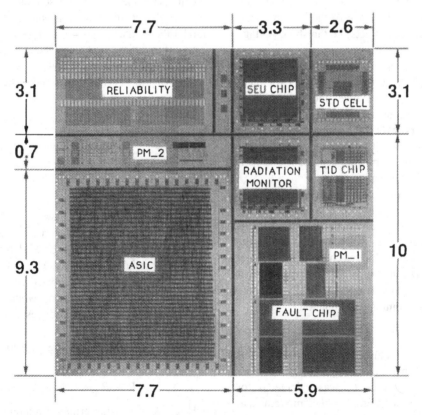

Figure 6-23. NASA/JPL test chip fabricated on MOSIS (N06J wafer run) including a comprehensive set of diagnostic test structures (dimensions in millimeters) [165].

experimental data showed the defect densities for n-pinholes (D_n) = 17 defects/cm², and for p-pinholes (D_p) = 5 defects/cm². The fault chip also contained matrix test structures used to assess the parameter variability, which is an indicator of the local control (e.g., ion implantation) obtained in the fabrication process. These matrix test structures included: (1) the linewidth/step matrix, (2) the contact matrix, and (3) the inverter matrix. The MOSIS run test results from the linewidth/step matrix showed good step coverages with poly 1, metal 1, and metal 2 step resistances being less than 3%. The contact matrix data showed contact resistance standard deviations (metal 1-p-poly, metal 1-n-poly, metal 1-n-diff) as somewhat higher than expected. The standard deviations for the inverter matrix are related to threshold voltage variations, and thus to the SEU sensitivity of the SRAM chip. These values were typical, and showed good local control of the ion implantation process which affects the threshold voltage variations.

The process monitor (PM) chips contained the split-cross bridge resistors for measuring the sheet resistance and linewidth of various layers, contact resistors, circular MOSFET capacitors, MOSFETs, and inverters. The test results of this MOSIS run showed that the contact resistances for the poly and n-diff contacts had a large spread.

The reliability chip was intended to characterize the electromigration in metal 1 or metal 2 and the metal diffusion contacts. In this MOSIS run, metal 2, which usually carries the highest current density, was characterized. Metal 2 electromigration results for the stress current density (J) and stress temperature (T) were fitted to Black's electromigration equation:

$$t_{50} = A_{50}J^{-n} \exp (E_a/kT).$$

The test structures used were metal 2 aluminum wires that had a length of L_c = 8 mm. From the test results, t_{50} was calculated for operating conditions of J_{op} = 0.2 mA/cm² and T_{op} = 125°C by using the following parameters:

$$E_a \text{ (eV)} = 0.49 \pm 0.04$$
$$n = 4.36 \pm 0.24$$

$$\sigma = 0.72 \pm 0.04$$
$$A_{50} = 0.0301.$$

Substituting in Black's equation,

$$t_{50} = 6014 \text{ years.}$$

6.7 RELIABILITY SCREENING AND QUALIFICATION

6.7.1 Reliability Testing

In the manufacturing of memory devices, the total yield of a product is given by

$$Y = \prod_{i=1}^{n} Y_i$$

where Y is the proportion of devices with no yield defects, and Y_i is the yield of each element of the device (e.g., transistors, dielectric layers, and metal interconnects). The Poisson random distribution of defects distribution is commonly used to model the element level yield, Y_i. However, because of variations in the defect density and clustering of defects, other distributions are also used.

For the Poisson distribution, the element yield is given by

$$Y_i = Y_{oi} \times e - \lambda_i \times A_i$$

where Y_{oi} = yield for nonrandom defects
  λ_i = defect density
  A_i = "critical area" for the product element.

During reliability defect modeling, defects of various sizes are placed on a design pattern to determine whether a given defect causes failure for reliability or yield. Integration of the convolution of defect size distribution and the probability of defect occurring where it would cause a reliability defect produces a "reliability critical area" (A_R) [166].

The reliability modeling equation is given by

$$R_i = R_{oi} \times e^{-\lambda}_i \times A_{R_i}$$

where R_i = product proportion having no reliability defects for element i (i.e., failures occur only from intrinsic or wearout mechanisms).

Therefore, $1-R_i$ represents the proportion of the product which may fail because of extrinsic failure mechanism (or infant mortality) and is the defect level (DL_i) for that mechanism.

The comparison of Poisson equations for the yield and reliability of a device gives

$$R_i = Y_i^{k_i}$$

where k_i represents the ratio between critical areas for reliability and yield, i.e.,

$$k_i = A_{R_i} / A_{Y_i}.$$

The purpose of reliability testing is to quantify the expected failure rate of a device at various points in its life cycle. The basic principle governing reliability testing of the semiconductor devices such as memories predicts that the failure rate for a group of devices will follow a bathtub-shaped curve as shown in Figure 6-1. As discussed earlier, the curve has three major regions as follows: (1) an infant mortality region with high failure rate, (2) a steady-state region which is useful life but may have random failures, and (3) a wearout region where the failures occur from intrinsic failure mechanisms. Reliability screens based on accelerated aging rely on the rapidly declining failure rate of infant mortality failure modes to reduce the early life failure rate. Burn-in, which is combined elevated temperature and voltage stress, is an example of these screens. The failure rate of semiconductor memory chips due to manufacturing defects can be improved significantly with burn-in.

Memory failure modes which can be accelerated by a combined elevated temperature and high-voltage stress are threshold voltage shifts (due to slow trapping, contamination with mobile ions, and surface charge spreading), the TDDB leading to oxide shorts, and data-retention degradation (charge loss and charge gain) for the EPROMs and flash memories [167]. The memory data retention and TDDB are both functions of oxide integrity, and an effective burn-in is required to weed out these infant mortality oxide-related defects. The oxide-hopping conduction between the floating gate and silicon substrate is typically less than 10^{-19} A or about 1 electron/s [168]. Depending on the biasing conditions, the electrons flow onto or off the floating gate, causing bit errors in the stored logic states. The oxide-hopping conduction is related to manufacturing defects, and this failure mechanism affects the infant mortality and random portions of the bathtub curve.

Table 6-10 lists some common failure mechanisms for n-channel MOS EPROMs, the

TABLE 6-10. A Listing of Some Common n-Channel MOS EPROM Failure Mechanisms, the Bathtub Region Most Affected by the Mechanism, and Corresponding Activation Energy. (From [168], with permission of Penton Publishing, Inc.)

Mode	Lifetime Region Affected (see Figure 6-1)	Thermal Activation Energy (eV)	Primary Detection Method
Slow trapping	Wear-out	1.0	High-temperature bias
Surface charge	Wear-out	0.5–1.0	High-temperature bias
Contamination	Infant/wear-out	1.0–1.4	High-temperature bias
Polarization	Wear-out	1.0	High-temperature bias
Electromigration	Wear-out	1.0	High-temperature operating life
Microcracks	Random	—	Temperature cycling
Contacts	Wear-out/infant	—	High-temperature operating life
Silicon defects	Infant/random	0.3	High-temperature operating life
Oxide breakdown/leakage	Infant/random	0.3	High-temperature operating life
Hot-electron injection	Wear-out	—	Low-temperature operating life
Fabrication defects	Infant	—	High-temperature burn-in
Charge loss	Infant/random/wear-out	1.4	High-temperature storage
Oxide-hopping conduction	Infant/random	0.6	High-temperature storage/burn-in

bathtub curve region most affected by each mechanism, and the corresponding activation energy (E_a) of the mechanism [168]. Thermal activation energy indicates the effect that the increased temperature has on frequency of the failure; the higher the activation energy, the greater the effect. Acceleration factors for the burn-in enable prediction of the expected operational life of a failed device which was exposed to certain voltage and temperature conditions over a compressed (shorter) period of time. Two types of acceleration factors for burn-in are: temperature (A_T) and electric field (A_{EF}).

Temperature acceleration factors are based on a commonly used model which assumes lognormal life distribution and uses the Arrhenius equation for reaction rates (R):

$$R = R_o \exp\left[\frac{-E_a}{kT}\right].$$

The temperature acceleration factor between two temperatures $(T_1$ and $T_2)$ is the ratio of time-to-fails at T_1 compared to that at T_2, and can be derived as

$$A_T = \frac{t_{f1}}{t_{f2}} = \exp\left[\frac{E_a}{k}\left(\frac{1}{T_2} - \frac{1}{T_1}\right)\right].$$

It is generally agreed that the TDDB data for intrinsic breakdown are lognormal in distribution, even though the activation energy values reported vary from 0.3 to 2.0 eV. The apparent activation energy is a function of the stressing electric field and dielectric strength of the material. For data-retention failures due to SiO_2 oxide defects, an E_a value of 0.6 eV has been reported [169]. In a study of charge loss through the oxide–nitride–oxide (ONO) layer, the activation energy has been reported as 0.35 eV [170]. Therefore, the final selection of E_a for computing A_T depends upon the application.

The electric field acceleration factor (A_{EF}) between two electric fields is the ratio of time-to-fails at first electric field (E_1) compared to that at second electric field (E_2). This is empirically derived by using test capacitors and actual memory devices to fit the following model:

$$A_{EF} = \frac{t_{f1}}{t_{f2}} = \exp\left[\frac{(E_1 - E_2)}{E_{EF}}\right]$$

where E_1 and E_2 are expressed in MV/cm. The E_{EF} is an acceleration rate constant which is related to the field parameter γ and can be expressed by

$$E_{EF} = \frac{1}{\ln 10^7}.$$

The value of γ has been reported to be in the range of 1.7–7 decades/MV/cm, and is found to be dependent on the temperature as expressed by the following equation:

$$\gamma = 0.4 \exp\left[\frac{0.07}{kT}\right].$$

Substituting values of E_{EF} and γ into the voltage acceleration factor equation gives

$$A_{EF} = \exp\left[(E_1 - E_2)\right] \ln 10^{0.4} \exp\left[\frac{0.07}{kT}\right].$$

The basic assumption is that burn-in is designed to detect the failures which are both temperature- and electric-field-sensitive. Therefore, the effects of both temperature and electric field should be taken into consideration. For example, assuming that a device fails at a time t_0 while operating at normal conditions (T_0, E_0), then under accelerated conditions, the time can be shortened to t_2 by exposing the device to a higher temperature T_1 $(>T_0)$. Then the combined temperature and electric field acceleration factor can be expressed as

$$A_{TEF} = \frac{t_o}{t_2} = A_T A_{EF}$$

$$= \exp\left[\frac{E_a}{k}\left(\frac{1}{T_0} - \frac{1}{T_1}\right) + \frac{(E_1 - E_0)}{E_{EF}}\right].$$

Once the A_{TEF} is known, the burn-in temperature and voltage stress levels can be varied to reach the desired values. A study was performed on known weak 1.5 μm CMOS EPROMs which exhibited charge loss by stressing them through the static burn-in using different temperature and dielectric field combinations [168]. For EPROMs, static burn-in stresses the gates of the cells while grounding the source and drain, which is in contrast to the dynamic burn-in where the selected cells are in a read mode. Three different temperature and electric field combinations were used to measure the V_{ccmax} margin which is the highest V_{cc} level wherein a programmed cell can still be read accurately.

Figure 6-24 shows a plot of rate of change in the V_{cc} margin (delta V_{ccmax} margin shift) plotted on a log time (seconds) scale for each of three cases [168]. Test results showed that the margin shift for each setup is different, but they have relatively constant slopes. The setup with higher temperature and electric field tends to induce a higher margin shift. To achieve the same level of margin shift, lesser stress time is needed at higher temperature and higher electric field stresses.

As discussed earlier in Section 6.3, two major reliability concerns for EPROMs are data-retention (or charge loss) characteristics and endurance performance over the extended program-erase cycling. Since charge loss has a high thermal activation energy (1.4 eV), a test involving a very high-temperature storage (e.g., 250°C) can determine where on the bathtub curve the devices begin to wear out. In a charge loss failure screen, the devices with a high percentage of programmed bits (typically more than 90%) are held in storage at 250°C for periods ranging from 168 to 1000 h. Intermediate data are taken to determine the shape of the failure rate curve. To evaluate the effect of

program-erase cycling on device wearout, the devices being screened are subjected to multiple programming and erase cycles before making supply voltage maximum and minimum measurements (V_{ccmax} and V_{ccmin}) that will support proper operation.

To produce reliable EPROMs, often "failure-mode screens" are used which are effective only on those EPROMs with latent defects due to a particular failure mechanism or mode. The bathtub curve shows that infant mortality and wearout failures each have an order of magnitude greater effect on the reliability compared to random failures. Therefore, the EPROM failure-mode screens are designed to isolate mechanisms that cause failures in these two regions of the bathtub curve. Infant mortality screens reduce a device population early life failure rate by prematurely aging each part just beyond the first change in the slope of bathtub curve. They do not lower the random failure rate, although they do reduce the cumulative failure rate, thereby improving the overall reliability. The wearout screens attempt to extend the random failure portion of the curve further to the right, thereby extending device useful life. Infant mortality screens have a negligible impact on the wearout failure rate. As opposed to infant mortality screening, where the objective is to eliminate a small freak population, the wearout screen should ensure that at least 99% of all parts would not fail during a specified lifetime.

In MOS DRAMs, as the gate dielectrics are scaled to thinner dimensions, it has been observed that the aging screens are not that effective in reducing the time-dependent failure rate of dielectric breakdown. The application of an aging screen requires an understanding of the time-dependent breakdown characteristics. The lognormal distribution described by the sigma and median time-to-failure (MTF) have been found to present the best fit to dielectric life test data. The experimental data have shown that there are certain distributions (with MTF and sigma) where aging screens are effective in reducing the failure rates, and other distributions where an aging screen would actually increase failure rates. The lognormal statistics predict that as the MTF and sigma become smaller, the

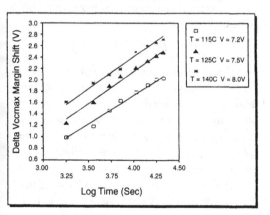

Figure 6-24. A 1.5 μm CMOS EPROM plot of rate of change in delta V_{ccmax} margin shift versus log time for different temperature and voltage combinations. (From [168], with permission of Penton Publishing Inc.)

failure rate declines more slowly. This means that the aging screens must rely on longer effective aging times. For example, in the case of ultrathin dielectrics (<200 Å), an aging stress that would cause degradation of intrinsic oxides requires a long aging time, and may be difficult to implement and not be cost-effective in a production environment. Therefore, the margin between required aging time and time-to-wearout defines the "window" which must be targeted by an aging screen.

A study performed by using an accelerated life test of screened production 64K DRAMs confirmed the effectiveness of using "nonaging" screens to enhance product reliability [171]. A nonaging screen which detects defective gates while minimizing stress on nondefective gates provides a better alternative than the aging screen. The physical mechanism of dielectric breakdown depends on the leakage current for failure activation. The DRAMs can take advantage of their charge-holding characteristics in order to detect the leakage current and implement a nonaging screen. In a typical DRAM cell, since the storage plate is held at higher voltage, dielectric failure would cause a stored "0" (low voltage) to become a "1" (high voltage). This unique failure mode allows dielectric leakage differentiation from other leakage modes. By testing charge-retention characteristics of each DRAM cell at elevated voltage across the storage dielectric, defective oxides can be easily detected. A nonaging screen applied to 64K DRAM production lot showed a greatly improved failure rate [172].

For a burn-in to be used as a reliability screen for memories, it has to be effective. The effectiveness of a specific burn-in is a function of the failure mechanisms and failure rate distribution of a given product. It is also a function of the burn-in time, temperature, electrical stress, and stress coverage. Therefore, the burn-in effectiveness for memories can be defined as the ratio between the failure probability of a product without burn-in to the failure probability with burn-in [173]. A convenient function to describe reliability failures as a function of time is

the Weibull hazard (or cumulative hazard function), and the effectiveness of a burn-in at time t can be mathematically expressed as

$$BI_{\mathrm{eff}}(t) = \frac{H_o(t)}{H_{BI}(t)}$$

where $H_0(t)$ = hazard without burn-in
$H_{BI}(t)$ = hazard with burn-in.

This ratio is always greater than 1 for decreasing failure rate, although it is a function of time in the field. Using the Weibull distribution, hazard with burn-in can be expressed as

$$H_{BI}(t) = \gamma \times [(t + t_{BI})^\beta - t_{BI}^\beta]$$

where γ is the scale parameter and β is the shape parameter.

Burn-in stress can be conveniently modeled as an acceleration to the time as follows:

$$t_{BI} = AF \times t$$

where t_{BI} = equivalent field time of the burn-in
AF = acceleration factor of the stress
t = time at stress condition.

It means that parts which have experienced burn-in stress of t_{BI} hours already completed have a corresponding fallout through that equivalent time $H(t_{BI})$ screened out from the population. After a burn-in which has 100% stress coverage, the conditional hazard function can be expressed as

$$H_{BI}(t) = H(t + t_{BI}) - H(t_{BI}).$$

If a burn-in process does not have 100% stress coverage, then the conditional hazard function equation would apply only to that portion of the device population which was subjected to burn-in. A partial stress coverage can apply to a portion of the semiconductor chip (e.g., 50% memory array cells of a memory chip are stressed). Then, the postburn-in hazard function for memory devices which do not have 100% stress coverage can be expressed as

$$H(t) = f \times H_{BI}(t) + (1 - f) \times H_0(t)$$

where f is the fraction of devices which experience stress, and therefore $(100 \times f)$ is the percentage of stress coverage. Knowing the performance of vendor product, both with and without burn-in, allows for calculation of the burn-in effectiveness.

A field data analysis was performed on some IBM bipolar SRAMs and time to failure (in arbitrary units) data plot fit to the conditional hazard function [173]. The failure distribution parameters for these memory devices were $\beta = 0.5$ and $t_{BI} = 1100$ equivalent hours, resulting in BI_{eff} of 1.8 at 3000 h. This SRAM had a failure mechanism that was sensitive to changes in states of memory cells when switching from 0 to 1, which produced electrical currents sufficiently high to cause failures. An effective burn-in was developed for this system that forced a memory cell array change of states at a rate of 10^7 switches/second (compared to 9×10^5 switches/second in the field application). Since these switches were the driving force for failure, the modified memory burn-in system was expected to provide an acceleration factor (AF) of 11. An independent analysis of the resulting SRAM devices reliability data, both in system tests and in the field, confirmed the existence of an AF of 11 through programmed electrical switches. Therefore, modeling of the characteristics of failure distribution of burned-in memory devices can be used to evaluate the relevant parameter estimates for the effective burn-in configuration.

A highly accelerated stress test (HAST) is often used for reliability evaluation of plastic-packaged memory devices by subjecting them to high temperatures (85°C or higher) and high relative humidity (85%) conditions (also called "85/85 testing") [174]. The objective is to accelerate the effect of humidity on corrosion of aluminum lines on a chip surface under electrical bias. The amount of current between two adjacent biased lines depends on the surface conductivity of the insulating surface between them. The Arrhenius model for burn-in acceleration relating junction temperature and failure rate is generally acceptable with some limitations, whereas for the HAST, there is no equivalent model with the same level of confidence

and verification linking failure rate and humidity stress. Still, HAST is often used by the manufacturers of plastic-packaged memories to provide a shorter time for product acceptance. It provides a good correlation with low-stress conditions, and offers the opportunity of establishing guaranteed reliability in humid conditions of applications.

A HAST study was performed by Intel Corporation using SRAMs, EPROMs, and DRAMs with various temperatures and relative humidity combination stresses [178]. Two CMOS DRAM memory sizes were used (64K and 256K), both manufactured using the same fabrication process. The EPROM data were collected using plastic-encapsulated 256K NMOS. Typical vendor SRAM material used for HAST evaluation was obtained from various commercial suppliers. The DRAM and SRAM data were collected using 5.5 V bias, whereas EPROMs were stressed at 5.25 V. Test results analysis showed two failure regimes: (1) the first is generally low-level random defect failures (below 25% cumulative) primarily occurring from hermiticity loss of the passivation, and (2) the second due to intrinsic moisture wearout of thin-film/passivation material which was reached only after thousands of hours of "85/85 testing."

The failure analysis showed quite similar results for all types of devices tested. The majority of moisture-related failures were due to single and multiple bit errors, followed by rows/columns and total array failures. There was no observable loss of metal in the bond pads or from metal lines. However, passivation irregularities were found in the failures taken from the earliest failure regime. In the case of the RAMs, passivation holes, cracks, and some physical damage were observed. The EPROM passivation anomalies were often in the form of large "bumps" of passivation. The passivation problem areas were found directly over or very near the failed cells. The test data showed that the results are sensitive to device processing, passivation integrity, and test apparatus cleanliness. The failure predictions are dependent upon which failure regime is used.

6.7.2 Screening, Qualification, and Quality Conformance Inspections (QCI)

MIL-STD-883, Method 5004 establishes screening procedures for microelectronic devices including memories to achieve levels of quality and reliability commensurate with intended application [175]. Most of the memory device manufacturers have screening flows which are either in direct compliance with this procedure, or have in-house reliability screens based on these test procedure requirements. Memories used for high-reliability military and space applications are usually procured to MIL-M-38510 requirements, which are identified as devices with Joint Army Navy (JAN) markings. MIL-M-38510 is the general specification for microcircuits which specifies requirements for the materials and process control, device qualification, quality conformance inspection (QCI), and screening, whereas the specific requirements such as pin identification, electrical characterization testing including functionals, burn-in schematics, and biasing conditions are given in the applicable MIL-M-38510 slash sheets.

MIL-M-38510 specifies two quality levels for microcircuits as "Class S" and "Class B," each with its own manufacturing and quality control inspection flows. MIL-M-38510 references MIL-STD-883, Method 5004 as a standard screening procedure for both Class S and B level screening. These screening flows include wafer lot acceptance requirements, nondestructive bond pull, internal and external visual inspection, temperature cycling, constant acceleration, particle impact noise detection (PIND), burn-in test with pre- and postelectrical measurements, seal leak, radiographic, and radiation latchup screening tests. Class S screening flow is more stringent (and costly) than Class B, and is used primarily in high-reliability applications. Table 6-11 shows MIL-STD-883, Method 5004 Screening Flow for Class S and B levels, each with its own percent defective allowable (PDA) requirements [175]. This table does not duplicate test paragraph references and notes

contained in Method 5004. For test method details including applicable test conditions and requirements, refer to MIL-STD-883, Method 5004.

MIL-STD-883, Method 1005 describes burn-in test conditions (A–F) such as: steady-state reverse-bias or forward-bias burn-in, static burn-in, dynamic burn-in, monitored burn-in or test during burn-in, and high-voltage cell stress test for memories. Steady-state reverse bias (condition A) is used for digital devices, mainly with the n-p-n inputs/outputs, and steady-state forward bias (condition B) for devices mainly with the p-n-p inputs/outputs. Steady-state burn-in for memories is most effective in identifying devices with ionic contamination, inversion, channeling, oxide defects (such as pinholes and shorts), metallization defects, surface-related defects, and wire bond problems. These defects may show up primarily as parametric drifts such as threshold voltage shifts, excessive leakage currents, and speed degradation which are detected during electrical measurements.

In static burn-in (Method 1015, condition C), the memory device is powered and the outputs are biased and loaded for maximum power dissipation. For a dynamic burn-in, all inputs including address lines are clocked with appropriate pulse sequences which produce larger current densities and higher chip temperatures than static burn-in. Increased current densities stress defects such as epitaxial and crystal imperfections, gate and field dielectrics, metallization conductors, and junction anomalies (including emitter–base shorts for bipolar memories). Many of these defects require localized thermal stresses to activate the associated failure mechanisms. However, the effectiveness of dynamic burn-in for screening contamination defects in memories is impaired because of a lack of constant electric field to provide preferential alignment of the ionic contaminants.

A subset of dynamic burn-in is "test during burn-in" which is primarily used for the DRAMs because of their long electrical test times. Test during burn-in is similar to dynamic burn-in, except that the devices are cycled with a functional test pattern (e.g., Checkerboard,

TABLE 6-11. MIL-STD-883, Method 5004 Screening flow [175].

	Class S		Class B	
Screen	Method	Req.	Method	Req.
Wafer lot acceptance	5007	All lots		
Nondestructive bond pull	2023	100%		
Internal visual	2010, test condition A	100%	2010, test condition B	100%
Temperature cycling	1010, test condition C	100%	1010, test condition C	100%
Constant acceleration	2001, test condition E (min) Y_1 orientation only	100%	2001, test condition E (min) Y_1 orientation only	100%
Visual inspection		100%		100%
Particle impact noise detection (PIND)	2020, test condition A	100%		
Preburn-in electrical measurements	Per device specification	100%	Per device specification	100%
Burn-in test	1015 240 hours @ 125°C min	100%	1015 160 hours @ 125°C min	100%
Interim (post burn-in) electrical measurements	Per device specification	100%		
Reverse bias burn-in*	1015, test condition A or C, 72 hours at 150°C min	100%		
Interim (post burn-in) electrical measurements	Per device specification	100%	Per device specification	100%
Percent defective allowable (PDA) calculation	5% (3% on functionals) @ 25°C	All lots	5%	All lots
Final electrical measurements @ 25°C, min and max rated operating temperatures	Per device specification including static, functional and switching tests	100%	Per device specification including static, functional and switching tests	100%
Seal tests Fine and Gross leak	1014	100%	1014	100%
Radiographic	2012, two views			

*The reverse bias burn-in is required only when specified in the applicable device specification, and is recommended only for certain MOS linear or other microcircuits where surface sensitivity may be of concern.

March, Galpat), while the device outputs are being monitored *in situ*. This allows for detection of the marginal/intermittent devices, as well as catastrophic failures which might escape detection during a standard dynamic burn-in.

The high-voltage cell stress tests are not true burn-in screens, although they are characterized as such because of their voltage, time, and temperature device-accelerated aging features. This stress test is used by the manufacturers as one effective means of identifying oxide defects in MOS DRAMs. A typical high-voltage cell stress involves cycling through all addresses using selected memory data patterns for 2 s at both logically high and low states, using 8.5 V V_{cc} (for a 5 V rated part, for example). A study has shown that when the high-voltage cell-stressed devices are returned to normal operating voltage, the random failure rate decreases. Thus, both high-voltage cell stress tests and tests during burn-in are widely used by memory manufacturers.

The reliability concerns associated with the fusible link (e.g., nichrome) PROMs such as structural imperfections in fuse fabrication, improperly blown fuses, and the regrowth of blown fuse links under some adverse environmental conditions were discussed in Section 6.3. The MIL-M-38510 specification for bipolar PROMs (e.g., MIL-M-38510/211 for 2K × 8 memories) requires a "freeze-out" test as 100% screen for Class S level devices that have nichrome as fusing links to verify the structural integrity of the fuses. This test is performed within 24 h after completion of burn-in and prior to the final electrical test. If more than 24 h have elapsed subsequent to 125°C burn-in exposure, the devices are preconditioned at 125°C for a minimum of 5 h immediately prior to the freeze-out test. When this test is performed, the 25°C final electrical parameters shall be completed within 96 h after the freeze-out test. The PROM freeze-out test sequence is as follows:

1. PROMs are connected with all address inputs either high, low, or open. All bit outputs are connected to separate, identical resistive loads. Figure 6-25 shows a typical freeze-out test bias configuration schematic used for MIL-M-38510/211 devices (2K × 8 PROMs) [176]. The bias is cycled 3 min "ON" and 3 min "OFF" throughout the duration of the test.

2. The device temperature is reduced to $T_c = -10°C \pm 2°C$ with the bias cycling operation maintained at this temperature for a minimum of 5 h cold soak.

3. The bias cycling is maintained while T_c is allowed to go to room temperature (by removal from the cold chamber or termination of forced cooling, but with no forced heating) and retained for a minimum of 19 h subsequent to the completion of 5 h cold

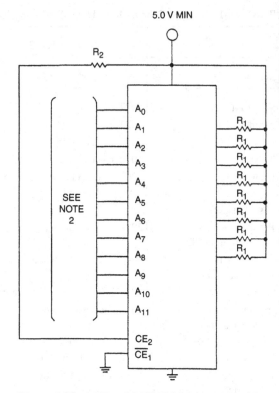

Figure 6-25. A 2K × 8 b PROM freeze-out test bias configuration per MIL-M-38510/211 [176].

NOTES:
1. All data bit outputs shall have separate identical loads ($R_1 = 4.7$ k$\Omega \pm 5\%$).
2. All address inputs shall be either high, low, or open.
3. $R_2 = 1$ k$\Omega \pm 5\%$.

soak. T_c shall not exceed 35°C during this period.

4. Then the bias is removed, and all devices are subjected to final electrical tests to establish the continuity of the nichrome fuses and removal of failed devices from the lot.

MIL-STD-883, Method 5005 establishes qualification and quality conformance inspection (QCI) procedures for microelectronic devices such as memories to assure that the lot quality conforms with the specification requirements. As in the case of screening Method 5004, qualification/QCI Method 5005 also specifies two quality levels: "Class S" and "Class B." For high-reliability memories procured for military and space applications as MIL-M-38510 devices, the full qualification and QCI requirements of MIL-STD-883, Method 5005 include Groups A, B, C, D, and E testing. Group A consists of electrical testing (dc and ac parametrics, functional, and dynamic tests) over a specified operating temperature range; Groups B and C combine various types of environmental stress testing and 1000 h life test at 125°C; Group D testing consists of package-related tests; and Group E tests are required for radiation-hardness assurance levels, as specified. For detailed test conditions and other applicable requirements, refer to MIL-STD-883, Method 5005.

MIL-M-38510 is being gradually substituted by a new military specification MIL-I-38535, General Specification for Integrated Circuits (Microcircuits) Manufacturing. This new specification, also referred to as the qualified manufacturer's listing (QML) specification, emphasizes the use of in-line process monitors and statistical process controls, self-audits, and continuous process improvements as quality management tools rather than end-of-the-line qualification/QCI testing. It also allows the manufacturers to eliminate several screening test requirements per MIL-STD-883, Method 5004 (see Table 6-11), if shown that proper SPC and in-process monitors are being used, and the failure rates for omitted tests are very low. This approach is favored by many memory manufacturers since it allows more flexibility and scope for rapid process improvements.

In order to maintain a source of qualified suppliers, the Defense Electronics Supply Corporation (DESC) has merged the QPL (or JAN) specification into the QML system. The QPL suppliers have been given a transitional certification with the provision that they demonstrate the capability to meet the MIL-I-38535 requirements within a reasonable time. The QML approach detailed in MIL-I-38535 emulates many of the best commercial practices (BCP) employed by high-volume microcircuit manufacturers in the computer, telecommunications, and automotive industries, producing plastic-encapsulated modules (PEMs). In general, the BCP includes the following [178]:

- A design, process, assembly, and test methodology which assures that the users' requirements are achieved consistently.
- A continuous process improvement with in-line controls, SPC monitors, self-audits, and a goal of eliminating the end-of-line testing and screens by reducing the product variability.
- In the microelectronic industry, the BCP is driven by the high-volume markets, and the QML system has no significant impact on the BCP.

REFERENCES

[1] *Hitachi IC Memory Data Book 1992, Reliability of Hitachi IC Memories.*

[2] E. B. Hakim, Ed., *Microelectronic Reliability: Reliability, Test and Diagnostics, Vol. 1*, ch. 2, "Failure mechanisms in microelectronic devices," G. L. Schnable. Dedham, MA: Artech House, 1989.

[3] W. C. Holton *et al.*, "A perspective on CMOS technology trends," *Proc. IEEE*, vol. 74, pp. 1646–1668, Dec. 1986.

[4] G. J. Hu, "A better understanding of CMOS latch-up," *IEEE Trans. Electron Devices*, vol. ED-31, pp. 62–67, Jan. 1984.

[5] R. R. Troutman, "Latchup in CMOS technologies," *IEEE Circuits Devices Mag.*, vol. 3, pp. 15–21, May 1987.

[6] A. H. Johnston and M. P. Baze, "Mechanism for the latchup window effect in integrated circuits," *IEEE Trans. Nucl. Sci.*, vol. NS-32, pp. 4018–4025, Dec. 1985.

[7] F. B. McLean and T. R. Oldham, "Basic mechanism of radiation effects in electronic materials and devices," HDL-TR-2129, U.S. Army Lab. Command, Harry Diamond Lab., Sept. 1987.

[8] G. C. Messenger and M. S. Ash, *The Effects of Radiation on Electronic Systems*. New York: Von Nostrand Reinhold, 1986.

[9] D. M. Brown et al., "Trends in advanced process technology—Submicrometer CMOS device design and process requirements," *Proc. IEEE*, vol. 74, pp. 1678–1702, Dec. 1986.

[10] N. D. Stojadinovic, "Failure physics of integrated circuits—A review," *Microelectron. Rel.*, vol. 23, pp. 609–707, 1983.

[11] M. H. Woods, "MOS VLSI reliability and yield trends," *Proc. IEEE*, vol. 74, pp. 1715–1729, Dec. 1986.

[12] G. L. Schnable and G. A. Schwartz, "In-process voltage stressing to increase reliability of MOS integrated circuits," *Microelectron. Rel.*, vol. 28, no. 5, pp. 757–781, 1988.

[13] H. A. Batdorf et al., "Reliability evaluation program and results for a 4K dynamic RAM," in *Proc. 16th Annu. Int. Rel. Phys. Symp. (IRPS)*, 1978, pp. 14–18.

[14] D. L. Crook, "Techniques of evaluating long term oxide reliability at wafer level," in *Proc. 1978 IEDM*, pp. 444–448.

[15] J. Klema et al., "Monitored burn-in of MOS 64K dynamic RAMs," *Semiconductor Int.*, vol. 7, pp. 104–108, Feb. 1984.

[16] S. K. Malik and E. F. Chace, "MOS gate oxide quality control and reliability assessment by voltage ramping," in *Proc. IEEE ITC 1984*, pp. 384–389.

[17] G. A. Schwartz, "Gate oxide integrity of NMOS transistor arrays," *IEEE Trans. Electron Devices*, vol. ED-33, pp. 1826–1829, Nov. 1986.

[18] I. C. Chen et al., "Oxide breakdown dependence on thickness and hole current enhanced reliability of ultra thin oxides," in *Proc. 1986 IEDM*, pp. 660–663.

[19] Y. Hokari et al., "Reliability of 6–10 nm thermal SiO_2 films showing intrinsic dielectric strength," *IEEE Trans. Electron Devices*, vol. ED-32, pp. 2485–2491, Nov. 1985.

[20] A. R. Leblanc et al., "Behavior of SiO_2 under high electric field/current stress conditions," in *Proc. 24th Annu. IRPS*, 1986, pp. 230–234.

[21] G. L. Schnable et al., "Passivation coatings on silicon devices," *J. Electrochem. Soc.*, vol. 122, pp. 1092–1103, Aug. 1975.

[22] S. Wolf and R. N. Tauber, *Silicon Processing for the VLSI Era, Vol. 1 (Process Technology)*. Lattice Press, 1986.

[23] V. Bhide and J. M. Eldridge, "Aluminum conductor line corrosion," in *Proc. 21st Annu. IRPS*, 1983, pp. 44–51.

[24] F. G. Will et al., "Corrosion of aluminum metallization through flawed polymer passivation layers: In-situ microscopy," in *Proc. 25th Annu. IRPS*, 1987, pp. 34–41.

[25] S. Dimitrijev et al., "Mechanisms of positive gate-bias stress induced instabilities in CMOS transistors," *Microelectron. Rel.*, vol. 27, no. 6, pp. 1001–1006, 1987.

[26] F. Fantini and C. Morandi, "Failure modes and mechanisms for VLSI ICs—A review," *Proc. IEEE*, vol. 132, part G, pp. 74–81, June 1985.

[27] E. H. Nicollian, "Interface instabilities," in *Proc. 12th Annu. IRPS*, 1974, pp. 267–272.

[28] E. R. Hnatek, *Integrated Circuits Quality and Reliability*, ch. 13, "Fabrication related causes." New York: Marcel Dekker, 1987.

[29] E. Pollino, Ed., *Microelectronics Reliability, Vol. II*, ch. 3, "Degradation mechanisms in insulating films on silicon," G. Baccarani and F. Fantini. Dedham, MA: Artech House, 1989.

[30] C. Hu et al., "Hot-electron induced MOSFET degradation—model monitor and improvement," *IEEE Trans. Electron Devices*, vol. ED-32, pp. 375–385, Feb. 1985.

[31] R. Petrova et al., "Hot carriers effects in short channel devices," *Microelectron. Rel.*, vol. 26, no. 1, pp. 155–162, 1986.

[32] A. G. Sabnis, "Hot-carrier damage mechanisms," *IEEE IRPS Tutorial Notes*, Apr. 1986.

[33] A. Yoshida et al., "Hot carrier induced degradation mode depending on the LDD structure in NMOSFETs," in *Proc. 1987 IEDM*, pp. 42–45.

[34] J. M. Soden and C. F. Hawkins, "Reliability of CMOS ICs with gate oxide shorts," *Semiconductor Int.*, May 1987.

[35] T. H. Ning et al., "Emission probability of hot electrons from silicon into silicon dioxide," *J. Appl. Phys.*, vol. 48, pp. 286–293, Jan. 1977.

[36] S. Tam et al., "Lucky-electron model of hot-electron injection in MOSFETs," *IEEE Trans. Electron Devices*, vol. ED-31, pp. 1116–1125, Sept. 1984.

[37] K. N. Quadar et al., "Simulation of CMOS circuit degradation due to hot-carrier effects," in *Proc. IRPS*, 1992, pp. 16–23.

[38] M. Noyori et al., "Characteristics and analysis instability induced by secondary slow trapping

in scaled CMOS devices," *IEEE Trans. Rel.*, vol. R-32, pp. 323–329, Aug. 1983.

[39] A. K. Sinha and T. E. Smith, "Kinetics of the slow-trapping instability at the Si/SiO$_2$ interface," *J. Electrochem. Soc.*, vol. 125, pp. 743–746, May 1978.

[40] D. S. Gardner *et al.*, "Interconnection and electromigration scaling theory," *IEEE Trans. Electron Devices*, vol. ED-34, pp. 633–643, Mar. 1987.

[41] S. K. Groothnius *et al.*, "Stress related failures causing open metallization," in *Proc. 25th Annu. IRPS*, 1987, pp. 1–8.

[42] G. L. Schnable *et al.*, "A survey of corrosion failure mechanisms in microelectronic devices," *RCA Rev.*, vol. 40, pp. 416–446, Dec. 1979.

[43] H. A. Schafft *et al.*, "Electromigration and the current density dependence," in *Proc. 23rd Annu. IRPS*, 1985, pp. 93–99.

[44] H. U. Schreiber, "Electromigration threshold in aluminum films," *Solid State Electron.*, vol. 28, pp. 617–626, June 1985.

[45] J. M. Towner, "Electromigration-induced short circuit failures," in *Proc. 23rd Annu. IRPS*, 1985, pp. 81–86.

[46] T. Wada *et al.*, "Electromigration in double-layer metallization," *IEEE Trans. Rel.*, vol. R-34, no. 1, pp. 2–7, 1985.

[47] J. McPherson, "VLSI multilevel metallization reliability issues," *IEEE IRPS Tutorial Notes*, pp. 2.1–2.39, 1989.

[48] J. T. Yue *et al.*, "Stress induced voids in aluminum interconnects during IC processing," in *Proc. 23rd Annu. IRPS*, 1985, pp. 126–137.

[49] J. R. Black, "Current limitations of thin film conductors," in *Proc. 20th Annu. IRPS*, 1982, pp. 300–306.

[50] H. H. Hoang *et al.*, "Electromigration in multi-layer metallization systems," *Solid State Technol.*, vol. 30, pp. 121–126, Oct. 1987.

[51] T. Hosoda *et al.*, "Effects of line size on thermal stress in aluminum conductors," in *Proc. IRPS*, 1991, pp. 77–83.

[52] E. I. Cole *et al.*, "Rapid localization of IC open conductors using charge-induced voltage acceleration (CIVA)," in *Proc. IRPS*, 1992, pp. 288–298.

[53] R. B. Camizzoli *et al.*, "Corrosion of aluminum IC metallization with defective surface passivation layer," in *Proc. 18th Annu. IRPS*, 1980, pp. 282–292.

[54] L. Gallace *et al.*, "Reliability of plastic-encapsulated integrated circuits in moisture environments," *RCA Rev.*, vol. 45, pp. 249–277, June 1984.

[55] T. Wada *et al.*, "A paradoxical relationship between width/spacing of aluminum electrodes and aluminum corrosion," in *Proc. 23rd Annu. IRPS*, 1985, pp. 159–163.

[56] J. J. Stephan *et al.*, "A review of corrosion failure mechanisms during accelerated tests," *J. Electrochem. Soc.*, vol. 134, pp. 175–190, Jan. 1987.

[57] M. R. Marks, "New thin plastic package crack mechanism induced by hot IC die," in *Proc. IRPS*, 1992, pp. 190–197.

[58] C. G. Shirley *et al.*, "Moisture-induced gold ball bond degradation of polyimide-passivated devices in plastic packages," in *Proc. IRPS*, 1993, pp. 217–226.

[59] R. L. Shook, "Moisture sensitivity characterization of plastic surface mount devices using scan acoustic microscopy," in *Proc. IRPS*, 1992, pp. 157–168.

[60] M. Mahalingam *et al.*, "Thermal effects of die bond voids," *Semiconductor Int.*, vol. 7, pp. 71–79, Sept. 1984.

[61] R. K. Shukla *et al.*, "A critical review of VLSI die-attachment in high reliability applications," *Solid State Technol.*, vol. 28, pp. 67–74, July 1985.

[62] S. I. Tan *et al.*, "Future analysis of die attachment in static random access memory (SRAM) semiconductor devices," *J. Electron. Mater.*, vol. 16, pp. 7–11, Jan. 1987.

[63] G. G. Harman, "Metallurgical failure modes of wire bonds," in *Proc. 12th Annu. Rel. Phys. Symp.*, 1974, pp. 131–141.

[64] T. Koch *et al.*, "A bond failure mechanism," in *Proc. 24th Annu. Rel. Phys. Symp.*, 1986, pp. 56–60.

[65] B. Selikson *et al.*, "A study of purple plague and its role in integrated circuits," *Proc. IEEE*, vol. 52, pp. 1638–1641, Dec. 1964.

[66] T. D. Hund *et al.*, "Improving thermosonic gold ball bond reliability," in *Proc. 35th Electron. Components Conf.*, May 1985, pp. 107–115.

[67] D. R. Edwards *et al.*, "VLSI packaging thermo-mechanical stresses," *IEEE IRPS Tutorial Notes*, pp. 8.1–8.39, 1988.

[68] J. G. Prendergast, "Reliability, yield and quality correlation for a particular failure mechanism," in *Proc. IRPS*, 1993, pp. 87–93.

[69] C. F. Hawkins and J. M. Soden, "Reliability and electrical properties of gate oxide shorts in CMOS ICs," Tech. Paper supplied by Sandia National Lab.

[70] S. Springer et al., "Gate oxide reliability and defect analysis of a high performance CMOS technology," Tech. Paper supplied by IBM Corp.

[71] C. Codella and S. Ogura, "Halo doping effects in submicron DI-LDD device design," in Proc. 1985 IEDM, pp. 230–233.

[72] A. W. Strong et al., "Gate dielectric integrity and reliability in 0.5-µm CMOS technology," in Proc. IRPS, 1993, pp. 18–21.

[73] W. Weber, "Dynamic stress experiments for understanding hot carriers degradation phenomenon," IEEE Trans. Electron Devices, vol. 35, pp. 1476–1486, Sept. 1988.

[74] A. G. Sabnis and J. T. Nelson, "A physical model for degradation of DRAMs during accelerated stress aging," in Proc. IRPS, 1983, pp. 90–95.

[75] J. A. van der Pol and J. J. Koomen, "Relation between the hot carrier lifetime of transistors and CMOS SRAM products," in Proc. IRPS, 1990, pp. 178–185.

[76] W. C. H. Gubbels et al., "A 40-ns/100-pf low power full-CMOS 256K (32K × 8) SRAM," IEEE J. Solid-State Circuits, vol. SC-22, no. 5, pp. 741–747, 1987.

[77] E. C. Cahoon et al., "Hot electron induced retention time degradation in MOS dynamic DRAMs," in Proc. IRPS, 1986, pp. 195–198.

[78] Y. Ohji et al., "Reliability of nano-meter thick multilayer dielectric films on poly-crystalline silicon," in Proc. IRPS, 1987, pp. 55–59.

[79] J. Kumagai et al., "Reduction of signal voltage of DRAM cell of trapped charges in nano-meter thick dual dielectric film (SiO_2/Si_3N_4)," in Proc. IRPS, 1990, pp. 170–177.

[80] H. Watanabe et al., "Stacked capacitor cells for high density dynamic RAMs," in Proc. 1988 IEDM, pp. 600–603.

[81] C. Duvvury, "Impact of hot carriers on DRAM circuits," in Proc. IRPS, 1987, pp. 201–206.

[82] D. A. Baglee et al., "Reliability of trench capacitors for VLSI memories," in Proc. IRPS, 1986, pp. 215–219.

[83] A. Nishimura et al., "Long term reliability of $SiO_2/SiN/SiO_2$ thin layer insulator formed in 9 µm deep trench on high boron concentrated silicon," in Proc. IRPS, 1989, pp. 158–162.

[84] T. Watanabe et al., "A 100 Å thick stacked $SiO_2/Si_3N_4/SiO_2$ dielectric layer for memory capacitor," in Proc. IRPS, 1985, pp. 18–23.

[85] P. C. Fazen, "Ultrathin oxide/nitride dielectrics for rugged stacked DRAM capacitors," IEEE Electron Device Lett., vol. 13, pp. 86–88, Feb. 1992.

[86] H. Ischiuchi et al., "Soft error rate reduction in dynamic memory with trench capacitor cell," in Proc. IRPS, 1986, pp. 235–238.

[87] T. Toyabe et al., "A soft error rate model for MOS dynamic RAMs," IEEE Trans. Electron Devices, vol. ED-29, pp. 732–737, 1982.

[88] E. Takeda et al., "Key factors in reducing soft errors in mega-bit DRAMs," in Proc. IRPS, 1987, pp. 207–211.

[89] D. Burnett et al., "Soft-error-rate improvement in advanced BiCMOS SRAMs," in Proc. IRPS, 1993, pp. 156–160.

[90] Z. Hasnain et al., "Building in reliability: Soft errors—A case study," in Proc. IRPS, 1992, pp. 276–280.

[91] Y. Ohno et al., "Soft-error study of DRAMs using nuclear microprobe," in Proc. IRPS, 1993, pp. 150–155.

[92] S. Onishi et al., "TEM analysis of failed bits and improvement of data retention properties in megabit-DRAMs," in Proc. IRPS, 1990, pp. 265–269.

[93] T. Katayama et al., "A new failure mechanism related to grain growth in DRAMs," in Proc. IRPS, 1991, pp. 183–187.

[94] P. Friedman and D. Burgess, "Reliability aspects of nichrome fusible link PROMs (programmable read only memories)," Tech. Paper supplied by Monolithic Memories, Inc.

[95] G. H. Parker et al., "Reliability considerations in the design and fabrication of polysilicon fusible link PROMs," Tech. Paper supplied by Intel Corp.

[96] A. Ito et al., "The physics and reliability of fusing polysilicon," in Proc. IRPS, 1984, pp. 17–29.

[97] R. E. Shiner et al., "Data retention in EPROMs," in Proc. IRPS, 1980, pp. 369–373.

[98] N. R. Mielke, "New EPROM data loss mechanisms," in Proc. IRPS, 1983, pp. 106–113.

[99] K. Wu et al., "A model for EPROM intrinsic charge loss through oxide-nitride-oxide (ONO)

interpoly dielectric," in *Proc. IRPS,* 1990, pp. 145–149.

[100] C. Pan, "Physical origin of long-term charge loss in floating gate EPROM with an interpoly oxide-nitride-oxide stacked dielectric," in *Proc. IRPS,* 1991, pp. 51–53.

[101] D. A. Baglee *et al.,* "Building reliability into EPROMs," in *Proc. IRPS,* 1991, pp. 12–18.

[102] S. A. Barker, "Effects of carbon on charge loss in EPROM structures," in *Proc. IRPS,* 1991, pp. 171–174.

[103] T. Miller *et al.,* "Charge loss associated with program disturb stresses in EPROMs," in *Proc. IRPS,* 1990, pp. 154–158.

[104] C. Dunn *et al.,* "Process reliability development for nonvolatile memories," in *Proc. IRPS,* 1993, pp. 133–145.

[105] P. Manos, "A self-aligned EPROM structure with superior data retention," *IEEE Electron Device Lett.,* vol. 11, pp. 309–311, July 1990.

[106] S. Mori *et al.,* "Reliability study of thin interpoly dielectrics for non-volatile memory application," in *Proc. IRPS,* 1990, pp. 132–144.

[107] S. Mori *et al.,* "Reliable CVD inter-poly dielectrics for advanced E & EE PROM," in *VLSI Symp. Dig. Tech. Papers,* 1985, pp. 16–17.

[108] N. Mielke *et al.,* "Reliability comparison of Flotox and textured polysilicon E^{2}PROMs," in *Proc. IRPS,* 1987, pp. 85–92.

[109] C. Papadas *et al.,* "Model for programming window degradation in FLOTOX EEPROM cells," *IEEE Electron Device Lett.,* vol. 13, pp. 89–91, Feb. 1992.

[110] R. B. Sethi *et al.,* "Electron barrier height change and its influence on EEPROM cells," *IEEE Electron Device Lett.,* vol. 13, pp. 244–246, May 1992.

[111] D. A. Baglee *et al.,* "The effects of processing on EEPROM reliability," in *Proc. IRPS,* 1987, pp. 93–96.

[112] T. Endoh *et al.,* "New design technology for EEPROM memory cells with 10 million write/erase cycling endurance," in *Proc. 1989 IEDM,* pp. 25.6.1–25.6.4.

[113] S. Aritome *et al.,* "Extended data retention characteristics after more than 10^4 write and erase cycles in EEPROMs," in *Proc. IRPS,* 1990, pp. 259–264.

[114] Y.-S. Kim *et al.,* "Low-defect-density and high-reliability FETMOS EEPROMs fabri-

cated using furnace N_2O oxynitridation," in *Proc. IRPS,* 1993, pp. 342–344.

[115] K. Naruke *et al.,* "Stress induced leakage current limiting to scale down EEPROM tunnel oxide thickness," in *Proc. 1988 IEDM,* pp. 424–427.

[116] S. Mori *et al.,* "ONO interpoly dielectric scaling for nonvolatile memory applications," *IEEE Trans. Electron Devices,* vol. 38, pp. 386–391, Feb. 1991.

[117] G. Verma *et al.,* "Reliability performance of ETOX flash memories," in *Proc. IRPS,* 1988, pp. 158–166.

[118] S. Haddad *et al.,* "Degradations due to hole trapping in flash memory cells," *IEEE Electron Device Lett.,* vol. 10, pp. 117–119, Mar. 1989.

[119] K. Tamer San *et al.,* "Determination of trapped oxide charge in flash EPROMs and MOSFETs with thin oxides," *IEEE Electron Device Lett.,* vol. 13, pp. 439–441, Aug. 1992.

[120] A. Brand *et al.,* "Novel read disturb failure mechanism induced by FLASH cycling," in *Proc. IRPS,* 1993, pp. 127–132.

[121] A. Roy *et al.,* "Substrate injection induced program disturb—A new reliability consideration for flash-EPROM arrays," in *Proc. IRPS,* 1992, pp. 68–75.

[122] T. C. Ong *et al.,* "Erratic erase in ETOX™ flash memory array," presented at the 1993 IEEE Symp. VLSI Technol., Kyoto, Japan, May 1993.

[123] D. E. Fisch *et al.,* "Analysis of thin film ferroelectric aging," in *Proc. IRPS,* 1990, pp. 237–242.

[124] A. K. Kulkarni *et al.,* "Fatigue mechanisms in thin film potassium nitrate memory devices," in *Proc. IRPS,* 1989, pp. 171–177.

[125] A. K. Kulkarni, "Thin film ferroelectric materials and devices," *IEEE IRPS Tutorial Notes,* Topic 8, 1990.

[126] G. W. Taylor, *J. Appl. Phys.,* vol. 38, p. 4697, 1967.

[127] B. Jaffe *et al., Piezoelectric Ceramics.* New York: Academic, 1971, pp. 10–13, pp. 77–78.

[128] C. Karan, IBM Tech. Rep., 1955.

[129] D. E. Fisch *et al.,* "Analysis of thin film ferroelectric aging," in *Proc. IRPS,* 1990, pp. 237–242.

[130] M. Takahashi *et al.,* "Space charge effects in lead zirconate titanate ceramics caused by the

addition of impurities," *Japan. J. Appl. Phys.,* vol. 9, p. 1236, Oct. 1970.

[131] H. Dederichs and G. Arlt, "Aging of Fe-doped PZT ceramics and domain wall contribution to the dielectric constant," *Ferroelec.,* vol. 68, p. 281, 1986.

[132] G. A. Rohrer *et al.,* "A new technique for characterization of thin film memory devices," *J. Vac. Sci. Technol.,* vol. 6, no. 3, pp. 1756–1760, 1988.

[133] A. Gregory *et al.,* "Thermal stability of ferroelectric memories," in *Proc. IRPS,* 1992, pp. 91–94.

[134] R. Moazzami *et al.,* "Electrical conduction and breakdown in sol-gel derived PZT thin films," in *Proc. IRPS,* 1990, pp. 231–236.

[135] Military Standard-721 (MIL-STD-721), "Definition of terms for reliability and maintainability."

[136] *Military Handbook-338 (MIL-HDBK-338): Reliability Design Handbook.*

[137] J. Kline, *Practical Electronic Reliability Engineering,* ch. 2, "Application of failure distributions to reliability." New York: Van Nostrand Reinhold, 1992.

[138] U.S. MIL-HDBK-217, "Reliability prediction of electronic equipment," version E, dated Oct. 27, 1986.

[139] Bellcore, TR-TSY-000332, "Reliability prediction procedure for electronic equipment," issue 2, July 1988.

[140] British Telecom, *Handbook of Reliability Data for Components Used in Telecommunication Systems,* issue 4, Jan. 1987.

[141] Nippon Telegraph and Telephone Corp., "Standard reliability table for semiconductor devices," Mar. 1985.

[142] National Center for Telecommunication Studies, "Collection of reliability data from CNET," 1983.

[143] Siemens Standard, SN29500, "Reliability and quality specification failure rates of components," 1986.

[144] J. B. Bowles, "A survey of reliability-prediction procedures for microelectronic devices," *IEEE Trans. Rel.,* vol. 41, Mar. 1992.

[145] S. Aur *et al.,* "Circuit hot electron effect simulation," in *Proc. 1987 IEDM,* pp. 498–501.

[146] B. J. Sheu *et al.,* "An integrated circuit simulator—RELY," *IEEE J. Solid-State Circuits,* pp. 473–477, Apr. 1989.

[147] Y. Lablebici and S. M. Kang, "A one-dimensional MOSFET model for simulation of hot carrier induced device and circuit degradation," in *Proc. IEEE Int. Symp. Circuits Syst.,* May 1990, pp. 109–112.

[148] C. Hu, "IC reliability simulation," in *Proc. IEEE Custom Integrated Circuits Conf.,* 1991, pp. 4.1.1–4.1.4.

[149] B. K. Liew *et al.,* "Reliability simulator for interconnect and intermetallic contact electromigration," in *Proc. IRPS,* Mar. 1990, pp. 111–118.

[150] M. S. Ash and H. L. Gorton, "A practical end-of-life model for semiconductor devices," *IEEE Trans. Rel.,* vol. 38, Oct. 1989.

[151] A. G. Sabnis *et al.,* "A physical model for degradation of DRAMs during accelerated stress aging," in *Proc. IRPS,* 1991, pp. 90–95.

[152] D. Wendell *et al.,* "Predicting oxide failure rates using the matrix of a 64K DRAM chip," in *Proc. IRPS,* 1984, pp. 113–118.

[153] T. R. Conrad *et al.,* "A test methodology to monitor and predict early life reliability failure mechanisms," in *Proc. IRPS,* 1988, pp. 126–130.

[154] H. A. R. Wegener *et al.,* "The prediction of textured poly floating gate memory endurance," in *Proc. IRPS,* 1985, pp. 11–17.

[155] C. Hu, "Lucky-electron modeling of channel hot electron emission," in *Proc. 1979 IEDM,* pp. 22–25.

[156] P. K. Ko *et al.,* "A unified model for hot-electron currents in MOSFETs," in *Proc. 1981 IEDM,* pp. 600–603.

[157] C. Fiegna *et al.,* "Simple and efficient modeling of EPROM writing," *IEEE Trans. Electron Devices,* vol. 38, pp. 603–609, Mar. 1991.

[158] D. L. Crook, "Evaluation of VLSI reliability engineering," in *Proc. IRPS,* 1990, pp. 2–11.

[159] C. Hong and D. Crook, "Breakdown energy of metal (BEM)—A new technique for monitoring metallization reliability at wafer level," in *Proc. IRPS,* 1985, p. 108.

[160] B. J. Root and T. Turner, "Wafer level electromigration tests for production monitoring," in *Proc. IRPS,* 1985, p. 100.

[161] D. Crook, "Techniques of evaluating long term oxide reliability at wafer level," in *Proc. 1978 IEDM*, p. 444.

[162] H. Geib *et al.*, "A novel test structure for monitoring technological mismatches in DRAMs processes," in *Proc. Int. Conf. Microelectron. Test Structures*, vol. 5, Mar. 1992, pp. 24–29.

[163] H. Wang *et al.*, "Improving hot-electron reliability through circuit analysis and design," in *Proc. IRPS*, 1991, pp. 107–111.

[164] M. G. Buehler *et al.*, "Fault chip for microcircuit characterization," Tech. Rep. dated Mar. 1988.

[165] M. G. Buehler *et al.*, "Design and qualification of the SEU/TD radiation monitor chip," JPL Publ. 92-18, Oct. 1, 1992.

[166] H. H. Huston, "Reliability defect detection and screening during processing—Theory and implementation," in *Proc. IRPS*, pp. 268–275.

[167] A. Suyko *et al.*, "Development of a burn-in time reduction algorithm using the principles of acceleration factors," in *Proc. IRPS*, 1991, pp. 264–270.

[168] S. Rosenberg, "Tests and screens weed out failures, project rates of reliability," *Electronics*, Aug. 14, 1980.

[169] R. E. Shiner *et al.*, "Data retention in EPROMs," in *Proc. IRPS*, 1980, pp. 238–243.

[170] K. Wu *et al.*, "A model for EPROM intrinsic charge loss through oxide-nitride-oxide (ONO) interpoly dielectric," in *Proc. IRPS*, 1990, pp. 145–149.

[171] W. K. Meyer *et al.*, "A non-aging screen to prevent wearout of ultra-thin dielectrics," in *Proc. IRPS*, 1985, pp. 6–10.

[172] D. L. Crook *et al.*, "The Intel 2164A HMOS-D III 64K dynamic RAM reliability report," RR-37, May 1983.

[173] H. H. Huston, "Burn-in effectiveness—Theory and measurement," in *Proc. IRPS*, 1991, pp. 271–276.

[174] D. Danielson *et al.*, "HAST applications: Acceleration factors and results for VLSI components," in *Proc. IRPS*, 1989, pp. 114–121.

[175] Military Standard 883D (MIL-STD-883D), "Military standard test methods and procedures for microelectronics."

[176] MIL-M-38510/211, "Military specification for bipolar Schottky, fusible link PROMs (2K × 8 bit)."

[177] P. Cappalletti *et al.*, "CAST: An electrical stress test to monitor single bit failures in flash-EEPROM structures," in *Proc. 13th Annu. IEEE Nonvolatile Semiconductor Memory Workshop*, Feb. 1994.

[178] E. B. Hakim *et al.*, "Beyond the qualified manufacturer list (QML)," in *Proc. IEEE Rel. and Maintainability Symp.*, 1995, pp. 362–369.

Semiconductor Memory Radiation Effects

7.1 INTRODUCTION

The space radiation environment poses a certain radiation risk to all electronic components on the earth-orbiting satellites and planetary mission spacecrafts. The irradiating particles in this environment consist primarily of high-energy electrons, protons, alpha particles, and cosmic rays. The weapon environment such as a nuclear explosion (often referred to as the "gamma dot") is characterized by X-rays, gamma neutrons, and other reaction debris constituents occurring within a short time span. This can cause latchup and transient upsets in integrated circuits such as memories. Although the natural space environment does not contain the high dose rate pulse characteristics of a nuclear weapon, the electronics systems exposed can accumulate a significant total dose from the electron and protons over a period of several years, depending upon the satellite orbital and inclination parameters. The radiation effects of charged particles in the space environment are dominated by ionization, which refers to any type of high energy that creates electron–hole (e–h) pairs when passing through a material. It can be either particulate in nature or electromagnetic. In addition to creating e–h pairs, the radiation can cause displacement damage in the crystal lattice by breaking the atomic bonds and creating trapping recombination centers. Both of these damage mechanisms can lead to degradation of the device performance.

The ionizing electromagnetic radiations of importance are the X-rays and gamma rays. Ionizing particulate radiation can be light uncharged particles such as neutrons; light charged particles such as electrons, protons, alpha, and beta particles; and heavy charged particles (heavy ions) such as iron, bromine, krypton, xenon, etc., which are present in the cosmic ray fluences. The cosmic rays originate from two sources, the sun (solar) and galactic sources outside the solar system. In the absence of any solar activity such as solar flares, cosmic radiation consists entirely of galactic radiation. Table 7-1 provides an overview of the various types of radiation and parameter degradation or circuit failures caused by them, in general [1].

Gamma rays (or X-rays) basically produce a similar kind of damage as light charged particles since the dominant mechanism is charge interaction with the material. Neutrons have no charge, and react primarily with the nucleus, causing lattice damage. Heavy charged particles create "tracks," or a volume of ionization within the substrate of the semiconductor devices such as memories. This charge can collect on circuit nodes and cause data loss through single-event upset (SEU) or transient upset.

The primary contributors to ionizing particle fluences are the Van Allen radiation belts in

TABLE 7-1. An Overview of Various Types of Radiation and Corresponding Transistor/Circuit Level Parameter Degradation and Failure. (From [1], with Permission of UTMC.)

Type of Radiation	Type of Degradation or Failure	
	Transistor Level	Circuit Level
Photons (X-rays and gamma rays)		
Total dose (Low dose rate) (High total dose)	• Threshold voltage shifts • Drive current shifts • Carrier mobility reduction • Parasitic leakage current	• Performance degradation • I/O parametric shift • Lost functionality • Data upset/loss
Transient (High dose rate) (Low total dose)	• Collapse of the depletion region • Burn-out of metal interconnects • Damage to junction region	• Latchup • Catastrophic failure • I/O glitches • High photocurrents
Charged Particles and Cosmic Rays		
Electrons and protons	• Threshold voltage shifts • Drive current shifts • Carrier mobility reduction • Parasitic leakage currents • Carrier recombination and trapping • Collapse of depletion region (protons only)	• Performance degradation or improvement • I/O parametric shifts • Lost functionality • Data upset (single bit/node, protons only)
Alpha particles	• Collapse of depletion region	• Data upset (single bit/node)
Heavy ions	• Collapse of depletion region	• Data upset (single or multibit/node) • Hard oxide failures • Latchup
Uncharged Particles		
Neutrons	• Junction leakage • Bipolar beta reduction • Carrier recombination and trapping	• No performance effects on CMOS • Improves latchup tolerance on CMOS circuits • Performance degradation on bipolar circuits

the earth's geomagnetic environments, solar flares, and cosmic rays. The energy of these charged particles varies from a few hundred kiloelectronvolts to more than 10 GeV. Electronic designers using semiconductor memories for space applications have to be concerned with two types of radiation damage induced by charge particle ionization. These are the total dose effects and single-event phenomenon (SEP).

The total dose effect is the cumulative ionization damage expressed in rads (material)

caused by all ionizing particles passing through a semiconductor device such as memory. For the MOS devices, the ionization traps positive charge in the gate oxide called the oxide traps, and produces interface states at $Si-SiO_2$ interfaces. These effects typically exhibit both a time and field dependence characteristic resulting in threshold voltage shifts, reduced speed, higher leakage currents, and possible loss of device functionality. In bipolar devices, the ionizing radiation causes a decrease in the gain and breakdown voltages, higher offset voltages, and leakage currents. The magnitude of these changes depends upon a number of factors such as the total radiation dose and its energy; dose rate; applied bias and temperature during the irradiation; types of transistors; and postirradiation annealing time, bias, and temperature.

The single-event phenomenon in semiconductor memories is caused by a high-energy particle (such as those present in cosmic rays) passing through the device. This can result in two types of errors: soft and hard. The soft errors (sometime also referred to as SEUs) can cause a random access memory cell logic state to change from "1" to "0" (bit-flip) and vice versa. These soft errors cause no permanent damage, and the device can be reprogrammed. However, the heavy ions can also cause hard errors as, for example, with the parasitic p-n-p-n devices in bulk CMOS technology. This can result in a latchup and device burnout unless the current is limited or the power is cycled off and on again. These total dose effects and SEP will be discussed in more detail in Section 7.2.

Integrated circuits such as memories can be radiation-hardened against the total dose effects and SEP by using special processing and design techniques. In bulk CMOS, the gate-oxide-hardening techniques include the use of doped SiO_2, double-layer oxide structures, new gate insulator material, as well as specially grown silicon dioxide films. The field oxides have also been hardened by doping, and using diffusion and ion implantation techniques. The oxide processing controls include the reduction of oxide growth temperature, and the elimination (or temperature reduction) of postoxidation annealing in nitrogen. Bulk CMOS technologies are susceptible to single-event latchup caused by heavy ions unless thin epitaxial layers are used. Other process parameters controlled for improved single-event hardness include the use of thin gate oxide and polycrystalline silicon resistors. Buried oxides which are fabricated using various silicon-on-insulator (SOI) technologies such as silicon-on-sapphire (SOS) are also being used to eliminate the possibilities of latchup and reduce the SEU sensitivities. At the memory circuit design level, special SRAM cells with cross-coupled resistors (or capacitors) are used to reduce the SEP susceptibility. These circuit design techniques for radiation hardness will be discussed in more detail in Section 7.3.

For electronic devices such as memories flown in space (satellites or planetary missions), radiation tolerance is assessed with respect to the projected total dose that would be accumulated in the mission-specific environments. This projected dose is the sum of absorbed dose contributions from all ionizing particles, and is calculated (in the form of dose–depth curves) by sophisticated environmental modeling based upon the orbital parameters, mission duration, and thickness of shielding on the spacecraft. To evaluate memory susceptibility to the SEU and latchup, the heavy ion fluence is translated to linear energy transfer (LET) spectra to calculate estimated bit-error rates. Although the total dose accumulated by a device can vary significantly with the amount of shielding (e.g., aluminum wall of the spacecraft), the SEU susceptibility from high-energy heavy ions does not change significantly with the shielding. The SEU effects have been studied on memories flown in space on some satellites such as the NASA Tracking and Data Relay Satellite System (TDRSS), the UoSAT family of research microsatellites, and the Combined Release Radiation Effects Satellite (CRRES). These will be discussed in more detail in Section 7.2.

For memories in space applications, radiation-hardness assurance testing is required for both total dose and SEP. Radiation-hardness assurance requirements, test methodology, and wafer lot radiation testing as well as the test structures will be discussed in more detail in Section 7.4.

7.2 RADIATION EFFECTS

7.2.1 Space Radiation Environments

This section discusses natural space environments in which most of the satellites operate in orbits ranging in altitudes from low earth orbits (150–600 km) to geosynchronous orbits (roughly 35,880 km). The radiation environment of greatest interest is the near earth region, about 1–12 earth radii (where $R_e = 6380$ km), which is mainly dominated by electrically charged particles trapped in the earth's magnetosphere, and to a lesser extent by the heavy ions from cosmic rays (solar and galactic). As the earth sweeps through the solar wind, a geomagnetic cavity is formed by the earth's magnetic field. The motion of the trapped charge particles is complex, as they gyrate and bounce along the magnetic field lines, and are reflected back and forth between the pairs of conjugate mirror points (regions of maximum magnetic field strength along their trajectories) in the opposite hemispheres. Also, because of the charge, the electrons drift in an easterly direction around earth, whereas protons and heavy ions drift westward. Figure 7-1 illustrates the spiral, bounce, and drift motion of the trapped particles [2]. Interplanetary space probes such as the Voyager (and Galileo to Jupiter) encounter ionizing particles trapped in the magnetosphere of other planets, as well as the solar flares and heavy ions from cosmic rays.

Electrons in the earth's magnetosphere have energies ranging from low kiloelectronvolts to about 7 MeV, and are trapped in the roughly toroidal region which is centered on the geomagnetic equator and extends to about 1–12 earth radii. These trapped electrons are differentiated by "inner zone" (<5 MeV) and "outer zone" (≈7 MeV) electron populations. The trapped protons originating mostly from the solar and galactic cosmic rays have energies ranging from a few megaelectronvolts to about 800 MeV. They occupy generally the same region as the electrons, although the region of highest proton flux for energies $E_p > 30$ MeV is concentrated in a relatively small area at roughly 1.5 R_e. The actual electron and proton flux encountered by a satellite is strongly dependent upon the orbital parameters, mission launch time, and duration. Electrons and protons from the trapped radiation belts on interacting with spacecraft materials produce secondary radiation (e.g., "bremsstrahlung" or braking radiation from the deceleration of electrons). This secondary radiation can extend the penetration range of primary radiation and lead to an increase in dose deposition. Incident electron and proton fluxes are typically calculated from the trapped radiation environmental models developed by the U.S. National Space Sciences Data Center (NSSDC).

The trapped particle fluxes responding to changes in the geomagnetic field induced by the solar activity exhibit dynamic behavior. In addition to the trapped geomagnetic radiation, another contribution to incident particle flux for an orbiting satellite is the transiting radiation from the solar flares and galactic cosmic rays. These solar energy particle events, usually occurring in association with the solar flares, consist

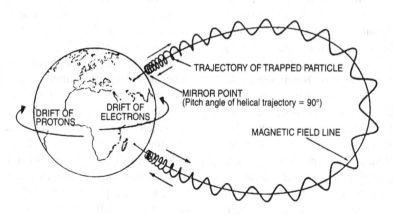

Figure 7-1. Motion of charged particles in earth's magnetosphere [2].

mainly of protons, some alpha particles (5–10%), heavy ions, and electrons. This solar flare phenomenon is categorized as an ordinary (OR) event or an anomalously large (AL) event. Particle fluxes from the solar flares can last from a few hours to several days. AL events, although occurring rarely, can cause serious damage (e.g., latchup, SEU) to nonhardened IC chips such as memories. For ordinary solar events, the relative abundance of helium ions can be between 5–10%, whereas ions heavier than He (e.g., carbon, oxygen, iron, etc.), referred to as the "heavy ions," are very small. However, the solar flare protons which contribute to the total ionizing dose radiation are not that significant a factor compared to the trapped radiation environment.

Another significant contribution to the transiting radiation is from cosmic rays originating from outside the solar system and consisting of 85% protons, 14% alpha particles, and 1% heavier ions. These galactic cosmic rays range in energy from a few megaelectronvolts to over 10 GeV per nucleon. The total flux of cosmic ray particles (primarily composed of protons) seen outside the magnetosphere at a distance of earth from the sun (1 AU) is approximately 4 particles/cm^2/s.

The particles from energetic solar flares (OR events) are heavily attenuated by the geomagnetic field at low altitude and low inclination orbits, such as U.S. Space Shuttle orbits (28.5° inclination). In a 500 km, 57° inclination orbit, some particle fluxes do penetrate. A characteristic of the geomagnetic field which is particularly significant to SEP (e.g., memory upsets) is the South Atlantic Anomaly (SAA), referring to an apparent depression of the magnetic field over the coast of Brazil where the Van Allen radiation belts dip low into the earth's atmosphere. This SAA is responsible for most of the trapped radiation in low earth orbits (LEOs). On the opposite side of the globe, the Southeast-Asian anomaly displays strong particle fluxes at higher altitudes. A polar orbit at any altitude experiences a high degree of exposure, and at geosynchronous orbit, geomagnetic shielding is rather ineffective.

The total dose absorbed by a semiconductor device such as a memory during a space mission depends not only on the radiation environment encountered, but also on the amount of shielding between the device and the incident flux. For small shield thicknesses (e.g., the spacecraft skin of 100 mil Al), the transmitted electrons and protons contribute roughly equally to the absorbed dose, whereas for thicker shields, the dose accumulates primarily from high-energy protons.

The number of single-event-induced upsets (SEUs) experienced by a semiconductor device such as a memory in a given radiation environment depends primarily on its threshold for upsets (usually given by the critical charge Q_c or critical LET) and the total device volume sensitive to ionic interaction. There are several models available for the prediction of upset/error rates due to cosmic ray events. An example is a popular CREME (Cosmic Ray Effects on MicroElectronics) model developed at the Naval Research Labs, which predicts a device upset/error rate based on a description of the radiation environment. These models will be discussed in more detail in Section 7.2.

The preceding sections discussed a natural space environment for the earth-orbiting satellites and planetary missions. The nuclear weapons radiation environment is of concern primarily for the strategic and tactical military weapons designers working on "operate-through" (nuclear event survivable) systems. Following a nuclear explosion, about 70–80% of the total energy of the material is initially emitted as thermal electromagnetic radiation in the soft X-ray region, 20–30% initially remains as kinetic energy of the weapon debris, and a small percentage is emitted as the energy of γ-rays and neutrons [3]. At higher altitudes (e.g., LEO of a satellite), where the air density is almost zero, the soft X-radiation may produce ionizing effects in the MOS devices. The total γ-ray energy fluence (also called the prompt γ-rays or gamma dot), roughly 10^8 rad/s, may cause transient photocurrent-induced upsets and latchup in the MOS devices. The neutrons are electrically neutral, and hence do not directly contribute to the ionizing dose. However, they may cause displacement damage. Radiation-hardening techniques used to harden the memories against SEU will be discussed in more detail in Section 7.3.

7.2.2 Total Dose Effects

In general, bipolar technology memories (e.g., RAMs, PROMs, etc.) are more tolerant to total dose radiation (ionizing damage) effects than the bulk MOS technology parts. In bipolar transistors, the principal factors responsible for performance degradation are trapped positive charges which build up in the passivating oxides surrounding silicon surfaces. This positive charge buildup results in the creation of surface states at these interfaces. Since a bipolar transistor depends on a junction for its operation, the effect of radiation degradation is not as severe as for the MOS transistor which depends on a surface and corresponding interfaces for operation. However, bipolar devices may be more susceptible to neutron damage.

For operation within a specified operating temperature range, the total dose radiation threshold for bipolar memories (e.g., TTL and I^2L devices) can exceed 100 krad (Si). At lower doses, the damage mechanisms are mainly due to surface and interface state generation. Dose rate effects on bipolar memories depend on whether the device is junction isolated (JI) or dielectrically isolated (DI). The substrate of junction-isolated devices can result in an n-p-n-p structure creating a parasitic SCR latchup path. To inhibit parasitic SCR action, often gold doping of the base and the collector region is used. Dielectrically isolated devices are essentially free from latchup and are used in a high-dose-rate environment.

MOS technology has been preferred by designers for military and space applications because of its lower static power requirements and superior noise margins. Also, the MOS devices, being majority carrier devices, are inherently hard to neutron irradiation to 10^{16} neutrons/cm^2. However, the nonhardened MOS devices are quite sensitive to total dose effects. For exam-

ple, commercial NMOS and CMOS devices may fail at doses as low as 1–2 krad (Si). Total dose effects on MOS devices will be discussed in more detail in this section. As mentioned earlier in the Introduction, the dominant effects resulting from the interaction of radiation with electronic devices are ionization damage or displacement damage. For CMOS devices, radiation damage is related to ionization effects, and consists of the buildup of trapped charge in the oxide which causes an increase in the number of bulk oxide traps and interface traps.

Ionizing radiation (charged particles) traversing through the bulk of material generate electron–hole (e–h) pairs within the silicon dioxide. Typically, the amount of energy required to create an e–h pair is roughly two–three times the bandgap energy. In silicon, approximately 3.5 eV is required to create an e–h pair. Therefore, the energetic electrons and holes which are created by incident ionizing radiation give rise themselves to a large number of e–h pairs in the solids. The significant effect of these e–h pairs created uniformly throughout the solid is the increase in bulk conductivity of the material. In an MOS structure, the insulating oxide (SiO_2) layer is most sensitive to ionizing radiation. As the radiation passes through the oxide, the energy deposited creates e–h pairs. In the first few picoseconds, some electrons and holes recombine; the fraction of recombination depends on the applied field and the energy and type of incident particles. This initial e–h pairs creation and almost instant recombination determine the actual charge (hole) yield in SiO_2 films, and thus the initial voltage shift.

After the e–h pairs creation and some instant recombination, these electrons and holes in the presence of an electric field will be free to diffuse and drift away from their points of generation. The electrons, being more mobile than the holes, are swept out of the oxide to be collected at the gate electrode. Some of the holes that escape initial recombination stay near their point of generation, acting as "oxide traps," causing a net positive charge. Other holes undergo a slow, stochastic "trap-hopping" process through the oxide. In the presence of positive applied gate bias, when the holes reach the SiO$_2$/Si interface, a fraction of them are captured in long-term trapping sites called "interface traps." These interface traps, which are localized states with energy levels within the Si bandgap, can cause small negative voltage shifts which may persist in time for few hours to several years. They may cause a degradation in mobility of the carriers in the channel of MOS transistors, which can lead to a reduction in the channel conductance. Figure 7-2 illustrates the process of charge generation and initial recombination, hole transport, and long-term trapping near the SiO$_2$/Si interface [4].

In addition to e–h pair generation, ionizing radiation can rupture chemical bonds in the SiO$_2$ structure. Some of these broken bonds may reform, whereas others can give rise to

electrically active defects that can serve as trap sites for carriers or as interface traps.

As the total dose to the device increases, the amount of energy absorbed by the device material also increases. Since the number of e–h pairs generated is directly proportional to the amount of energy absorbed, the ionization damage in the form of oxide traps and interface traps also monotonically increases. The total dose received by a device such as a memory is measured in units of rads (Si) or rads (SiO$_2$), a unit equal to 100 ergs absorbed per gram of Si (or other material specified).

Therefore,

1 rad (Si) = 100 ergs absorbed/gram of Si

= 10^5 J absorbed/gram of Si.

Since the density of silicon is 2.42 g/cm^3, this corresponds to a deposition of 2.42 × 10^5 J/cm^3. A generation constant "g" can be defined as the amount of energy delivered per unit volume of the material divided by the energy required to create an e–h pair. This gives the number e–h pairs created per unit volume per rad (Si).

Thus, for Si,

$$g = 4.2 \times 10^{13} \ e\text{–}h \ \frac{\text{pairs}}{\text{cm}^3}.$$

The generation rate for the e–h pairs is given by $G = g \times d\gamma/dt$.

The radiation hardness of an MOS memory is determined by the rate at which the two types of traps (oxide and interface) build up with increasing levels of total dose radiation. For commercial memories (unhardened devices), the radiation-generated trapped holes may be greater than 50% of the number generated within the oxide layer, and the number of interface traps may also increase, although not to the same degree as the oxide traps. In contrast, for the radiation-hardened space application memories, the increase in oxide traps and interface traps is relatively insignificant.

Figure 7-3(a) shows a schematic cross-section of an n-channel MOSFET to illustrate the effect of total dose exposure. It shows the radiation-induced charging (positive) of the thin gate

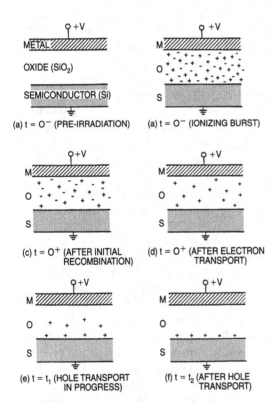

Figure 7-2. Schematic illustration of processes of charge generation and initial recombination, hole transport, and long-term trapping near SiO$_2$/Si interface [4].

oxide region, which generates an additional space-charge field at the Si surface. This additional buildup of charge can cause voltage offsets, shifts in turn-on voltage of the devices which can cause circuit degradation and failure. As shown, for sufficiently large amounts of trapped positive charge, the device may be turned on even for zero applied gate bias ($V_G = 0$).

The effect of radiation-induced charge components studied using the model of an MOS capacitor has shown that the oxide-trapped charge shifts the C–V curve in a negative direc-

tion, whereas the interface traps tend to "stretch out" the curve. Figure 7-3(b) shows a plot of the drain current versus gate voltage for an n-channel MOS (NMOS) transistor pre- and post-irradiation. As in the case of an MOS capacitor, the curve has shifted in the negative direction, which implies that a less positive charge is required to turn on the transistor.

The effect of ionizing radiation damage is different for the p-channel (PMOS) devices than for the NMOS. In a PMOS transistor, since the gate voltage is negative, the hole charges are trapped near the gate–oxide interface. As such, the PMOS transistors are more radiation-resistant than the NMOS transistors in which the hole charges are trapped near the Si–SiO$_2$ interface due to the positive gate bias voltage. The effect of ionizing radiation (gamma) on the n- and p-channel transistors is shown in Figure 7-4 [4]. The net effect of gamma radiation on a CMOS device threshold voltage is that the n-channel transistors become easier to turn on, and can actually become depletion mode (i.e., remain "on" with zero gate bias), whereas the p-channel transistors become more difficult to turn on.

Ionizing radiation damage causes changes in the memory circuit parameters such as the standby power supply currents, input and output voltage level thresholds and leakage currents, critical path delays, and timing specification

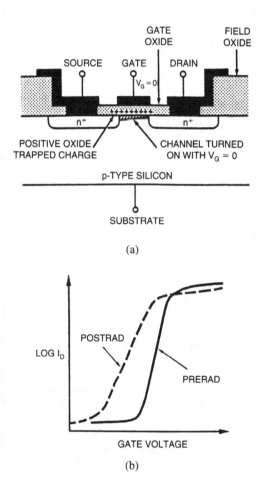

(a)

(b)

Figure 7-3. (a) Schematic cross-section of an n-channel MOSFET, postirradiation, showing the basic effect of charging of gate oxide. (b) Plot of the drain current (logarithm) versus gate voltage for an N-MOS transistor, pre- and postirradiation.

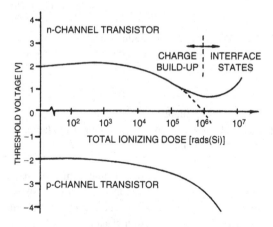

Figure 7-4. Curves showing effect of ionizing radiation (gamma) on the n- and p-channel transistors [4].

degradations (e.g., access time for the RAMs). Increasing radiation doses can cause progressive loss of function and eventual failure of an MOS device. This effect is illustrated in Figure 7-5, which shows a series of parallel drain current (I_D) versus gate voltage (V_G) for an n-channel MOS device which has been subjected to increasing dose levels (1–4) [5]. The zero (0) curve represents preirradiation level data. The remaining curves (1–4), each corresponding to a specific dose level, represent the onset of a particular failure mechanism and its main degradation effect, along with typical values of negative threshold voltages ($-\Delta V_t$) as listed in Table 7-2 [5]. For space applications, a radiation dose limit or "maximum acceptable dose" (D_A max) for any CMOS device can be defined as the dose at which a significant failure mechanism appears for that circuit. For example, in Figure 7-5 (and corresponding Table 7-2), this may be at a dose level of 2–5 krad (Si) that causes the threshold voltage of the n-channel to

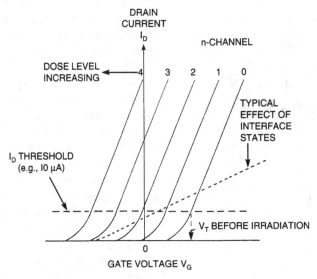

Figure 7-5. Typical drain current (I_D) versus gate voltage (V_G) for an n-MOS device subjected to increasing radiation dose levels (1–4) [5].

TABLE 7-2. Dose Levels, Failure Mechanism, Main Degradation Effect, and Typical Threshold Voltage Shifts Corresponding to Figure 7-5, Curves (1–4) [5]

Dose Level in Fig. 7-5	Failure Mechanism Number	Main Degradation Effect	Symbol	Dose rad (Si)	$-\Delta V_T$ volts
				Typical Values for CMOS LSI	
1	1	Minor "noise immunity reduction"; possibly minor loss in switching speed	NIR	8×10^2	0.2
2	2	Sharp quiescent current increase due to "V_T of n-channel crossing zero"	VTNZ	5×10^3	1
3	3	Switching speed reduction	SSR	1×10^4	2
4	4	Change of logic state impossible: "Logic failure"	LF	3×10^4	4

cross zero (VTNZ), which can lead to large increases in quiescent current (I_{DD}).

The increase in quiescent supply currents (I_{DD}) does not only depend on the total dose level, but also on the rate at which the dose is received. For n-channel transistors, the threshold voltage shift is less at lower dose rates, and hence a lower increase in supply current due to trapped hole annealing effects. After irradiation, this leakage current tends to decrease with time as the n-channel threshold voltage increases. Figure 7-6 illustrates this dose rate effect in I_{DD} versus total dose krad(Si) curve for a 1 kb SRAM subjected to six different dose rates [6]. Input voltage switching levels, both CMOS and TTL (e.g., V_{IL}, V_{IH}), which are closely linked with the threshold voltage of the transistors in the input stage, are also affected by the radiation. Similarly, the output drive levels, which are also closely related to the individual transistor performance, are also degraded. In DRAMs, the increase in leakage currents with higher radiation doses can lead to a decrease in the hold time (or the minimum refresh period) [7], [8].

In general, ionizing radiation tends to reduce the operating margin for an MOS circuit. For example, in SRAMs, the operating margins for sense amplifier circuitry tend to decrease with increasing dose levels [9]. Radiation-induced leakage currents can reduce the logic "1" level and increase logic "0" to eventually cause functional failures from reduced operating margins. The effect of radiation on the CMOS memory timing parameters, such as the access time or minimum write pulse width time, is highly dependent upon the manufacturing technology and fabrication process variables. For example, in the MOS fabrication processes which cause very little mobility degradation with radiation, the negative threshold voltage shift for the n-channel transistors may actually lead to improved performance in some parameters, although degradation in power supply leakage currents may still occur [10].

For those processes which have a large interface trap buildup with irradiation (and thus higher mobility degradation), the circuit speed tends to degrade with radiation, and rebound may occur on postirradiation anneal. Another variation in the radiation response may cause the access time to be improved at low doses, but degrade at high total dose. Figure 7-7 illustrates an example of this type of response for a 4 kb CMOS SRAM [11]. Normalized read access time is plotted as a function of the radiation dose. The solid curve shows measurements

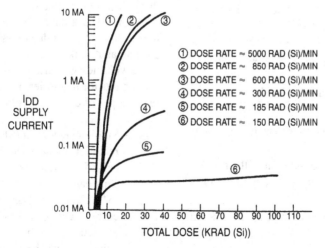

Figure 7-6. I_{DD} versus total dose krad (Si) curve for a 1 kb CMOS RAM subjected to six different dose rates. (From [6], with permission of IEEE.)

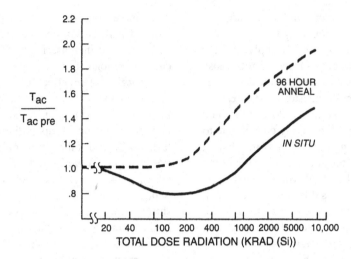

Figure 7-7. Normalized read access time (T_{ac}) as a function of radiation dose, and post 96 h annealing (unbiased) curve for a 4 kb CMOS RAM. (From [11], with permission of IEEE.)

made *in situ* (at the radiation source), whereas the dashed curve represents timing measurements made 96 h after removal from the radiation source with the unbiased parts held at room temperature.

To summarize, the parametric degradations in threshold voltages and operating margins, leakage currents, and variations in timing parameters for commercial MOS memories (unhardened) in response to total dose radiation depend on various factors, such as:

- Manufacturing processes and fabrication variables
- Total dose irradiation levels
- Dose rates
- Circuit biasing conditions during irradiation
- Time between completion of the irradiation and measurement of circuit parameters
- Test temperatures (during irradiation and testing).

Total dose testing for radiation hardness assurance (RHA) verification for the memories used in space applications is performed in accordance with MIL-STD-883, Method 1019, which will be discussed in more detail in Section 7.4.

7.2.3 Single-Event Phenomenon (SEP)

The single-event phenomenon (SEP) induced by the charged particles penetrating through the microelectronic circuits is a major concern in space environments. The SEP refers to the ionizing effect due to a single high-energy particle as it strikes the sensitive nodes (or sensitive volume) within the electronic device. In semiconductor memories, it can cause three different types of effects, as follows:

- Single-event upset (SEU) is the ionizing radiation-induced logic change in the cell of a storage device such as a RAM. This can cause a bit-flip from logic "1" to "0" and vice versa because of the high current level or excess charge collected at a sensitive bit node in the memory circuit. This can be a "soft error," which means the logic reversal is temporary, and the cell can become functional again upon a reset during the next logic cycle or memory write operation.

- Transient radiation effects are those that depend on the ionizing dose rate delivered to a circuit rather than the total amount of dose delivered. A higher dose rate creates a greater number of e–h pairs, both in the insulator layer and the silicon substrate. High photocurrents generated can also alter the contents of memory elements to cause soft errors without permanent damage to the circuit. However, this effect, caused by a burst of ionizing radiation from an extended source, is referred to as a "transient upset," as compared to an SEU caused by a single energetic ion.

- Single-event latchup (SEL) or a "hard error" may occur when a single charged particle causes a latched low-impedance state by turning on the parasitic SCR structure (p-n-p-n or n-p-n-p) in bulk CMOS devices. It can also result in a "snap-back" effect, which is a result of photocurrent-induced initiation of avalanching in the drain junction of an n-channel MOS transistor. If the induced parasitic current levels are sufficiently high, they can cause permanent device failures such as a junction burnout. These parasitic current states are highly dependent upon a circuit layout and the device geometry.

The impact of SEU on the memories, because of their shrinking dimensions and increasing densities, has become a significant reliability concern. In the 1970s, the soft errors in random access memories from the SEP were first reported in operating satellites [12], [13]. The satellites at GEO orbits and in regions outside the earth's radiation belts can experience upsets due to heavy ions from the galactic cosmic rays or solar flares. It is estimated that natural, cosmic ray, heavy ion flux has approximately 100 particles/cm²/day. In this environment, many devices can upset at a rate of about 10^{-6} upsets/bit-day. For example, for a memory size of 1 Mb (10^6 b),

this rate equates to 1 upset/day. The SEUs can also occur from proton-induced nuclear reactions, even though the energy loss rates by direct ionization from protons are too low to upset most devices. SEUs can have significant effects within the heart of proton belts, which may roughly have $10^7 - 10^9$ protons/cm²/day with energies above 30 MeV (which is approximately the minimum energy required to penetrate a typical spacecraft). It is estimated that about 1 in 10^5 protons has a nuclear reaction with the bulk silicon.

The rate of energy loss, and therefore the density of generated charge, depends on several factors such as energy, mass, and charge of ionizing particle. The energy deposition rate (dE/dX), the stopping power, or the linear energy transfer (LET) are measured in units of MeV/[cm²/gm] (or MeV/μm). An important consideration in device SEU susceptibility is the critical charge, which is defined as the minimum charge that must be deposited in a device to cause an upset. It is equivalent to the number of stored electrons representing the difference between a stored "0" and a stored "1" [14]. For example, in a DRAM operating at 5 V, the largest voltage swing which can be operated without losing the stored information is roughly 2.5 V. If the storage capacitance is 50 fF, then

$$Q_{crit} = C_S \times \Delta V_{crit}$$
$$= 50 \times 10^{-15} F \times 2.5V = 125fC$$
$$\approx 7.8 \times 10^5 \text{ electrons.}$$

Because of the scaling and reduced feature sizes, the advanced high-density memories have smaller area capacitors and hence lower critical charges. Q_c is device-dependent, and can vary from 0.01 to about 2.5 pC.

In general, for the unhardened devices, the critical charge decreases with reduced feature sizes. Figure 7-8(a) shows a plot of SEU critical charge (Q_c) versus feature size (l) for a broad range of technologies [15]. This curve shows a lack of dependence on the device technology, and seems to follow the l^2 scaling rule. Figure 7-8(b) shows the SEU rate as a function of Q_c for various sensitive region transistor dimensions of a CMOS SRAM [15].

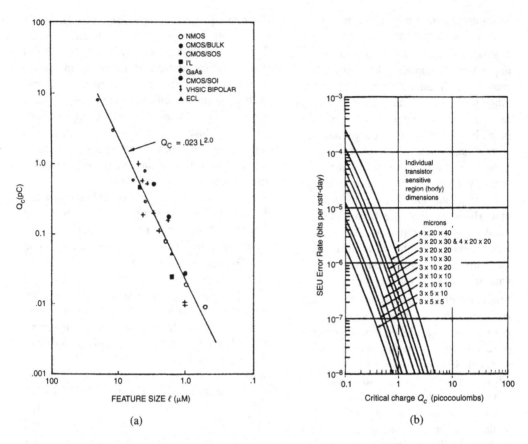

Figure 7-8. (a) A plot of SEU critical charge (Q_c) versus feature size (l) for various technologies. (b) A plot of SEU error rate versus Q_c for various sensitive region transistor dimensions of a CMOS SRAM [15].

As a heavy ion transits through the sensitive volume of a device, it creates $e-h$ pairs along its ionization track. The original charge collection modeling postulated that the charge deposited in a depletion layer of the affected device would be collected rapidly by drift and the remaining charge more slowly by diffusion [16]. An effect not taken into consideration by this model, and which made the SEU a more serious threat, was the field-funneling effect illustrated in Figure 7-9 [17], [18]. This shows the effect of an alpha particle striking a node in a circuit array, with the solid curve representing the expected charge collection through the diffusion mechanism. However, taking the funneling phenomenon into consideration, more charge would be collected at the struck node (shown by the dashed curve) and less at the surrounding nodes.

If Q_{crit} of this circuit is within the indicated range, an upset would occur because of the funneling effect which increases charge collection at the target node. Funneling is a drift process which directs the deposited charge onto the target node, whereas the diffusion can distribute the deposited charge over several nodes. The funneling requires significant conductivity differential between the ion track and the surrounding semiconductor (i.e., high injection conditions). The circuit-hardening techniques for SEU are aimed at neutralizing the effect of this charge collection mechanism. These techniques will be discussed in more detail in Section 7.3.

The heavy ions or transient radiation (high dose rate burst) can induce latchup at certain internal circuit nodes which can produce

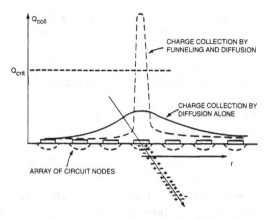

Figure 7-9. Schematic of charge collection profiles for a target node from the diffusion, and combination of diffusion with funneling effects. (From [18], with permission of IEEE.)

high photocurrent, resulting in large voltage drops in the isolation wells, substrate, and metallization. A study was performed to investigate the topological factors that affect the latchup by using test structures fabricated with two different CMOS foundries [123]. Test results showed that the geometry of isolation well contacts had a large effect on the photocurrent responses because of the effect of lateral resistance of the well. The experiments that delayed the application of bias until well after the initial radiation pulse showed that the latchup would still occur, which implied that the triggering mechanism does not depend on the fast transition times of the radiation pulse.

The latchup tests on a bulk CMOS technology 64K SRAM revealed a latchup window that was caused by the presence of two different internal latchup paths with different radiation triggering levels and holding voltages. It was shown that the power supply current surge characteristics were an important diagnostic tool to select the radiation test levels to minimize the interference from latchup windows.

The insulating substrate for silicon-on-insulator (SOI) technologies such as silicon-on-sapphire (SOS) offers considerable improvement in SEU and SEL hardness. The SOI RAMs (see

Chapter 2, Section 2.2.5) have three–four orders of magnitude lower SEU error rates than for bulk CMOS memories. The SOI memories are SEL-immune since the insulating substrate eliminates the parasitic p-n-p-n path.

7.2.3.1 DRAM and SRAM Upsets. The DRAMs store information in storage cells through the presence or absence of charge on a capacitive node. This charge can be upset in the presence of alpha particles produced in the decay of trace radioactive materials found in the packaging of some DRAMs, which can cause soft errors (see Chapter 2, Section 2.3.5). The relationship between the electrical states of a RAM cell and the logic states it represents is determined by the read circuitry of the cell. For DRAMs, the critical charge is considered to be the difference between the lesser of a logical "1" or a "0" and the reference cell (logical $V/2$) rather than the absolute value of difference (in charge content) between a "1" and a "0." However, in reality, the component cells have a certain range of critical charges determined by the length of the refresh cycle [19]. This implies that the cells most recently refreshed have maximum critical charge. Since the DRAMs may be upset by charge collection through diffusion, several storage cells may be affected by a single-particle strike ionization track [20–22]. This is also known as multiple upsets (from a single-particle strike), and to minimize its possibility of occurrence in a given application, some DRAMs are designed so that the adjacent cells will carry opposite charge states.

In comparison to a DRAM with a capacitive storage cell, a SRAM stores information in a latch, which is basically a cross-coupled pair of inverters. In MOS SRAMs, the storage node consists of a combination of the gate node of one inverter coupled with the output node of the other inverter in that cell. For an SEU to occur, the induced perturbation must exceed the charge restoration capability of this cross-coupled configuration. By a similar argument, the cell can be hardened against SEU by protecting the charge at one inverter's gate. For CMOS SRAMs which have very low static power consumption, the critical charge as a measure of

SEU susceptibility is generally valid (as in DRAMs), provided the duration of charge collection from an SEU pulse is short compared to the write time of the cell.

In general, SRAMs radiation-hardened for both total dose and SEU are preferred for critical space applications. Commercial (unhardened) SRAMs are susceptible to low-dose radiation tolerance, multiple-bit upsets, permanently "stuck" bits, and a possibility of latchup.

The cosmic-ray environment is specified in terms of particles with a given energy loss (LET spectrum). This LET spectrum can be combined with device dimensions to find the probabilities of different amounts of energy deposition, and then to calculate the number of events that will deposit a charge greater than Q_c. Table 7-3 shows the SEU and latchup results of some 1 Mb (128K × 8 b) and 256 kb (32K × 8 b) SRAMs ground-tested by the Aerospace Corporation [22], [23]. It lists the threshold LET values and saturation cross-sections for both the SEU and latchup. These devices were tested at the Lawrence Berkeley Laboratory 88 in cyclotron facility, using various ion species such as xenon (603 MeV), krypton (380 MeV), copper (290 MeV), and hydrogen (20, 30, and 55 MeV). SEU and latchup cross-sections are calculated by the following formula [24]:

$$\sigma = \left(\frac{N}{F}\right) \sec \theta$$

TABLE 7-3. SEU and Latchup Test Results of Some 1 Mb and 256 kb SRAMs. (From [22], [23] with Permission of IEEE.)

Device Type	Manufacturer (Die Type if not Mfr's.)	Technology	Organization	Feature Size	SEU LET$_{Th}$	SEU X-Sec	Latchup LET$_{Th}$	Latchup X-Sec
HM658128LP[†]	Hitachi	CMOS/NMOS	128K × 8	≈1 μm	4	1 × 10^0	72	3 × 10^{-6}
MSM8128SLMB[†]	Mosaic (Hitachi)	CMOS/NMOS	128K × 8	≈1 μm	4	1 × 10^0	72	3 × 10^{-6}
MT5C1008C-25[~]	Micron Technology	CMOS/NMOS	128K × 8	≈1 μm	4	2 × 10^0	None	None
CXK581000P	Sony	CMOS/NMOS	128K × 8	1 μm	4	7 × 10^{-2}	50	2 × 10^{-5}
CXK58255P	Sony	CMOS (6-Trans)	32K × 8	0.8 μm	10	1 × 10^{-1}	25	8 × 10^{-4}
MT5C2568[~]	Micron Technology	CMOS/NMOS	32K × 8	≈1 μm	3	9 × 10^{-1}	None	None
UPD43256A[†]	NEC	CMOS/NMOS	32K × 8	≈1 μm	3	4 × 10^{-1}	None	None
84256	Fujitsu	CMOS/NMOS	32K × 8	≈1 μm	3	1 × 10^0	30	3 × 10^{-5}
S32KX8	Seiko	CMOS/NMOS	32K × 8	1.3 μm	3	1 × 10^{-1}	30	1 × 10^{-3}
IDT71256[†~]	IDT	CMOS/NMOS	32K × 8	1.3 μm	≈3	2 × 10^{-1}	None	None
XCDM62256[†]	RCA (Seiko)	CMOS/NMOS	32K × 8	1.3 μm	≈3	5 × 10^{-1}	40	1 × 10^{-3}
EDH8832C[†]	EDI (Mitsubishi)	CMOS/NMOS	32K × 8	1.3 μm	≈3	4 × 10^{-1}	20	2 × 10^{-3}
OW62256[†]	Omni-Wave (Hitachi)	CMOS/NMOS	32K × 8	1.3 μm	≈3	4 × 10^{-1}	None	None

[†]Tested for proton-induced single-event upset.

[~]These devices incorporate an epitaxial layer.

(1) Threshold LETs given in MeV/(mg/cm^2).

(2) Saturation cross-sections given in cm^2/device.

(3) "None" indicates a cross-section <10^{-7} cm^2/device at an LET of 100 MeV/(mg/cm^2).

where N = number of errors (counting multiple-bit errors separately)

F = beam fluence in particles/cm^2

θ = angle between the beam and chip surface normal.

A lower LET threshold implies that the device is more sensitive to an SEU or latchup. A recent study also showed that the devices are more susceptible to proton-induced upsets if the heavy ion-induced LET threshold is less than about 8 MeV/(mg/cm^2) [25].

The SEU rates (upsets/bit-day) were calculated for four unhardened memory technologies using a 10% cosmic ray environment for geosynchronous orbits [26]. The plot showed SEU probabilities to be lowest for the CMOS/SOS and CMOS/SOI technologies because of their lower charge collection volumes. The CMOS devices using the proper thickness epitaxial layer had slightly higher upset rates, but they were better than the bulk devices (with no epitaxial layer). The bipolar devices showed a broad range of sensitivities because of various technologies and geometries. Also, the very high-speed bipolar memories which use small amounts of charge have higher upset rates than the slower devices. The memories in low earth orbit experience a reduced cosmic ray flux. However, the upset rates can be higher for devices which are sensitive to protons.

7.2.3.2 SEU Modeling and Error Rate Prediction.
The SEU threshold (or vulnerability) is commonly measured in terms of critical charge (Q_c) and critical LET. Q_c is primarily correlated to circuit design characteristics, whereas the critical LET is a function of both circuit design and technology. Q_c can be determined from computer simulation, whereas the critical LET is found experimentally by bombarding the device with various ion species (e.g., in a cyclotron). The LET is basically a measure of how much energy (or charge) is deposited per unit distance within the device for the first micrometer of track length. The critical LET values can be different for devices fabricated with identical designs and processes, but on different substrates. Also, the particles with the same initial (incident) LET values may not necessarily deposit the same total amount of charge within a device since, in general, they do not have the same track lengths.

There are several aspects of memory SEU susceptibility which can be evaluated by computer analysis, such as the charge collection and transport simulations, critical charge simulations, and predicted error rates in standard environments. The heavy ions and device interaction simulations can be used to model the effects of particle hits on sensitive devices. A wide variety of software tools are available to do these charge collection simulations. Sometimes these simulations are performed by using tabulated values of range [27], [28], and stopping power versus ion energy [27], [29] instead of using the ion charge deposition models. Some other options available are the use of Monte Carlo simulations and Poisson-solver equations for modeling a charge deposition process in the circuit [30–32].

The critical charge can be independently used as a measure of SEU susceptibility in DRAMs since they have isolated cells for the charge collection process and active devices are not involved in logic state storage. The SRAMs are upset if a single event deposits sufficient charge at a sensitive circuit node to "write" the cell [33]. A commonly used software tool to simulate critical charge is the SPICE model, which is available in several versions [34–36]. MOS transistors fabricated on insulating substrates (e.g., CMOS/SOS and CMOS/SOI) require special simulation models, even though the radiation concerns are not as high as those for bulk CMOS devices [37–39].

For a semiconductor memory device, the SEU rate in a given cosmic ray space environment depends primarily on its upset threshold expressed as a critical charge (or critical LET) and the total device volume sensitive to ionic interaction. The sensitive volume is associated with the area around sensitive nodes, e.g., n-channel and p-channel drain nodes of the "OFF" devices in a CMOS SRAM cell. The devices with small feature size l are more sensitive to

SEU than those with a larger feature size. In general, the critical charge Q_c is proportional to l^2. This proportionality factor has been combined with the number of particles in a space environment capable of delivering Q_c to a given circuit, and project its SEU susceptibility to further scaling [40]. In a device, multiple nodes may be sensitive. The DRAMs are susceptible to multiple errors from single events when critical charge for upset is transported through diffusion to the storage nodes near the event site. For both SRAMs and DRAMs, the soft-error rate predictions for a sensitive volume (or a group of nodes for DRAMs) are made by estimating the number of events capable of delivering charge in excess of Q_c for the upset. These predictions require environmental models.

The most commonly used environmental models for SEU by cosmic rays are as follows:

- CRIER (cosmic ray-induced error rate) code [41]
- CRUP (cosmic ray upset) model
- CREME (cosmic ray effects on microelectronics) [42].

The CREME model is commonly used to calculate upset/error rates given the description of the radiation environment including the effects of trapped particles and spacecraft shielding, the circuit's critical charge, and sensitive volume. The modeling performed is based upon a spacecraft's orbit, calculating the LET spectra, and then the error rate. The upset/error rates from the CREME model compare favorably with the actual spaceflight dosimetry data [43]. The CRUP model saves the user from the complexity of a detailed description of an environment as required by the CREME model. It is designed to operate in one of the following three standard environments: solar minimum, solar maximum, and Adams' 90% worst case [44].

Cosmic-ray-induced upsets can be calculated from the following integral equation:

$$N_e = \frac{S}{4} \int_{L_0}^{L_{max}} \phi(L) \, C\left(\frac{\Delta E}{\rho L}\right) dL$$

where S = surface area

ΔE = critical energy for upset

L = linear energy transfer

$L_0 = L$ (minimum) that causes upset

$\phi(L)$ = differential LET distribution

$C(\Delta E/\rho L)$ = integral path distribution.

This integration is performed using either approximations or exact forms.

The heavy nucleii like iron are the most important components of the space environments, and the particle spectra are usually presented in the form of differential energy spectra. The actual space environment includes several other elements, and so their contributions must also be added. However, the upset rate calculations do not use the energy spectra, but the LET spectrum, i.e., the number of particles as a function of their rate of energy loss. Another approach for SEU rate calculations, the integral LET spectrum, is used. Figure 7-10 shows the integral energy loss spectrum at a geosynchronous orbit. The curve labeled 100% corresponds to solar maximum conditions, and the environment is always worse. The 10% curve combines solar minimum cosmic rays and solar activity such that the environment is worse only 10% of the time. The 0.03% curve corresponds to a large solar flare. The curves labeled GPS and TIROS-N represent the predicted environments for two satellites that were used during the calculation of their respective upset rates.

The single-event error rate for a device can also be estimated using a "figure of merit" approach which can be useful as an approximation for comparing the device sensitivity. The upset/error rate for a part is calculated from the probability that a particle with LET value $\geq$LET$_{crit}$ would be interrupting a given sensitive volume. Using the figure of merit, the error rate at a sensitive node can be approximated as

$$\text{error rate} = 5 \times 10^{-10}\left(\frac{\sigma L}{L_c^2}\right)$$

$$= 5 \times 10^{-10} \frac{A \cdot B \cdot C^2}{Q_c^2}$$

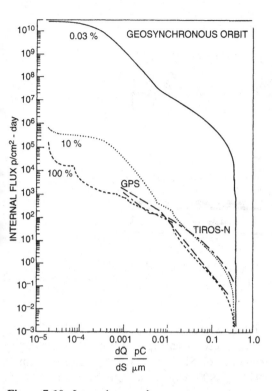

Figure 7-10. Integral energy loss spectra at a geosynchronous orbit for three different
environments: 100% curve representing solar maximum, 10% curve for a
combination of solar minimum cosmic
rays and solar activity, and 0.03%
curve corresponding to a large solar
flare. (From [45], with permission of
IEEE.)

where $\sigma L = A \cdot B$ = surface area of the sensitive node

$$L_c = \text{LET}_{\text{crit}} = Q_c/C$$

Q_c = critical charge in picocoulombs
(pC)

C = collection length at that node for
single-event track (usually
depth) in micrometers.

The error rate may be estimated by sum-
ming up the results at each sensitive node. This
algorithm is quite often used to estimate the
worst case error rate (referred to as "Adams'
90% worst case environments"), which is la-

beled as the 10% curve in Figure 7-10. The
SEUs caused by proton-initiated nuclear reactions (during solar flares or in the heart of radiation belts) can be estimated by using the
Bendel–Petersen one-parameter approach and
experimental determination of the proton upset
cross-sections at one proton energy [46].

An important technique for SEU modeling and error rate prediction is through circuit
simulation of single-particle events. For example, the SEU response of n^+–AlGaAs/GaAs
heterostructure field effect transistors (HFETs—
also known as SDHTs, HEMTs, MODFETs, and
TEGFETs) and HFET SRAMs was evaluated by
measuring their response to focused 39 kV electron beam pulses [124]. Figure 7-11 shows a
cross-sectional diagram of the enhancement
mode HFETs (gate length = 1 μm, gate width
= 17.5 μm). The HFET layer structure was
grown on semi-insulating GaAs substrate using
molecular beam epitaxy (MBE) techniques. The
gate metallization was WSi, and the drain and
source metallization was Ni–Au–Ge. In addition, all metallization lines and bond pads were
isolated from the GaAs substrate by an insulating layer.

Initially, focused *e*-beam pulses were used
to measure and model the HFET drain and gate
SEU responses. Test results showed that the
drain response was most sensitive to *e*-beam
pulses applied between the source and drain,
while the gate response was significant only
when the *e*-beam pulses were located at the end
of the gate. Using these data, current sources
were used to simulate single-particle events in
circuit simulations of an HFET memory cell.
The simulations showed that this HFET memory cell is most vulnerable to an SEU hit in the
area between the drain and source region of the
OFF pull-down HFET. The test results showed
that measurements on individual HFETs and circuit simulation of the SEU hits may be used to
predict the SEU response of HFET memories.

7.2.3.3 SEP In-Orbit Flight Data.   The
SEU modeling discussed in Section 7.2.3.2 is an
important tool for device error rate prediction in
a given radiation environment. Ground-based
testing in a simulated environment using heavy

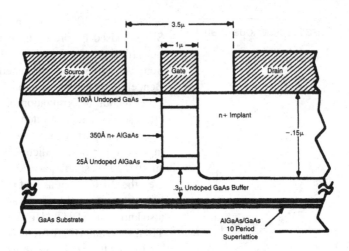

Figure 7-11. Cross-section of an n^+–AlGaAs/GaAs HFET. (From [124], with permission of IEEE.)

ions and proton beams (see Section 7.4.2) provides useful information on the relative sensitivity of the devices. However, it is important to verify ground test results obtained by observation of actual device behavior in orbit. In the 1970s, soft errors in random access memories from SEP were first reported in operating satellites [12], [13]. A number of flight anomalies in the DOD, NASA, and commercial satellites (spacecraft systems and instruments) have been attributed to this phenomenon. However, in a flight system, it is often difficult to establish a precise cause–effect relationship between the energetic particles (protons and heavy ions) and bit errors experienced because of several other sources such as noise, power line variations, and spacecraft discharge that could result in a device's anomalous behavior. Since those early SEPs reporting, a systematic recording of SEU on satellites was initiated, and several in-orbit experiments were designed to study the effect of radiation particle environments on various semiconductor device types, including the memories. Some of those SEP in-orbit flight data and experiments are discussed in the following sections:

(1) NASA Tracking and Data Relay Satellite Systems (TDRSS): This satellite system was designed by NASA to provide communications and high-data-rate transmissions for low-earth-

orbiting satellites and platforms such as the Space Telescope and Space Shuttle. A systematic recording of SEUs on TDRSS satellites from 1984 to 1990 allowed some correlations to be drawn between the upsets and space environments [47]. The anomalous responses observed in the TDRSS attitude control systems (ACS) were traced to bit upsets in its RAMs. The ACS contained four pages of RAM with 256 bytes/page, and each page consisted of two static bipolar 93L422 (256 × 4 b) RAM chips. About half of the total SRAM was in use. Table 7-4 shows the TDRSS ACS RAM usage and SEU history [47]. A series of large solar flares in August–October 1989 caused a substantial increase in the TDRS-1 SEU rate, which resulted in either operational anomalies or checksum errors.

This experience with a high SEU rate on TDRS-1 resulted in ACS system redesign for subsequent TDRSS series satellites. The SEU-sensitive static bipolar 93L422 was replaced with a hardened CMM5114 CMOS/SOS device. No SEUs were observed in the upgraded ACS system. Ground testing for SEU showed effective LET for CMM5114 to be much higher than that of 93L422 (58.0 versus 1.5 MeV/[mg/cm^2]).

(2) National Space Development Agency (NASDA) Satellites: The in-orbit observation and measurement of SEP was also made by NASDA of Japan through its Engineering Test

TABLE 7-4. TDRSS ACS RAM Usage and SEU History. (From [47], with Permission of IEEE.)

	Used	1984	1985	1986	1987	1988	1989	1990
Page 0	93%	10	17	22	24	23	14	5
Page 1	57%	71	102	157	131	106	114	57
Page 2	53%	83	254	277	269	258	266	101
Page 3	0%	174	262	302	269	200	259	94
Total	51%	338	635	758	693	587	653	257
SEUs/Chip/Day*		0.238	0.360	0.414	0.368	0.274	0.355	0.129
Estimated Total*		696	1048	1208	1076	800	1036	376

*Based on Page 3 SEU count.

Satellites (ETS-V) in geostationary orbit [48]. This spacecraft included a RAM soft-error monitor (RSM) which recorded measurements of the SEU or SEL occurring in eight 64 kb CMOS SRAMs. The objective of this RSM chip was to: (1) measure the rate of SEU/SEL occurrences, (2) measure the number of bits losing functionality because of hard errors, and (3) monitor the total current supplied to RAMs in order to detect any deterioration of the devices due to the total dose effects. The period of data acquisition was about 180 weeks from November 22, 1987 to June 13, 1991, and included periods of solar flares.

Test data for ETS-V showed an increase of SEL during the solar flares, and not much of an increase in SEU (except during the SEL occurrences). Another set of data was collected through command memory (decoder system) used in Marine Observation Satellite (MOS-I), a medium-altitude satellite. Test results showed that the number of SEL or SEU events followed a Poisson distribution, although there was little variation in the distribution before and after the solar flares. For MOS-I satellites which used TTL RAM devices (93419, 64 × 9 b), approximately one SEU occurred every day from protons trapped in the radiation belts. The upset rate increased during solar flare, and was attributed to heavy particles in the flare. The SEU data for MOS-I spacecraft agreed with the experimental ground data. However, there was some discrepancy observed for on-orbit SEL data and the CREME model calculations based on Cf-252 and accelerator testing.

(3) UoSAT Satellites: The UoSAT family of research microsatellites (50–60 kg) were designed by the University of Surrey, England, and flown in low-earth, near-polar, sun-synchronous orbits [49]. One of the objectives of these experiments was to observe single-event, multiple-bit upset (MBU) in the devices, and to verify the integrity of error-protection coding schemes. These satellites were also equipped with a cosmic-ray effects and dosimetry (CREDO) payload which allowed measurement of the radiation environment inside the spacecraft, thus making possible the correlation of SEU activity with solar proton events and long-term changes in the radiation environments [50].

Table 7-5(a) shows UoSAT orbital parameters [49]. UoSAT-2 contained a digital communication experiment (DCE) to provide experience with dense semiconductor memory systems and as a testbed for block coding schemes used to protect on-board bulk memory systems (RAMDISKs). Table 7-5(b) shows the details of memory devices used in the UoSAT family of microsatellites [49].

The memory systems on board were protected against accumulated errors due to SEU by coding schemes as follows: (1) Hamming EDAC code implemented in hardware, and (2) two types of block coding schemes in hardware.

It was observed that with the possible exception of one memory device type, all other

TABLE 7-5. UoSAT Satellites. (a) Orbital Parameters. (b) Memory Devices used. (From [49], with Permission of IEEE.)

Satellite	Launch	Orbit	Incl.	Status
UoSAT-1	Oct. 1981	—	—	Decayed
UoSAT-2	Mar. 1984	674 × 657 km	97.9	Active
UoSAT-3	Jan. 1990	802 × 784 km	98.6	Active
UoSAT-4	Jan. 1990	803 × 787 km	98.6	Failed
UoSAT-5	July 1991	775 × 764 km	98.5	Active

(a)

System	Manufacturer	Device	Size	Type (Quantity)	Bits Monitored
UoSAT-2 OBC	Mostek	MKB4116J-83	16K × 1	NMOS DRAM (12)	196,608
	Texas	TMS4416-15NL	16K × 4	NMOS DRAM (3)	
	Texas	TMS4416-20L	16K × 4	NMOS DRAM (3)	393,216
UoSAT-2 DSR	Toshiba	TC5516AP-2	2K × 8	CMOS SRAM (96)	786,432
UoSAT-2 DCE	Harris	H6564	16K × 4	Hybrid (3)	
		(=48 × HM6504	4K × 1	CMOS SRAMs)	147,456
	Harris	MI6516-2	2K × 8	CMOS SRAM (2)	
	Harris	MI6516-9	2K × 8	CMOS SRAM (5)	18,432
	Hitachi	HM6264LP-12	8K × 8	CMOS SRAM (8)	524,288
	Hitachi	HM6116L-3	2K × 8	CMOS SRAM (8)	
	Hitachi	HM6116-3	2K × 8	CMOS SRAM (8)	245,760
UoSAT-3 OBC-0	EDI	EDH8832C-15KMHR	32K × 8	CMOS SRAM (1)	
	EDI	EDH8832P-85KMHR	32K × 8	CMOS SRAM (1)	393,216
	IDT	IDT71256-S45CRE	32K × 8	CMOS/epi SRAM (2)	368,640
UoSAT-3 PCE CPU	Hitachi	MSM832TLI-10	32K × 8	HMP Package (8)	
		(=1 × HC62256	32K × 8	CMOS SRAM)	1,572,864
	NEC	D43256AC-10L	32K × 8	CMOS SRAM (4)	786,432
	EDI	EDH8832C-15KMHR	32K × 8	CMOS SRAM (4)	786,432
UoSAT-3 RAMDISK	Hitachi	HMS628128JILP-15	128K × 8	HMP Hybrid (8)	
		(=4 × HM62256LFP-12T	32K × 8	CMOS SRAMs)	8,388,608
	Hitachi	HMS626256SILP-12	256K × 8	HMP Hybrid (8)	
		(=8 × HM62256LFP-10T	32K × 8	CMOS SRAMs)	16,777,216
	Mitsubishi	HMS628256SILP-12	256K × 8	HMP Hybrid (4)	
		(=8 × MSM256AFP-70	32K × 8	CMOS SRAMs)	8,388,608
UoSAT-5 OBC-186	Hitachi	MSM8128VLI-10	128K × 8	HMP package (8)	
		(=1 × HC628128	128K × 8	CMOS SRAM)	4,194,304
UoSAT-5 RAMDISK	NEC	MS8256RKXLI-12	256K × 8	HMP Hybrid (4)	
		(=8 × D43256AGU-10L	32K × 8	CMOS SRAMs)	8,355,840
	NEC	MS81000RKXLI-12	1M × 8	HMP Hybrid (8)	
		(=4 × D431000GW-10L	128K × 8	CMOS SRAMs)	66,846,720
	Sony	MS81000RKXLI-12	1M × 8	HMP Hybrid (4)	
		(=4 × CXK581000M-10L	128K × 8	CMOS SRAMs)	33,423,360

(b)

HMP = Hybrid Memory Products.

devices were susceptible to upset in the proton environment of the South Atlantic Anomaly (SAA). However, the correlation with CREDO data showed that solar proton events have only a marginal effect on the error rates for CMOS devices, although they do affect NMOS

DRAMs. The older 16 and 64 kb CMOS exhibited error rates of about 5×10^{-7} errors/bit-day, whereas for 256 kb parts, the error rates were close to 1×10^{-6} errors/bit-day. Table 7-6 shows UoSAT memory device observed failure rates [49].

TABLE 7-6. Memory Device Observed Failure Rates. (a) UoSAT-2. (b) UoSAT-3. (c) UoSAT-5. (From [49], with Permission of IEEE.)

Device	SEU	MBU	SEU Rate (10^{-6} SEU/Bit-Day)	Uncertainty +/−	% MBU
MKB4116	982	—	3.72	6%	—
TMS4416	7081	419	13.4	2%	6%
TC5516	370	—	0.338	10%	—
HM6264	266	6	0.437	12%	2%
HM6116	173	8	0.606	15%	5%
MI6516	4	0	0.191	{2–9}	0%

Notes:

1) The uncertainties represent the 95% confidence limits. Where the number of events observed is small, the Poisson 95% bounds are shown in { }.

2) Multiple-Byte Upset (in this context) necessarily implies Multiple-Bit Upset (MBU). However, MBU does *not* necessarily imply an MBU/Byte Error.

(a)

Device	SEU	MBU	SEU Rate (10^{-6} SEU/Bit-Day)	Uncertainty +/−	% MBU
EDH8832	86	—	1.38	21%	—
IDT71256	10*	—	0.198	{7–17}	—
HC62256	545	—	0.973	8%	—
D43256	445	—	1.59	9%	—
EDH8832	315	—	1.13	11%	—
HM62256 /MSM256	15635	87	1.10	2%	0.6%

Notes:

1) *160 errors in the "soft" patch have been ignored.

(b)

Device	SEU	MBU	SEU Rate (10^{-6} SEU/Bit-Day)	Uncertainty +/−	% MBU
HC628128	695	—	0.337	7%	—
D43256	3095	13	1.14	4%	0.4%
431000	14306	7	0.656	9%	0.5%
581000	88	0	0.00803	11%	0%

(c)

For the majority of multiple-bit upsets, the outcome was an upset occurring in two separate bytes. The ratio of one-bit upset occurring within a single-byte to double-byte upset was roughly 200:1 (i.e., 0.5% of all events were MBUs). The lower density CMOS SRAMs showed a larger proportion of the MBUs (i.e., a few percent of the total number of events) occurring mainly outside the SAA, and were attributed to the heavy ions. Although the results of ground-based testing showed that some of the devices used were susceptible to MBU/byte errors, there was no conclusive evidence of the MBU affecting bits within a byte for any of the SRAMs. This implies that a simple Hamming code (12, 8) can offer good error protection, even with byte-wide devices. The block coding schemes developed for these satellites provided 99.5% protection (error detection).

(4) Combined Release and Radiation Effects Satellite (CRRES): The CRRES launched in July 1990 contained a number of different experiments to measure the radiation environments of space [51]. The CRRES orbit was highly elliptical (348 × 33,582 km) with an 18.2° inclination and 9.87 h period. The elliptical nature of the orbit caused the spacecraft to pass through the heart of the earth's radiation belts twice per orbit, while spending 90% of its orbit time beyond the inner proton belts (L-shells $> 2R_e$, where R_e is the earth's radius) to near geosynchronous altitudes. This eccentric orbit made it easier to separate SEUs into those caused by proton-initiated nuclear reactions versus those caused by direct ionization with energetic cosmic rays above the radiation belts.

CRRES contained a microelectronics package (MEP) space experiment which had 40 different device types, including RAMs, PROMs, and EAROMs. The MEP computer system was designed to report approximately every second all SEUs according to the device identification and logical address in the test memory and time tag the event with the MEP clock. This internal clock was synchronized with a satellite clock and converted to universal time (UT).

The data were analyzed from MEP for all 1067 CRRES orbits for measurement of the SEU rate for several different device types, and included the effects of protons from a large solar flare that occurred on March 23, 1991. Table 7-7 shows several different device types, including static RAMs which do not require refresh. Each

TABLE 7-7. CRRES MEP Devices with Block (Bk) Identification [2], Board Location, and the Total Number of Bits Tested for SEU. The Column Labeled "No. Bits" is the Total Number of Bits Tested for SEU. CMOS=Complementary Metal–Oxide–Semiconductor, SOS=Silicon-on–Sapphire, RP=Rad-Pak Total Dose Shielding Package, and CML=Current Mode Logic.

Bk	Bd[1]	Device	No. Bits
01	1	1K×1 GaAs RAM	4096
03	1	1K×1 GaAs RAM	4096
04	1	CMOS version of 8085 μP	98304
06	1	16K×1 CMOS/SOS RAM	98304
07	1	16K×1 CMOS RAM	196608
0A	1	6504 CMOS 4K×1 RAM (RP)	40960
0B	1	6504 CMOS 4k×1 RAM	40960
0E	1	1×3 CMOS/SOS gate array	12
10	3	6504 CMOS 4K×1 RAM (RP)	40960
11	3	6504 CMOS 4K×1 RAM	40960
12	3	6641 512×8 CMOS PROM	40960
13	3	6616 2K×8 CMOS PROM	163840
14	3	21L47 4K×1 NMOS RAM	40960
15	3	92L44 4K×1 NMOS RAM	32768
16	3	32×8 CML RAM	1024
17	3	8K×9 RAM	262144
18	3	2K×4 4-port CMOS RAM	32768
19	3	8K×8 CMOS RAM (2K×8)	65536
1A	3	256×8 CMOS RAM	12288
1B	3	93L422 256×4 bipolar RAM	4096
1C	3	93422 256×4 bipolar RAM	4096
1D	3	82S12 256×9 bipolar RAM (×8)	8192
23	2	71681 4K×4 CMOS/NMOS RAM	98304
24	2	6116 2K×8 CMOS/NMOS RAM	98304

[1]Bd=board number. Each board is duplicated (A and B) in the MEP for redundancy to increase reliability. Half of the test parts are on side A and half on side B. Boards 1A and 1B, in the first layer, at the front, are closest to the outside environment and receive the maximum exposure to radiation. Boards 2A and 2B are in the second layer, and boards 3A and 3B are in the third layer toward the back, furthest from the outside environment and just ahead of the control electronics.

block contained four–ten different devices split equally into two parts of a redundant MEP bus [52]. The listing shows the block identification number, the board position, the device description, and the total number of bits tested for SEU.

Table 7-8 shows the SEU rate, number of SEUs/kilobit/day, for the following four periods: (1) before orbits 4–585 (238 days), (2) during orbits 586–590 (2 days), (3) just after orbits 591–600 (4 days), and (4) after orbits 601–1067 (197 days) the March 23, 1991 flare [52]. The overall rate (O) is given for the number of devices where no SEUs were observed.

The CRRES MEP data analysis shows that the SEU rates are function of dynamics of the radiation environments including effects of solar flares. The effect of solar flares varied widely from one device type to another. GaAs SRAMs showed a different response to proton environments than the silicon RAMs. The SEUs, while CRRES was in the proton belts, were more common than those seen at higher altitudes. Also, multiple SEUs occurred more frequently than expected, but mostly in the proton belts. The solar flare did not affect either the SEU rate or the proton flux in the belts. The

TABLE 7-8. CRRES MEP SEU Rate (SEUs/kilobit/day) Before, During, Just After, and After the March 23, 1991 Solar Flare for All L-Shell Values with the Overall Rate (O) for Blocks with No Upsets. The Blanks Refer to Either No Calculated Overall Rate or No Measured SEUs Before, During, Just After, or After the Flare. (From [52] with permission of IEEE.)

Bk	O	Before	During	Just After	After
01		0.0295	2.38	0.238	1.65
03		8.14	177.	10.6	7.42
04	0				
06		0.000509	0	0	0
07	0				
0A	0				
0B		0.000174	0	0	0
0E	0				
10		0	0	0	0.000496
11		0	0	0	0.000248
12	0				
13	0				
14		0.0719	0	0.0358	0.0766
15		0.126	0.0596	0.0894	0.162
16		5.27	15.2	4.29	9.51
17		0.0540	0.223	0.0969	0.0102
18	0				
19		0.0117	0	0	0.0141
1A	0				
1B		3.45	5.72	2.50	3.87
1C		4.54	7.86	3.58	4.82
1D		0.913	2.14	0.715	0.960
23		0.00577	0	0.00497	0.00413
24		0.00318	0	0.00993	0.0122

flare did have a brief but significant effect on the SEU rate in the intermediate L-shell region. The largest increase in SEU rate occurred in L-shell values > 4, where very little geomagnetic shielding exists. A large variation in SEU rates can be avoided by selecting memory devices from the same wafer batch lot.

7.2.4 Nonvolatile Memory Radiation Characteristics

As discussed in Chapter 3, the earliest nonvolatile semiconductor memories were fabricated in MNOS (metal–nitride–oxide–semiconductor), which is similar to MOS technology except that the gate oxide is replaced by a stack of dielectrics. The dielectric stack is equivalent to the floating gate of the floating-gate technology (e.g., FAMOS) and serves as a charge-trapping region. There are several variations of MNOS technology, including SNOS (silicon–nitride–oxide–semiconductor) and SONOS (silicon–oxide–nitride–oxide–semiconductor), all of which may be referred to as SNOS since basic operation is similar in all cases. Figure 7-12 shows the cross-section of an SNOS memory transistor with typical thicknesses for various layers of the memory stack [53].

An exposure to ionizing radiation can cause a net loss in the trapped charge in SNOS transistors, thus affecting the retention charac-

teristics of the memory. Figure 7-13, showing irradiation effects on SNOS threshold voltages for two logic states of "1" and "0" as a function of the retention time, illustrates this effect [54]. The various plots are retention curves for identical SNOS transistors written into different initial states by changing the programming time. The magnitude of initial threshold voltage (V_{th}) depends on the number, duration, and amplitude of the write/erase pulses. In the curves shown, the larger V_{th} corresponds to parts which were programmed with larger duration pulses, smaller V_{th} curves with smaller time duration pulses, and the center curve for a virgin (never been written or erased) SNOS transistor for which the nitride contains no injected charge. It can be seen that the V_{th} shift caused by a given total dose irradiation depends upon the dose rate as well as the initial threshold voltage, and hence the charge trapped in nitride prior to irradiation. This charge can be represented by a model which yields a simple analytical solution that can accurately predict radiation-induced V_{th} shifts of SNOS transistors for a wide range of initial threshold voltages.

Radiation test results have shown that SNOS transistors can retain data for ten years after irradiation to 500 krad (Si), and are able to function even after receiving a high total dose [> 1 Mrad (Si)] [54–56]. However, the peripheral circuitry associated with SNOS transistor-

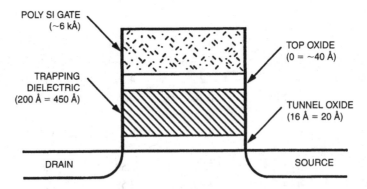

Figure 7-12. Cross-section of an SNOS memory transistor showing typical thicknesses for various layers of the memory stack [53].

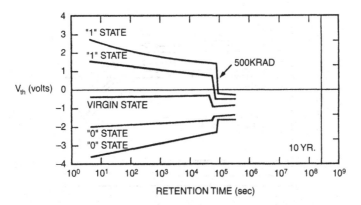

Figure 7-13. Irradiation effects on SNOS threshold voltage for transistors written into different initial states as a function of retention time. (From [54], with permission of IEEE.)

based memory circuits may degrade the overall radiation characteristics. Hence, the radiation hardness (both dose rate upset and total dose) of the SNOS memories for write/erase operations is limited by the hardness of the peripheral circuitry. Neutron radiation has no significant effect for SNOS devices beyond that observed for MOS transistors. The SNOS transistors are not prone to "soft errors" by high-energy heavy ions. However, under certain conditions, the SNOS transistors are susceptible to permanent damage, "hard errors" from high-energy particles [57]. For an SNOS device, the write and erase operations induce a high field across the memory stack. An exposure to ionizing radiation during the presence of this field can create a conductive path discharging the capacitor formed between the gate and the substrate. This discharge energy can be high enough to cause local heating, and form an electrical short between the gate and the substrate which may prevent the transistor from being written or erased [58].

In general, the radiation characteristics of unhardened MOS masked ROMs and fusible-link-based PROMs are the same as for MOS devices. Bipolar PROMs have a higher total dose radiation tolerance compared to CMOS PROMs. Radiation testing of floating-gate nonvolatile memories such as EPROMs and EEP-

ROMs have shown them to be susceptible to parametric and functional degradation. Since floating-gate devices such as EPROMs (or UVPROMs) can be erased by ultraviolet illumination, it would be expected that exposure (and hence erasure) may occur in the ionizing radiation environment also. A charge loss occurs over a period of time, causing a shift in the memory threshold voltage which eventually results in a loss of memory content in the programmed cells. The amount of voltage (or charge) on the floating gate above the minimum needed to differentiate a logic "1" from a logic "0" is defined as the cell margin voltage.

A UVPROM cell is programmed via hot-electron injection from the depletion region onto the floating gate. Ideally, under normal operating conditions, the charge is supposed to remain on the floating gate permanently. However, in actual practice, the charge leaks off the floating gate and tunnels into oxide where recombination takes place. Additional charge loss can occur on exposure of UVPROMs to ionizing radiation that has a sufficient amount of energy required to free the remaining charge. Figure 7-14 shows the threshold voltage shift of a floating-gate transistor as a function of total dose irradiation for transistors that were programmed with different initial threshold voltages, and a comparison with model calculations [59].

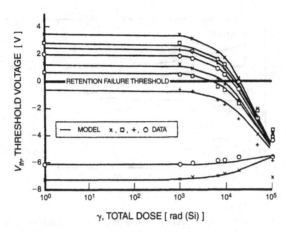

Figure 7-14. Threshold voltage shifts of a floating-gate transistor
as a function of total dose irradiation for transistors
programmed with different initial threshold voltages.
(From [59], with permission of IEEE.)

Neutron damage for floating-gate devices is the same as for MOS devices, which show very little sensitivity to neutrons. Neutron testing performed on Intel 64 and 128 kb HMOS UVEPROM samples showed that the range of neutron fluences for which memory bit failures were observed ranged from 8.4×10^{12} to 9.4×10^{12} n/cm^2 [60]. The neutron threshold for functional failures was slightly less than 10^{12} n/cm^2 for both UVEPROM types. Total dose failures were observed on roughly 20 krad (Si) exposure. Since the radiation response is highly dependent upon the device geometry and process variations, the thresholds may shift for different devices, and from lot to lot.

In general, the memories (including floating gate) on exposure to ionizing radiation may experience the following three types of observable degradation effects:

- Damage in the peripheral circuitry (hard errors)
- Upset of sense amplifiers/readout circuitry (soft errors)
- Data loss by transfer of charge from the floating gate.

The first of these degradation mechanisms is common to all MOS devices, and the second

is applicable to RAMs. The third type of defect, which is qualitatively different from the "soft errors" observed in RAMs, can be called a "firm error" [61]. This "firm error" has been defined as the change of data occurring as a result of charge transfer from a floating gate on exposure to ionizing radiation. Unlike soft errors which are caused by single ionizing particles, charge transfer to a floating gate is a cumulative effect of the ionizing radiation. The carriers collected on a floating gate may come from two sources. One source is the ionizing radiation creation of carriers in SiO$_2$ which lies between the floating gate and another electrode or the substrate when they are at different potentials. The second source is electrons excited in the floating gate which have enough kinetic energy to surmount the potential barrier between the conduction bands in Si and SiO$_2$. This "firm error" can cause a read error which can be corrected by rewriting the affected cell (or cells).

Total dose radiation testing on some floating-gate EEPROMs (SEEQ 28C256) have shown observed failures to be mode-dependent, occurring at roughly 33 krad (Si) for read operation and 9.5 krad (Si) for write operation [62]. During the write mode, failures occur earlier at higher dose rates such as 11 rad (Si)/s compared to a lower dose rate of 0.1 rad (Si)/s. The prompt dose and latchup testing showed the average threshold as 3.8×10^8 rad (Si)/s and 7.7×10^8 rad (Si)/s, respectively. Total dose radiation testing on Westinghouse SONOS EEPROMs (W28C64) has demonstrated these parts as functional to 1 Mrad (Si) with reprogramming at 100 krad (Si) intervals. No latchup has been observed in the heavy ion or prompt dose environments. The SEU test data show LET thresholds of 60 MeV/[mg/cm^2] for data latch errors. Hard errors were observed during the WRITE mode testing with krypton (and all heavier) ions.

7.3 RADIATION-HARDENING TECHNIQUES

Radiation hardness of the memories is influenced by a number of factors, both process- and design-related. CMOS is the dominant technology used in the fabrication of radiation-hardened memories, both in bulk process and CMOS/SOS (or CMOS/SOI). Radiation-hardening concerns and issues for the memories can be listed as follows:

- Gate oxide dielectric
- Field oxide region for bulk CMOS, and the back channel for SOS/SOI
- Latchup for bulk CMOS
- SEU.

Radiation-hardness process-related issues are discussed in Section 7.3.1, and circuit design issues, both total dose and SEP, in Section 7.3.2. Section 7.3.3 discusses radiation-hardened memory characteristics (Example).

7.3.1 Radiation-Hardening Process Issues

Several advanced processing techniques including E-beam and X-ray lithography, reactive ion etching (RIE), and other plasma processes which involve the use of high-energy photons and charged particles may result in process-induced ionizing radiation damage effects. In MOS devices, it means a buildup of positive charge in the oxide and an increase in the interface traps. This process-induced radia-

tion damage is not significant from the reliability point of view, as long as it can be completely removed by a subsequent annealing process. Thermal annealing is the most commonly used process. However, the applicable thermal annealing temperature may be limited in a given technology, resulting in an incomplete healing of the radiation damage. Studies have shown that the use of hydrogen during thermal anneal can significantly improve its effectiveness. There are other techniques, including the use of RF plasma environments, to remove radiation-induced charges and traps in silicon dioxide and at the SiO$_2$/Si interface. Some of the other process-related factors which affect the radiation response are substrate effects, gate oxidation and gate electrode effects, postpolysilicon processing, and field oxide hardening.

7.3.1.1 Substrate Effects. For bulk CMOS devices grown on different substrates such as <100> and <111>, studies have shown some intrinsic differences in radiation response, although not major ones. In general, the silicon surfaces should be free of contaminants before oxidation in order to produce consistently radiation-hardened oxides. The surface damage caused to silicon wafers by improper cleaning or polishing may result in a variability of the radiation response. Several studies have shown that postprocessing densities of silicon surface defects such as stacking faults and edge dislocations are associated with increased radiation-induced charge in the MOS oxides [63], [64]. The surface contaminants may also degrade radiation hardness. Also, the doping concentration affects the initial threshold (or flatband voltage), i.e., V_{th} of an n-channel MOS transistor can be increased by the substrate doping.

7.3.1.2 Gate Oxide (Dielectric) Effects. Total dose radiation response of the MOS devices such as memories is highly dependent on charge-trapping properties of the gate dielectric such as gate oxide thickness, oxide growth temperature and anneals, and the presence of impurities in gate oxide. The number of e–h pairs generated within the gate oxide of an MOS transistor under the gate electrode is directly proportional to its volume (and hence its thickness). Also, the number of holes distributed throughout the oxide thickness or trapped at the Si/SiO$_2$ interface will produce a threshold voltage shift linearly in proportion to the oxide thickness, $d_{ox}{}^n$. This power law dependence has been observed for gate oxide thicknesses typically in the range of 20–100 nm. The experimental value of n has been reported to range from 3 to 2 [65], [66]. For thicker oxides, the value of n more closely approaches 1 [67]. For thinner oxides (<20 nm), very little radiation damage has been observed. Clearly, significant gains in radiation hardness may be achieved by reducing the oxide thickness.

Another issue related to radiation response variability is the growth of oxides in dry oxygen versus those grown in wet oxygen or steam ambient. There have been conflicting studies, some of which claim that oxides grown in dry oxygen were harder, whereas others claim steam-grown oxides to have better radiation response. Test results published in the literature show that the hardness of steam-grown or dry-oxygen-grown oxides can be made comparable if the processing conditions for each are stabilized. For the dry-oxygen-grown oxides, the optimum temperature is about 1000°C, whereas for the steam oxides, it is in the range of 850–925°C.

Several studies have shown that postoxidation annealing of the gate oxide at temperatures greater than 925°C can significantly degrade the radiation hardness [64], [65], [68], [69]. At high temperatures, the hardness continues to degrade as the annealing time is increased. The postoxidation annealing ambient is usually nitrogen. However, no significant effects in radiation-hardness characteristics have been reported for samples annealed in argon instead of nitrogen at temperatures less than 950°C.

The contaminants or impurities that may be introduced during the oxide growth process affect the radiation response of the device. This approach has also been used for hardening the devices. For example, in the late 1960s and early 1970s, a phosphosilicate glass layer was deposited on top of the gate oxide, which improved the radiation response at positive bias, but degraded it under negative bias. The concept of dual dielectric structures used in the MNOS (and SNOS) memories has shown little functional degradation to very high total dose levels. For the development of radiation-hardened technologies, the use of other dielectrics, e.g., silicon oxynitride, has also been investigated [70]. Another promising insulator explored was aluminum oxide which, on irradiation, exhibited very small shifts under either positive or negative bias [71], [72]. However, for aluminum oxide insulators, the temperature-dependent charge injection for fields greater than roughly 1×10^6 V/cm caused long-term stability problems. The most practical approach to radiation hardening of gate dielectric has been the use of a pure, undoped thermally grown SiO$_2$ layer.

7.3.1.3 Gate Electrode Effects.

The most commonly used gate electrode materials have been metal (aluminum) and polysilicon. Studies have shown that metal gate devices have better radiation characteristics than ones with polysilicon gates [64], [73]. The polysilicon gates are preferred because of their enhanced density and performance. Figure 7-15 shows the radiation-induced threshold voltage shifts for the n- and p-channel transistors processed in a metal gate and in polysilicon gate technology [73]. However, it has been shown that it is not the gate material itself which is responsible for variation in the radiation response, but the processing differences between the two technologies. For the metal gate technology, the gate oxide is grown very near the end of the wafer processing sequence, whereas for the polysilicon gate technology, the dielectric is grown fairly early in the process.

Gate dielectrics made of refractory materials such as tungsten, molybdenum with lower resistivity than polysilicon are also being used. Studies with tungsten on tungsten silicide on top of silicon dioxide gate dielectrics have shown a significantly reduced number of interface traps [74]. It has been shown that this reduction is related to the stress induced in the oxide film layer by overlaying films [75]. Therefore, the use of

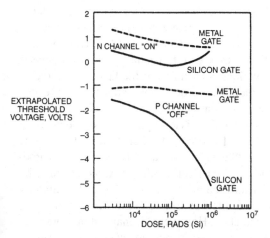

Figure 7-15. Radiation-induced threshold voltage shifts for n- and p-channel transistors processed in a metal gate and in poly-silicon gate technology. (From [73], with permission of IEEE.)

refractory materials for gate electrodes appear to improve the radiation hardness.

7.3.1.4 Post-gate-Electrode Deposition Processing.

After the growth of gate oxide, the temperature and duration of high temperature for subsequent processing steps should be minimized, for optimizing the radiation characteristics [64], [76]. Studies have shown that after gate oxidation, the processing temperatures should be kept below 900°C. Ion implantation, often through the gate oxide, is commonly used in the processing sequence to control the threshold voltage of the MOS transistors and for doping the source and drain regions of the transistors. In general, this ion implantation has been found to degrade the radiation hardness [77]. This damage is not removed by the anneals, at least up to 850°C.

Some metal deposition processes can cause damage that can be subsequently annealed through sintering of the final metallization which is performed to reduce the postprocessing interface charge and improve the contact between the metal and underlying layers. Studies have shown that sputtering of the metal directly onto the gate oxide causes a radiation-hardness degradation

compared to the metal deposition by the thermal evaporation process [78], [79]. However, this degradation due to the sputtering of metal can be avoided if the gate oxide is covered by a polysilicon electrode. The time and temperature of the sintering process are also critical parameters which can affect the stress levels in the oxides, and need to be controlled.

One of the final wafer fabrication processing steps is the deposition of a passivation layer as protection from contamination and handling damage. Phosphorous-doped glass has been the most commonly used passivating layer material. Another approach has been the use of silicon nitride deposited by plasma-enhanced chemical vapor deposition (PECVD). Studies have shown that this process can significantly degrade the radiation characteristics [80], [81].

7.3.1.5 Field Oxide Hardening.

In addition to hardening of the gate oxides, the field oxide also has to be hardened. Field oxides for the bulk devices are typically 0.8–1 μm thick. One of the major failure modes for bulk MOS devices in a radiation environment is the field inversion of p-substrate, and for n-substrates, the radiation-induced threshold voltage shifts drive the surface deeper into accumulation. This radiation-induced field inversion usually occurs below the interconnections, particularly where the gate electrode (deposited over the gate oxide) protrudes into the field oxide region, resulting in a parasitic shunt across the device. Several techniques have been used for field oxide hardening. One of them has been the use of closed geometry layout in which the field region between the drain and source is eliminated, although it places severe limitations on the circuit layout. Another option has been the use of a diffused guardband to place a heavy doped region in the gate field transition such that the field threshold voltage is several hundred volts, i.e., well above the radiation-induced shift. However, with shrinking device geometries, penalties for both of these approaches are severe. Currently, the most popular approach is to use hardened field oxides by employing techniques similar to those for the hardened gate oxides [82].

7.3.1.6 Bulk CMOS Latchup Considerations.

The bulk CMOS circuits contain parasitic bipolar n-p-n and p-n-p transistors which can form a p-n-p-n type of SCR structure in parallel with each CMOS inverter, as shown in Figure 7-16. The vertical n-p-n transistor is formed by n^+ source–drains (S/D) into the p-well (emitter), the p-well itself (base), and the n-substrate (collector). The lateral p-n-p structure is formed by p^+ S/D into the substrate (emitter), the substrate (base), and the p-well (collector). The latchup path is characterized by the value of the holding current, I_h, which is the minimum current required to maintain the latched state. During the normal IC operation, this p-n-p-n structure stays in its normal high-impedance state with n-p-n beta (β) being above 30, whereas the lateral n-p-n β is roughly 0.1. However, a transient radiation pulse (in a gamma dot environment) or a heavy ion can forward-bias this p-n-p-n structure by forcing it into a low-impedance state (or a latchup). For this latchup to occur, the closed-loop gain of this p-n-p-n structure must exceed unity. Under these conditions, the radiation-induced latchup can be destructive if no current limiting is used. This can occur for gamma dot levels as low as 1×10^9 rad (Si)/s.

In bulk CMOS devices, "latchup windows" have also been observed, which means that the device initially latches up at a given transient dose rate, then reaches a point (on the higher dose rate curve) when it no longer latches up, but then again experiences a latchup as the dose rate further increases [83]. The optimum fix for bulk CMOS latchup would be a generic solution independent of the layout and process variability. A process solution which is effective in preventing latchup under certain design rule considerations is through control of the minority carrier lifetime [84]. For 5 μm design rule CMOS devices, the parasitic transistor's β product will be less than unity, provided the minority carrier lifetime is less than 10 ns. This minority carrier lifetime can be controlled with neutron irradiation. However, this may not be a practical solution, and is not adequate for 3 μm or smaller design rules.

The most common process-oriented approach for latchup prevention is the use of an epitaxial (epi-) layer over a heavily doped substrate. This provides a shunt for the p-n-p emitter that will prevent forward bias from an IR drop. This technique has proved quite effective for technologies scaled up 1.5 μm. However, the epi-layer thickness should be minimized to reduce the shunt resistance. It is also recommended that designers use enough well and substrate connections to reduce the parasitic resistances.

A system approach implemented for handling latchup typically consists of either current limiting or power cycling. Current limiting, as the name implies, prevents parasitic SCR from attaining current levels that are necessary to sus-

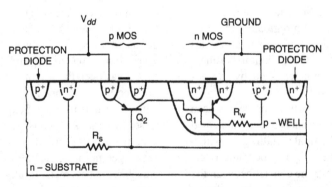

Figure 7-16. A typical p-n-p-n type structure consisting of a vertical n-p-n and a lateral p-n-p parasitic transistor in a bulk CMOS inverter.

tain latchup. Power cycling turns the device "OFF" on detecting latchup conditions (e.g., current exceeding a certain threshold value).

7.3.1.7 CMOS SOS/SOI Processes.
CMOS SOS/SOI technologies utilize insulator isolation as opposed to junction isolation for bulk technologies, and therefore offer a substantial advantage in latchup, transient upset, and SEU characteristics. The SOI devices have a "gate electrode" on both sides of the back channel oxide (usually the silicon substrate), whereas the SOS substrate is a complete insulator. Figure 7-17 shows a cross-sectional view of an n-channel SOI transistor and the location of radiation-induced backchannel leakage current [85]. This leakage current is caused by the positive charge accumulation in sapphire near the surface region of the silicon–sapphire interface and is process-dependent. The sensitivity of this back silicon surface/insulator interface can be significantly altered by the processing conditions utilized in its formation.

Several other process-related factors that are important in radiation response and need to be controlled are: silicon growth rate, ambient of anneal performed before the growth of epitaxial silicon layer, and control of surface contamination. Studies have shown that a hydrogen anneal with a burst growth of the silicon layer produces the best silicon/sapphire interface with the lowest radiation-induced charge trapping [86]. Also, the optimization of silicon islands on sapphire through laser annealing can improve channel mobilities with no radiation hardness degradation [87]. During the gate oxidation process, the wafers that are oxidized in steam at 875°C have superior backchannel properties (and hence better radiation response) than those processed in dry oxygen at 1000°C [88].

Techniques such as deep boron implants into the silicon island have been used to increase the threshold of the backchannel leakage path and improve radiation hardness [89], [90]. This selective boron implant has also been used to increase radiation hardness of the soft sidewall devices. In the SOS devices, sidewalls where the gate electrode covers the gate oxide in the channel width region may be of <111> orientation. The higher interface state densities with this

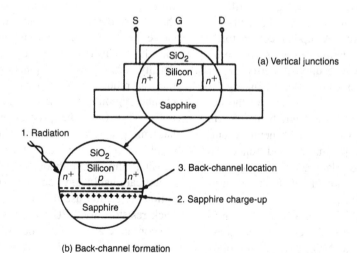

(a) Vertical junctions

1. Radiation

3. Back-channel location

2. Sapphire charge-up

(b) Back-channel formation

Figure 7-17. (a) Cross-sectional view of an n-channel SOI transistor. (b) Location of radiation-induced backchannel leakage current at silicon–sapphire interface. (From [85], with Permission of IEEE.)

<111> orientation can make this edge more prone to radiation-induced edge leakage. This effect is minimized by proper selection of the boron doping level along the device edge. There are also other process techniques available for hardening the sidewalls [91].

7.3.1.8 Bipolar Process Radiation Characteristics.

In general, bipolar devices are less sensitive to ionizing radiation than MOS circuits since their operation does not depend upon the surface potentials, except for some isolation techniques. However, they operate with both majority and minority carriers, and therefore exhibit higher sensitivity to neutrons. Commercial bipolar devices, especially ECL and CML technologies, are quite hard to both neutron levels and total dose. Typical neutron failure levels for these devices are in the range of 1×10^{14}–1×10^{15} neutrons/cm^2, and the total dose can exceed 1 Mrad (Si).

The major bipolar processes identified by their isolation techniques are standard buried collector (SBC) oxide isolation and the dielectric isolation (DI). Oxide isolation has been the most popular choice for bipolar circuits because of its high packing density. The DI technology, which does not have much commercial potential, was developed as a radiation-hardened bipolar process to minimize the photocurrents and eliminate latchup. A bipolar device tolerance to neutron irradiation can be improved by reducing its dependence upon minority carrier lifetime, which is accomplished primarily by minimizing its base width. In general, the devices fabricated with a base width of 0.5 μm would be tolerant to 1×10^{14} neutrons/cm^2 without significant parametric and functional degradation. Neutron damage in a bipolar device anneals as a function of both time and emitter current density [92].

The bipolar devices in a total dose environment are sensitive to the emitter–base leakage currents (as are the MOS devices), although their degree of tolerance is higher. A bipolar device latchup can be prevented by the use of DI technology. A system level solution to bipolar latchup is the use of power cycling upon detection of a transient event.

7.3.2 Radiation-Hardening Design Issues

Section 7.3.1 discussed the effects of various IC (memory) fabrication steps on the radiation response, and some radiation-hardened technologies. However, radiation hardening process considerations require certain tradeoffs between meeting the desired hardness goals versus long-term memory reliability. For example, the radiation hardness of MOS devices may be substantially improved by reducing the gate oxide thickness. However, thin gate oxide devices are susceptible to failure mechanisms such as the time-dependent dielectric breakdown (TDDB) in the presence of high dielectric fields and pinhole defects. This means that a combination of reduced gate oxide thicknesses and defects can lead to higher failure rates and lower yield. Therefore, the gate oxide thickness optimization should take into consideration long-term memory reliability.

The radiation hardening of a memory process also involves achieving a proper balance between the desired hardness goals and device performance. For hardened technologies, the threshold voltages of n-channel MOS transistors are established higher (relative to those for commercial technologies) to compensate for the postirradiation threshold shifts which can cause the devices to go into the depletion mode [93], [94]. This is done by increasing the doping density, which increases junction capacitances that can reduce the mobility of electrons in the channel region and cause performance degradations. Therefore, the process-level radiation-hardening approach requires an appropriate combination of gate oxide thicknesses, doping density, and preirradiation threshold voltage specifications. In addition, several other modifications (e.g., the use of guardbands, and feedback resistors for SEU hardness) compared to the commercial baseline process can lead to additional process complexities. A suitable mix of the process-related approach and optimized design is integral to radiation hardening [98]. Memory design considerations for radiation hardness, both total dose and SEP, are discussed in the following sections.

7.3.2.1 Total Dose Radiation Hardness.
Radiation-induced failure mechanisms can be classified into three categories: (1) power-supply-related failures, (2) logic level failures, and (3) timing failures [95]. Power-supply-related failures can occur if the entire chip leakage current through the n-channel devices exceeds the maximum allowable supply current. This failure mechanism is relatively simple and does not require detailed circuit analysis. Logic level failures may occur if the n-channel leakage currents coupled with reduced p-channel drive prevent the signal at a critical node from reaching a valid HIGH logic level. Logic level failures depend on the n-channel versus p-channel transistor transconductance ratios. Timing failures will occur at the nodes where the reduced p-channel drive capability leads to the degradation of a "low-to-high" transition, and have the same worst case irradiation biases as logic level failures.

The design of high-performance radiation-hardened devices requires circuit characterization and assessment which is usually done through simulation. SPICE is the commonly used circuit simulator which can generate multiple sets of device parameters to accurately model pre- and postirradiation characteristics. These pre- and postirradiation parameters also depend upon the biasing configuration. Figure 7-18 shows a schematic of a simple CMOS RAM cell irradiated under the worst case bias-ing conditions [96], [97]. Threshold voltage shifts will be greatest for the n-channel transistor $N1$ and p-channel transistor $P1$. A circuit simulation requires the designer to define each transistor with a unique model that would accurately simulate its characteristics. This may be done by creating at least three sets of device models to cover the worst, nominal, and best case processing parameters [97]. Also, the designer must define a realistic set of voltage and temperature conditions for the design simulation.

Figure 7-19 shows an example of a simple SRAM circuit that will be used to illustrate total dose radiation-hardening design techniques [96]. This SRAM uses the technique of address transition detection (ATD) to precharge the RAM bit lines to a high state before a read or write operation. As shown in the upper half of the ATD circuit, a low-to-high transition on the address input A_0 will generate an active low precharge pulse, ATD_1. Similarly, in the lower half of the same ATD circuit, a high-to-low transition will generate an active low precharge pulse, ATD_2. All precharge signals are inverted, then NOR-gated to create a single, active, low precharge pulse. This precharge pulse will activate p-channel pull-up devices ($P10$ and $P20$) to charge the bit lines to a high state. Using this simulation result, the preirradiation minimum pulse width required to fully charge the bit-line capacitance can be found. This time to

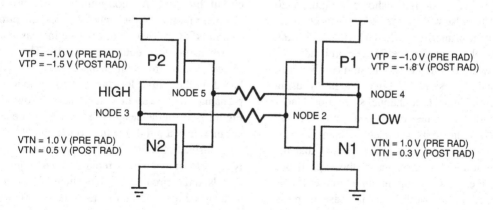

Figure 7-18. Schematic of a CMOS RAM cell irradiated under worst case biasing conditions.
(From [96], IEEE tutorial, with permission of IEEE.)

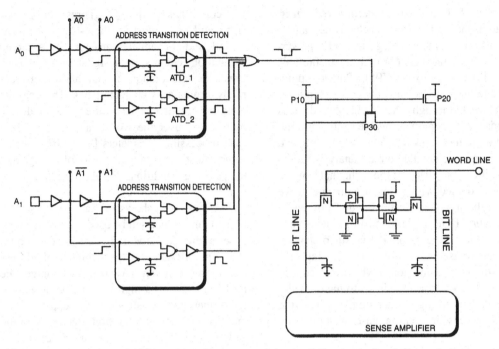

Figure 7-19. A simple SRAM circuit to illustrate total dose radiation-hardening design techniques. (From [96], with permission of IEEE.)

precharge the bit lines increases with irradiation due to a decrease in the drive of the precharge p-channel devices. Therefore, the postirradiation precharge pulse width should also increase to compensate for the additional time required for precharge.

A more practical three-inverter ATD circuit for memory design is shown in Figure 7-20 [96]. The pulse width generated by this circuit is dependent primarily on the capacitance $C1$ and p-channel drive of device $P1$. The preirradiation pulse width of this inverter design can be made identical to a single-inverter design by using proper p-channel to n-channel size ratios. Under postirradiation conditions, the drive of $P1$ will be reduced due to its degraded p-channel threshold voltage and mobility. This means that the time required for charging of the capacitance will increase, resulting in an increase in the precharge pulse width. This increase in pulse width is adequate for the precharging of memory bit lines under postirradiation conditions.

Therefore, it is considered a radiation-tolerant design, although with an increased chip area.

The next stage of the precharge circuit and its associated timing diagrams are shown in Figure 7-21 [96]. This circuit inverts all ATD signals, performs a NOR operation on them, and generates a single precharge pulse. Since the circuit has a series configuration and lower channel mobility of p-channel devices, the p-to-n ratio NOR transistors have to be increased as shown to produce equal drives [96]. Thus, by selecting the proper n-channel device width based on the output loading, the corresponding p-channel width can be calculated as shown in Figure 7-21. In the design of radiation-hardened circuits, it is a good practice to use the NAND gates instead of NORs. Therefore, the four-input NOR gate shown can be replaced by an equivalent NAND circuit which occupies less chip area, has higher radiation tolerance, and produces a more symmetrical output wave under postirradiation conditions.

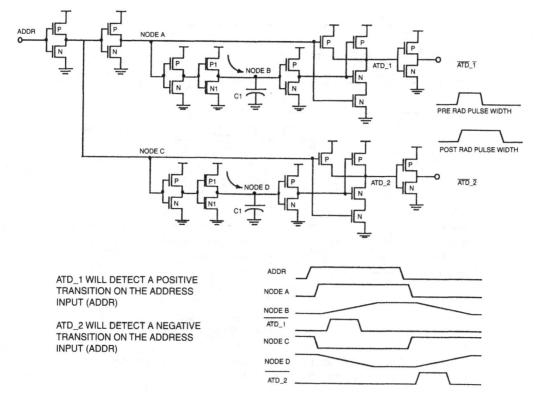

ATD_1 WILL DETECT A POSITIVE
TRANSITION ON THE ADDRESS
INPUT (ADDR)

ATD_2 WILL DETECT A NEGATIVE
TRANSITION ON THE ADDRESS
INPUT (ADDR)

Figure 7-20. A three-inverter address transition detector (ATD) circuit radiation tolerant design. (From [96], with permission of IEEE.)

Memory cells are quite sensitive to imbalances in the threshold voltages due to biasing conditions during irradiation. Therefore, simulations should be performed on the memory cells to ensure their postirradiation stability during read or write operations. Once the simulated memory cell has been determined to be stable with balanced postirradiation device threshold voltages, the cell must be evaluated for the effects of imbalanced thresholds due to worst case biasing conditions during irradiation. For the circuits which depend on threshold voltage matching, bias-related shifts are very critical. Therefore, if the imbalanced memory cell read margin simulation is not performed during the design phase, the end result may be lower than expected radiation failure limit.

For designing radiation-hardened memories, the postirradiation performance that a given fabrication process provides should be first characterized. Then, given these postirradiation parametric shifts, the basic design elements (e.g., transistors) can be structured to provide radiation tolerance. In addition to the radiation-hardening approach, the normal design variables such as temperature effects, circuit fan-out, and process variations also have to be taken into account. After the circuit design and simulations are complete, some additional layout constraints are imposed to enhance latchup protection and total dose radiation performance. Since oxides usually trap a net positive charge on exposure to ionizing radiation, p-type surfaces are susceptible to inversion. Therefore, a heavy p^+ guardband surrounds all p-wells and between all n-channel transistors, and the hardened gate oxide of the n-channel devices is extended from transistors to the guardband under the polysilicon gate to prevent the p-well from inverting.

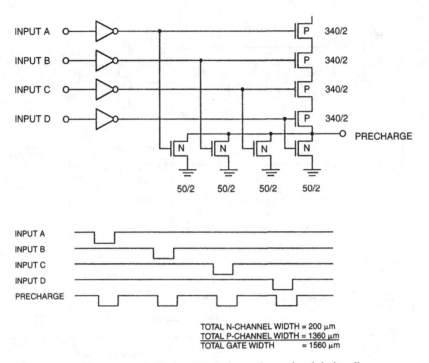

TOTAL N-CHANNEL WIDTH = 200 μm
TOTAL P-CHANNEL WIDTH = 1360 μm
TOTAL GATE WIDTH   = 1560 μm

Figure 7-21. A precharge circuit for ATD design and associated timing diagrams. (From [96], with permission of IEEE.)

Figure 7-22 shows an example flowchart for a methodology of evaluating the worst case total dose radiation environment failure levels for CMOS circuits [95]. The basic approach consists of first identifying the most sensitive nodes in the circuit by approximate calculations. In the following step, all critical subcircuits surrounding the sensitive nodes are extracted; the irradiation and postirradiation operating points are selected to produce worst case conditions at the sensitive nodes. Finally, detailed simulations of subcircuits are performed using the appropriate postirradiation device parameters to characterize the radiation tolerance of the entire circuit.

Figure 7-23 shows a schematic of a single column of a basic six-transistor cell memory circuit using n-channel access transistors whose gates are driven by the word lines (not shown). In accordance with the flowchart methodology discussed above, the bit lines are identified as the most sensitive nodes for both logic level and timing faults due to the large bit-line capacitance and a number of potential leakage paths. Then, following the rules for sensitive destina-

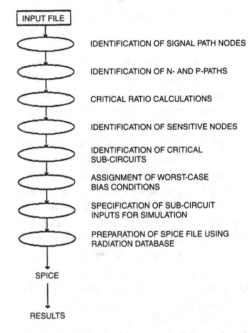

Figure 7-22. Flowchart for a methodology of evaluating worst case total dose radiation environment failure levels. (From [95], with permission of IEEE.)

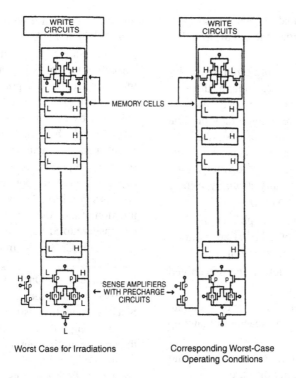

Figure 7-23. Schematic of one column of a six-transistor cell memory. (From [95], with permission of IEEE.)

tion nodes, the memory cell transistors providing n-paths and p-paths through the access transistors are assigned the maximum parameter shifts. The n-channel access transistor (assumed active) is assigned the OFF-bias parameter shift, and the remaining access transistors the ON-bias parameter shifts. For logic level failure simulations, the active memory cell is assigned a HIGH state, and the remaining cells a LOW state to produce maximum leakage current at the sensitive node. For timing failure simulations, all memory cells are assigned a LOW state initially, and then the active cell is flipped before performing a read operation.

In general, for designing memory circuits which are tolerant of the effects of total dose radiation, the following effects are usually taken into consideration:

• n-channel transistor drive increases with increasing total dose, whereas p-channel transistor drive decreases.

• Threshold shifts are more severe for n-channel transistors which are biased "ON" and p-channels which are biased "OFF" during irradiation.

• n-channel transistor subthreshold leakage current increases with increasing total dose irradiation.

• "Body effect" (when the source and substrate are at different potentials) also influences threshold, especially for n-channel transistors.

• Mobility degradation due to total dose radiation influences the current drive of both n- and p-channel transistors.

Some of the good design practices recommended by the manufacturers of radiation-hardened memories for space applications are listed below:

• Synchronous circuits are preferred over asynchronous designs. If asynchronous design is unavoidable, then care must be taken to desensitize the circuit to radiation.

- Static circuits are preferred over dynamic designs, even though there is some area penalty (because of extra transistors required) since they provide an additional measure of design margin in the radiation environment. An active transistor controls the logic voltage state of each node, typically at supply voltage rails (in CMOS devices).

- Differential sense amplifiers for increased speed should be biased symmetrically when not in operation.

- RAMs should be left in the precharged state when deselected.

- Use of frequent well and substrate straps, large n^+ to p^+ spacing for latchup immunity.

- Uniform distribution of power and ground pads around the chip.

- Minimize wired-OR circuits.

- Use split gate transistors to maximize gain while minimizing drain diffusion area.

- Minimize the number of transistors used in series (especially p-channel), limiting to no more than three in series.

- Minimize the use of transmission gates, limiting to just one transmission gate in series, and use full complementary pairs.

- Limit fan-out to approximately four critical paths. Use conservative drive-to-load ratios.

7.3.2.2 Single-Event Upset (SEU) Hardening.

There is a continuing trend toward decreasing the minimum feature size of memories in order to minimize the parasitics for higher performance and maximize the circuit packing density. This scaling affects the radiation hardness of circuits. In an SEU environment, the critical charge (Q_c) for upset has been shown to decrease with scaling, and the scaled designs are relatively insensitive to specific device technology. When a SRAM is hit by a high-energy particle (single ion), the predominant self-correcting recovery mechanism versus upset reinforcing feedback response determines whether the cell will experience a logic upset. The charge collected by this single event creates a voltage transient which is roughly proportional to the target node capacitance. Figure 7-24 is the schematic representation of a bulk CMOS memory cell and corresponding n-well technology cross-section [96]. As shown, the reverse-biased junction of the "OFF" transistors $P1$ and $N2$ will be sensitive to SEU.

Bulk CMOS memories have been hardened to SEU by increasing the delay in the feedback inverters. One of the most common approaches to hardening a SRAM cell against SEU is through resistive hardening, which involves using polysilicon decoupling resistors in the cross-coupling segment of each cell. A combination of these resistors with gate capacitance creates an RC time constant between the target node and feedback inverter. However, the use of high-value resistors increases the switching time constant, which slows the circuit response and increases the minimum cell write time. This effect may worsen at low temperatures because lightly doped, high-resistivity polysilicon has a negative temperature coefficient [99]. Therefore, the selection of proper value resistors may require several iterative simulation cycles during the memory design for SEU hardening. The designer has to specify the dimensions of critical volume (e.g., drain diffusion of an n-channel transistor), the critical charge required to cause a cell upset, and the cosmic ray environment. Then, by using one of several environmental models mentioned earlier (CREME, CRIER, etc.), the number of particles that have sufficient LET $\geq Q_{crit}$ in the given sensitive volume can be calculated.

This procedure can be used to generate a plot of critical charge versus error rate as shown in Figure 7-25 [96]. This curve specifies the amount of Q_c that would be required to produce a specific error rate. For example, the error rate of 5×10^{-9} errors/bit-day corresponds to a Q_c

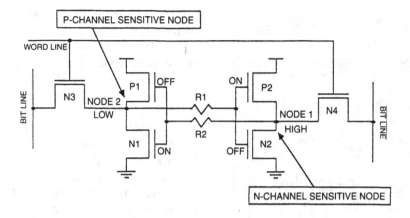

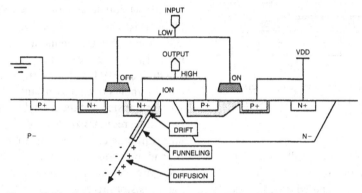

Figure 7-24. Schematic representation of a bulk CMOS memory cell and corresponding n-well technology cross-section. (From [96], with permission of IEEE.)

of 2.0 pC. Therefore, the cell feedback resistor values must be selected to withstand a 2.0 pC "hit." It should be noted that Q_c corresponding to a given error rate is memory-cell-design-specific. Typical SEU modeling steps for a given error rate corresponding to a Q_c of 2.0 pC required for the calculation of feedback resistor values can be summarized as follows:

- The memory cell feedback resistors are arbitrarily set to low values.
- Memory cell transistors are set up with worst case bias-dependent threshold voltages.
- SEU simulation is performed with V_{cc} = 4.5 V and at 125°C to yield the worst

case drive for the "on" P-channel transistor.

- If these conditions cause a cell upset, feedback resistor values must be increased for longer *RC* delay.
- This simulation process is repeated until the proper value of the resistor that would prevent upset is found.
- After the proper resistor value has been determined, the complete memory cell and peripheral circuitry must be simulated.

As the memory cell resistance increases (it is highest at the lowest temperature), the write time also increases. The read time is not

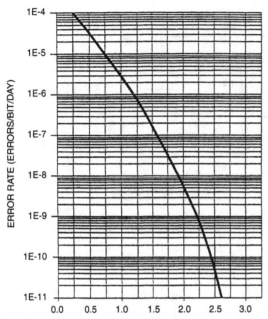

Figure 7-25. A plot of critical charge (Q_c) versus error rate. (From [96], with permission of IEEE.)

high ρ polysilicon resistors [100]. Therefore, a certain tradeoff has to be made between the SEU hardness (desired bit-error rate in a given environment) and the write time specifications.

The resistive load SEU hardening technique has been implemented in several variations. Figure 7-27(a) shows a schematic of a conventional SRAM cell with feedback resistors (R_F). On the left side, the voltage transient shown is for a strike on the "off" p-channel drain node, and represents the case where the cell recovers from the strike [101]. On the right side, the voltage transient at the unstruck node represents the case where SEU occurs. Figure 7-27(b) shows a variation on this approach by an LRAM cell schematic in which the decoupling resistors are used only to protect against short n-channel transients, and longer persisting pulses are reduced in magnitude by a voltage divider [101]. This technique is based upon measurements of time constants of ion-induced voltage transients at the n- and p-channel drains (sensitive regions) of SRAM transistors. The large differences between the time constants for n- and p-channel drains is the basis for the LRAM concept which allows the use of feedback resistors five–ten times smaller than cur-

affected by the use of these feedback resistors. Figure 7-26 shows the read and write cycle times for a memory with and without the use of

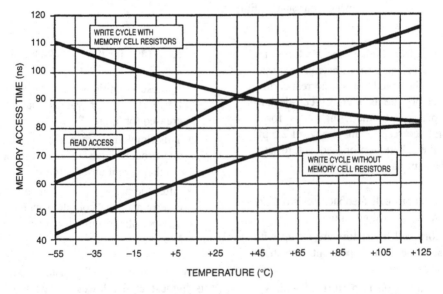

Figure 7-26. Read and write cycle times for a memory with and without the use of cell resistors. (From [100], with permission of IEEE.)

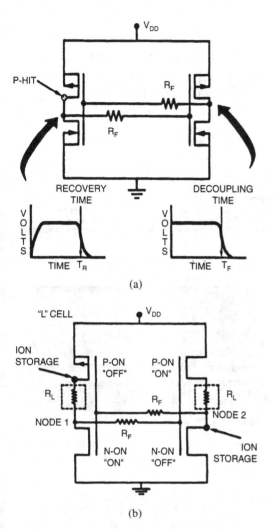

Figure 7-27. (a) Schematic of a conventional SRAM cell with feedback resistors (R_F). (b) Schematic of an LRAM cell. (From [101], with permission of IEEE.)

rently used in SRAMs with equivalent SEU tolerance.

The LRAM cell shown uses two sets of resistors: (a) R_F to decouple two inverters and protect against strikes on n-channel transistors, and (b) R_L as a voltage divider between the "on" n-channel transistor for limiting the voltage amplitude at nodes 1 and 2 to values below the switching point. This LRAM cell design uses $R_F = R_L$.

Another example of resistive load-hardening technique uses a logic configuration which separates p-type and n-type diffusions nodes within the memory circuit [102]. Figure 7-28 shows a schematic of an improved SEU tolerant RAM cell design which uses the following three basic concepts: (1) use of redundancy in the memory circuit to maintain a source of un-corrupted data after an SEU occurrence, (2) data in the uncorrupted section to provide feedback for corrupted data recovery, and (3) a particle-hit-inducing current which always flows from an n-type diffusion to p-type diffusion [102]. The RAM cell shown consists of two storage structures which have complementary transistor pairs $M6/M7$ ($M16/M17$) inserted between power supply V_{DD} (V_{SS}) and n-type (p-type) memory structure. This enables dc paths to be discon-nected, reducing the static power consumption not only during normal operation, but also during the SEU recovery, and improves recovery time. The output is divided into a p-driver and an n-driver so that full SEU protection can be achieved when two RAM cells are combined to

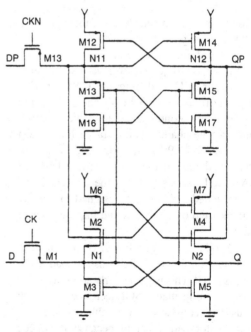

Figure 7-28. Schematic of an improved SEU toler-ant RAM cell design. (From [102], with permission of IEEE.)

form a flip-flop. SPICE simulation performed on this cell showed a recovery time of less than 2 ns from peak to midvoltage of 2.25 V, and under worst case conditions (V_{dd} = 5.5 V, T_j = −55°C, and three σ parameters), drew about 2 nA current.

A major concern with the use of polysilicon intracell resistors is their sensitivity to variations in process controls (e.g., doping concentrations) and operating conditions such as variations in temperature. The scaled (high-density) memories which exhibit greater SEU sensitivity require larger intracell resistors that can slow down memory write-cycle time. One technique offered to mitigate those effects is the use of gated resistors (instead of passive), which are actively clocked polysilicon resistors built in two polysilicon layers and separated by a thin interlevel thermal oxide layer. These high-resistance polysilicon resistors are clocked into a low-resistance state during write-cell operations to minimize the write time delays. Figure 7-29(a) shows a schematic of a gated-resistor-hardened CMOS RAM cell design in which the first polysilicon layer forms the conducting plane of the device and the second level of polysilicon forms the gate electrode of the gate resistor [99]. These poly-2 gate electrodes are tied to the row word line of the cell array.

The main objectives for the use of gated resistors for SEU hardening are: (1) to provide adequate off-state channel resistance, and (2) to maximize transconductance to achieve maximum on-state channel conductance for highest speed. Figure 7-29(b) shows the cell layouts of unhardened and gated-resistor-hardened CMOS SRAM cell macros [99]. In the hardened cell version, gated resistors are designed to overlay the thick field oxide region and impose no area penalty.

Another technique for SEU hardening is charge partitioning (CP) for resistive-load MOS SRAMs [103]. This is particularly appropriate for the silicon-on-insulator (SOI) technologies which are usually considered unsuitable for conventional resistor load hardening because of reduced storage capacitances. The MOS SRAMs are susceptible to the depletion of stored charge on the vulnerable node by a single-event strike. If a por-

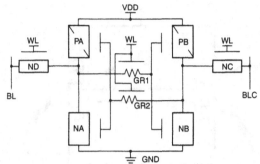

Legend:
R(?) -> Gated Resistors WL -> Wordline
P(?) -> PFETs in N-well at VDD BL -> Bitline
N(?) -> NFETs in P-substrate at GND BLC -> Bitline Complement

(a)

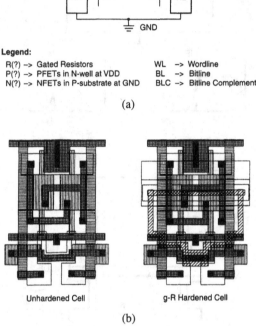

Unhardened Cell g-R Hardened Cell

(b)

Figure 7-29. (a) Gated-resistor-hardened CMOS SRAM cell design. (b) SRAM cell layouts: unhardened versus gated-resistor-hardened. (From [99], with permission of IEEE.)

tion of the charge representing stored information could be partitioned from the sensitive region, then this charge can be used to partially recharge the node after single-event decay. This is the basic concept of clock partitioning, as illustrated in Figure 7-30 [103]. The contact to each transistor gate is made within the high-resistance polysilicon load region, creating two series resistors. Although the resistors shown appear similar to conventional cross-coupled resistors in hardened SRAMs, or drain load resistors in *L*-hardened CMOS cells, the charge resupply mechanism is different from feedback decoupling. For the

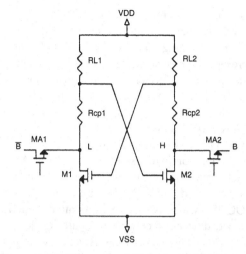

Figure 7-30. Schematic of CP-hardened RMOS RAM cell. (From [103], with permission of IEEE.)

RMOS SRAM SEU hardening, the addition of extrinsic cell capacitance has been recommended [104]. However, in SOI, this is unnecessary because of their much smaller junction areas (and associated capacitances); the gate capacitance can be several factors larger than parasitic drain capacitances. This technique, used in simulation of the RMOS SOI SRAM with CP resistors of 60K and 100K, has shown a gross error rate (CREME model) of 1.8×10^{-8} and 4.5×10^{-12} errors/bit-day, respectively [103].

To summarize, the memory design for SEP hardness requires extensive simulation for the integration and optimization of the process and circuit design techniques. Some of the process and design techniques typically used by manufacturers for high-reliability space applications are as follows:

Process Techniques

• Thin gate oxide.
• Use of p-well, twin-tub process with thin epitaxial layer, low well resistance, and high doping concentrations.
• Polycrystalline silicon resistors in cross-coupled cell segments.
• SOS/SOI substrates.

Design Techniques

• Special SRAM cells with cross-coupled resistors and cross-coupled capacitors.
• Use of high-drive circuits.
• Special flip-flops with additional transistors at critical nodes for partial or fully redundant designs. Other techniques include replacement of the NOR gates with NANDs, and transmission gates with clocked inverters.

Silicon bipolar and GaAs FET SRAMs have proven to be more difficult to harden with respect to SEU as compared to silicon CMOS SRAMs [125]. This is due to the fundamental property of bipolar and JFET (or MESFET) device technologies which do not have a high-impedance, nonactive isolation between the control electrode and the current or voltage being controlled. All SEU circuit-level-hardening techniques applied at the local cell level must use some type of information storage redundancy such that information loss on one node due to an SEU event can be recovered from information stored elsewhere in the cell. For example, in CMOS SRAMs, this can be achieved by the use of simple cross-coupling resistors, whereas in bipolar and FET technologies, no such simple approach is possible. The circuit simulations for proposed SEU-hardened bipolar and FET designs consume considerably more cell area than simple unhardened RAM cells, and have about twice the power dissipation of conventional SRAMs. An alternative approach to local cell level SEU hardening is the use of system level redundancy and error-correction techniques.

7.3.3 Radiation-Hardened Memories Characteristics (Example)

An example of bulk CMOS radiation-hardened memory is a 256K SRAM designed and fabricated in 0.8 μm enhanced radiation-hardened (RHCMOS-E) technology developed by Loral Federal Systems [105]. According to the manufacturer, this 256K SRAM has been used as a standard evaluation circuit (SEC) for further scaling to produce 1 Mb radiation-hardened SRAM using 0.5 μm ultra large scale integration (ULSI) technology CMOS process [106].

This 0.5 μm CMOS process supports a 3.3 V power supply, includes 0.5 μm contacts, 1.6 μm pitch, three levels of metal and one level of polysilicon. The 1 Mb SRAM is designed to be configurable to a 128K × 8, 256K × 4, or a 1 M × 1 organization. This fully asynchronous base design contains all logic circuitry, external data steering multiplexers and drivers to obtain the desired configuration through electrical programming at the wirebond pad level. There are 16 redundant column and 16 redundant row addresses available for replacement with fuse programming.

The 1 Mb SRAM chip uses a segmented precharge architecture which allows for lower power dissipations, a simplified timing scheme,

and a static page mode capability. The memory array is segmented into 16 sections, with each section containing 256 rows and 256 columns, representing physical address space for each bit. Each section is further divided into 8 blocks, containing 256 rows and 32 columns. Figure 7-31 shows the block diagram of this 1 Mb RH SRAM [106]. Table 7-9 shows major technology and process features [106].

Radiation hardness for this 1 Mb SRAM is achieved through an enhanced field oxide process based on local oxidation of silicon (LOCOS), which is used to eliminate radiation induced device-off current and parasitic leakage currents between devices. A radiation hardened gate oxide, and overall lower processing temperatures, control the radiation induced threshold voltage shifts. Thin epitaxial substrates and a retrograde n-well created by high

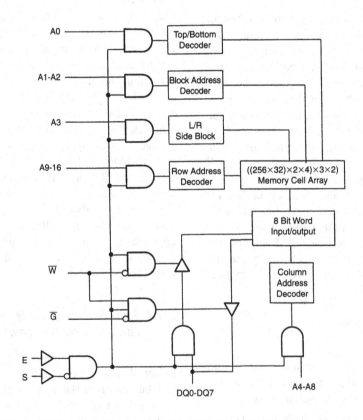

Figure 7-31. Block diagram of a 1 Mb RH SRAM. (From [106], with permission of Loral Federal Systems.)

TABLE 7-9. 1 Mb SRAM Radiation Hardened (a) Technology, and (b) Process Features. (From [106], with permission of Loral Federal Systems.)

Technology	0.5 μm fully-scaled CMOS	Contact size	0.5 μm × 0.5 μm
Organization	128K × 8, 256K × 4, 1M × 1	Diffusion space: p+ to n+	2.3 μm
Operation	Asynchronous, Synchronous	Diffusion space: same type	0.8 μm
Cell Size	10.0 μm × 8.9 μm	Polysilicon line width/space	0.6 μm/0.5 μm
Chip Size	12.0 mm × 11.6 mm	1st metal line width/space	0.8 μm/0.8 μm
Read/Write		2nd metal line width/space	0.9 μm/0.9 μm
Time (WC)	30 ns	3rd metal line width/space	2.0 μm/2.0 μm
Redundancy	16 Rows, 16 Columns	Via size	0.9 μm/0.9 μm
Power Supply	3.3 V ±5%		
I/O Levels	TTL/CMOS	(b)	
Package	40 pin flat pack		

(a)

energy ion implantation guarantee latchup immunity by lowering parasitic resistances and degrading parasitic bipolar gains. These features also limit photocurrent collection at circuit nodes, which enhances transient upset and SEU levels.

This 1 Mb SRAM has been subjected to various radiation testing environments such as total dose, prompt dose, and SEU testing. Total dose irradiation was performed at 200 rad (Si)/s up to exposures of 4 Mrad(Si) and 10 Mrad(Si), followed by 100°C biased annealing. During irradiation and annealing, the devices were held under worst case input (static high) with checkerboard data pattern loaded in the memory cells. Figure 7-32 shows the total dose effects on 1 Mb SRAM (a) address access time, and (b) active and standby currents [106]. No significant shifts in address

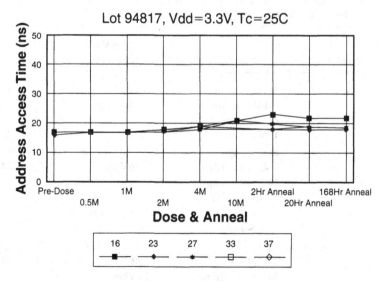

Figure 7-32. Total dose and anneal test results for 1 Mb RH SRAM. (a) Address access time.

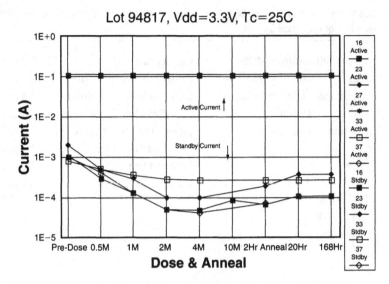

Figure 7-32 (Cont.). (b) Active and standby currents. (From [106], with permission of Loral Federal Systems.)

access time are observed. The reduction of standby current with radiation indicates that the initial leakage is caused by the aggregate effect of p-channel transistors conduction.

The SEU testing was performed at Brookhaven National Laboratory in both static and dynamic operating modes. No upset was detected up to an LET of 120 MeV/mg/cm² and a fluence of 1E7 particles/cm² at 125°C and worst case power supply. No latchup was observed. Table 7-10 summarizes test results under various radiation environments.

TABLE 7-10. A Summary of Test Results for 1 Mb RH SRAM Under Various Radiation Environments. (From [106], with Permission of Loral Federal Systems.)

Radiation Environment	Samples	Results	Voltage(Vdd)/ Temp.(Ta)	Test Facility/ Test Date
Total dose	5	No fail <10Mrad(Si)/s or after anneal of 168hrs @100°C	3.3V±5% −55 to 125°C	LORAL Gamma Source/ June '94
Prompt dose	5	No data upset <4E9rad(Si)/s No transient upset <1E9rad(Si)/s	3.15V for upset 3.46V for Latchup	Crane/ May '94
Prompt dose (survivability)	5	No fail <2E13rad(Si)/s	3.96V & 125°C	ARL/ June '94
Single event Phenomena (SEP)	5	LET >120MeV/mg/cm2 SER<1E-12error/ bit-day	3.15V for Upset 3.46V for Latchup	Brookhaven/ May '94

7.4 RADIATION HARDNESS ASSURANCE AND TESTING

Section 7.2.1 discussed: (1) space radiation environments, near earth (mainly the trapped radiation belts) which can produce large-dose buildup over several years of satellite mission time, although low-dose rates and simultaneous annealing effects limit the damage to some extent; (2) space probe environments, including the effects of the earth's radiation belts, solar flares, solar wind, and cosmic rays over long mission times; and (3) nuclear weapons environments (gamma dot) in which the total dose is delivered within a very short time period. Military Handbooks 279 and 280 define total dose hardness and neutron fluence level (respectively) assurance guidelines for semiconductor devices and microcircuits [107], [108]. Radiation hardness assurance requires establishing mission requirements, and formulating a plan to assure verification of those requirements through proper part selection, and lot sample radiation testing.

7.4.1 Radiation Hardness Assurance

The objective of the radiation hardness assurance program is to define the mission total dose radiation requirements based on the environmental models available which were discussed earlier, and to generate the dose–depth curves. These curves are then used for spacecraft shielding analysis, and to predict the total dose radiation that the spacecraft systems, subsystems, and components would experience over the mission lifetime. This detailed analysis provides the designers with an estimate of the total dose that semiconductor devices such as memories populating the board would receive. The component level environment R_{SPEC} is usually expressed in terms of dose, dose rate, and LET flux. The second factor is to define part failure level (accept/reject) criteria which, in the case of the memories, may mean functional failure or bit-flips. The third factor is part radiation response data for each environment in order to determine a failure threshold R_F in that environ-

ment. This R_F can be a nominal value, a worst case lower bound, or with a statistical distribution, depending on the part categorization criteria which are established to determine part suitability.

Three categories of parts are: (1) unacceptable, (2) acceptable with hardness assurance required (hardness-critical), and (3) acceptable with no hardness assurance required (hardness-noncritical). In a traditional approach, the part categorization criteria are based upon the radiation design margin (RDM), which is defined as the normal part failure level R_F divided by the radiation specification level R_{SPEC}. The method used to determine values of RDM which fall into three categories are based upon statistical variation (for a significant sample size) in part failure levels, the confidence and part failure rates required from the system. A commonly used value of RDM for a conservative approach is 2.

Figure 7-33 shows a block diagram of the traditional approach to memory selection and a hardness assurance program for space applications [109].

Acceptable memories may (or may not) require hardness assurance based upon a review of: (1) technology maturity and assessment, (2) the existence of a radiation characterization database, and (3) the identification of a part-specific failure mechanism such as a susceptibility to latchup. Memories which are characterized as hardness-noncritical can be used with no further evaluation. Some technologies (and device types) can be classified as hardness-noncritical-based on a general knowledge of radiation effects (e.g., technologies with no p-n-p-n paths such as CMOS/SOS are free of latchup). All MOS technologies may be considered hardness-noncritical for displacement damage (neutrons), and as such, no neutron testing may be required.

For memories that are hardness-critical, the most widely used technique for hardness assurance which can be applied for both part qualification and flight lot acceptability evaluation is lot sample radiation testing. This can be for the total dose effects or SEP, both of which are discussed in Section 7.4.2. In general, for hardness

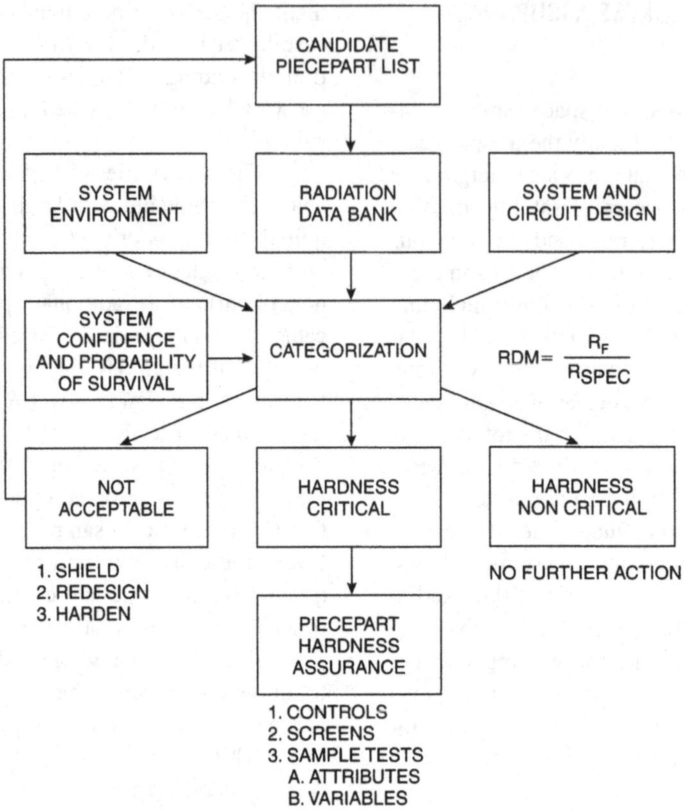

Figure 7-33. Block diagram of traditional approach to memory selection
and hardness assurance program for space applications.
(From [109], with permission of IEEE.)

assurance lot sample testing, applicable methods
are: (1) attribute testing, which involves pass/fail
decisions on each sample; and (2) variable testing to determine parametric degradation and the
functional failure level of each sample in a given
environment. For space applications, a variable
test data approach is recommended.

For a space mission, if a hardness-critical
memory fails lot sample total dose radiation
testing (i.e., it has an RDM of <2), the options
available are to: (1) provide shielding at the
board or component level to reduce total dose
accumulation, (2) change the part failure definition, and (3) substitute with a harder part.
Section 7.3 discussed radiation-hardening techniques, including process- and design-related issues that can be used as an evaluation criterion

for the selection of suitable memories. If the
memory substitutes do not meet system performance requirements (i.e., the lot sample radiation testing shows them to be below a specified
RDM), then the spacecraft shielding at the box,
board, or component level has to be evaluated.
The box and board level protection requires
shielding analysis using the models based upon
the radiation environment and spacecraft geometry. At the component level, designers may occasionally use "spot shielding" consisting of
aluminum foil (or other shielding material) of
appropriate thickness.

Another component of the total dose
shielding approach limited to a few part types
has been the use of devices packaged in RAD-
PAK™. These are specially designed packages

with shielding material that can attenuate proton dosage and mitigate the total dose effects. However, they do not offer protection from the high-energy particles (heavy ions) that can penetrate through the package and cause bit upsets.

A solution to building SEU tolerance is often through a circuit redesign which includes error detection and correction and fault tolerance. These topics were discussed in Chapter 5. For the use of memories in space applications, the hardness assurance issues can be summarized as follows:

- For a sufficient number of parts available, lot sample radiation testing variable data are used to characterize the radiation response. Typically, overtesting is performed at two or three times the level required by the system. The safety margins (such as RDM) are used to compensate for uncertainties inherent in the space environment modeling and a lack of detailed knowledge of the device response. Total dose, neutron, and SEP test procedures and guidelines are discussed in Section 7.4.2.

- Extrapolation of lot sample, ground radiation testing data to the space environment should also take into consideration effects such as "rebound" which, for some device types, may exhibit ionization damage effects at low-dose space rates. This requires a comprehensive analysis of the radiation environment, device response, and mechanisms causing failure.

- Wafer level radiation testing and special test structures are also used for hardness assurance. This approach is discussed in Section 7.4.3.

7.4.2 Radiation Testing

Semiconductor devices such as memories have a radiation response which is a function of technology, fabrication process, and design characteristics. Therefore, the radiation sensitivity, especially of unhardened devices, can vary from lot to lot. As such, actual radiation testing is necessary to characterize the device response to assure that it meets mission requirements. However, radiation testing on ground is complicated by the fact that it differs significantly from the actual space environments, which is quite complex, as discussed in Section 7.2.1. In space, there is a large variability in the environmental parameters: particle type, energy, and time dependency. An ionizing dose is delivered steadily over a long period of time (mission periods usually ranging from one to ten years) at very low dose rates from a number of ion species at various energy levels. The ideal situation would be to put the test devices in actual space environments and measure their radiation response. However, this approach is not practical, and is limited to research types of space satellites that were discussed in Section 7.2.3.3.

Therefore, for all practical purposes, the space environment has to be simulated on the ground for testing semiconductor devices. Although ground testing is a simulation of space environments, the test results are generally valid, as long as the parameters responsible for the degradation effect are known and measured. The major objectives of ground-based radiation testing are as follows:

- Identify the mechanism of the radiation interaction with the device material, and establish the correlation of these effects with the device failure.

- Characterize the response of specific device types and technologies for use in space applications.

- Evaluate the device susceptibility to radiation and conformance to mission (application) requirements.

Radiation sources include radioactive isotopes such as Co-60, X-ray tubes, and particle accelerators. In addition to radiation sources, the radiation response of semiconductor devices strongly depends upon biasing conditions and temperature during irradiation. Therefore, device modeling and testing are performed for worst case biasing conditions.

A study was performed to identify the bias conditions that lead to worst case postirradiation speed and timing response for SRAMs [126]. The study was based on the following four assumptions: (1) the SRAM cell layout is symmetrical, (2) any initial preference for the "1" or "0" state is small compared to the radiation-induced threshold voltage shifts at levels of interest, (3) the n-channel transistors dominate the SRAM cell response, and (4) the SRAM cell is the largest contributor to changes in the overall circuit response. These assumptions are valid for many hardened memory designs, as well as commercial technologies. The test results showed that the worst case postirradiation response (in terms of relevant failure mechanisms) is exhibited by devices irradiated with the SRAM cells biased in a particular state (e.g., logic "1"s) and then switched to the opposite state (i.e., logic "0"s) before postirradiation anneal.

Radiation testing for memories can be for: (1) total dose effects, which is the damage caused by ionization energy through cumulative absorption by the sensitive material; (2) SEU (or soft errors), simulated by the high-energy, single-particle environment; (3) transient response for dose-rate effects (latchup); and (4) neutron effects. Radiation testing for each of these effects is discussed in the following sections.

7.4.2.1 Total Dose Testing.

Total dose testing is performed to measure the damage caused by ionization energy absorbed by the sensitive material. Radiation induces gradual parametric degradation (changes in threshold voltages and leakage currents) and eventual function failures. The rate of degradation may depend upon the dose rate (flux). Total dose testing can be performed in the following two ways:

- Exposing a sample lot of devices to a number of fixed, incremental dose level steps, and after each radiation step, making electrical measurements before returning the devices to the irradiation facility for the next dose level. This is also referred to as "not in-flux testing."

- "In-flux testing," in which the device performance is continually monitored while it is being irradiated.

Not in-flux testing is most widely used for the memories since it is easier to perform by exposure in an irradiation fixture followed by electrical characterization with a standard tester. In-flux testing is more complex because of special interfacing (irradiation source chamber to tester) requirements. However, it is useful in some applications that require higher measurement accuracy in functional failure levels.

7.4.2.1.1 IRRADIATION SOURCES. Irradiation sources commonly used for the total dose effects (ionization damage) investigation through ground simulation testing can be listed as follows:

- Gamma rays through radioactive isotopes such as cobalt-60 (or ^{60}C) and cesium-139 (Cs-139)

- X-rays

- Electrons

- Protons.

Each of these sources exhibits a specific energy spectrum and affects the test device differently. Co-60 is the most commonly used gamma-ray irradiation source for the measurement of ionization effects on silicon devices. It emits photons of energy 1.173226 and 1.332483 MeV, and has a half-life of 5.27 years. It has been established through various studies and experimentally that a rad (Si) deposited by gamma rays produces quantitatively an equivalent response in SiO_2 films with respect to the oxide traps (ΔV_{OT}) and interface traps (ΔV_{IT}) creation, as for the space environment protons, electrons, and bremsstrahlung (secondary electrons). In a typical irradiator, the Co-60 source is sealed in a steel jacket and placed inside a thick lead shield chamber. The devices to be irradiated are plugged into sockets on the fixture board, which is connected with wires to the power supply and biasing source located outside the chamber. The distance of the source or source exposure window can be varied to get different dose rates such as 10^5–100 rad (Si)/h or less. For example, a typical high-radiation orbit mission of ten years may average 200 krad total dose (including a 2:1 radiation design margin, RDM). Then a "one-shot" total dose exposure to that level (200 krad) at a dose rate of 100 krad/h may take only 2 h. However, the total dose testing is usually performed at various steps (cumulative), followed by electrical measurements at each of those steps. Figure 7-34 shows a schematic of a typical arrangement for Co-60 gamma-ray source testing setup, test levels, etc. [110].

X-rays (even low-energy) can also be used to simulate the space environment provided they can be accurately dosed and introduced into the active region of the device. The main advantage of using X-rays as an irradiation source is the low cost, wider distribution, and higher safety standards available for X-ray equipment. For commercial applications, 10 keV X-ray sources are often used for wafer level testing of radiation-hardened memories before they are packaged. This is a cost-effective approach for process validation and evaluation of

radiation response of the devices before expensive packaging and final inspection. However, analysis of X-rays test results should take into account effects such as dose enhancements and the absorption coefficient in comparing them to higher energy radiation source test results.

Electron beams can act as sources of ionization, and the electron sources include dc accelerators such as the Dynamitron. Another useful source is the Van de Graaff accelerator, which may be (typically) designed to operate at particle energies between 0.1 and 5 MeV. The dose rates can be varied by altering the beam current and beam focus, or sweeping the beam. A variation in the beam current from 10 nA to 10 A in a 20 mm diameter beam may yield particle fluxes from about 2×10^{10}/cm^2/s to 2×10^{13}/cm^2/s, which roughly corresponds to dose rates ranging from 600 to 600,000 rad/s. These dose rates are several orders of magnitude (about 106) times higher than the typical space dose rates of 10^{-5}–10^{-3} rad/s. Linear accelerators (LINACs) provide electrons of higher energies, typically 4–40 MeV (or even higher) in rapid, square pulses (e.g., pulse widths of a few

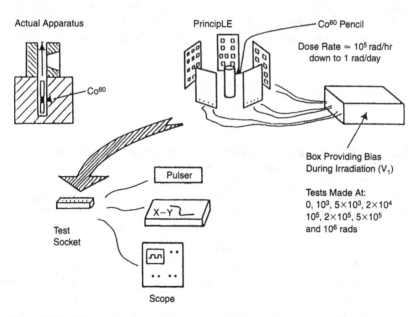

Figure 7-34. Schematic of a typical arrangement for testing with Co-60 gamma rays [110].

microseconds and repetition rate of up to 120 pps). However, these are quite expensive machines and have a high test time overhead.

Protons in space environments (heart of radiation belts) cause ionization damage primarily through nuclear interactions. Ground testing with protons is done mainly for displacement damage studies. For the acceleration of protons to energies above 15 MeV, the most commonly used machine is the cyclotron, and there are several of them spread around the country. Table 7-11 summarizes total dose radiation sources and some of their characteristics compared to natural space environment.

7.4.2.1.2 TEST PROCEDURE. Ground testing for total dose radiation testing is usually performed at dose rates which are typically several orders of magnitude higher than the actual space-like dose rates. The major concerns related to higher dose rate ground testing are: (1) the residual trapped hole density per rad (SiO_2) is quite high, and (2) rebound failures cannot be measured. In a low dose rate space environment, a significant amount of hole detrapping occurs (annealing effects). Therefore, the higher dose rate in ground testing represents the worst case for failure mechanisms dependent on the oxide traps, although they do not accurately simulate the interface traps (time-dependent) effects.

Total dose testing on semiconductor devices such as memories is performed per MIL-STD-883, Method 1019, which defines the requirements for effects from a Co-60 gamma-ray source [111]. This procedure also provides an accelerated aging test for estimating low dose rate effects which may be important for some device types that exhibit time-dependent effects. This test procedure includes specifications and requirements for the following:

- Test apparatus including radiation source, dosimetry system, electrical test instruments, and test circuit board.
- Test procedures including sample selection and handling, dosimetry measurements, radiation levels and dose rates, and temperatures.
- Electrical performance measurements, test conditions including bias and loading, and postirradiation procedures.
- MOS accelerated aging test procedures and requirements.

7.4.2.2 Single-Event Phenomenon (SEP) Testing.

As discussed earlier, the SEP (SEU and SEL) natural space environment concerns are related to: (1) protons from the earth's trapped radiation belts and solar flares, and

TABLE 7-11. Total Dose Radiation Sources and Some of their Characteristics Compared to Natural Space Environment. (From [109], with Permission of IEEE.)

Source	Type	Energy	Dose Rate
^{60}Co	Photon	1.25 MeV	$10^{-4} - 10^3$ rads (Si)/sec
^{137}Cs	Photon	.67 MeV	$10^{-3} - 10^2$ rads (Si)/sec
X-ray	Photon	$10 - 300$ KeV	$10^3 - 10^4$ rads (Si)/sec
LINAC	Electron	$1 - 50$ KeV	$10^3 - 10^{11}$ rads (Si)/sec (pulse mode)
Natural Space Environment	Electrons Protons Protons	1 MeV 100 MeV	$10^{-4} - 10^3$ rads (Si)/sec

(2) heavy ions from solar flares and galactic cosmic rays. SEP testing for space applications is performed in accordance with the general guidelines established by the EIA/JEDEC Standard EIA/JESD57 [112]. This test method defines the requirements and procedures for earth-based single event effects (SEE) testing of integrated circuits and is valid only when using a Van de Graaff or cyclotron accelerator. The SEP applicable to memories includes upset caused by a single-ion strike that may cause soft errors (one or more simultaneous reversible bit-flips) or hard errors (irreversible bit-flips) and latchup.

An SEP test consists of irradiating a given device with a prescribed heavy ion beam of known energy and flux in such a way that the number of SEUs or latchup can be detected as a function of the beam fluence (particles/cm²). For both SEU and latchup, the beam LET, which is equivalent to the ion's stopping power dE/dx (energy/distance), is the basic measurement variable. The SEU error rate is expressed in terms of critical charge (Q_c) related to circuit design and device cross-section (geometry of the sensitive volume). Q_c can be represented by a threshold LET, which is the minimum charge per unit length that must be deposited in sensitive volume to cause upset or latchup. A full device characterization requires irradiation with beams of several different LETs, which requires changing the ion species, energy, or angle of incidence with respect to the chip surface. The end product of SEP testing is a plot of the device upset rate (or cross-section) as a function of beam LET. Figure 7-35 shows the curves for some commonly used ion energy ranges and their corresponding LET values [110].

SEU testing is performed using the in-flux test method, and requires a source of energetic heavy ions. The cost, availability, and ion/energy capabilities are all important considerations when selecting a test facility. Some of the

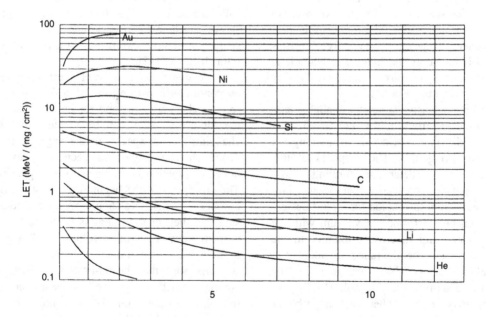

Figure 7-35. The curves for some commonly used ion energy ranges and their corresponding LET values [110].

commonly used sources for conducting SEP tests are listed below:

- Cyclotrons
- Van de Graaff accelerators
- Synchrotrons
- Fission sources
- Alpha emitters.

The cyclotrons provide the greatest flexibility since they can supply a number of different ions (including alpha particles) at a finite number of different energies. This allows the selection of ions with an adequate penetration range in the device. The cyclotrons used for SEU testing may have accelerating potentials of up to 300 MeV (or even higher) and use gaseous ion species such as krypton, argon, oxygen, and neon. The disadvantages associated with the use of cyclotrons are test expenses and down times associated with ions source replacement and changes of ion energy.

Van de Graaff accelerators are often used in the testing of sensitive devices with low LET thresholds for which lower atomic weight (Z) ions with continuously variable energy are required. These machines offer a relatively faster change of ion species, and are less expensive to operate than cyclotrons. However, they have limited energy range and function. Synchrotrons are very high-energy machines. An example is the European Synchrotron Radiation Facility (ESRF) in Grenoble, France.

A less expensive simulator available uses fission products from californium Cf-252 that have LET distributed primarily in the range of 41–45 MeV/[mg/cm^2]. The purposes of this simulator are to determine limiting cross-sections, and as a characterization tool to establish upset/latchup thresholds for circuit design. Irradiations are performed in vacuum with the IC (memory) chip in close proximity to the source. LET can be degraded to about 15–20 MeV/[mg/cm^2], which may not be low enough to determine the threshold (LET$_{crit}$) of some memory devices. Other practical problems which restrict the accuracy of this method are the low range of fission fragments and the sput-

tering of californium from the source by fission fragments contaminating the vacuum system. An improved Cf-252 method has been suggested which uses time-of-flight techniques, the response of ultrathin scintillation detectors to Cf-252 fission fragments, and the properties of fission fragments to reduce uncertainties in Cf-252 testing [113]. The major advantages of CASE (californium assessment of single event effects) are its low cost, simplicity, and flexibility. The complete facility is contained within a single bell jar, so that with normal precautions taken for handling radioactive sources, it may be used in any laboratory environment.

There are other techniques that use short pulses of highly focused laser beams as a research tool for the investigation of single-particle ionization effects [114], [115].

The SEU testing is performed in an evacuated chamber. The delidded device in a test socket is mounted on a platform which can be rotated so that the angle of incidence between the ion beam and the chip surface can be changed. A memory device under test (DUT) mounted on a text fixture is connected through interfacing cables to a power supply and tester outside the chamber which are used to electrically bias and test the device. Figure 7-36(a) shows a schematic overview of an SEU test setup, and Figure 7-36(b) shows the typical vacuum chamber [112].

A device such as a RAM is tested for SEU by continuously writing it with a fixed pattern (e.g., a checkerboard) and reading it to detect the bit errors. LET can be varied either by changing the ion species or the angle of incidence. The changing of ion specie takes several hours, and cyclotron time gets to be expensive. Therefore, before the beginning of testing, the number of ion species should be selected to assure that their energy range is adequate to traverse the device-sensitive volume at all angles of incidence. The particle flux (or beam current) is varied such that the bit-error rate (BER) is compatible with the speed of the tester to ensure that all errors are counted. A common practice to increase the LET through a sensitive volume is by varying the angle of incidence so that

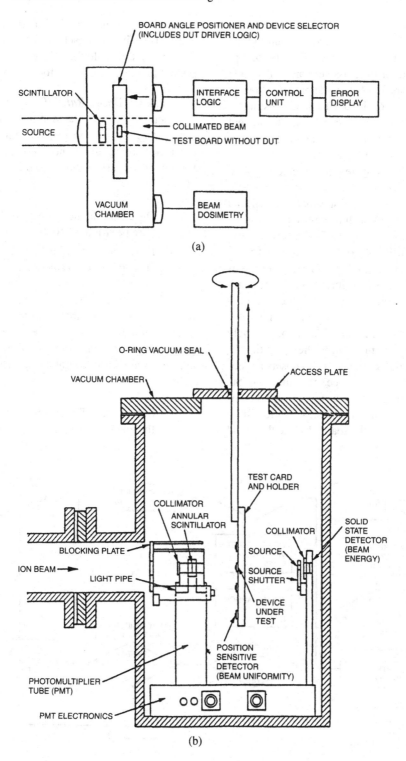

Figure 7-36. (a) Schematic overview of an SEU test setup. (b) A typical vacuum chamber [112].

$LET_{eff} = LET \times \sec \theta$. At each effective value of LET_{eff}, the BER is measured by counting the statistically significant number of errors (bit upsets). Then the error cross-section is the total number of errors divided by the particle fluence.

For heavy ion-induced latchup testing of memories, the test circuit is designed to prevent latchup-induced burnout, and the power is interrupted after each latching event.

The SEP testing requires a carefully laid out test plan that includes the scope of the test, overall test objectives, test schedules, special conditions, logistics such as device preparation, and tester checkout. Figure 7-37 shows an example of the SEP test flow diagram used for the NASA Galileo program [116]. Some of the steps shown were performed during the preliminary phase of the testing to ensure that the beam behaved as expected by testing parts of known SEU sensitivity. The variation in part supply voltage allows examination of the upset threshold as a function of voltage. If no upsets are observed at normal incidence, the angle of the beam with respect to chip surface is varied to increase LET_{eff} through the sensitive region. If an upset occurs readily, the search for the threshold

continues by using lower LET particles. The objective is to find the upset cross-section and corresponding threshold for SEU events in order to predict the SEU error rates for the mission.

7.4.2.3 Dose Rate Transient Effects.
Transient effects are related to the weapon environment for which the general guidelines and test procedure requirements are specified in the following two test methods from MIL-STD-883 [111]:

- Dose Rate Induced Latchup Test Procedure per Method 1020
- Dose Rate Upset Testing of Digital Microcircuits per Method 1021.

Transient effects testing requires ionizing radiation simulators that can test microelectronic components to high-level X-rays and gamma rays. These photon spectrum simulators are classified into several categories as listed below:

- Soft X-ray (SXR) simulators that use the energy from pulsed electrical gener-

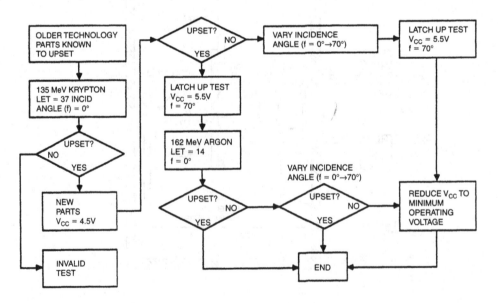

Figure 7-37. SEP test flow diagram for NASA Galileo Mission. (From [116], with permission of IEEE.)

ators to heat small volumes of plasma to kilovolt temperatures. The soft X-rays, in comparison to hard X-rays, have longer wavelengths, lower energies (<10 keV), and are more easily absorbed than the hard X-rays. The most commonly used spectra for testing are those of aluminum, neon, krypton, and argon.

- Moderate/hard X-ray simulators are either low- or moderate-energy bremsstrahlung sources. Low-energy bremsstrahlung sources test the effects of low-fluence, moderate-energy X-rays. Moderate-energy bremsstrahlung bombards a metallic foil with high-current electron beams with energies in the 0.5–2.0 MeV range to produce photons.

- Gamma-ray simulators produce bremsstrahlung and X-rays in the 0.1–10 MeV range, and are used to study transient radiation effects on electronics (TREE) produced by ionizing radiation. The Defense Nuclear Agency (DNA) facilities Aurora and TAGS are classified as gamma-ray simulators.

- Electron beam sources are used to study radiation effects often at higher fluences than are available from various X-ray and gamma-ray simulators.

The dose-rate-induced latchup test procedure (MIL-STD-883/Method 1020) specifies general requirements such as:

- Test plan including radiation pulse width(s), radiation dose(s) per pulse, and dose rate angles(s); total dose limit for each device type and failure criteria; methods to detect latchup; worst case biasing conditions; functional test requirements and output monitoring; etc.

- Test apparatus including radiation source, dosimetry system, latchup test system device interfacing, bias and functional test circuit, and temperature control.

- Test procedure details such as characterization testing and analysis, general requirements for production testing, and detailed test sequence.

Dose rate upset testing of digital microcircuits including memories (MIL-STD-883/ Method 1021) specifies general requirements and guidelines. The objective of this test is to find the dose rate at which any of the following occurs: (1) a transient upset, (2) stored data or logic upset, and (3) dynamic upset. In addition to general requirements such as test circuit and apparatus, this test method also defines several interference sources that need to be considered, approaches for state vector selection, transient upset criteria, radiation exposure, and the test sequence for upset threshold testing.

7.4.2.4 Neutron Irradiation.

The neutron irradiation test is performed on semiconductor devices such as bipolar memories to determine their susceptibilities in neutron environments. Neutron irradiation is performed per MIL-STD-883, Method 1017 [111]. The objectives of this test are: (1) to detect and measure the degradation of critical parameters as a function of neutron fluence, and (2) to determine if device parameters are within specified limits after exposure to a specified level of neutron fluence. The radiation sources specified are a TRIGA reactor or a fast burst reactor. The operation may be in either pulse or steady-state mode.

7.4.3 Radiation Dosimetry

Dosimetry is the process of measuring the amount of radiation to which a sample is exposed in a given time. A lack of proper radiation dosimetry technique can invalidate the test data. In total dose effect radiation testing, the objective is to accurately measure the ionization energy deposited in silicon dioxide and silicon. The RAD (radiation absorbed dose) and, in equivalent MKS units, gray (Gy) are the units of energy deposition. A RAD is defined as the energy absorbed by sample material when 100 ergs/g have been deposited.

$$\frac{100 \text{ ergs}}{\text{g(Si)}} = 1 \text{ RAD (Si)}$$

$$= 0.01 \text{ Gy (Si)}.$$

There are numerous radiation dosimetry techniques in existence. One that is most commonly used for total dose testing is thermoluminescent dosimetry (TLD). This uses hot-pressed polycrystalline lithium fluoride (LiF) chips which absorb a portion of radiation energy in the form of electrons that remain trapped in the LiF lattice for a long time at room temperature. When these are heated to about 150°C, the electron energy is released as light. This thermoluminescence emission is measured by a cool photomultiplier tube, and the signal emitted over a given swept temperature range is integrated. The integrated charge determined is roughly proportional to the radiation exposure over several orders of magnitude. Therefore, with calibration, the charge indicates the dose received.

The major advantages of TLD chips are their small size in comparison to the smallest ionization gauge and their long-term retention of dose information through trapped electrons. ASTM has approved a standard for the use of TLD in radiation testing: ANSI/ASTM E 668-78 [117].

Electron and proton beam accelerators require measuring of the radiation fluence for each exposure. For charged particle beams, a block of metal with appropriate thickness can stop all particles, and the flow of resultant charge can be measured. This basic concept is known as a Faraday Cup, and its further refinements are used to control many electron and proton accelerators in conjunction with single-particle counters collecting scattered radiation. The major refinements are the evacuation of air around the cup electrode, and shaping of the cup as a long thin cavity which does not allow secondary electrons to escape.

7.4.4 Wafer Level Radiation Testing and Test Structures

7.4.4.1 Wafer Level Radiation Testing.
Military specification MIL-M-38510, the general specification for microcircuits including memories, specifies four levels of total dose radiation hardness assurance (RHA levels M, D, R, and H corresponding to 3, 10, 100, and 1000 krad (Si), respectively) and a neutron fluence of 2×10^{12} n/cm^2 corresponding to each of these four levels [118]. Qualification and quality conformance inspection (QCI) of devices such as memories for total dose testing and neutron irradiation at wafer level is specified in accordance with MIL-STD-883, Method 5005, Group E [111]. The devices (in dice form) to be tested are selected from each wafer lot as specified and packaged. However, recently, this approach of wafer lot to wafer lot radiation hardness assurance testing is being replaced by a Qualified Manufacturers List (QML) methodology specified in MIL-I-38535, the General Specification for Manufacturing of Integrated Circuits (Microcircuits) [119]. In this new approach applicable to all silicon technology ICs including memories, radiation hardness is assured by proper control of all manufacturing sequences from design, fabrication processes to assembly, rather than end-of-the-line wafer product testing. This is done by identifying key technology parameters for the design, processing, and assembly, and bringing them under statistical process control (SPC). The SPC controls can be implemented through process monitors (PMs) and appropriate test structures, as well as wafer level test systems, to map the test structure and memory chip response across the wafer. Statistical and deterministic approaches are used to correlate the test structure data to memory chip performance including the use of circuit simulators.

Figure 7-38 shows a block diagram of a typical wafer level test system built around the 10 keV ARACOR X-ray irradiator (including wafer prober), HP 4062 tester for test structure parametric measurements, HP 8112 pulse generator, and HP 82000 tester (50 MHz) for parametric and functional IC measurements with interfacing to an HP 375 workstation [120]. This wafer level system is capable of making dc I–V and quasi-static C–V parametric measurements. The system shown has been used to generate detailed wafer maps (e.g., n-channel threshold voltage shifts, ΔV_{th} following irradiation) for

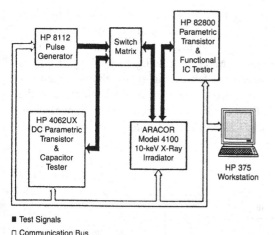

■ Test Signals
□ Communication Bus

Figure 7-38. Block diagram of a wafer level test
system. (From [120], with permission
of IEEE.)

test structures and delay chains from Honeywell's Yield and Circuit Reliability Analysis
Test (YCRAT) wafers. It can perform
current–voltage and charge pump measurements (to calculate the density of interface
traps) on transistors pre- and postirradiation.

7.4.4.2 Radiation Test Structure. Radi-
ation test structures are used at the wafer (or
chip level) as process monitors for implementation of SPC and hardness assurance. In a good
circuit design methodology, the critical paths
are first identified along with soft/hard cells.
Then soft circuits in critical paths are candidate
test structures and are prioritized according to
their position in the critical path. Analysis and
test simulations are used to define the test structures and measurement points. The test structure
data must relate to critical paths in the final design. Some actual test structures are given as examples to discuss their process characterization
goals and requirements.

*(1) Texas Instruments R2D3 Test Bar
[121]:* The purpose of the R2D3 test bar was to
develop a test vehicle for evaluation of process
controls, electrical and layout design rules, and radiation effects on bipolar microcircuits of VHSIC
level complexity. The test bar was processed

using 1.25 μm design rules for all radiation effect
test structures, and used an approach that included
the following: (1) targeting a specific radiation
environment/mechanism with each structure,
(2) emphasis on the radiation effect being measured, (3) systematic parametric variation
measurements for design rule validation, and
(4) grouping of structures for testing efficiency at
the wafer probe level and packaged devices.

This test bar consisted of nine square sections, of which two (sections 4 and 5) were devoted to the radiation test structures for the
definition of radiation-related design rules and
the investigation of basic failure mechanisms.
Section 4 contained test structures to investigate
the photoresponse, latchup, and SEU. Photocurrent measurement diodes used were large
enough to provide measurable signals at doses as
low as 10^5 rad (Si)/s. Also, the devices with significant differences in aspect ratio were included
to evaluate the amount of lateral photocurrent
collection. Latchup structures were used to evaluate both p-n-p-n paths within the isolation regions and through the substrate. These structures
were also useful in evaluating secondary photocurrents in parasitic transistors. For SEU
charge collection studies, an expanded transistor
structure containing all doping levels was used.

Test bar section 5 contained test structures
to investigate the total dose effects. In the
process used, most of the devices are thick field
oxide MOS transistors, with the source and drain
defined by the adjacent, buried layers and the
gate formed by first-level metallization. A summary of the total dose effects studied and the test
structures used for investigation is given below.

- Buried Layer to Buried Layer Leakage
  - Field oxide MOSFETs and buried
    layer to buried layer structures with-
    out field plates
  - Field oxide MOSFETs with trench
    isolation
- Collector-to-Emitter Sidewall Leakage
  and Gain Degradation
  - Standard transistors
  - Walled emitter transistors

• Resistor Dedgradation (Increase)

—Standard resistors

—Resistors with field plates.

(2) NASA/JPL Radiation Monitor Chips [122]: A single-event upset and total dose (SEU/TD) radiation monitor chip was developed, along with three other diagnostic chips

(fault chip, process monitor, and reliability chip) for NASA space applications. The objective was to develop a custom radiation monitor with the same CMOS processing technology which is used in the fabrication of spaceborne computers and signal processors. A block diagram of the SEU/TD radiation monitor chip is shown in Figure 7-39(a) [122]. It consists of the following

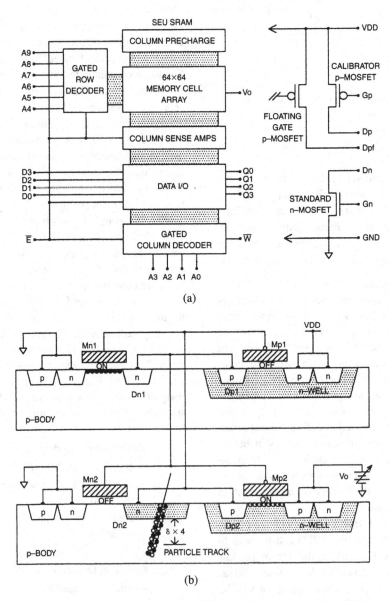

(a)

(b)

Figure 7-39. (a) A block diagram of SEU/TD radiation monitor chip. (b) Schematic cross-section of SEU RAM showing a particle track [122].

devices: (1) SEU SRAM, (2) standard n-MOSFET, (3) calibrator p-MOSFET, and (4) floating-gate MOSFET.

The floating-gate MOSFET experiences channel conductance shifts due to the radiation-induced gate charge. Total dose effects are monitored by measuring the floating-gate drain current and comparing it to the drain current from the calibrator MOSFET to determine the radiation-induced floating-gate charge. The 4 kb SRAM was designed to monitor the heavy ion upset rate. It has a conventional six-transistor cell which was modified to add an offset voltage, V_o. This allows the sensitivity of the cell to particle upset to be adjusted externally. Also,

the drain $Dn2$ was enlarged to increase the cell cross-section to particle capture. SRAM sensitivity is fixed so that cells can be upset by particles with LET > 2.88 MeV/[cm^2/mg]. Figure 7-39(b) shows a cross-section of an SEU RAM showing a particle track through diode $Dn2$ which can initiate a bit-flip [122]. The n-MOSFETs ($Mn1$ and $Mn2$) and p-MOSFETs ($Mp1$ and $Mp2$) are also shown in the schematic.

The radiation monitor chips along with diagnostic chips were fabricated in a 1.6 μm n-well, double-level metal CMOS process. Figure 7-40 shows a photomicrograph of the SEU/TD radiation monitor chip [122].

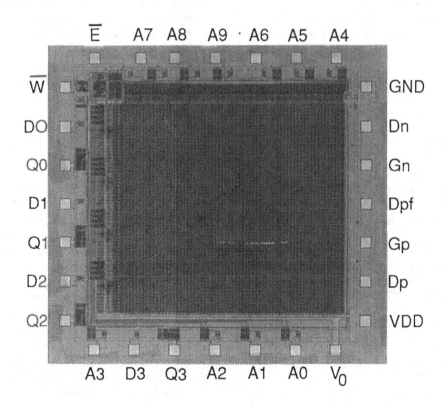

Figure 7-40. Photomicrograph of SEU/TD radiation monitor chip (3.12 × 3.29 mm^2) [122].

REFERENCES

[1] UTMC Tech. Description Notes.

[2] W. N. Spjeldvik and P. L. Rothwell, "The earth's radiation belts," AFGL-TR-83-0240, Air Force Geophysics Lab., Hanscom AFB, MA, Sept. 29, 1983.

[3] S. Glasstone and P. J. Dolan, *The Effect of Nuclear Weapons,* 3rd ed. Washington, DC: U.S. Dept. of Defense and U.S. Dept. of Energy, 1977.

[4] J. R. Srour, "Basic mechanisms of radiation effects on electronic materials, devices and integrated circuits," DNA Tech. Rep. DNA-TR-82-20, Aug. 1982.

[5] *ESA Radiation Design Handbook Draft,* 1987.

[6] W. E. Abare *et al.,* "Radiation response of two Harris Semiconductor radiation hardened 1K CMOS RAMs," *IEEE Trans. Nucl. Sci.,* vol. NS-29, no. 6, p. 1712, 1982.

[7] A. G. Sabnis *et al.,* "Influence of encapsulation films on properties of Si/SiO$_2$ interface of MOS-structures when exposed to radiation," in *Proc. 1988 IEDM,* p. 244.

[8] D. K. Myers *et al.,* "Radiation tolerant memory selection for Mars Observer Camera," *IEEE Trans. Nucl. Sci.,* vol. NS-34, no. 6, p. 1467, 1987.

[9] T. P. Haraszti, "Radiation hardened CMOS/SOS memory circuits," *IEEE Trans. Nucl. Sci.,* vol. NS-25, no. 6, p. 1187, 1978.

[10] R. H. Passow *et al.,* "Characterization summary for radiation hardened 16K × 1 SRAM," *IEEE Trans. Nucl. Sci.,* vol. NS-33, no. 6, p. 1535, 1986.

[11] B. L. Gingerich *et al.,* "Total dose and dose rate characterization of epi-CMOS radiation hardened memory and microprocessor devices," *IEEE Trans. Nucl. Sci.,* vol. NS-31, no. 6, p. 1332, 1984.

[12] D. Binder *et al.,* "Satellite anomalies from galactic cosmic rays," *IEEE Trans. Nucl. Sci.,* vol. NS-22, p. 2675, 1975.

[13] J. C. Pickel and J. T. Blandford, Jr., "Cosmic ray induced errors in MOS memory cells," *IEEE Trans. Nucl. Sci.,* vol. NS-25, p. 1166, 1978.

[14] J. F. Ziegler and W. Langford, *Science,* vol. 206, p. 776, 1979.

[15] E. L. Petersen and P. W. Marshall, "Single event phenomenon in the space and SDI arenas," Naval Research Lab. (NRL) Tech. Presentation.

[16] S. Kirkpatrick, "Modeling diffusion and collection of charge from ionizing radiation in silicon devices," *IEEE Trans. Electron Devices,* vol. ED-26, pp. 1742–1753, 1979.

[17] C. M. Hsieh, P. C. Murley, and R. R. O'Brien, *IEEE Electron Device Lett.,* vol. EDL-2, p. 103, 1981; and *IEEE Trans. Electron Devices,* vol. ED-30, p. 686, 1983.

[18] F. B. McLean and T. R. Oldham, "Charge funneling in N- and P-type substrates," *IEEE Trans. Nucl. Sci.,* vol. NS-29, pp. 2018–2023, Dec. 1982.

[19] A. B. Campbell *et al.,* "Investigation of soft upsets in MOS memories with a microbeam," *Nucl. Inst. Methods,* vol. 191, pp. 437–442, 1981.

[20] G. A. Sai-Halasz *et al.,* "Alpha-particle-induced soft error rate in VLSI circuits," *IEEE Trans. Electron Devices,* vol. ED-29, pp. 725–731, 1982.

[21] J. M. Biogrove *et al.,* "Comparison of soft errors induced by heavy ions and protons," *IEEE Trans. Nucl. Sci.,* vol. NS-33, pp. 1571–1576, 1986.

[22] R. Koga *et al.,* "SEU test techniques for 256K static RAMs and comparison of upsets by heavy ions and protons," *IEEE Trans. Nucl. Sci.,* vol. 35, p. 1638, 1988.

[23] R. Koga *et al.,* "On the suitability of non-hardened high density SRAMs for space applications," *IEEE Trans. Nucl. Sci.,* vol. 38, Dec. 1991.

[24] W. A. Kolasinski *et al.,* "Heavy ion induced upsets in semiconductor devices," *IEEE Trans. Nucl. Sci.,* vol. NS-32, pp. 159–162, 1985.

[25] J. G. Rollins, "Estimation of proton upset rates from heavy ion test data," *IEEE Trans. Nucl. Sci.,* vol. 37, pp. 1961–1965, 1990.

[26] J. H. Adams *et al.,* "Cosmic ray effects on microelectronics, Part I: The near earth particle environment," NRL Memo. Rep. 4506, 1981.

[27] L. C. Northcliffe and R. F. Schilling, *Range and Stopping Power Tables for Heavy Ions.* Orlando, FL: Academic, 1970, Nuclear Data A7.

[28] U. T. Littmark and J. F. Ziegler, *Handbook of Range Distributions for Energetic Ions in all Elements, Vol. 6.* Elmsford, NY: Pergamon, 1980.

[29] J. F. Ziegler, *Handbook of Stopping Cross-Sections for Energetic Ions in All Elements, Vol. 5.* Elmsford, NY: Pergamon, 1980.

[30] G. A. Sai-Halasz and M. R. Wordeman, "Monte-Carlo modeling of the transport of ionizing radiation created carriers in integrated circuits," *IEEE Electron Device Lett.,* vol. EDL-1, pp. 211–213, 1980.

[31] H. L. Grubin *et al.,* "Numerical studies of charge collection and funneling in silicon devices," *IEEE Trans. Nucl. Sci.,* vol. NS-31, pp. 1161–1166, 1984.

[32] J. P. Kreskovsky and H. L. Grubin, "Simulation of charge collection in a multilayer device," *IEEE Trans. Nucl. Sci.,* vol. NS-32, pp. 4140–4144, 1985.

[33] J. C. Pickel and J. T. Blandford, Jr., "Cosmic-ray-induced errors in MOS memory cells," *IEEE Trans. Nucl. Sci.,* vol. NS-25, pp. 1166–1171, 1978.

[34] S. E. Diehl *et al.,* "Error analysis and prevention of cosmic-ion-induced soft errors in static CMOS RAMs," *IEEE Trans. Nucl. Sci.,* vol. NS-29, pp. 2032–2039, 1982.

[35] S. E. Diehl *et al.,* "Single event upset rate predictions for complex logic systems," *IEEE Trans. Nucl. Sci.,* vol. NS-31, pp. 1132–1138, 1984.

[36] R. L. Johnson *et al.,* "Simulation approach for modeling single event upsets on advanced CMOS SRAMs," *IEEE Trans. Nucl. Sci.,* vol. NS-32, pp. 4122–4127, 1985.

[37] Y. A. El-Mansy *et al.,* "Characterization of silicon-on-sapphire IGFET transistors," *IEEE Trans. Electron Devices,* vol. ED-24, pp. 1148–1153, 1975.

[38] H. K. Lim and J. G. Fossum, "Threshold voltage of thin-film silicon-on-insulator (SOI) MOSFETs," *IEEE Trans. Electron Devices,* vol. ED-30, pp. 1244–1251, 1983.

[39] H. K. Lim and J. G. Fossum, "A charge-based large-signal model for thin-film SOI MOSFETs," *IEEE Trans. Electron Devices,* vol. ED-32, pp. 446–457, 1985.

[40] J. C. Pickel, "Effect of CMOS Miniaturization on cosmic-ray-induced error rate," *IEEE Trans. Nucl. Sci.,* vol. NS-29, pp. 2049–2054, 1982.

[41] J. C. Pickel and J. T. Blandford, Jr., "CMOS RAM cosmic-ray-induced error rate analysis," *IEEE Trans. Nucl. Sci.,* vol. NS-28, pp. 3962–3967, 1981.

[42] J. H. Adams, Jr. *et al.,* "LET spectra in low earth orbit," *IEEE Trans. Nucl. Sci.,* vol. 33, pp. 1386–1389, 1989.

[43] J. K. Letaw and J. H. Adams, Jr., "Comparison of CREME model LET spectra with spaceflight dosimetry data," *IEEE Trans. Nucl. Sci.,* vol. NS-33, pp. 1620–1625, 1986.

[44] J. H. Adams, Jr., "The variability of single event upset rates in the natural environments," *IEEE Trans. Nucl. Sci.,* vol. NS-30, pp. 4474–4480, 1983.

[45] E. L. Petersen *et al.,* "Calculation of cosmic-ray-induced soft upsets and scaling in VLSI devices," *IEEE Trans. Nucl. Sci.,* vol. NS-29, pp. 2055–2063, 1982.

[46] W. L. Bendel and E. L. Petersen, "Proton upsets in orbit," *IEEE Trans. Nucl. Sci.,* vol. NS-30, pp. 4481–4485, 1983.

[47] D. C. Wilkinson *et al.,* "TDRS-1 single event upsets and the effect of the space environment," *IEEE Trans. Nucl. Sci.,* vol. 38, p. 1708, Dec. 1991.

[48] T. Goka *et al.,* "The ON-ORBIT measurements of single event phenomenon by ETS-V spacecraft," *IEEE Trans. Nucl. Sci.,* vol. 38, Dec. 1991.

[49] C. I. Underwood *et al.,* "Observations of single-event upsets in non-hardened high density SRAMs in sun-synchronous orbit," *IEEE Trans. Nucl. Sci.,* vol. 39, Dec. 1992.

[50] C. S. Dyer *et al.,* "Radiation environment measurements and single event upset observations in sun-synchronous orbit," *IEEE Trans. Nucl. Sci.,* vol. 38, pp. 1700–1707, 1991.

[51] A. Campbell *et al.,* "SEU flight data from the CRRES MEP," *IEEE Trans. Nucl. Sci.,* vol. 38, pp. 1647–1654, Dec. 1991.

[52] A. Campbell *et al.,* "Single event upset rates in space," *IEEE Trans. Nucl. Sci.,* vol. 39, Dec. 1992.

[53] P. V. Dressendorfer, "An overview of advanced nonvolatile memories technologies," IEEE Nucl. and Rad. Effects Conf. Short Course, July 15, 1991.

[54] P. J. McWhorter *et al.,* "Radiation response of SNOS nonvolatile transistors," *IEEE Trans. Nucl. Sci.,* vol. NS-33, p. 1414, 1986.

[55] P. J. McWhorter *et al.,* "Retention characteristics of SNOS nonvolatile devices in a radiation environment," *IEEE Trans. Nucl. Sci.,* vol. NS-33, Dec. 1986.

[56] H. A. R. Wegener et al., "Radiation resistant MNOS memories," IEEE Trans. Nucl. Sci., vol. NS-19, no. 6, p. 291, 1972.

[57] J. J. Chang, "Theory of MNOS memory transistor," IEEE Trans. Electron Devices, vol. ED-24, no. 5, p. 511, 1977.

[58] T. F. Wrobel, "On heavy ion induced hard-errors in dielectric structures," IEEE Trans. Nucl. Sci., vol. NS-34, no. 6, p. 1262, 1987.

[59] E. S. Snyder et al., "Radiation response of floating gate EEPROM memory cells," IEEE Trans. Nucl. Sci., vol. 36, no. 6, pp. 2131–2139, 1989.

[60] G. D. Resner et al., "Nuclear radiation response of Intel 64-Kbit and 128-Kbit HMOS ultraviolet erasable programmable read only memories (UVEPROMs)," IEEE Trans. Nucl. Sci., vol. NS-32, Dec. 1985.

[61] J. M. Caywood et al., "Radiation induced soft errors and floating gate memories," in Proc. IRPS, 1983, pp. 167–172.

[62] J. F. Wrobel, "Radiation characterization of a 28C256 EEPROM," IEEE Trans. Nucl. Sci., vol. 36, Dec. 1989.

[63] H. L. Hughes and E. E. King, "The influence of silicon surface defects on MOS radiation sensitivity," IEEE Trans. Nucl. Sci., vol. NS-23, no. 6, p. 1573, 1976.

[64] H. Borkan, "Radiation hardening of CMOS technologies—An overview," IEEE Trans. Nucl. Sci., vol. NS-24, no. 6, p. 2043, 1977.

[65] W. R. Dawes et al., "Process technology for radiation-hardened CMOS integrated circuits," IEEE J. Solid State Circuits, vol. SC-11, no. 4, p. 459, 1976.

[66] C. R. Vishwanathan et al., "Model for thickness dependence of radiation charging in MOS structures," IEEE Trans. Nucl. Sci., vol. NS-23 no. 6, p. 1540, 1976.

[67] J. R. Adams et al., "A radiation hardened field oxide," IEEE Trans. Nucl. Sci., vol. NS-24, no. 6, p. 2099, 1977.

[68] K. M. Schlesier et al., "Processing effects on steam oxide hardness," IEEE Trans. Nucl. Sci., vol. NS-23, no. 6, p. 1599, 1976.

[69] A. Pikor et al., "Technological advances in the manufacture of radiation-hardened CMOS integrated circuits," IEEE Trans. Nucl. Sci., vol. NS-24, no. 6, p. 2047, 1977.

[70] F. F. Schmidt et al., "Radiation-insensitive silicon oxynitride films for use in silicon oxynitride films for use in silicon devices—Part II," IEEE Trans. Nucl. Sci., vol. NS-17, no. 6, p. 11, 1970.

[71] A. Waxman et al., "Al_2O_3-silicon insulated gate field effect transistors," Appl. Phys. Lett., vol. 12, no. 3, p. 109, 1968.

[72] F. B. Micheletti et al., "Permanent radiation effects in hardened Al_2O_3 MOS integrated circuits," IEEE Trans. Nucl. Sci., vol. NS-17, no. 6, p. 27, 1970.

[73] T. V. Nordstrom et al., "The effect of gate oxide thickness on the radiation hardness of silicon-gate CMOS," IEEE Trans. Nucl. Sci., vol. NS-28, no. 6, p. 4349, 1981.

[74] R. K. Smeltzer, "Refractory gate technology for radiation hardened circuits," IEEE Trans. Nucl. Sci., vol. NS-27, no. 6, p. 1745, 1980.

[75] K. Kasama et al., "Mechanical stress dependence of radiation effects in MOS structures," IEEE Trans. Nucl. Sci., vol. NS-33, no. 6, p. 1210, 1986.

[76] S. N. Lee et al., "Radiation-hardened silicon gate CMOS/SOS," IEEE Trans. Nucl. Sci., vol. NS-24, no. 6, p. 2205, 1977.

[77] K. G. Aubuchon et al., "Radiation-hardened CMOS/SOS LSI circuits," IEEE Trans. Nucl. Sci., vol. NS-23, no. 6, p. 1613, 1976.

[78] P. S. Winokur et al., "Optimizing and controlling the radiation hardness of a Si-gate CMOS process," IEEE Trans. Nucl. Sci., vol. NS-32, no. 6, p. 3954, 1985.

[79] H. L. Hughes, "Radiation hardness of LSI/VLSI fabrication processes," IEEE Trans. Nucl. Sci., vol. NS-26, no. 6, p. 5053, 1979.

[80] R. F. Anderson et al., "Degradation of radiation hardness in CMOS integrated circuits passivated with plasma-deposited silicon nitride," IEEE Trans. Nucl. Sci., vol. NS-26, no. 6, p. 5180, 1979.

[81] A. G. Sabnis et al., "Influence of encapsulation films on the properties of Si/SiO_2 interface of MOS structures when exposed to radiation," in Proc. 1981 IEDM, p. 244.

[82] J. R. Adams et al., "A radiation hardened field oxide," IEEE Trans. Nucl. Sci., vol. NS-24, p. 2099, Dec. 1977.

[83] J. L. Azarwicz et al., "Latch-up window tests," IEEE Trans. Nucl. Sci., vol. NS-29, p. 1804, Dec. 1982.

[84] B. L. Gregory et al., "Latchup in CMOS integrated circuits," IEEE Trans. Nucl. Sci., vol. NS-20, p. 293, Dec. 1973.

[85] J. L. Repace et al., "The effect of process variations on interfacial and radiation-induced charge in silicon-on-sapphire capacitors," *IEEE Trans. Electron Devices,* vol. ED-25, p. 978, 1978.

[86] D. Neamen, "Rapid anneal of radiation-induced silicon-sapphire interface charge trapping," *IEEE Trans. Nucl. Sci.,* vol. NS-25, no. 5, p. 1160, 1978.

[87] A. Gupta et al., "Co^{60} radiation effects on laser annealed silicon on sapphire," *IEEE Trans. Nucl. Sci.,* vol. NS-28, no. 6, p. 4080, 1981.

[88] J. L. Peel et al., "Investigation of radiation effects and hardening procedures for CMOS/SOS," *IEEE Trans. Nucl. Sci.,* vol. NS-22, no. 6, p. 2185, 1975.

[89] R. A. Kjar et al., "Radiation-induced leakage current in n-channel SOS transistors," *IEEE Trans. Nucl. Sci.,* vol. NS-21, no. 6, p. 2081, 1974.

[90] B. L. Buchanan et al., "SOS device radiation effects and hardening," *IEEE Trans. Electron Devices,* vol. ED-25, no. 8, p. 959, 1978.

[91] T. Ohno et al., "CMOS/SIMOX devices having a radiation hardness of 2 Mrad (Si)," *Electron. Lett.,* vol. 23, no. 4, p. 141, 1987.

[92] H. H. Sander et al., "Transient annealing of semiconductor devices following pulsed neutron irradiation," *IEEE Trans. Nucl. Sci.,* vol. NS-13, pp. 53–62, Dec. 1966.

[93] A. London et al., "Establishment of a radiation hardened CMOS manufacturing process," *IEEE Trans. Nucl. Sci.,* vol. NS-24, no. 6, p. 2056, 1977.

[94] E. E. King, "Radiation-hardening static NMOS RAMs," *IEEE Trans. Nucl. Sci.,* vol. NS-26, no. 6, p. 5060, 1979.

[95] B. L. Bhuva et al., "Simulation of worst-case total dose radiation effects in CMOS VLSI circuits," *IEEE Trans. Nucl. Sci.,* vol. NS-33, p. 1546, Dec. 1986.

[96] J. J. Silver et al., "Circuit design for reliable operation in hazardous environments," Tutorial Short Course, IEEE 1987 NSREC.

[97] R. E. Martina, *CMOS Circuit Design.* New York: Wiley, 1984, pp. 127, 114.

[98] W. S. Kim et al., "Radiation-hard design principles utilized in the CMOS 8085 microprocessor family," *IEEE Trans. Nucl. Sci.,* vol. NS-30, p. 4229, Dec. 1983.

[99] L. H. Rockett, Jr., "Simulated SEU hardened scaled CMOS SRAM cell design using gated resistors," *IEEE Trans. Nucl. Sci.,* vol. NS-29, p. 2055, Dec. 1982.

[100] E. L. Petersen et al., "Calculation of cosmic ray induced soft upsets and scaling in VLSI devices," *IEEE Trans. Nucl. Sci.,* vol. NS-29, p. 2055, Dec. 1982.

[101] H. T. Weaver et al., "An SEU tolerant memory cell derived from fundamental studies of SEU mechanisms in SRAM," *IEEE Trans. Nucl. Sci.,* vol. NS-34, Dec. 1987.

[102] S. Whitaker et al., "SEU hardened memory cells for a CCSDS Reed Solomon encoder," *IEEE Trans. Nucl. Sci.,* vol. 38, pp. 1471–1477, Dec. 1991.

[103] L. W. Messengil, "SEU-hardened resistive load static RAM," *IEEE Trans. Nucl. Sci.,* vol. 38, Dec. 1991.

[104] P. M. Carter et al., "Influences on soft error rates in static RAMs," *IEEE J. Solid-State Circuits,* vol. SC-22, no. 3, pp. 430–436, 1987.

[105] R. Brown et al., "QML qualified 256K radiation-hardened CMOS SRAM," paper supplied by Loral Federal Systems.

[106] N. Haddad et al., "Radiation hardened ULSI technology and design of IM SRAM," Paper supplied by Loral Federal Systems.

[107] MIL-HDBK-279, *Military Handbook Total-Dose Hardness Assurance Guidelines for Semiconductor Devices and Microcircuits.*

[108] MIL-HDBK-280, *Military Handbook Neutron Hardness Assurance Guidelines for Semiconductor Devices and Microcircuits.*

[109] Nick Van Vonno, "Advanced Test Methodologies," Short Course, IEEE 1995.

[110] *ESA Radiation Design Handbook ESA-PSS-01-609,* issue 1, Jan. 1987.

[111] MIL-STD-883, "Test methods and procedures for microcircuits."

[112] EIA/JEDEC Standard EIA/JESD57: "Test procedures for the measurement of single-event effects in semiconductor devices from heavy ion irradiation."

[113] M. D. Weeks and J. S. Browning, "An improved CF-252 method of testing microelectronics for single events," presented at the 11th Annu. IEEE Ideas in Sci. and Electron. Exposition and Symp., Albuquerque, NM, May 1989.

[114] S. P. Buchner *et al.*, "Laser simulation of single event upsets," *IEEE Trans. Nucl. Sci.*, vol. NS-34, p. 1228, Dec. 1987.

[115] A. K. Richter *et al.*, "Simulation of heavy charged particle tracks using focussed laser beams," *IEEE Trans. Nucl. Sci.*, vol. NS-34, p. 1234, Dec. 1987.

[116] P. A. Robinson, "Packaging, testing and hardness assurance," IEEE 1987 NSREC Tutorial: Section 6, Radiation Testing and Dosimetry.

[117] ANSI/ASTM E668-78, "Standard practice for the application of thermoluminescence dosimetry (TLD) systems for determining absorbent dose in radiation hardness testing of electronic devices."

[118] Military Specification MIL-M-38510, "General specification for integrated circuits."

[119] Military Specification MIL-I-38535, "General specification for manufacturing of integrated circuits (microcircuits)."

[120] M. R. Shaneyfelt *et al.*, "Wafer-level radiation testing for hardness assurance," *IEEE Trans. Nucl. Sci.*, vol. 38, p. 1598, Dec. 1991.

[121] J. Salzman (TI), R. Alexander *et al.*, "Radiation effects test structures on the Texas Instruments R2D3 test bar," presented at the DNA/Aerospace Corp. Workshop on Test Structures for Radiation Hardening and Hardness Assurance, Feb. 19, 1986.

[122] M. G. Buehler *et al.*, "Design and qualification of the SEU/TD radiation monitor chip," Jet Propulsion Lab., California Inst. Technol., Pasadena, CA, JPL Publ. 92-18.

[123] A. H. Johnston *et al.*, "The effect of circuit topology on radiation-induced latchup," *IEEE Trans. Nucl. Sci.*, vol. 36, pp. 2229–2237, Dec. 1989.

[124] R. L. Remke *et al.*, "SEU measurements on HFETs and HFET SRAMs," *IEEE Trans. Nucl. Sci.*, vol. 36, pp. 2362–2366, Dec. 1989.

[125] J. R. Hauser, "SEU-hardened silicon bipolar and GaAs MESFET SRAM cells using local redundancy techniques," *IEEE Trans. Nucl. Sci.*, vol. 39, pp. 2–6, Feb. 1992.

[126] D. M. Fleetwood and P. V. Dressendorfer, "A simple method to identify radiation and annealing biases that lead to worst-case CMOS static RAM post irradiation response," *IEEE Trans. Nucl. Sci.*, vol. NS-34, pp. 1408–1413, Dec. 1987.

8

Advanced Memory Technologies

8.1 INTRODUCTION

Chapter 3 discussed nonvolatile semiconductor memories such as ROMs, PROMs, floating-gate technology EPROMs (or UVEPROMs), and EEPROMs including flash memories that retain information even when the power is temporarily interrupted or a device is left unpowered for indefinite periods of time. In addition to the semiconductor memories, other commonly used nonvolatile storage devices are magnetic memories that use magnetic core and plated wire, magnetic bubbles and permalloy thin films, and permanent magnetic storage media such as magnetic tapes and disks. Another mass memory storage medium being developed is the use of optical memories that are characterized by write-once-read-many (WORM) operations.

In semiconductor memories, special nonvolatile random access memory (NOVRAM) or shadow RAM configurations are available that combine on the same chip a SRAM array and a backup EEPROM array of equal bits. On detection of power failure, or when the voltage drops below a certain critical value, the SRAM memory contents are transferred to EEPROM. These NOVRAM cells are larger compared to EEPROM cells, and thus have lower density and a higher price compared to EEPROMs and battery-backed RAMs. However, a newer generation of NOVRAMs based on DRAMs and

MNOS are being developed for high-density nonvolatile memory storage applications.

In recent years, an area of interest in advanced nonvolatile memories has been the development of thin-film ferroelectric (FE) technology to build ferroelectric random access memories (FRAMs) as substitutes for NOVRAMs. The high dielectric constant materials such as lead zirconate titanate (PZT) thin film can be used as a capacitive, nonvolatile storage element similar to trench capacitors in DRAMs. The basic principle governing the data storage mechanism is the use of the electrical polarization (or hysteresis) property of ferroelectric films versus applied voltage. This ferroelectric film technology can be easily integrated with standard semiconductor processing techniques to fabricate FRAMs which offer considerable size and density advantages over core and plated wire memories. Also, they have a higher WRITE/ERASE endurance life cycle and lower cost per bit than the equivalent NOVRAMs [1]. The storage of data by polarization instead of electric charge makes FRAM technology intrinsically radiation-hard, and thus a good candidate for space and military applications. Gallium arsenide ferroelectric random access memory (FERRAM) prototypes have been fabricated that use GaAs EJFET (enhancement-junction field-effect transistor) with ferroelectric capacitor memory elements [2].

Another technology development for non-volatile storage is the magnetoresistive (MR) memories (MRAMs) which use a magnetic thin-film sandwich configured in two-dimensional (2-D) arrays [4]. The magnetic storage elements are formed by a sandwiched pair of magnetic film strips which are separated by an isolation layer to prevent exchange coupling. The storage mechanism is based upon the magnetoresistive material's resistance that changes due to the presence of a magnetic field. The MR technology has characteristics such as nondestructive readout (NDRO), very high-radiation tolerance, higher WRITE/ERASE endurance compared to FRAMs, and virtually unlimited power-off storage capability. Some 16K MRAM prototypes have been fabricated. However, further technology development has been limited by process uniformity issues with the magnetoresistors. Another variation on this technology is the design and conceptual development of micromagnet-Hall effect random access memory (MHRAM) where information is stored in small magnetic elements [5]. A high storage density of 256 kb–to 1 Mb can possibly be achieved since a unit cell consists only of two transistors and a micromagnet-Hall (MH) effect element as compared to a conventional six-transistor SRAM cell.

A listing of few sources, performance characteristics of FRAMs, MRAMs, and MHRAM technologies relative to EEPROM and flash EEPROM technologies for comparison purposes is shown in Table 8-1 [5].

Memory storage, volatile or nonvolatile, usually refers to the storage of digital bits of information ("0"s and "1"s). However, in the last few years, analog nonvolatile data storage has also been investigated using EEPROMs and FRAMs in applications such as audio recording of speech and analog synaptic weight storage for neural networks. These are discussed in Section 8.4. The latest research in advanced memory technologies and designs includes solid-state devices that use quantum–mechanical effects, e.g., resonant-tunneling diodes (RTDs). The RTDs can be used by themselves, or as the base–emitter diode of a resonant-tunneling hot-electron transistor (RHET). These RTDs and some other experimental memory devices such as a GaAs n-p-n-p thyristor/JFET memory cell and single-electron memory are briefly discussed in Section 8.6.

TABLE 8-1. A Listing of Few Sources, Performance Characteristics of FRAMs, MRAMs, and MHRAMs Relative to EEPROMs and Flash EEPROMs Technologies for Comparison Purposes

Type of Nonvolatility	EEPROM	Flash EEPROM	FRAM	MRAM	MHRAM
Sources	Hitachi	Intel	Ramtron	Honeywell	JPL
Part Number	HN58C1001	28F016	FM1208S/ 1608S	HC7116	Experimental Prototypes
Address Access Time/ Read Cycle Time[1]	150 ns	85 ns	200/250 ns	250 μs	100 ns
Write Cycle Time[1]	10 ms	75 ns	340/500 ns	400 ns	100 ns
Capacity (bits)	1M	16M	4K/64K	16K	256K/1M
Erase (or Read)/ Write Endurance (cycles)	10,000 (page mode)	$10^5 - 10^6$	10^8	$>10^{15}$	Very high
Data Retention (years)[2]	10	10	10	>10	Very high
Total Dose Rad (Si)	Variable[3]	Variable[3]	IE6	IE6	IE6

[1] Address access time/Read cycle time and Write cycle time are specified by the manufacturers' for certain test conditions and specific modes.

[2] Data retention period specification are estimates based upon accelerated life testing performed by the manufacturer.

[3] Limited total dose testing performed in some sample devices have shown them to start degrading parametrically and funtionally between 20–30 krad and higher doses.

8.2 FERROELECTRIC RANDOM ACCESS MEMORIES (FRAMs)

8.2.1 Basic Theory

FRAMs utilize the ferroelectric effect, which is the tendency of dipoles (small electrically asymmetric elements) within certain crystals to spontaneously polarize (align in parallel) under the influence of an externally applied electric field, and these elements remain polarized after the electric field is removed. Then reversing the electric field causes spontaneous polarization (i.e., alignment of the dipoles) in the opposite direction. Thus, ferroelectric materials have two stable polarization states, and can be modeled as a bistable capacitor with two distinct polarization voltage thresholds. Since no external electric field or current is required for the ferroelectric material to remain polarized in either state, a memory device can be built for storing digital (binary) data that does not require power to retain stored information.

Ferroelectric (FE) films that are used as storage elements have relative dielectric constants which are a few orders of magnitude higher than that of silicon dioxide (e.g., 1000–1500 versus 3.8–7.0 of some typical DRAM capacitors). A ferroelectric capacitor using lead zirconate titanate (PZT) film can store a larger charge, e.g., 10 μC/cm^2 compared to an equivalent sized SiO$_2$ capacitor that may store only 0.1 μC/cm^2. Ferroelectric films such as the PZT remain ferroelectric from -80 to $+350°$C, well beyond the operating temperature of existing silicon devices. Also, the FE film processing is quite compatible with conventional semiconductor wafer processing techniques.

The earliest nonvolatile FRAM was developed by the Ramtron Corporation by introducing a device which used a ferroelectric thin film of potassium nitrate and lead zirconate titanate (PZT) [6]. The storage element was a capacitor constructed from two metal electrodes and a thin FE film inserted between the transistor and metallization layers of a CMOS process. Figure 8-1 shows a typical hysteresis $I-V$ switching loop for the PZT film and operating characteristics [7], [8].

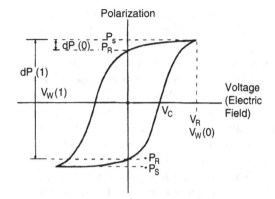

Figure 8-1. Schematic of a typical hysteresis curve of a ferroelectric capacitor and operating characteristics. (From [8], with permission of IEEE.)

For positive voltages greater than the coercive voltage (V_c) applied to the ferroelectric capacitor, the film is polarized in the positive direction to a saturation value of P_s. The coercive voltage (V_c) is defined as the value where polarization reverses and the curve crosses the X-axis. On removal of the applied voltage, the polarization relaxes to a value P_r, called the remnant polarization. This positive polarization value can be arbitrarily defined as a "0." On application of negative voltage to the ferroelectric film, the resulting polarization is in the negative direction, reaching a saturation value of $-P_s$ and a remnant (or relaxed) polarization of $-P_r$. This can be arbitrarily defined as a "1." Therefore, there are two stable states at zero voltage, defined as a "0" and a "1" for nonvolatile storage.

For a read operation of this ferroelectric storage element, a positive voltage is applied to the capacitor. If stored data are a "0," a small current equal to the increase in polarization from P_r to P_s is observed. If stored data are a "1," then some current is created by the change in polarization from $-P_r$ to P_s. The current differential between two memory states is sensed to produce memory output. Therefore, the greater the difference between the $dP(1)$ and $dP(0)$, the higher is the signal-to-noise ratio of the memory output. The read voltage is always in the same direction; the stored value of "0" is read as a low current, and "1" as a high current. Also, the value of "1" must be rewritten after the read operation, which means reversing the voltage and rewriting negative polarization.

There are hundreds or even thousands of ferroelectric materials out of which only a few are suited for switching applications and can be successfully integrated into baseline semiconductor processing. The earliest materials used to demonstrate the ferroelectric effects included thin ferroelectric films of bismuth titanate $(Bi_4Ti_3O_{12})$ deposited on the bulk semiconductor silicon by RF sputtering [9]. This material was used to fabricate a metal-ferroelectric semiconductor transistor (MFST) which utilized the remnant polarization of a ferroelectric film to control the surface conductivity of the bulk semiconductor substrate and perform a memory function. Materials such as barium titanate $(BaTiO_3)$ have been extensively studied [10]. A major problem with some ferroelectric materials has been fatigue, i.e., the polarization tends to decrease and disappear after a number of reversal cycles. For a ferroelectric material to be commercially feasible, it should have good retention and endurance characteristics, high dielectric breakdown for submicron layers, a Curie temperature above the highest storage temperature, a signal charge of at least 5 $\mu C/cm^2$, and nanosecond switching speeds [11].

The Ramtron Corporation, for their FRAM products, uses PZT as the preferred ferroelectric material that satisfies all of the above-stated requirements. The PZT thin-film deposition methods include sol-gel processing, metallo-organic decomposition (MOD), chemical-vapor deposition (CVD), and sputtering. Figure 8-2 shows a cross-section of a PZT film sandwiched between two metal electrodes to form a storage capacitor incorporated with standard semiconductor processing technology [11]. The bottom electrode, the PZT layer, and a top electrode are placed above the underlying CMOS circuitry before the source, drain, and gate contacts are cut, and before the interconnection (aluminum metallization) step. One–three additional masking operations are required for this integration of FE film into a standard CMOS process. The scaling to higher density FRAM memories is feasible because of high signal charge per unit area.

8.2.2 FRAM Cell and Memory Operation

Figure 8-3(a) shows a schematic of a single-ended FRAM cell that uses one transistor and one capacitor cell similar to that of a standard DRAM or EEPROM cell [6]. During a read operation, the cell element is polarized and charge transferred is compared to a referenced cell or other fixed level to determine whether a "1" or a "0" was stored. Figure 8-3(b) shows the dual-element differential sense amplifier approach used by the Ramtron Corporation [6].

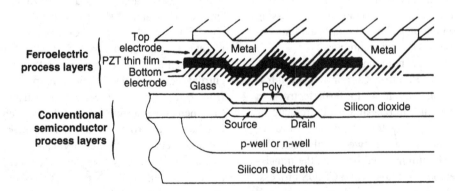

Figure 8-2. Cross-section of a PZT film sandwiched between two metal electrodes to form a storage capacitor incorporated with standard semiconductor processing technology. (From [11], with permission of IEEE.)

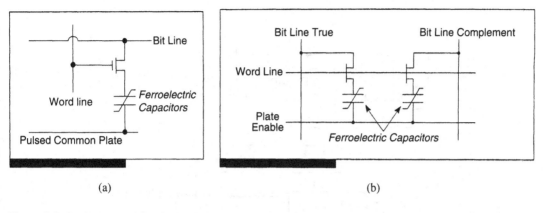

Figure 8-3. FRAM schematic of (a) a single-element memory cell, and (b) a dual-element memory cell. (From [6], with permission of Ramtron Corporation.)

Two nonvolatile elements are integrated in every memory cell, each polarized in the opposite direction. For a memory cell state read operation, both nonvolatile elements are polarized in the opposite direction. A differential sense amplifier which is connected to bit lines measures the difference between the amount of charge transferred from the two cells, and sets the outputs accordingly. The differential sense approach enhances reliability since the common mode variations in nonvolatile elements are cancelled out. However, this scheme requires dual nonvolatile elements, two access devices and two bit lines, which result in lower density compared to single-element memory cell devices.

The earliest proof of the concept FRAMs developed included an experimental 512 b nonvolatile memory, externally controlled device (ECD). Figure 8-4 shows a block diagram of this 512 b ECD [12]. The memory cells utilized a pass transistor and ferroelectric capacitor combination configured into a double-ended, differential sense architecture. The data bit consists of a word line (WL) which controls two pass transistors, a bit line (BL) and its complement ($\overline{BL}$) to collect charge from the capacitors, and a common drive line (DL) to actively drive the capacitors. A sense amplifier is connected between BL and $\overline{BL}$.

For a WRITE operation, the sense amplifier is set to the desired state, driving BL and $\overline{BL}$ to opposite voltage values of V_{drive} and ground.

Then the drive line is pulsed such that the high drive line against the grounded bit line writes the $Q(0)$ state. When the drive line drops to ground level after the pulse operation, the second capacitor has a $Q(1)$ written in it by its high bit-line voltage. The logic state written into the capacitor or BL represents polarity of the stored datum.

During a READ operation, a voltage step is applied to the drive line with the bit lines floating and sense amplifier OFF. Since the capacitors are in opposite states, BL and $\overline{BL}$ will collect different amounts of charge and produce a voltage differential of a polarity determined by the stored data.

Although FRAMs have demonstrated very high write endurance cycle times, the ferroelectric capacitors depolarize over time from read/write cycling. The rate of depolarization is a function of temperature, intrinsic stress, WRITE voltage, and ferroelectric material composition. The search for high-density nonvolatile memories has led to the development of DRAM-like architectures with thin-film dielectrics as direct replacements for the oxide–nitride–oxide (ONO) structures. A ferroelectric nonvolatile RAM (FN-VRAM) has been developed which normally operates as a conventional DRAM while utilizing the hysteresis loop of ferroelectric materials for nonvolatile operation [13]. This device uses metal–ferroelectric–metal (MFM) fabricated with lead zirconate titanate as the ferroelectric

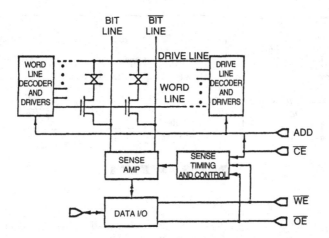

Figure 8-4. Block diagram of a 512 b ECD. (From [12], with permission of IEEE.)

material. This cell can be operated in either a nonvolatile mode or a DRAM mode.

The FNVRAM cell with a one-transistor DRAM cell and a ferroelectric capacitor are shown in Figure 8-5 [13]. Two different biasing schemes are used. In the first scheme, the cell is always held at half supply voltage ($V_{DD}/2$). During DRAM operation, the storage node is either at $V_{DD}/2$ or V_{DD} such that the ferroelectric capacitor is not cycled between opposite polarization states. In the second biasing scheme, the cell plate is held at V_{DD} during DRAM operation, and the storage node is either at 0 V or V_{DD}. On power failure (or command), a nonvolatile or store operation is executed in which the cell logic state is read and written back as one of the two permanent polarization states of the ferro-electric (FE) film. In the first scheme, if the DRAM datum is zero, the word line is selected and the bit line grounded so that the FE film is now polarized in one direction (nonvolatile zero). In the second scheme, a DRAM zero is stored as a nonvolatile zero by selecting the word line, driving the bit line to V_{DD}, and grounding the cell plate. Thus, the FE film is cycled between opposite polarization states during nonvolatile store/recall operations, and not during the DRAM read/write operations. The DRAM read/write cycling operations do cause some degradation in capacitor polarization. However, this FNVRAM cell is expected to tolerate orders of magnitude higher nonswitching read/write cycles than the 10^{12} cycles demonstrated for the PZT films [14].

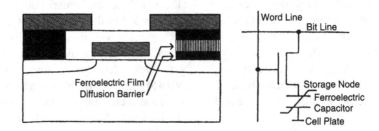

Figure 8-5. FNVRAM cell with one-transistor DRAM cell and a ferro-electric capacitor. (From [13], with permission of IEEE.)

8.2.3 FRAM Technology Developments

An example of a commercial FRAM is Ramtron Corporation's FM 1208S/1608S series 4K/64K series devices which combine the flexibility of a low-power SRAM with the non-volatile characteristics of an EEPROM. These are fabricated in a 1.5 μm silicon gate CMOS technology with the addition of integrated thin-film ferroelectric capacitors. These operate from a single + 5 V power supply, and are TTL/CMOS compatible on all the input/output pins. FM 1208 has 4K memory organized as 512×8 b and utilizes the JEDEC standard byte-wide SRAM pinout. This device is specified for 250 ns read access, 500 ns read/write cycle time and 10 year data retention without power.

Figure 8-6(a) shows a functional diagram of an FM 1208S FRAM which uses dual-memory capacitor cells [15]. Figure 8-6(b) shows the read cycle timing, and (c) the write cycle timing [15]. This FRAM operates synchronously using the $\overline{CE}$ signal as a clock. Memory read cycle time (t_{RC}) and write cycle time (t_{WC}) are both measured between the falling edges of $\overline{CE}$. The $\overline{CE}$ signal must be active for time t_{CA}, and a minimum precharge time t_{PC} is required to precharge the internal buses between operations.

The memory latches the address internally on the falling edge of $\overline{CE}$. The address data must meet a minimum setup time t_{AS} and hold time t_{AH} relative to a clock edge. Read data are valid for a maximum access time t_{CE} after the beginning of the read cycle. The $\overline{OE}$ signal is used to gate the data to I/O pins. The $\overline{WE}$ signal must be high during the entire read operation. For a write operation, $\overline{WE}$ must be stable for time t_{WP} prior to the rising edge of $\overline{CE}$. The data must be valid on I/O pins for time t_{DS} prior to the rising edge of $\overline{WE}$ and hold time t_{DH} after $\overline{WE}$. Also, the $\overline{OE}$ signal must disable the chip outputs for time $t_{0\,Hz}$ prior to placing data on I/O pins to prevent a data conflict.

In order to obtain high-density FRAMs, a one transistor/one capacitor cell with a stacked capacitor structure is necessary, for which the following key technology developments are required [44]:

- A well-defined planarization process before fabricating the capacitor to obtain a smooth capacitor structure.
- Since the barrier metal layer between the capacitor bottom electrode and the underlying layer is easily oxidized during the annealing, causing increased contact resistance, the oxygen diffusion through the bottom electrode should be eliminated.
- The use of a dry etching process to obtain the fine patterning since the plasma etch causes capacitor degradation.

A 0.5 μm design rule ferroelectric process has been used to fabricate a memory cell size of 10.5 μm^2 with a capacitor size of 1.5×1.5 μm^2, suitable for a 4 Mb NVDRAM. This process uses PZT (Pb(^{2r}1–x^{Ti}x)O$_3$) as the ferroelectric material. After fabricating the MOS transistors, n$^+$-doped polysilicon is filled at the contact holes. A planarized poly-Si plug is formed by using chemical mechanical polishing (CMP). The capacitor module consists of PZT (0.25 μm)/Pt/TiN/Ti (0.1/0.2/0.03 μm). The Pt/TiN/Ti multielectrode is deposited by dc magnetron sputtering, and the PZT film is coated by the conventional sol-gel method. After patterning the PZT/Pt/TiN/Ti layers, the TiO$_2$ and SiO$_2$ films are deposited, and subsequently the contact holes are opened on the PZT film. The top electrode consisting of Ti (0.02 μm)/Pt (0.1 μm) is formed. After fabricating the top electrode, the contact holes are opened on Si and the top electrode, and then AlSiCu/TiW lines are formed.

8.2.4 FRAM Reliability Issues

Ferroelectric memories' major reliability issues were discussed in Chapter 6 (see Section 6.3.5). Thermal stability, fatigue, and aging of the ferroelectric RAMs are the key reliability concerns. A study was performed on the FE film capacitors to evaluate the effects of thermal excursions on the remnant polarization level in polycrystalline films [21]. The reliability of ferroelectric nonvolatile memory is limited by

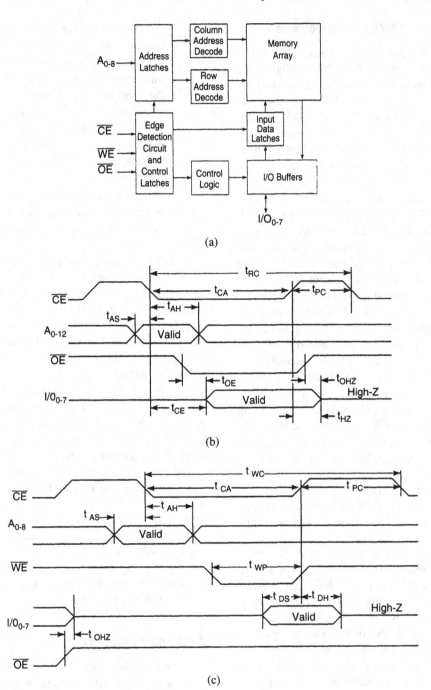

Figure 8-6. FM 1208S FRAM memory. (a) Functional diagram. (b) Read cycle timing.
(c) Write cycle timing. (From [15], with permission of Ramtron Corporation.)

degradation in the value of suitable remnant polarization. Typical FRAMs available on the market are specified for the data storage and operating temperature range of 0–70°C. The elevated temperature excursions (e.g., to 125 and 175°C) were found to cause a reduction in the retained polarization. A severe degradation in signal strength is experienced when the FE capacitors are thermally cycled from -25 to $+125°C$. Negative thermal excursions have no effect on the retained polarization.

Fatigue is induced by cycling, and is an important characteristic of the FE films since the nonvolatile memories experience polarization reversal with every read and write cycle. The fatigue characterization testing and plot of some 20×20 μm^2 capacitors at 6 MHz using high-current cycling boxes showed a characteristic knee at 10^{12} cycles, beyond which a steep degradation of the polarization with the log (cycling) occurred.

Aging is the loss of remnant polarization with time, and may increase with temperature and fatigue. An aging test is performed by leaving a poled capacitor in one state for a specified time, and then characterization of the write/read behavior through the measurement of remnant polarization. Test results on some FE capacitors indicated that both the fatigue (up to 10^{10} cycles) and high temperature ($125°C$) had very little effect on the aging rate.

8.2.5 FRAM Radiation Effects

In general, the FE capacitors and memories made from thin-film PZT have shown high radiation tolerance characteristics suitable for space and military applications. Radiation testing has also been performed on CMOS/FE integrated devices, including the memories, to determine whether the additional ferroelectric processing causes significant degradation to the baseline CMOS process. Test results for some of these studies are summarized below:

- In a total dose radiation-induced degradation of thin-film PZT ferroelectric capacitors study performed by Sandia National Laboratories, the FE capacitors fabricated through different processes were irradiated using X-ray and Co-60 sources to 16 Mrad (Si) dose levels [16]. The capacitors were characterized for retained polarization charge and remnant polarization both before and after radiation. It was observed that the radiation hardness was process-dependent. Three out of four processes for the FE capacitors showed less than 30% radiation-induced degradation for exposures up to the 16 Mrad (Si) level. However, for one of the processes, significant degradation was observed above 1 Mrad (Si). The radiation-induced degradation appeared to be due to the switching characteristics of FE material, and showed recovery on postirradiation biased anneal.

- In a study conducted by Harry Diamond Laboratories and Arizona State University, thin-film sol-gel capacitors were irradiated to 100 Mrad (Si) with 10 keV X-rays [17]. Test results showed that the PZT FE films fabricated by the sol-gel techniques degrade very little in terms of amount of switched energy available, up to 5 Mrad (Si) exposure. Above the 5 Mrad (Si) exposure, some distortion in the hysteresis loop was observed, and there is a gradual loss of switched charge out to 100 Mrad (Si). The type and degree of hysteresis loop distortion depended upon the polarization state and the applied bias during irradiation. Some of the postradiation-induced damage could be annealed by repeated bias cycling of the FE capacitor.

- Another evaluation was performed by Arizona State University in which sol-gel derived PZT thin films were irradiated to a total dose of 1 Mrad (Si) under open-circuit bias conditions [18]. Test results showed that the irradiation changed the magnitude of the internal bias fields and produced hysteresis curve distortion, depending on the polarization state of the capacitor before irradiation. The postirradiation electrical cycling makes the hysteresis loop symmetric initially, but eventually causes fatigue effects. Static $I–V$ measurements showed that irradiation changes the switching current, but does not degrade the leakage current. Irradiation does not significantly affect the data-retention and fatigue properties, except for polarization reduction.

- Neutron irradiation (displacement damage) effects on the PZT thin films for nonvolatile memory applications were evaluated by Harry Diamond Laboratory [19]. Test results showed that the PZT films were quite resistant to switched charge loss, showing a reduction of only 10% at the 10^{15} neutron/cm^2 flux level. In comparison, the retained polarization measured by pulse techniques showed a greater loss of remnant polarization which saturated at the lowest fluence measured $(1 \times 10^{13}$ neutrons/cm$^2)$. However, none of the PZT film devices "failed" at the fluences at or below 1×10^{15} neutrons/cm^2. The endurance characteristics were unchanged by neutron irradiation. A comparison of pre- and postneutron irradiation fatigue data indicated that the neutron damage was apparently permanent and not annealed by the read/write cycling, as was the case with ionizing radiation damage [17].

- A radiation evaluation of commercial ferroelectric (FE) nonvolatile memories was made by Harry Diamond Laboratories and the National Semiconductor Corporation [1]. Test devices included

4K nonvolatile memories, 8 b octal latches (with and without FE), and some process control test chips that were used to establish a baseline characterization of the radiation response of the CMOS/FE integrated devices. A major objective was to determine whether the additional FE processing caused a significant degradation to the baseline CMOS process. The test results showed 4K memories and octal latches failing functionally at total dose exposures between 2–4 krad (Si). Analysis showed that the radiation response of FRAM test devices was dominated by the underlying baseline, unhardened, commercial CMOS process. No significant difference was observed between the radiation response of devices with and without the FE film in this commercial process.

- The large polarization charge density of the ferroelectric capacitor memory cells makes them relatively insensitive to single-event upset (SEU) from the heavy ions [20]. Some experimental radiation tests, including displacement damage, ionizing dose, and SEU (Cf-252), were run on the ferroelectric capacitors for a comparison of theoretical (estimated) and experimental (measured) radiation data. These test results are summarized in Table 8-2.

TABLE 8-2. Radiation Test Results for Comparison of Theoretical and Experimental Radiation Data. (From [20], with Permission of IEEE.)

Radiation Stress	Displacement Damage, n/cm^2	Ionizing Dose, rads (Si)	Ionizing Dose rads (Si)/s	Fission Fragments Cf-252, particles/cm^2
Theoretical Threshold (Estimated Level Threshold)	5×10^{16}	$>10^8$	2.5×10^{13}	Probably No Damage, No SEU at any level
Experimental Data (Measured Level at which No Damage is observed)	10^{14}	10^7	1.2×10^{11}	At 6×10^6 Particles/cm^2, No Damage and No SEU

8.2.6 FRAMs versus EEPROMs

In nonvolatile memory applications, the EEPROM is an established technology that is widely used, whereas FRAM fabrication is in the development stages. The FRAMs appear to have a good future potential. However, there are some major reliability concerns such as process uniformity control, thermal stability, fatigue, and aging of the ferroelectric RAMs. Some of the key differences in the basic storage mechanisms and operating characteristics of the two technologies can be summarized as follows:

- FRAMs utilize magnetic polarization storage techniques, in contrast to the EEPROMs' charge tunneling mechanism, and use a 5 V supply for all internal operations instead of the 12–15 V required by conventional EEPROM technologies.

- For FRAM cell programming, an electric field is applied for less than 100 ns to polarize nonvolatile elements, whereas for standard EEPROMs, 1 ms or more are needed to generate sufficient charge storage at the gate element.

- In EEPROMs, charge tunneling across the insulating oxide layers degrades characteristics of the oxides, causing excessive trapped charge or catastrophic breakdown. Therefore, typically, EEPROM devices are guaranteed for 10,000–100,000 write cycles. According to the manufacturer's reliability test data, the FRAMs which utilize the polarization phenomenon can provide up to 10 billion (10^{10}) write endurance cycles.

8.3 GALLIUM ARSENIDE (GaAs) FRAMs

Section 8.2 discussed FRAMs based on ferroelectric film capacitors integrated with conventional silicon CMOS processing. Ferroelectric element processing also has been combined with GaAs technology to produce ferroelectric nonvolatile memory (or FERRAM) prototypes with 2K/4K bit density levels [21]. A common design approach for FE processing combined with the GaAs JFET has been the placement of a capacitor alongside the transistor structure, as shown in Figure 8-7(a) [8], [22]. This arrangement allows the transistor fabrication to be completed after the FE capacitor has been deposited. Thus, the temperature-sensitive step of contact metal deposition and dopant diffusion can be performed after the high-temperature annealing of the FE film.

Another major design consideration is the use of a single-capacitor cell data sensing scheme, or a dual-capacitor differential sense approach. The single storage element results in a small cell size which allows for doubling of the memory density for same chip area. However, the disadvantage of the single cell is the reduction in signal level for the sense amplifier and the need for a reference sensing scheme. The dual-element approach works well even with smaller remnant polarization values, and does not need a separate reference. Figure 8-7(b) shows a block diagram of an experimental FERRAM memory circuit which can be externally switched for operation in either 2K (256 word × 8b) single-cell mode, or 4K (512 word × 8b) dual-cell operation. The reference capacitors are ignored when the dual-cell option is selected. Since the readout operation is destructive, an automatic rewrite circuitry is included in the design.

The memory array is divided into four quadrants, each containing 64 word × 8 b for a 2K and 128 word × 8 b for a 4K density mode. Each of these quadrants contains a different size FE capacitor for the determination of optimum capacitor size. For the single-cell devices, each quadrant has different size reference capacitors, which allows for various reference capacitor versus memory cell capacitor ratios. The physical size of FE capacitors and their compatibility with semiconductor processes used are the major limitations toward the development of practical memory circuits with commercial yields. Another critical control parameter is the uniformity of FE properties of thin film which must be maintained over the entire chip surface and wafer area.

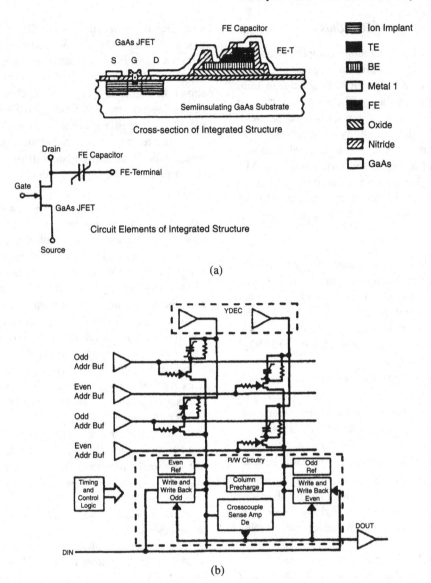

Figure 8-7. (a) Ferroelectric element processing combined with GaAs JFET technology. (b) Block diagram of FERRAM memory circuit. (From [8], with permission of IEEE.)

8.4 ANALOG MEMORIES

A conventional electronic data memory, volatile or nonvolatile, usually refers to the storage of digital bits of information as logic "0"s and "1"s. Analog signal data storage also requires digitization using analog-to-digital (A/D) converters and some digital signal processing. A recent development of the "analog memory" concept refers to direct analog storage without intermediate digital manipulation. A company called Information Storage Devices, Inc. (ISD) has developed a single-chip voice message system ISD1200/1400 series family which includes several device types that can provide 10–20 s of high-quality recording and playback of the human voice without transforming the sound into digital form [23], [24]. These chips are based upon ISD's patented direct analog storage technology (DAST ™) in a CMOS EEPROM process. For example, the analog memory ISD 1400 has an array of 128,000 storage cells, which is the amount of storage equivalent to 1 Mb of standard digital EEPROM (128K × 8 b). This equivalence is based upon each analog sample stored and read out with 8 b digital resolution.

According to ISD, this nonvolatile analog storage is accomplished by using the EEP-ROMs, which are inherently analog memories on a cell-by-cell basis because each floating gate can store a variable voltage. The EEP-ROMs discussed in Chapter 3 used a floating gate for charge storage corresponding to digital "1"s and "0"s, through a charge transport mechanism called Fowler–Nordheim tunneling. The charge on the floating gate creates an electric field that modifies the conductivity of the channel, making it either strongly conductive or nonconductive. When high voltage is applied, the oxides conduct sufficiently via the electron tunneling to charge or discharge the floating gate. The tunneling current is a sharp exponential function of the bias voltage across the oxide [36]. The polarity and location of the bias volt-

age applied control the charging and discharging of the gate. When the bias value is reduced (or the device is unpowered), the oxides behave as excellent insulators, retaining the charge stored on the floating gate for several years. The voltage on the floating gate can be "read" without disturbing the charge (or stored voltage). The sensed value of a cell's conductivity corresponds to the value of the analog level stored.

In digital memories, high erase and programming voltages are applied to cause large amounts of tunneling current to write a logic "1" or a "0" to the floating gate. In contrast, the analog EEPROMs take advantage of intermediate conducting values between the strongly conductive and nonconductive cell. This allows for denser analog memory arrays than digital memories. For example, an analog signal sample voltage which requires eight digital cells for encoding can be stored in a single analog EEP-ROM cell. This technology also eliminates the need for A/D and D/A converters since the input, storage, and output are all in the analog domain.

Figure 8-8 shows a block diagram of the ISD 1200/1400 series single-chip voice record/playback system utilizing analog memory storage technology [37]. The chip includes automatic gain control circuitry, sampling clocks,

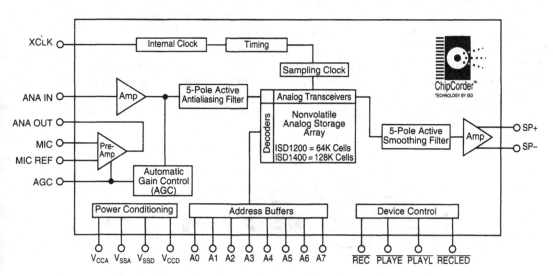

Figure 8-8. Block diagram of ISD 1200/1400 single-chip voice record/playback system utilizing analog memory system technology. (From [37], with permission of ISD.)

analog transceivers, decoders, oscillators, filters, and speaker amplifiers. Each sample is stored as a precisely metered charge on the floating gate of a standard EEPROM. For "playback," the conductance of each storage transistor is measured to determine the analog level stored in the cell. As the storage array is read out in sequence at the original sampling rate, each cell produces a corresponding voltage level. The signal is reconstructed from these voltage samples and amplified. The major audio applications for this analog memory include products such as answering machines, pagers, and cellular phones. An advanced version of this chip developed is the ISD 2500 series suitable for 60–90 s messaging applications, and offers 5.3, 6.4, and 8.0 kHz sampling frequencies.

In addition to the recording of speech, other applications being developed for the analog memories include offset compensation in op-amps [26], analog recording of speech [27], and analog synaptic weight storage for neural networks [28]. For all of these applications, the EEPROMs utilize a tunneling mechanism for both programming and erasing. The programming of an analog value into a floating-gate EEPROM requires the iterative application of a programming pulse and reading of the device output. The output signal during the read operation is compared to the desired output value, and the iteration process stops when the two values closely match.

In contrast to the above approach, a new vertical-injection punchthrough-based MOS (VIPMOS) EEPROM structure has been developed that can be programmed in a single step without high control gate pulses [25], [29]. This circuit operates from a single +5 V supply, and the output voltage is also continuously available during the programming. Figure 8-9 shows a schematic of this VIPMOS structure, which is based on an NMOS transistor [25]. The source of the transistor has been omitted, the gate is left floating, and with a control gate placed on top of the floating gate. An n-type buried injection is formed under the floating gate by high-energy implantation.

During normal operation, the ENABLE input is connected to ground, and only the supply voltage and reference current have to be supplied to the circuit. In the program mode, the ENABLE input is connected to the supply voltage, and programming current has to be supplied. This program current has to be made dependent on the difference of the measured and desired output potential. After the desired output potential is reached, programming can be disabled. This analog programmable voltage source can be used to set a multiplication factor in programmable filters, or to set weight factors in synaptic connections for neural networks.

Analog nonvolatile memory devices based on thin-film ferroelectric capacitors have also been proposed for applications such as synaptic connections in neural networks. Electronic hardware implementation of the neural networks often requires a large number of reprogrammable synaptic connections. The conventional approach required the use of hybrid analog–digital (A/D) converters, RAMs, and multiplying DACs that resulted in a design complexity which limited the number of synapses available due to power dissipation and size considerations. The floating-gate EEPROMs have a life cycle endurance which is

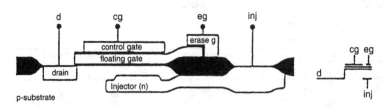

Figure 8-9. VIPMOS EEPROM structure and circuit schematic. (From [25], with permission of IEEE.)

often limited to 100,000 cycles. An experimental thin-film ferroelectric nonvolatile analog memory has been developed which enables nondestructive readout, electrical and optical interfaces with input data, and ac coupling techniques [3]. In contrast to FRAMs that have binary storage mode and destructive readout as discussed is Section 8.2, these memories can be "read" without destroying the stored analog data.

These memory devices use a ferroelectric (FE) thin film sandwiched between two metal electrodes as the storage element. These FE memory capacitors can be fabricated in a vertical or lateral configuration, as shown in Figure 8-10 [3]. In the vertical configuration, two FE thin films are connected to three electrodes. The upper film serves as the memory element of a memory capacitor, and the lower film serves as the dielectric element of a "readout" (interrogation) capacitor. In the lateral configuration, the two capacitors lie adjacent to each other, both sharing the same dielectric film and a common electrode.

Memory write operation is performed by the application of a voltage pulse of suitable amplitude and duration across the pair of electrodes to vary the remnant polarization of the capacitors. During the read operation, an ac voltage is applied across the readout capacitor to generate an elastic wave via the inverse piezoelectric effect. This elastic wave induces an ac voltage across the memory capacitor. The amplitude and phase of this ac voltage are proportional to the magnitude and direction of the remnant polarization.

These devices have a potential for fairly large-scale integration as synaptic elements of an electronic neural network.

8.5 MAGNETORESISTIVE RANDOM ACCESS MEMORIES (MRAMs)

Magnetoresistive (MRAMs) are based upon the principle that a material's magnetoresistance will change due to the presence of a magnetic field. A resistor made out a of common ferromagnetic material will have a resistive component (R) which varies as the square of the cosine of the angle between the magnetization (M) and the current. This means that for a larger angle, the resistance component is large, and for a smaller angle, the resistance component is small. This magnetoresistive effect is utilized in the fabrication of Honeywell MRAMs which use a magnetic thin film sandwich configured in two-dimensional arrays in RICMOS™ (radiation insensitive CMOS), n-well bulk CMOS technology [32]. This technology incorporates additional magnetoresistive processing to produce nonvolatile MRAM prototypes of 16K and 64K densities, and with 256K and 1 Mb circuits under development.

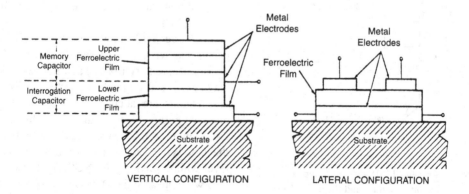

Figure 8-10. Ferroelectric memory and "readout" (interrogation) capacitors in vertical or lateral configurations for neural networks [3].

Figure 8-11(a) shows the magnetoresistive bit structure for Honeywell MRAMs [30], [31]. The magnetic storage elements formed by a sandwiched pair of magnetic film (Permalloy™) stripe are separated by an isolation layer to prevent exchange coupling. These electrically isolated film strips serve the dual function of sensing for reading and supplying a magnetic field for wiring. These films are patterned in two-dimensional arrays in which each row of bits forms a sense line, and each column forms a word line. A nonvolatile memory bit is stored at the intersection of a word line and a bit line, which allows individual access to each bit in the array.

Figure 8-11(b) shows a simplified schematic of a true differential sensing circuit

used by Honeywell MRAMs [31]. The word line has two values of current which can be applied, depending on the access cycle. A write operation uses the "write" value of the word current in combination with the appropriate polarity of the sense current to store a digital "1" or "0" to the bit at the selected bit intersection. It is the magnetic field from the combined currents which forces the desired memory state. The bit logic states ("1" or "0") are controlled by sense current polarity, or direction. A read operation uses the "read" value of the word current in combination with the sense current. When the applied currents create a magnetic field which opposes the stored magnetic flux state of a bit, a large magnetic rotation is produced. If the applied currents create a magnetic field that coincides with the stored state, a much smaller magnetic rotation results. Due to the magnetoresistivity of the sandwich bit structure, the magnetic rotation causes differing values of the bit resistance which can be electrically sensed as a stored logic "1" or "0." This is a nondestructive operation.

Figure 8-12 shows a functional block diagram of the Honeywell 16K × 1 HC7167 MRAM [31]. It has both low-power-mode operation and a power-down mode. This MRAM has a specified total ionizing dose radiation hardness of 1×10^6 rad (SiO$_2$), and can be accessed before and after exposure to a transient ionizing radiation pulse of ≤1 μs duration up to 1×10^8 rad (Si)/s under the specified operating conditions. The MRAM bit cells are immune to soft errors. These MRAMs require only a single +5 V supply, and operate over the full military temperature range.

Figure 8-13 shows the truth table, read cycle, and write cycle timing diagrams. Read operations are synchronous and nondestructive, leaving the cell logic state unaffected. The address inputs are latched into the chip by the negative-going edge of the Not Chip Select (NCS) input. A read cycle is initiated by holding the Not Write Enable (NWE) input at a high level while switching the NCS input low. Valid data will appear at the Data Output (Q) pin after time TSLQV. Data out is released to a high-impedance state by a positive transition on NCS

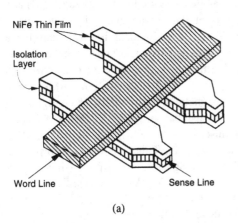

NiFe Thin Film

Isolation Layer

Word Line   Sense Line

(a)

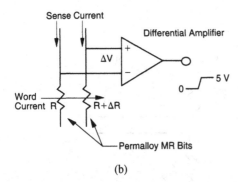

Sense Current

Differential Amplifier

ΔV

5 V
0

Word Current R   R+ΔR

Permalloy MR Bits

(b)

Figure 8-11. Honeywell MRAMs (a) magnetoresistive bit structure, and (b) magnetoresistive sensing circuit. (From [31], with permission of Honeywell, Inc.)

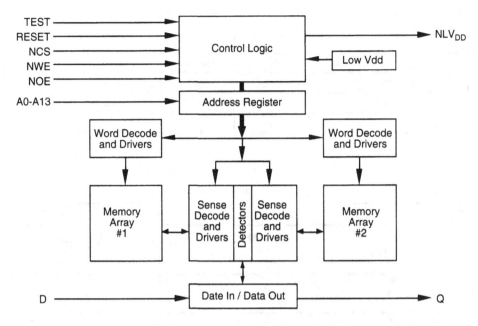

Figure 8-12. Functional block diagram of Honeywell 16K × 1 HC7167. (From [31], with permission of Honeywell, Inc.)

until the next active read access cycle. Once the address hold time (TSLAX) condition has been satisfied, the addresses may be changed for the next cycle. The NCS input must be held high for the internal array precharge time (TSHSL) before the next read or write cycle begins.

During a write cycle, the resistance of the magnetoresistive element (MRE) is altered by the application of two orthogonal currents, a sense current and a word current. The direction of the sense current will alter the resistance of the MRE to store a logic "1" or "0" state. A write cycle is initiated by holding the NWE input at a low level while clocking the NCS input also to a low level. The addresses and Data In (D) inputs are latched on-chip by the negative-going edge of the NCS input. After the write time (TSLSHW) has expired, the NCS signal should return to the high state. The read cycle time (TSLSLR) for these devices is specified as 600 ns minimum, and the write cycle time (TSLSLW) as 200 ns minimum under the worst case operating conditions.

An experimental 1 Mb chip has been designed using 1.5 × 5 μm² M-R double-layer

memory elements and bipolar circuitry based on 1.25 μm optical lithography [33]. Three masking steps are required to define the M-R element array, of which two steps are part of normal semiconductor processing. The total chip area for this design is 8.5 × 9 mm². The storage as well as the sensing element is a ferromagnetic thin film (NiFeCo) which is also magnetoresistive. Key design parameters are the number of elements in series on a sense line, the worst case supply voltage, the required number of read pulses for a specified signal-to-noise (S/N) ratio, first and second metal pitch, sense and word line currents, and M-R element array yield. Figure 8-14 shows the M-R element array geometry [33].

In this design and layout, the readback signal is quite small (≈0.45 mV) and is in the presence of significant noise, so that the S/N ratio is low and the error rate is high. A special amplifier is needed to handle the small signal, and a multiple-sample read scheme is used which results in an unacceptably long read time of 3 μs and a write time of 0.2 μs.

A fast-access (<100 ns), nonvolatile, radiation-hard, high-density (>10⁶ b/cm²),

NCS	NWE	NOE	RESET	MODE	Q
L	H	L	L	Read	Data Out
L	L	X	L	Write	High Z
H	X	X	L	Deselected	High Z
X	X	X	H	Low Power	High Z

Notes

X: VI=VIH or VIL

NOE=H: High Z output state maintained fc
       NCS=X or NWE=X

(a)

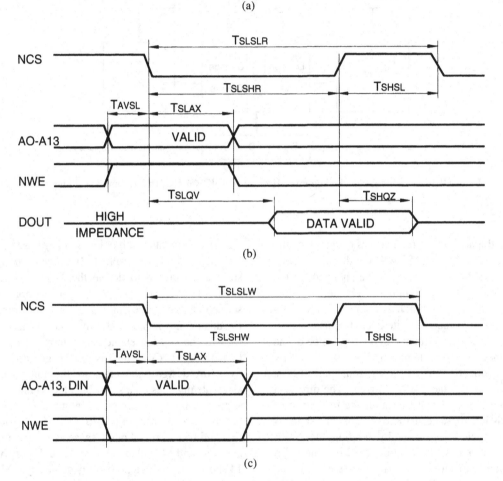

(b)

(c)

Figure 8-13. Honeywell 16K × 1 MRAM HC7167. (a) Truth table. (b) Read cycle. (c) Write cycle timing diagrams. (From [31], with permission of Honeywell, Inc.)

experimental 64 kb micromagnet-Hall effect random access memory (MHRAM) has been designed for space applications [34]. In this device, the information is stored magnetically in small magnetic elements (micromagnets) which allow unlimited data-retention time, a large

number of write/rewrite cycles ($>1 \times 10^{15}$), and inherent radiation hardness to total dose and SEU. For this high-speed MRAM, a highly magnetoresistive and high-resistivity material (not necessarily ferromagnetic), e.g., Bi or InSb, is used as a sensing element, and two ferromag-

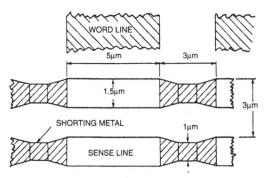

Figure 8-14. 1 Mb M-R element array geometry. (From [33], with permission of IEEE.)

netic materials, e.g., CoCr and permalloy, are used to form the information storage element. The M-R element is sandwiched between two ferromagnetic elements. The three layers are separated from each other by insulators.

Figure 8-15(a) show an MHRAM cell structure's top and side views [5]. The M-H element consists of a ferromagnetic element (micromagnet), shown in the diagram as the right-slanted area, and a Hall effect sensor, shown as the shaded area. The nonvolatile storage function is achieved with the micromagnet having an in-plane uniaxial anisotropy and in-plane bipolar remnant magnetization states. The information stored in the micromagnet is detected by a Hall effect sensor. As shown in the side view of the micromagnet, the direction of magnetization is pointing toward the right (north), and a current flowing from terminal 3 to 4 (east) in the Hall sensor would produce a Hall voltage across terminals 1 and 2, with terminal 2 being positive with respect to terminal 1. When the direction of magnetization is reversed in the micromagnet and pointing toward the left (south), the same current in the Hall sensor produces a Hall voltage with the same magnitude, but terminal 1 becomes positive. This polarity reversal between the two terminals can be easily detected by a differential sense amplifier.

Figure 8-15(b) shows the organization of a 2 × 2 b MHRAM [5]. The M-H elements are

incorporated in the gating transistors matrix, with micromagnets shown as rectangles and Hall sensors as the shaded regions. As an example, consider cell M-H 21 located at the intersection of the second row and first column. To read its contents, signals $RS2$ (Row Select 2) and $CS1$ (Column Select 1) & Read become high. This means that transistors $Q7$, $Q9$, and $Q13$ are turned on by $RS2$, which sends a current through and produces a Hall voltage at every Hall sensor in the second row. This Hall voltage is amplified by the sense amplifier at the corresponding column. However, since only transistor $Q1$ is turned on by $CS1$ & Read, therefore the signal output from sensor 21 is connected to the final output V_{out}. Although the transistor $Q8$ is also turned on when the second row is selected, no current flows through it since none of the other transistors $Q2$, $Q3$, $Q10$, and $Q11$ is turned on.

For cell M-H 21's write operation, signals $RS2$ and $CS1$ & Write become high. To write a logic "1," transistors $Q3$, $Q8$, and $Q11$ are turned on, whereas for the logic "0," transistors $Q2$, $Q8$, and $Q10$ are turned on. The stored bit value then determines the sense of current through the conductor over the magnetic element, and therefore the sense of in-plane magnetization.

For MRAMs, one of the critical memory performance parameters is the "read-out" signal levels. The output voltage of the Hall sensor (V_{out}) is given by

$$V_{out} = \mu V_x B_z \frac{W}{L}$$

where μ is the Hall electron mobility, V_x is the voltage drop across the Hall sensor, B_z is the coercivity of the micromagnet, and W/L is the width-to-length ratio of the Hall sensor. It can be seen by the above equation that the output voltage is proportional to μ. A Hall sensor film for a memory cell should satisfy the following requirements: (1) it must have an electron mobility $\geq 10^4$ cm^2/s·V for desired sensitivity and response, and (2) its thickness must be less than 1 μm to limit the readout current and size of the transistor needed to switch the readout current [38].

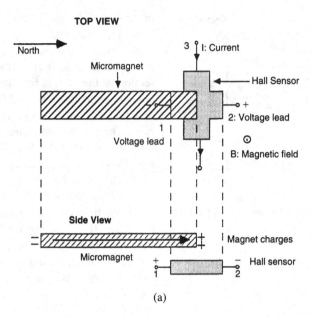

(a)

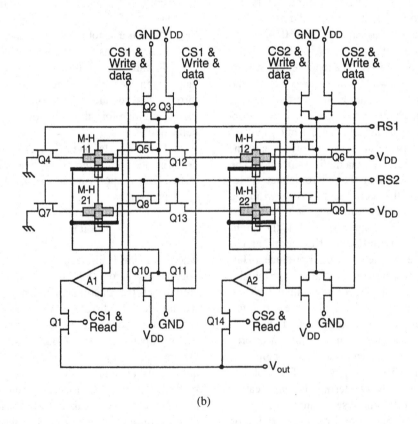

(b)

Figure 8-15. (a) An MHRAM cell structure, top and side views. (b) A 2 × 2 b
MHRAM organization [5].

Improved Hall effect sensors with adequate response (± 10 mV) and high speed have been made for the MRAMs using InAs as the micromagnet material. A single-crystal InAs cannot be easily formed directly on silicon substrate because of the lattice constant mismatch between the two materials. A solution to smoothing out the transition between mismatched crystal lattices has been to deposit the InAs film by molecular beam epitaxy (MBE) on an intermediate, electrically insulating buffer layer that consists of sublayers of materials with intermediate lattice constants. Also, improved readout circuitry has been designed for the 2 × 2 b MHRAM organization shown in Figure 8-15(b), which incorporates additional transistors to eliminate current shunts that can cause read-out errors [39].

8.6 EXPERIMENTAL MEMORY DEVICES

8.6.1 Quantum–Mechanical Switch Memories

Quantum–mechanical switch-based devices such as resonant-tunneling diodes (RTDs) and resonant-tunneling hot-electron transistors (RHETs) are being investigated for possible development of gigabit memory densities [35]. These devices are based upon the negative resistance property (or negative differential conductance) which causes a decrease in current for an increase in voltage. Figure 8-16(a) shows the characteristic I–V curve for an RHET device which exhibits normal resistance characteristics at low voltage, similar to a normal transistor. However, when the collector voltage reaches a certain threshold, the RHET device shows negative resistance, and the collector current drops off with an increase in the collector voltage.

This effect has been used in the development of a SRAM cell that uses two RTDs and one ordinary tunnel diode (TD) for the complete cell. This cell configuration produces considerably higher densities than the devices fabricated with a conventional six-transistor SRAM cell. Figure 8-16(b) shows the equivalent circuit for an RTD-based SRAM cell, and (c) shows a dia-

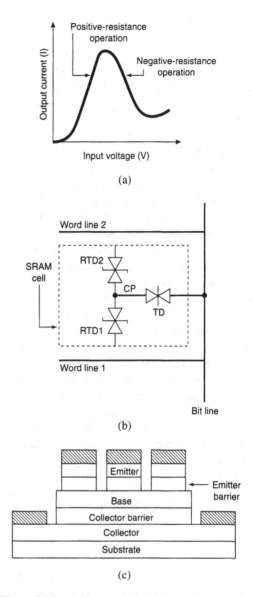

Figure 8-16. (a) Characteristic I–V curve for a RHET device. (b) Equivalent circuit for an RTD-based SRAM cell [35]. (c) Cross-section of a multiemitter RHET [42].

gram of the operating characteristics of the RTDs used. The intersection of two RTD curves, labeled $S1$ and $S2$, is the stable operating points, whereas $S3$ is unstable. The application of appropriate voltages to both ends of the RTDs makes the center point (CP) of a SRAM cell circuit assume either one of the two stable states. The application of a toggle signal to the tunnel diode (TD) which exceeds its threshold voltage makes the CP switch to the other stable state. This three-diode cell with two stable logic states has the potential for scaling up to gigabit SRAM densities.

However, the logic circuit using conventional RHETs has many resistors which can limit the speed and scale of integration. A multiemitter RHET approach has been developed which uses separate emitters and emitter barriers, and a common base, collector barrier, and collector. Figure 8-16(c) shows the cross-section of a multiemitter RHET [42]. The transistor turns on when the maximum potential difference between the emitters exceeds a certain threshold voltage—the sum of the emitter barrier forward and reverse turn-on voltages. As the transistor turns on, electrons in the low-potential emitter are injected into the base layer, where most of the electrons surmounting the collector barrier reach the collector electrode. The electrons which are scattered in the base region pass through the emitter barrier, and reach the high-potential emitter electrode which acts as a sink for the scattered electrons, similar to the base electrode in a single-emitter RHET. The emitter electrode, depending on its potential, works as both an emitter and a base.

These multiemitter RHETs have been used to form a logic family including a NAND/NOR gate, a NOR gate, an AND gate, an EX-NOR gate, a data selector, latches, and flip-flops. The fabrication of these devices is less complex than for the single-emitter RHETs because the transistor does not need a base contact. Since the multiemitter RHETs can also function as a SRAM cell, the logic circuits and SRAM cells can be integrated using the same device structure.

8.6.2 A GaAs n-p-n-p Thyristor/JFET Memory Cell

An experimental GaAs homojunction storage device has been reported, which is based upon the merger of a JFET sense channel with a p-n-p storage capacitor. This approach has been used to fabricate an integrated GaAs n-p-n-p thyristor-junction field-effect transistor (JFET) structure which displays a memory characteristic by storing charge on the thyristor reverse-biased junctions [40]. The device can be electrically programmed and erased through a single terminal. A buried p-channel, which also functions as the thyristor anode, is used to read the stored charge nondestructively over a small range of applied drain voltages (±1.5 V). The thyristor memory cell cross-section is shown in Figure 8-17. The n-p-n-p storage element is vertically integrated with the p-layer charge sense channel. A positive gate-to-source voltage less than the breakdown voltage value is applied to program a

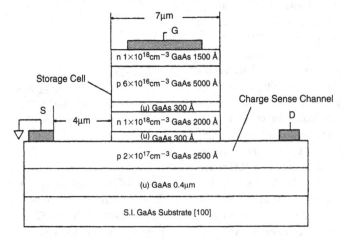

Figure 8-17. Cross-sectional view of GaAs n-p-n-p thyristor-JFET memory cell [40].

logic "1" (WRITE). This applied voltage forward-biases the middle p-n junction, which causes majority carrier depletion from the floating p- and n-epilayers. Immediately following the WRITE pulse, carriers are redistributed such that all of the junctions become reverse-biased, establishing the charged state of the device. No external voltages are required to maintain the stored charge.

The READ operation is performed by monitoring the JFET drain current by using a small drain voltage while the source and gate are grounded. To erase the stored charge, a negative gate-to-source voltage greater than the forward breakover voltage is applied, switching the transistor into its forward-biased state, which effectively neutralizes the stored depletion charge.

8.6.3 Single-Electron Memory

A team of researchers from Hitachi and Cambridge University's Cavendish Laboratory in the United Kingdom have used the quantum tunneling effects in GaAs material at extremely low temperatures (0.1 K) to demonstrate that, in principle, one bit of information can be stored by one electron in a semiconductor memory. The development is based on the device structure that used side-gated structures in delta-doped GaAs material to form a multiple-tunnel junction (MTJ), as shown in Figure 8-18(a) [41]. This design allows control over single electrons through a phenomenon known as the "Coulomb blockade effect," illustrated in Figure 8-18(b). A WRITE operation is performed by feeding single electrons from the MTJ into a capacitor. When the device is within the Coulomb blockade regime, no electrons can enter or exit the memory node. When the memory voltage reaches the boundary of the Coulomb blockade regime, one electron enters or leaves, and the electron state reverts to the Coulomb blockade regime. After another discrete increase in the node voltage, the boundary of the Coulomb blockade regime is again reached, and another electron can be transferred. To discharge the node, the added electrons must be removed one at a time, which requires a definite voltage change. This makes it possible to have two elec-

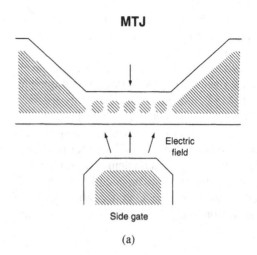

MTJ

Electric field

Side gate

(a)

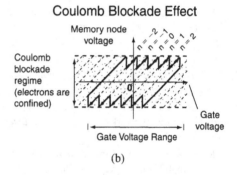

Coulomb Blockade Effect

Memory node voltage

Coulomb blockade regime (electrons are confined)

$n = -2$ $n = -1$ $n = 0$ $n = 1$ $n = 2$

Gate voltage

Gate Voltage Range

(b)

Figure 8-18. (a) A memory device that uses side-gated structures in delta-doped GaAs material to form a multiple-tunnel junction (MTJ). (b) Illustration of Coulomb blockade effect [41].

tron states for the same applied voltage, thus enabling "1" and "0" memory states to be defined. Also, if the structures can be scaled to less than 5 nm, the memory could operate at room temperature. This experimental memory device has the potential of scaling to multigigabit DRAMs.

8.6.4 Neuron-MOS Multiple-Valued (MV) Memory Technology

In real-time image processing applications, the analog signals are taken into a 2-D array of image sensors, converted to digital data, and then processed by a digital computer.

However, due to the increasing number of pixels and enormous volumes of binary information, a new scheme for data storage and computation is needed to perform completely parallel processing of analog or multiple-valued (MV) information at the hardware level [43]. An experimental RAM technology has been developed that quantizes the analog input data and stores them as MV data for intelligent data processing. This implementation of intelligent functions by a memory cell has been studied by using unique circuit configurations of the neuron MOS (νMOS) transistor, a multi-

functional device that stimulates the action of biological neurons. These development concepts were experimentally verified by the fabrication of test devices in a standard double-polysilicon CMOS process. The analysis on a four-valued system with 0.25 μm design rules, a planar capacitor, and a double-poly/double-metal process showed the gain cell area to be only 2.13 μm², which is less than three times the area of the 1-T cell for a 256 Mb DRAM. Thus, an eight-valued MV DRAM may store more bits per area than the current generation DRAMs.

REFERENCES

[1] J. M. Benedetto et al., "Radiation evaluation of commercial ferroelectric nonvolatile memories," IEEE Trans. Nucl. Sci., vol. 38, Dec. 1991.

[2] D. L. Harrington et al., "A GaAs nonvolatile memory," in Proc. IEEE GaAs IC Symp., 1991.

[3] H. Stadler et al., "Ferroelectric memory capacitors for neural networks," NASA Tech. Brief, vol. 15, no. 4, item 125.

[4] J. M. Daughton, "Thin film magnetic RAM devices," Appl. Notes, Honeywell, Inc.

[5] H. Stadler et al., "Nonvolatile high density, high speed, micromagnet-Hall effect random access memory (MHRAM)," in Technology 2001, NASA Conf. Publ. 3136, vol. 1, Dec. 1991.

[6] Ramtron Corp., "Nonvolatile ferroelectric technology and products," Tech. Rep. and Appl. Notes.

[7] J. T. Evans and R. Womack, "An experimental 512-bit nonvolatile memory with ferroelectric storage cell," IEEE J. Solid-State Circuits, vol. 23, pp. 1171–1175, Oct. 1988.

[8] W. A. Geideman et al., "Progress in ferroelectric memory technology," IEEE Trans. Ultrason., Ferroelec., Freq. Contr., vol. 38, Nov. 1991.

[9] S.-Y. Wu, "A new ferroelectric memory device, metal-ferroelectric-semiconductor transistor," IEEE Trans. Electron Devices, vol. ED-21, pp. 499–504, Aug. 1974.

[10] R. P. Feynman, The Feynman Lectures on Physics. Reading, MA: Addison-Wesley, 1963.

[11] F. P. Gnadinger, Ramtron Corp., "High speed nonvolatile memories employing ferroelectric

technology," in Proc. IEEE Conf. VLSI and Comput. Peripherals, 1991.

[12] J. T. Evans and R. Womack, "An experimental 512-bit nonvolatile memory with ferroelectric storage cell," IEEE J. Solid-State Circuits, vol. 23, pp. 1171–1175, Oct. 1988.

[13] R. Moazzami et al., "A ferroelectric DRAM cell for high-density NVRAMs," IEEE Electron Device Lett., vol. 11, pp. 454–456, Oct. 1990.

[14] W. I. Kinney et al., "A nonvolatile memory cell based on ferroelectric storage capacitors," in Proc. 1987 IEDM, p. 850.

[15] Ramtron Specialty Memory Products 1993 Data Sheets.

[16] J. R. Schwank et al., "Total-dose radiation-induced degradation of thin film ferroelectric capacitors," IEEE Trans. Nucl. Sci., vol. 37, Dec. 1990.

[17] J. M. Benedetto et al., "The effect of ionizing radiation on sol-gel ferroelectric capacitors," IEEE Trans. Nucl. Sci., vol. 37, Dec. 1990.

[18] S. C. Lee et al., "Total-dose radiation effects on sol-gel derived PZT thin films," IEEE Trans. Nucl. Sci., vol. 39, Dec. 1992.

[19] R. A. Moore et al., "Neutron irradiation on PZT thin films for nonvolatile-memory applications," IEEE Trans. Nucl. Sci., vol. 38, Dec. 1991.

[20] G. C. Messenger et al., "Ferroelectric memories: A possible answer to the hardened nonvolatile question," IEEE Trans. Nucl. Sci., vol. 35, Dec. 1988.

[21] A. Gregory et al., "Thermal stability of ferroelectric memories," in Proc. IRPS, 1992.

[22] W. A. Geideman, "Single event upset immune GaAs memories," in *Government Microcircuits Appl. Conf. (GOMAC), Dig. Papers.*

[23] J. Bond, "Analog memory—A revolution in electronics," *Test and Measurement World,* Mar. 1991.

[24] H. G. Willett, "New memory goes analog," *Electron. Buyer's News,* issue 739, Feb. 4, 1991.

[25] K. Hoen *et al.,* "A nonvolatile analog programmable voltage source using the VIPMOS EEPROM structure," *IEEE J. Solid-State Circuits,* vol. 28, July 1993.

[26] E. Sackinger *et al.,* "An analog trimming circuit based on a floating-gate device," *IEEE J. Solid-State Circuits,* vol. 23, pp. 1437–1440, Dec. 1988.

[27] T. Blyth *et al.,* "A non-volatile storage device using EEPROM technology," in *IEEE ISSCC Tech. Dig.,* 1991, pp. 92–93.

[28] M. Holler *et al.,* "An electrically trainable artificial neural network (ETANN) with 10240 floating gate synapses," in *Proc. Int. Annu. Conf. Neural Networks,* 1989, pp. 191–196.

[29] R. C. M. Wijburg *et al.,* "VIPMOS, A novel buried injector structure for EPROM applications," *IEEE Trans. Electron Devices,* vol. 38, pp. 111–120, Jan. 1991.

[30] Honeywell Solid State Electronics Center 1993 Products and Services Data Book

[31] *Honeywell Non-Volatile Magnetic Memory Technology Description and Data Sheets.*

[32] J. M. Daughton, "Thin film magnetic RAM devices," Honeywell, Inc., 1988.

[33] A. V. Pohm *et al.,* "The design of one megabit non-volatile M-R memory chip using 1.5 μm $\times$ 5 μm cells," *IEEE Trans. Magn.,* vol. 24, p. 3117, 1988.

[34] H. Stadler *et al.,* "Fast magnetoresistive random-access memory," *NASA Tech. Brief,* vol. 15, no. 4, item 124 from JPL Invention Rep. NPO-17954/7452.

[35] D. Gabel, "Tunnel to high density," *Electron. Buyer's News,* issue 833, Dec. 14, 1992.

[36] F. Goodenough, "IC holds 16 seconds of audio without power," *Electron. Design,* Jan. 31, 1991.

[37] *Information Storage Devices ISD 1200/1400 Data Sheets,* Dec. 1993.

[38] H. Stadler *et al.,* "Improved Hall-effect sensors for magnetic memories," *NASA Tech. Brief,* Sept. 1993 (NPO-18628).

[39] H. Stadler *et al.,* "Improved readout for micromagnet/Hall effect memories," *NASA Tech. Brief,* Sept. 1993 (NPO-18627).

[40] D. L. Hetherington *et al.,* "An integrated GaAs n-p-n-p thyristor/JFET memory cell exhibiting nondestructive read," *IEEE Electron Device Lett.,* vol. 13, pp. 476–478, Sept. 1992.

[41] "Single-electron memory," Semiconductor International, May 1993.

[42] M. Takatsu *et al.,* "Logic circuits using multiemitter resonant-tunneling hot-electron transistors (RHETs)," in *IEEE ISSCC Tech. Dig.,* 1994, pp. 124–125.

[43] R. Au *et al.,* "Neuron-MOS multiple-valued memory technology for intelligent data processing," in *IEEE ISSCC Tech. Dig.,* 1994, pp. 270–271.

[44] S. Onishi *et al.,* "A half-micron ferroelectric memory cell technology with stacked capacitor structure," in *Proc. 1994 IEDM,* pp. 34.4.1–34.4.4.

9

High-Density Memory Packaging Technologies

9.1 INTRODUCTION

The most common high-volume usage semiconductor random access memories and nonvolatile memories discussed in the preceding chapters are monolithic dies sealed in plastic and ceramic packages. There are two technologies for mounting first-level packages to the mounting platform: "through-the-hole" (also called insertion mount) and surface mount technology (SMT). Through-the-hole is a mature technology, although it lacks some of the desirable features of the SMT, such as small size and high packaging density, lead pitch of 50 mil or less, flexibility of mounting packages on both sides of the board, and easy process automation that can reduce the packaging costs and improve yields. However, SMT technology has disadvantages, such as more expensive substrates than for through-the-hole packages, the use of higher temperatures during soldering which may compound existing thermal problems, and it involves the use of finer linewidths and spaces that increase the potential for signal cross-coupling.

For high-reliability military and space applications, hermetically sealed ceramic packages are usually preferred because of their specified operating temperature range of -55 to $+125°C$. In the 1960s, through-the-hole, dual-in-line packages (DIPs) were introduced, and have been used quite extensively for packaging memories with pin configurations ranging from 16 to 40 pins. The ceramic DIP packages are available in several versions such as the CERDIP, side-brazed solder-sealed, and leadless chip carriers (LCCs). The mass memories for high-volume, commercial usage are supplied in plastic dual-in-line (PDIP) packages, which are nonhermetic. However, process improvements in plastic packaging including die passivation, coating the chip with hermetic material such as silicon-gel, have been developed which offer better contamination control and higher moisture resistance.

For through-the-hole assemblies requiring high-density layouts on the PC board, zigzag-in-line packages (ZIPs) are used which have a body shape similar to the PDIP, but which have leads only along one edge. This reduces the board "footprint" of a ZIP package to 100 mil compared to 300 mil for a "slimline DIP." A variation on this is the single-in-line package (SIP) which is similar to the ZIP, except that the leads are in a straight line. The pin grid array (PGA) is another insertion mount technology in use for high-pin-count devices. The erasable, programmable read-only memories (EPROMs) discussed in Chapter 3 require quartz windows in the packages for ultraviolet exposure during the erase operation. Therefore, these are supplied in either a CERDIP, or a side-brazed package with a quartz window for the die exposure.

The SMT is preferred in building mass memory computer and telecommunications systems because of the high packaging density requirements. The SMT packages lead spacing range from 50 mil for small outline (SO) packages to 20 and 25 mil for the miniature SO packages. The use of these packages in single-sided or double-sided memory board configurations can result in space savings and memory compaction by a factor of two or more. One version of the SMT package developed for memory chips which has the leads spreading outwards like "gull wings" is referred to as a small-outline package (SOP) or a small-outline IC (SOIC). It has 50 mil spacing between the leads, is thinner than a DIP package, and is widely used for 64K and 256K SRAM commercial applications. Another SO package commonly used for the memories has "J-bend" leads (SOJ) with 50 mil spacing between the leads that are curved back under the package. The SOJ packages, which are slightly thicker than the SOP packages, are preferred for larger, multimegabit DRAM chips. These packages have been standardized by the EIA JEDEC committee. A variation on the SOP packages has been the introduction of a thin, small outline package (TSOP) for applications such as smart memory cards, where high density but low packaging height are required. It has the same pinout as a standard dual-in-line package, but may have 25 mil lead spacing instead of 50 mil.

The SMT packages in chip carrier configurations are available in various sizes, lead types, and package styles. These are more suitable for high-pin-count chips used in the wordwide bus (e.g., 16 b and higher) for mass memory system applications. In addition to higher board density, the chip carrier configurations can offer faster switching times because of reduced conductor lengths and shorter interconnections. Three common types of chip carriers are flatpacks (FPs) or quad flatpacks (QFPs), ceramic leadless chip carriers (LCCs), and plastic leaded chip carriers (PLCCs). Flatpacks are rectangular chip carriers packages with gull wing leads on all four sides. The LCCs can be mounted in sockets, but are usually soldered directly to the substrate. The PLCCs are cheaper and lighter than the ceramic versions. They have J-bend leads, and can be attached using sockets, a motherboard, or directly onto the PC board. They are being widely used for very high-pin-count packages and for assembly mounting into the multichip modules (MCMs).

Figure 9-1 shows the schematic outlines and pin configurations for some of the commonly used packages for memories in both the insertion mount and surface mount technologies. The choice of packaging for a memory device depends upon the application requirements. In addition, the following other packaging technology attributes have to be considered: I/O pitch, maximum I/O count available, packaging efficiency and cost, thermal characteristics, package profile, performance and testability, ease of visual inspection, and reworkability. Table 9-1 lists commonly used insertion mount and surface mount technology packages and their attributes [1].

The increasing requirements for denser memories in computers, communications, and consumer electronics have led to further compaction of packaging technology through conventional hybrid manufacturing techniques and MCMs. A major objective of these techniques is to shorten the interconnection path (and hence the signal delay), and at the same time reduce the PC board size. Several interconnection technologies have been developed for use in the assembly of MCMs. Wire-bonding technology, used to make interconnections from a memory die to another die, substrate, lead, or a lead frame, is the oldest and most well-established technology. A tape automated bonding (TAB) assembly is also being widely used along with the SMT to interconnect the active elements to leads or the lead frame of a package, or to interconnect the active elements directly to a substrate. Flip-chip interconnection technology has been widely used by companies such as IBM and AT&T. In this technology, the chip is connected to the substrate by reflowing the solder bumps deposited on the area array metal terminals (located on the chip) to the matching footprints of solder wettable terminals on the substrate. GE and Texas Instruments have

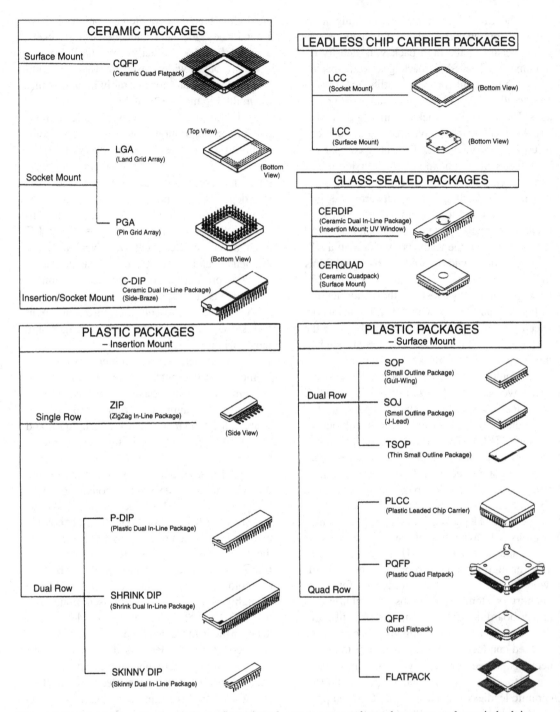

Figure 9-1. Schematic outlines and pin configurations for some commonly used memory packages in both insertion mount and surface mount technologies.

TABLE 9-1. Insertion Mount and Surface Mount Technology Package Attributes and Selection Criteria [1]

Attribute/Mounting Technology	Insertion Mount			Surface Mount					
Attribute/Package Type	SIP	DIP	PGA	SO	FP	QFP	LCC	LLCC	LGA
I/O pitch (mils):									
common	50	100	100	50	50	50	50	50	50
minimum	50	50	50	30	10	10	8	8	25
Typical maximum I/O count	24	64	>600	64	84	400	256	300	>600
Packaging efficiency	poor	poor	excellent	fair	fair	good	good	good	excellent
Package cost	low	medium	high	low	medium	medium	medium	medium	high
Heat removal*	poor	poor	good	poor	fair	fair	good	good	good
Package profile	high	high	high	low	low	low	medium	low	low
Performance	fair	fair	excellent	good	good	good	excellent	excellent	excellent
Testability	good	good	fair	good	good	good	good	good	fair
Ease of visual inspection	excellent	excellent	fair	excellent	fair	fair	fair	poor	poor
Reworkability	good	good	fair	excellent	good	good	excellent	excellent	fair

SIP = Single Inline Package
SO = Small Outline Package
LCC = Leaded Chip Carrier
* Assume no heat sinks
DIP = Dual Inline Package
FP = Flat Pack
LLCC = LeadLess Chip Carrier
PGA = Pin Grid Array
QFP = Quad Flatpack
LGA = Land Grid Array

developed high-density interconnect (HDI) MCM technology that connects complex bare chips with a multilayer overlay structure.

A commonly used multichip module configuration for DRAMs is single-in-line memory modules (SIMMs). An example is the 9 Mb DRAM module built from nine 1-Mb DRAMs in SOJ packaging. These SIMMs can be expanded by adding several rows and building a 2-D memory array. For space applications, high-density SRAM modules have been fabricated by using available radiation-hardened SRAM chips, and assembling them through the conventional hybrids or MCM techniques. For example, a 1 Mb radiation-hardened (RH) SRAM module can be assembled by using 16 64-kb chips, or 4 256-kb chips along with some additional control circuitry. In a conventional hybrid assembly, wire bonding is used for interconnection. An alternative approach would require the use of MCM assembly in conjunction with other interconnection technologies.

Several variations on MCM technology have evolved for memories with the goal of improving the storage densities while lowering the cost per bit. Since the packaging size reduction usually reduces the cost and drive performance, the goal is to minimize the device footprint. The density of chip packaging expressed as the "silicon efficiency" is determined by the ratio of silicon die area to PC board area. In a very high-speed integrated circuit (VHSIC) development program for the Advanced Spaceborne Computer Module (ASCM) developed by Honeywell Inc., the memory MCMs were fabricated using the following two technologies: (1) MCM-D using thin-film multilayer copper/polyimide interconnects on an alumina ceramic substrate mounted on a perimeter-leaded cofired ceramic flatpack, and (2) MCM-C which used a multilayer cofired ceramic substrate in an alumina package. In the chip-on-board (COB) technology, the bare memory chip (or die) is directly attached to a substrate, or even to a circuit board (such as FR 4 glass epoxy). IBM (now LORAL Federal Systems) has developed VLSI chip on silicon (VCOS) MCMs which combine high-density silicon interconnect technology

with the flip-chip, and the C4 (controlled-collapse chip connect) attach process. These planar 2-D MCM memory technologies will be discussed in Section 9.2.

An extension of 2-D planar technology has been the three-dimensional (3-D) concept in which the memory chips are mounted vertically prior to the attachment of a suitable interconnect. This stacking of memory chips, also called "memory cubing," was pioneered by Irvine Sensors Corporation (ISC) and, currently being used by several other companies such as TI and Thomson CSF, has resulted in an improvement of packaging densities by several orders of magnitude. Figure 9-2 shows a comparison between the board area (packaging density) for various packaging technologies such as TSOPs, wire-bonded MCMs, 3-D TSOPs, and 3-D bare die relative to the equivalent density DIPs on a PCB [2]. It shows that the 3-D bare die MCMs provide significantly higher silicon density of typically 435% versus 10–20% for the DIPs. These 3-D MCM technologies by various companies and some of their applications will be discussed in Section 9.3.

The MCM defects and failures can occur due to materials including the substrate, dice, chip interconnections, and manufacturing processes variations; a lack of proper statistical controls (SPCs) during fabrication and assembly; inadequate screens and device qualification procedures; and a lack of proper design for testability (DFT) techniques. The availability of "known-good-die" (KGD) and "known-good-substrate" (KGS) are important prerequisites for high-yield MCMs, and minimize the need for rework/repair. Process verification and quality assurance tests for the MCM substrates, interconnection technologies, module level screening and qualification, and other reliability issues are discussed in Section 9.4. The MCM design for testability techniques such as the boundary scan, level-sensitive scan design (LSSD) for IBM Federal System's VLSI chip on silicon (VCOS) is also discussed in Section 9.4.

Another application for high-density memory bare chip assembly has been the development of memory cards suitable for notebook

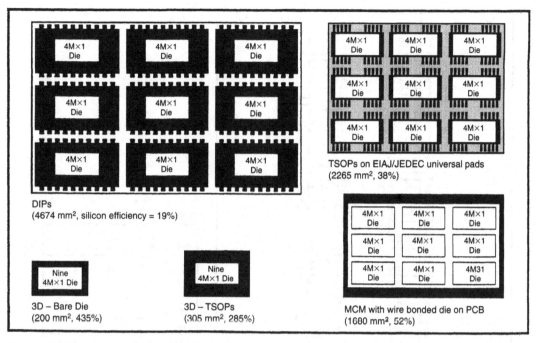

Figure 9-2. Comparison of board areas (packaging densities) of various memory packaging technologies [2].

computers and mobile communications. These are lightweight plastic and metal cards containing memory chips and supporting circuitry intended to serve as alternatives for the traditional hard disks and floppy drives. They offer considerable advantages in size, power consumption, speed, and weight. These cards integrate multiple-memory technologies such as SRAM, DRAM, PSRAM, OTPROM, EPROM, MROM, and flash memories. These memory cards are discussed in Section 9.5.

The future direction is to produce mass memories of very high bit densities ranging from tens of megabytes to several hundreds of gigabytes by integrating cubing technology into MCMs that would include fault-tolerant schemes to make them more reliable. These mass memories have applications in high-capacity solid-state recorders for space applications, high-speed rate buffers, and block storage cards. A potential application is the fabrication of a high-density hermetic configuration board (HDHC), which is briefly discussed in Section 9.6.

9.2 MEMORY HYBRIDS AND MCMs (2-D)

This section gives several examples of commercially available memory modules. Plastic-packaged memories hold a high share of the commercial market because of lower material and processing costs. Several concepts for plastic packaging of LSI memory modules have been developed, such as lead-on-chip (LOC) by IBM and chip-on-lead (COL) by Hitachi [3]. In military and space applications, high-density memory custom MCMs were developed by Honeywell Inc. for the Advanced Spaceborne Computer Module (ASCM) program, and VLSI chip on silicon (VCOS) processing by IBM as an extension of very high-speed integrated circuit (VHSIC) technology.

9.2.1 Memory Modules (Commercial)

Figure 9-3 shows the packaging configuration and pin diagram of a commercially available, 168-pin dual in-line, memory module,

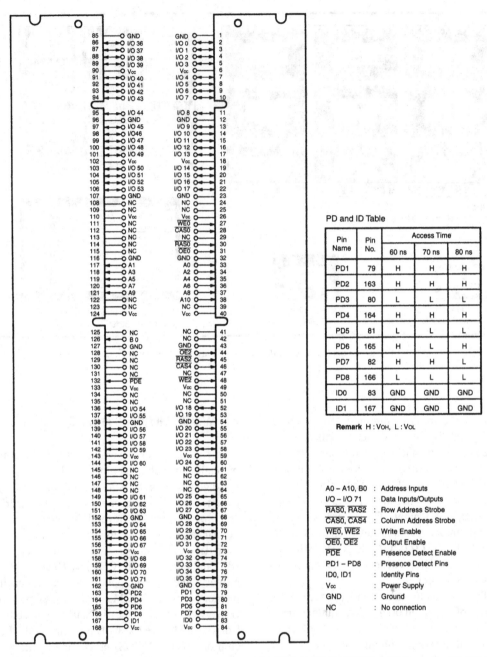

Pin Name	Pin No.	Access Time		
		60 ns	70 ns	80 ns
PD1	79	H	H	H
PD2	163	H	H	H
PD3	80	L	L	L
PD4	164	H	H	H
PD5	81	L	L	L
PD6	165	H	L	H
PD7	82	H	H	L
PD8	166	L	L	L
ID0	83	GND	GND	GND
ID1	167	GND	GND	GND

PD and ID Table

Remark H : VOH, L : VOL

A0 – A10, B0	:	Address Inputs
I/O – I/O 71	:	Data Inputs/Outputs
RAS0, RAS2	:	Row Address Strobe
CAS0, CAS4	:	Column Address Strobe
WE0, WE2	:	Write Enable
OE0, OE2	:	Output Enable
PDE	:	Presence Detect Enable
PD1 – PD8	:	Presence Detect Pins
ID0, ID1	:	Identity Pins
Vcc	:	Power Supply
GND	:	Ground
NC	:	No connection

Figure 9-3. Packaging configuration and pin diagram of NEC MC-424000LAB72F, a 4 M-word × 72-DRAM module, 168-pin, dual in-line, socket type. (From [4], with permission of NEC Inc.)

socket type (NEC MC-424000LAB72F). This is 4 M-word × 72-DRAM module assembled with 18 pieces of 4 M-word × 4-fast page mode DRAM [4]. It is designed to operate from a single 3.3 V power supply. Decoupling capaci-

tors are mounted in the power supply lines for noise reduction. The DRAM refreshing is performed using $\overline{CAS}$-before-$\overline{RAS}$ refresh, $\overline{RAS}$-only refresh, and hidden refresh cycling. It is supplied in three different versions with the following maximum access times: 60 ns, 70 ns, and 80 ns.

Figure 9-4 shows a block diagram of the NEC MC-424000LAB72F DRAM module.

Another example of a high-density (36 Mb) memory module is the Hitachi HB56D136B, which is a 1 Mword × 36 b DRAM module consisting of eight pieces of 4 Mb DRAMs and four

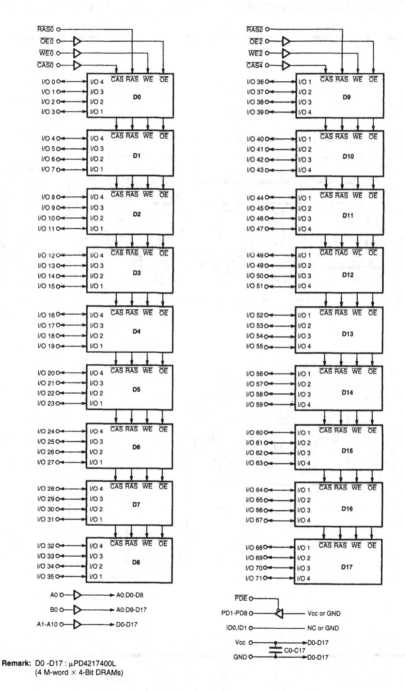

Figure 9-4. Block diagram of NEC MC-424000LAB72F DRAM module (From [4], with permission of NEC Inc.)

pieces of 1 Mb DRAMs, all of them sealed in SOJ packages and contained within a 72-pin single-in-line (SIP) package of the socket type. This is supplied in three different versions with access times of 80, 100, and 120 ns maximum. It operates from a single +5 V supply and has fast page mode capability. Figure 9-5 shows a physical outline of this memory module and its packaging dimensions [5]. Figure 9-6 shows a block diagram of the Hitachi HB56D136B DRAM [5].

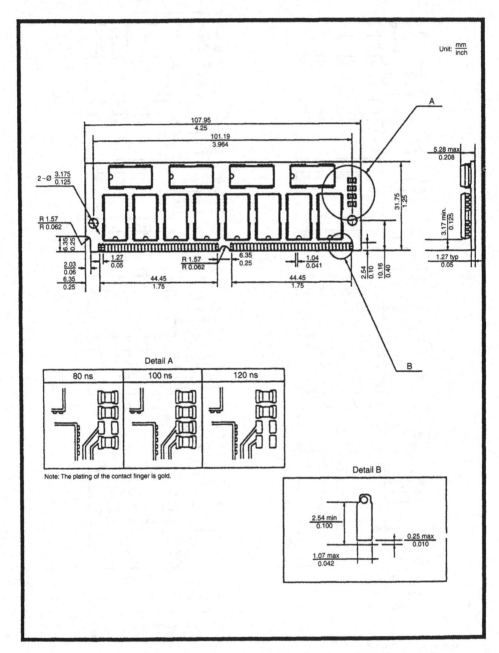

Figure 9-5. Physical outline and packaging dimensions of Hitachi 36 Mb DRAM module (HB56D136B). (From [5], with permission of Hitachi America, Ltd.)

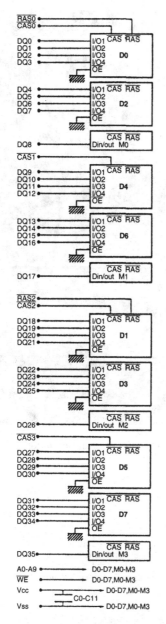

Figure 9-6. Block diagram of Hitachi 36 Mb DRAM module (HB56D136B). (From [5], with permission of Hitachi America, Ltd.)

9.2.2 Memory MCMs (Honeywell ASCM)

The Honeywell ASCM is a general-purpose Advanced Spaceborne Computer Module based on MIL-STD-1750A [6] processor design with very high throughput and large memory requirements. It uses three types of MCMs. One MCM contains Honeywell's Generic VHSIC Spaceborne Computer (GVSC) five-chip set. The second MCM is a memory module with 64K × 8 b architecture and containing nine 8K × 8 SRAMs. The third MCM is a cache memory containing six 8K × 8 SRAM dies and a cache controller chip. These MCMs have been fabricated in both copper/polyimide thin-film multilayer (MCM-D) and cofired ceramic technology (MCM-C).

Figure 9-7 shows a photograph of the memory MCM-C designed as an integrated cofired ceramic module with package dimensions of 1.6 × 2.5 in^2 and 122 leads [2]. It contains nine SRAM dies and one memory line driver. SRAM die sites are designed to accept either a 64K or 256K SRAM die. It has four signal routing layers (versus two in MCM-D), and 13 total metal layers counting the wire-bond and seal ring layers. Figure 9-8 shows a photograph of cache memory MCM-D with package dimensions of 1.6 × 2.5 in^2 and 244 leads. It contains six memory ICs and a cache controller IC [2]. These MCMs use epoxy die bond technology for die attach which permits rework (replacement of a defective die).

9.2.3 VLSI Chip-on-Silicon (VCOS) Technology

The MCM technology called VLSI chip-on-silicon (VCOS) developed by IBM Federal Systems combines high-density silicon interconnect technology with the flip-chip, C4 (controlled-collapse chip connect) die attach. VCOS technology provides high I/O chip capacity and low thermal resistance by the area array of C4 solder bumps that permit wiring directly beneath the chip [7]. The VCOS package consists of a multilayer ceramic substrate, a silicon interconnect wafer (SIW) on which all the active elements are attached, IC chips (or dies), and a protective cover to provide hermeticity. The SIW is fabricated with the same process as the 1.0 μm (or 0.8 μm) CMOS chip design.

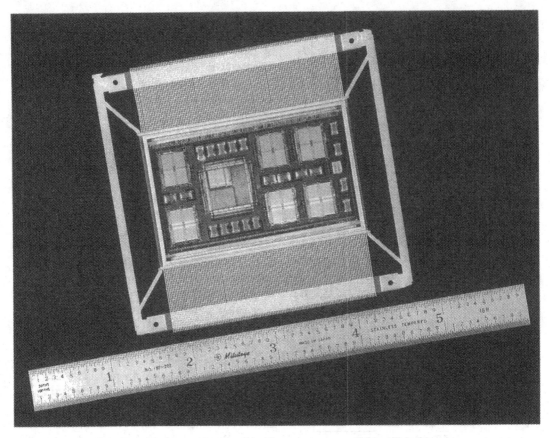

Figure 9-8. Photograph of Honeywell cache memory MCM-D. (From [2], with permission of Honeywell Inc.)

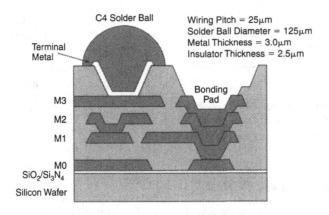

Figure 9-9. (a) VCOS interconnecting silicon substrate cross-section. (From [8], with permission of IBM Federal Systems.)

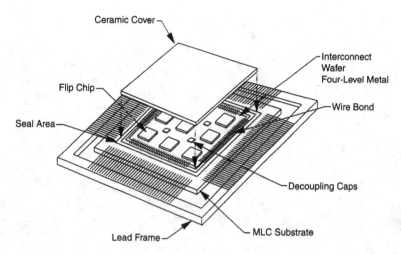

Figure 9-9 (Cont.). (b) Overview of a typical 308-lead VCOS package. (From [8], with permission of IBM Federal Systems.)

synchronous operation using on-chip address latches. Figure 9-10 shows the physical (silicon carrier) layout with 18 memory chips, a buffer chip, and 12 capacitors [9]. Figure 9-11 shows a photograph of the VCOS 1 Mb memory module [9].

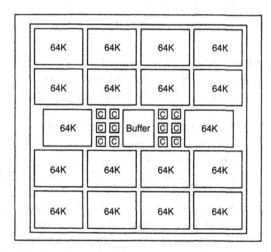

Figure 9-10. Physical (silicon carrier) layout of VCOS 1 Mb memory module. (From [9], with permission of IBM Federal Systems.)

9.3 MEMORY STACKS AND MCMs (3-D)

Three-dimensional (3-D) memory packaging and interconnectivity, which can be considered as an extension of 2-D MCM technology, provides higher packaging density and improved electrical performance because of lower interconnect parasitics. The 3-D approach also reduces the substrate size, module weight, and volume. The footprints for 3-D packages vary from a single die to MCMs. Three fundamental packaging techniques have been developed for 3-D memory chip assemblies: packaged chips, bare chips, and MCMs. There are other experimental techniques such as stacked wafers and folded flex circuitry that have not been commercialized yet [10].

The packaged chip 3-D stacking approach includes both standard and custom packages. For example, Thomson-CSF mounts memory TSOPs to copper lead frames, stacks the packages, and then molds the stack in plastic. This stack is sawed into a cube, and the edges of the stack are interconnected by laser-patterning a metal layer deposited on the surface of the cube. Mitsubishi stacks TSOP packages and solders the leads to a pair of PC boards on each side of the stack. For high-reliability hermetic applications, Dense-Pac

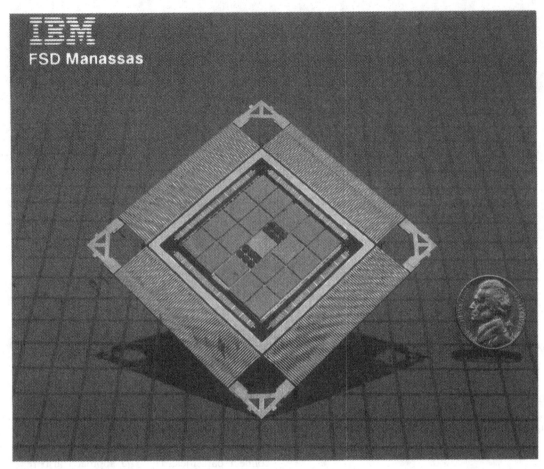

Figure 9-11. Photograph of a VCOS 1 Mb memory module in 308-lead package. (From [9], with permission of IBM Federal Systems.)

Microsystem offers stacked ceramic leadless chip carrier (LCC) packages. Harris Corporation has developed a low-temperature cofired ceramic (LTCC) tub that can hold two memory chips, with the tubs stacked in such a way so as to hermetically seal each layer. To increase the packaged chip density even further, Dense-Pac Microsystem has developed a nonhermetic stackable package that uses a two-layer LTCC substrate in which the IC is glued to the ceramic and interconnected with the wire bonding [11].

In addition to packaged chip stacking, the other commonly used 3-D memory configurations involve the use of bare dies and MCMs. Four generic types of 3-D packaging techniques currently being used by several manufacturers

are: layered die, die stacked on edge, die stacked in layers, and vertically stacked modules. These four techniques are briefly described as follows:

• Layered die is a single multifunctional die containing artwork from several unique dies fabricated onto thin layers of recrystallized silicon with thin dielectric layers separating them. The base IC is produced on a monocrystalline silicon wafer using standard wafer processing techniques. Figure 9-12(a) shows the details of a three-layered die structure [1]. In this technology, still under development, the major areas under investigation are the maintenance of proper

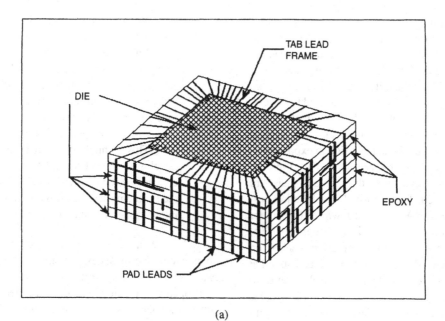

(a)

(b)

Figure 9-12. Schematics of (a) three-layered die structure, and (b) die stacked on edge of a 3-D memory module and its edge cross-section [1].

orientation of crystal growth and improvements in methods to minimize damage to the sub-IC layers resulting from excessive heat during recrystallization. Thermal mismatch can interfere with smooth crystal growth.

• The die stacked on edge approach uses a group of cut dies that are attached to-

gether, face to back, and that utilize some methods of pad alignment to provide a flip-chip mountable block of die. The manufacturing process begins with a wafer containing uncut memory die from an IC manufacturer. This method of die assembly requires relocation of all I/O pads to one side of the device. A method commonly used involves laying down a layer of metallization over the existing die for routing traces from the die pads to the side. The dies are tested using wafer probes to identify rejects. The wafers are lapped and thinned to a desired thickness. An individual die is laminated, often with epoxy, to form a die block which is then lapped on the contact side to expose electrical interconnects for the flip-chip mounting. Then the contact pads are formed. The deposited metallization provides I/O lines onto which the solder bumps are deposited. Figure 9-12(b) shows a schematic of a die stacked on edge 3-D memory module and its edge cross-section [1]. In die stacked on edge technology, some of the problem areas which can cause defects are: nonplanarity at the wafer level, the die pad relocation process, misalignment during die lamination, poor metallization attachment to the die pad as a result of contamination, and incorrect photomasking on etching.

- The die stacked in layers approach produces a 3-D memory module similar to the die stacked on edge, with an orientation rotation of 90°, which requires a different electrical interconnect scheme. Typically, the dies are mounted conventionally, and the pads are extended beyond the die by using TAB interconnections or wire bonding on tape. This is an emerging 3-D technology, and the reliability problems associated with this are similar to that with the die stacked on edge. This process uses TAB (instead of metallization to expose the die pads), which has an advantage over the die stack on edge by allowing individual screening of the die (e.g., burn-in) prior to the module assembly.

- The stacked module 3-D packaging approach utilizes vertical interconnection of either single-memory chips or the MCMs. The process flow requires die mounting to a single chip or MCM by using any of the standard interconnect technologies, such as wire bond, TAB, flip-chip, or HDI. The fabrication of 3-D MCM requires additional interconnect pads on both sides of the substrate for connection to the neighboring modules. The vertical interconnects are formed by elastometric connectors through-the-hole pins and surface mount pins. This technology is currently under development, and has all the reliability concerns associated with the MCMs plus the problems associated with the vertical interconnect MCM assemblies.

In 3-D memory packaging, vertical interconnection is a critical design issue. Several vertical interconnection techniques have been developed which can be broadly classified as peripheral connections or array connections. The peripheral connections require that all signals be routed to the edge of the stack, and then routed from layer to layer. The array connection techniques allow for more vertical channels than the peripheral connections. Also, many array connection techniques offer flexibility by allowing custom placement of the vertical connections for each layer. The several vertical interconnection techniques developed include thin-film metal deposition and patterning, solder connections, laser-machined conductors, metal pins, and TAB tape.

The advantages of 3-D memory stacks include reduced volume, lower line capacitance and drive requirements, and reduced signal propagation delay times which allow mass memory densification. Some of these 3-D memory stacks and MCM technologies from various manufacturers are discussed as examples in the following sections.

9.3.1 3-D Memory Stacks (Irvine Sensors Corporation)

In 1985, the Irvine Sensors Corporation (ISC) began exploring 3-D die-on-edge stacking technology as a more efficient alternative to the existing planar IC packaging approach. Subsequently, two memory chip stacking technologies were developed: short stacks and full stacks [12]. The short stack puts 4–16 memory chips in a "stack-of-pancakes" configuration to produce a thick IC with typical dimensions of 1.25 × 0.65 × 0.025 cm³. The full stack can place up to 100 memory chips in a "loaf-of-bread" configuration, with typical dimensions of 2.5 × 1.25 × 0.65 cm³. This full stack is typically interconnected to the next higher level of assembly by bump-bonding to a substrate or a wire-bonding into a deep, custom-designed package. The short stack is designed for wire-bonding or TAB interconnection into the standard packages, SMT, or chip-on-board mounting.

The ISC process for 3-D memory modules begins by procuring chips to be stacked in wafer forms, and then applying a new metallization layer to reroute the chip leads to the edge of a device. The wafer is then tested and thinned to 7 mil or more. Then the IC chips are diced out and laminated to form short stacks or large stacks. The stacked chips and stacking fixture are placed in an oven and baked at the curing temperature. The "face" of the stack is then plasma-etched to ex-pose thin-film metal leads, which are typically 1 μm thick and 125 μm wide. After sufficient etching, several layers of polyimide are deposited over the stack face to allow for a depth coverage somewhat greater than the length of the exposed metal thin-film leads. This polyimide serves as a passivating layer between the silicon chips and metal pads/bus lines deposited later in the process. The pads and bus lines on the stack face are formed by lift-off photolithography and sputter deposition of Ti-W/Au. This pad to thin-film metal lead interconnect is called the "T-connect."

The memory short stacks were developed by ISC for applications that did not have the head room required for full stack custom packages. These short stacks can use up to 16 memory dies in a standard package. The top layer is a ceramic "cap chip" which provides bond pads suitable for the wire bonds, TAB, SMT, or chip-on-board.

Figure 9-13 shows a photograph of a full stack 3-D memory module [13]. For achieving interconnect redundancy and to maximize the

Figure 9-13. Photograph of ISC full stack 3-D memory module. (From [13], with permission of ISC.)

heat dissipation in full stacks, solder bumps are placed at the bottom of the cubes so that the heat can be dissipated through the metal bumps into the substrate. The 3-D full stack applications include the main memories for large computer systems, cluster memories for parallel processors, and cache memories for high-performance disk systems. The use of 100 dies full stack can provide better than 50:1 improvement in the packaging density of main memory boards. For cache memory applications, full 3-D stack implementation on the disk controller board using 16 Mb DRAM chips can provide over 200 Mb of on-line, high-speed data buffering.

9.3.1.1 4 Mb SRAM Short Stack™ (Example).
Figure 9-14 shows (a) the pinout diagram, and (b) the block diagram of an ISC 4 Mb SRAM short stack module which uses four 128K $\times$ 8 b SRAMs and is organized as 512K words $\times$ 8 b [14]. Four chip enable inputs $\overline{CE1}$, $\overline{CE2}$, $\overline{CE3}$, and $\overline{CE4}$ are used to enable the short stacks' four memory chips independently. SRAM memory write operation is performed when write enable ($\overline{WE}$) and one of the four chip enable ($\overline{CE}$) inputs are low. Data are read when $\overline{WE}$ remains

HIGH, and $\overline{CE}$ and output enable ($\overline{OE}$) are held low. Figure 9-15 shows a photograph of this 4 Mb SRAM short stack module [14].

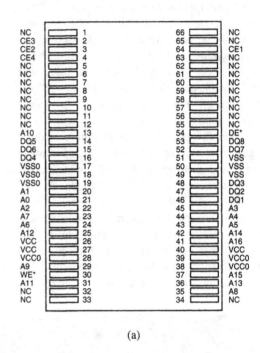

(a)

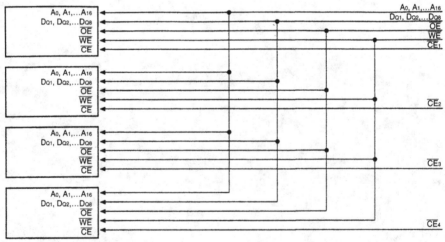

NOTE: Vcc, Vco, Vss, Vsso are not shown
Voltage on Vcc Supply Relative to Vss is −1V to +7V
Short Circuit Output Current is 50 mA

(b)

Figure 9-14. A 4 Mb SRAM short stack module. (a) Pinout. (b) Block diagram. (From [14], with permission of ISC.)

wafer level or the creation of solder bumps, as is the case with ISC and TI designs [16], [17]. It uses standard, off-the-shelf chips and six sides of the cubes as a routing surface instead of a single side, which reduces the cost of mounting substrates. The Thomson CSF 3-D memory cube basic processing sequence can be summarized as follows:

1. The gold wires are bonded using an automated wire-bonding machine to interconnect the "chips on tape," which allows burn-in and testing of each chip prior to stacking.

2. The n "chips on tape" sets are glued to the epoxy layer.

3. The sets of wires between the chips and tape are trimmed with a diamond saw so that only the chips and their wires are preserved.

4. The gold wire cross-sections are metallized with a chemical deposit.

5. The side conductors are drawn with a laser.

The memory cubes (or blocks) can be mounted in ceramic or metal hermetic packages for space and military applications. However, these can also be mounted on epoxy substrates to reduce costs. Thomson CSF has built memory cubes consisting of eight 256K SRAM memory chips as technology characterization specimens. Memory cubes are being developed for assembly into ultradense memory cards with commercial and military applications.

9.3.3 3-D Memory MCMs (GE-HDI/TI)

GE-HDI technology, which was originally applicable to 2-D MCMs, was extended to 3-D through a program for memory densification by the application of z-plane lithography (MDAZL). This was a collaborative effort between ISC and Texas Instruments (TI), sponsored by the Strategic Defensive Initiative Organization (SDIO) [18]. The goals of the MDAZL program were aimed at ISC to package four memory chips in a short stack (with a form

factor of a single chip but thicker) and GE to integrate those test assemblies into a hybrid wafer scale integration (WSI) process. The MDAZL approach used a modified configuration of the ISC 3-D traditional stacking process (discussed earlier), in which the edges of individual chips are perpendicular to the substrate and the "memory cube" is interfaced to a mounting substrate by a solder bump array laid down on the interconnect face. In the MDAZL approach, the stack consists of four chips (or more, if needed) and a top surface ceramic interconnect cap chip. The MDAZL stack and cap are mounted in a planar fashion, similar to the ISC short stack configuration shown in Figure 9-15. These short stacks are then integrated into the GE-HDI process to build single-stack (or multiple-stack) modules.

The MDAZL concept was demonstrated for quad-stack modules which have a memory density four times the capacity of a single-chip layer HDI. A single MDAZL HDI layer using a 16 Mb DRAM die currently available can have a gigabit of storage capacity per 2×2 in^2 module.

TI, under contract with the Defense Advanced Research Projects Agency (DARPA), developed the 3-D HDI for memory packaging as an extension of its silicon-on-silicon (SOS) 2-D HDI process. This was done by combining identically sized 2-D HDI modules into a very compact assembly through direct stacking. One of the goals of this multiphase program was to package 4 G-bytes of memory into 100 in^3. Figure 9-16 shows a simplified sequence of this 3-D stacking approach [19]. The electrical contacts are formed on the edges of individual HDI modules that are combined into a 3-D assembly. After the HDI module stack is laminated together, new HDI patterned overlays are created that interconnect the edges of individual layers together. This results in a robust 3-D package that can survive severe thermal and mechanical storage environments.

The key features of this 3-D packaging approach are TAB and a multilayer polyimide interconnect substrate. The TAB lead frame provides both a mechanical and electrical connection between the memory chip and the

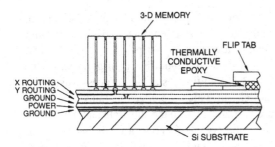

Figure 9-16. Simplified schematic of 3-D memory HDI cross-section. (From [19], with permission of TI.)

interconnect substrate through a butt-lead solder joint [20]. The polyimide interconnects are manufactured on silicon wafers where polyimide is used as the dielectric and aluminum as the circuit metallization. An application of this 3-D memory technology, discussed in the following section, is the solid-state recorder (SSR) for military and space applications.

9.3.3.1 3-D HDI Solid-State Recorder (Example).

TI 3-D HDI MCM packaging technology was used for the development of a solid-state recorder (SSR) for DARPA. The system requirements called for a high-density, low-power, and high-reliability digital data recorder with initial storage capacity of 1.3 Gb expandable to 10.4 Gb. This SSR had data storage and retrieval rate specifications of up to 10 Mb/s (continuous) over a single serial data interface, and a demonstrated capability of performing extensive self-test and reconfiguration operations.

The basic SSR architecture consists of several gigabit memory units (GMUs), each containing 1.3 Gb of SRAM; memory driver unit (MDU) ASICs to provide the interface among the address, data, and control signals of the MBUS and high-density memory modules; and a memory interface unit (MIU) which contains all the logic necessary to support up to eight GMUs within the SSR. The MIU is further partitioned into two major functional blocks: a system interface unit (SIU), and a system control unit (SCU). Figure 9-17 shows (a) the GMU architecture, and (b) the MIU architecture [21].

Figure 9-18 shows a photograph of the GMU which measures approximately $4 \times 6 \times 0.4$ in^3 and weighs only 0.5 lb [22]. It requires 1.4 W of power in its fully active mode. This GMU uses flip-tape automated bonded (TAB) and consists of 1280 SRAMs (1 Mb each), ten CMOS ASICs, four buffers, and some discrete components, all placed on high-density substrates. Seven individual substrates are used, of which five are known as the memory substrates. These five substrates are identical, each containing 32 3-D stacks, two ASICs for memory interface and control purpose, and 16 discrete components. The remaining two, known as buffer substrates, contain the buffers, discrete components, and provide connection to an 8 b parallel bus.

The MIU which provides interface and control functions required by the SSR system is a standard multilayer printing wiring board (PWB) with through-hole and surface mount components. It includes a microcontroller (TI TMS 370), a processor interface controller (PIC), an address generation unit (AGU), a bus interface controller (BIC), and an error control unit (ECU). Several 3-D stacks are mounted on the MIU card to provide storage for the SSR's error and configuration data.

9.3.4 3-D Memory Stacks (nCHIPS)

A new, low-cost process for stacking memory chips up to four high on an MCM substrate has been developed by nCHIP, Inc. [23]. This technology is particularly useful when utilized with a thin-film interconnection substrate (MCM-D) which typically leads to 2–4× reduction in the manufacturing cost of the MCM substrate. The resulting increase in routing density leads to much more efficient use of the high-density interconnection capability of an MCM-D substrate. This technology, referred to as "laminated memory," is derived from a conventional dicing, die attach, and wire-bond process. It enables chips to be stacked "flat" with very close spacing (~0.4 mm chip pitch), so that a short (two–four chip) stack could be produced without increasing the overall package thickness.

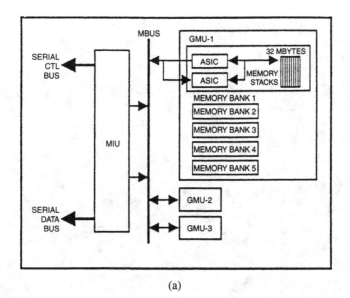

(a)

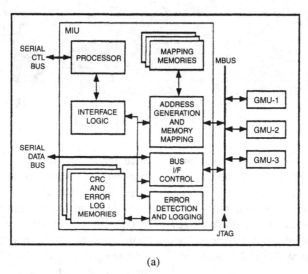

(a)

Figure 9-17. Solid-state recorder (SSR). (a) GMU architecture.
(b) MIU architecture. (From [21], with permission
of TI.)

Figure 9-19 shows a cross-sectional view of a laminated memory three-high stack architecture [23]. The basic approach consists of a repeated sequence of die attach and wire bonding. The bottom edges of upper chips are mechanically beveled or notched on all sides which contain wire bonds such that the wire bonds on the lower chips are not physically contacted when the upper chips are attached. To achieve a minimum stack height, the chips are prethinned, typically in the range of 0.3–0.4 mm. The first (lowest) thinned memory chip in each stack is epoxy-bonded to the MCM substrate using a conventional die attach process. After epoxy curing, all chips are wire-bonded to the substrate using a low-height wire-bond loop profile

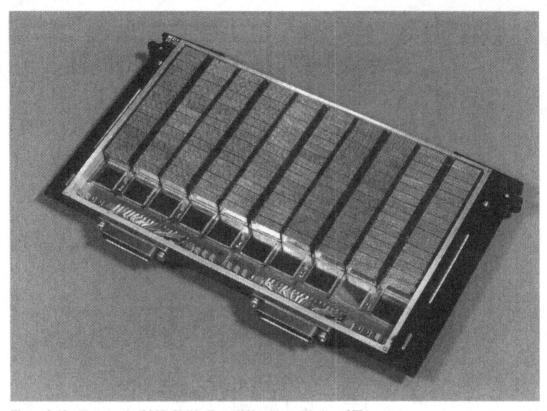

Figure 9-18. Photograph of SSR GMU. (From [22], with permission of TI.)

(ideally <4 mil). After the wire bonding is completed, the next tier of chips is attached using electrically insulating adhesive, typically 2 mil thick. After the adhesive is cured, the wire-bonding process is repeated on the upper chip. In this case, the bond wire looping parameters will be somewhat different due to the increased

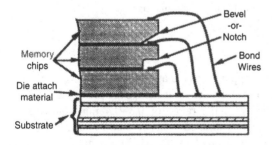

Figure 9-19. Cross-sectional view of a laminated memory three-high stack architecture. (From [23], with permission of IEEE.)

wire height, and in general, the substrate bond pads will be slightly farther away from the chip edge than the first chip.

High-accuracy automated die attach equipment is required for manufacturing the "laminated memory" modules in order to avoid damaging the underlying wire bonds. The chip beveling or notching step, which is required on all upper level memory chips (and optional at the bottom level), is performed when chips are provided in wafer form. In case the memory chips are available only in die form, a special fixture can be fabricated to perform the beveling operation.

A multichip memory module was designed using the laminated memory process for a two-high stack. Then fully functional modules were fabricated using 32K × 8 SRAM chips from two different suppliers. No failures were observed on a few sample memory modules that were subjected to 50 cycles of thermal shock.

However, more extensive reliability testing is needed to further characterize this laminated memory technology.

9.4 MEMORY MCM TESTING AND RELIABILITY ISSUES

Memory MCM technology from the testability and reliability points of view lies somewhere between the chips and a board level system. Therefore, a well-defined test strategy and reliability screens are important for high-yield and reliable MCMs. The availability of "known-good-die" (KGD) fully tested and burned in for memory and control logic functions is essential for minimizing the rework and repair of final MCM assembly. The yield of bare die currently available for MCM assembly may vary widely, depending upon the extent of wafer level testing, device geometry and complexity, and the semiconductor manufacturing process maturity. In general, the effect of a memory die yield on a projected MCM yield can be expressed by the following equation [24]:

$$Y_{MCM} = Y_C^{(NC)}$$

where Y_C = chip yield

NC = number of chips in MCM assembly.

This means that even with memory (or other logic control) die yields as high as 97%, the chance of finding a defective chip in a 20-chip MCM assembly is about 50:50. In addition to the KGD requirements, known-good-substrate (KGS) is also an important issue. Some other strategies for high-yield and reliable MCMs include the use of process monitors (PMs) and other reliability test structures, design for testability (DFT) and robust circuit design techniques, high fault coverage testing and accurate and efficient diagnostics. The MCM test structures have been developed for process verification and monitoring in terms of material characterization, geometric tolerances, thermal

management and stress evaluation, and assembly techniques [25].

In general, the memory MCM testing may be divided into two categories: (1) an assembly test to verify the connectivity of components to each other and to their substrate, and (2) performance testing to verify the functional and parametric specifications (dc and ac) [26]. Many of the MCM faults may not show up until the MCMs are driven "at speed." Since the MCMs are also assemblies with a requirement for in-manufacture repairability, testers are needed that should also have diagnostic capabilities. The Advanced Research Projects Agency (ARPA) specifications for MCM testing require test equipment capability >100 MHz. However, the use of a high-speed tester may not be the optimum solution for a memory MCM with several complex dies for which the development of an appropriate functional test vector set could become a prohibitive task. In this case, a test strategy that combines a KGD or wafer level testing with more limited functional testing may be a more viable and cost-effective approach.

The MCMs require a large number of test vectors which are generated during the design and simulation cycles. A current trend is to include the boundary scan architecture in MCM designs to facilitate testing in conformance with the JTAG IEEE 1149.1 Standard and to increase the fault coverage. For devices like the memory chips that may not include scan circuitry, boundary scan techniques can still be implemented to perform testing through the surrounding JTAG circuitry. Memory testing requires the capability to perform read/write operations at system speeds, and the performance of periodic memory read operations to check data retention. The boundary scan architecture requirements for the serialization of control signals and the data line may increase memory access time delays by an order of magnitude. Therefore, even though the boundary scan methodology may be inadequate for memory MCM testing, it is an important element of overall test strategy. For example, if the MCM contains a gate array for memory control logic, then the module architecture can take on several different configurations, depending upon system applications. In this case, testability requirements may include the following: a built-in-test (BIT) for memory chips implemented through the gate array, and a scan-path and boundary scan test for the gate array.

Some MCM manufacturers are using logical synthesis tools which automatically include JTAG circuitry in the design. ATE manufacturing companies such as Teradyne have developed automatic test generation software for boundary scan devices that can generate interconnect test patterns to provide 100% pin-level fault coverage on the MCMs with full boundary scan networks [26]. Since the JTAG boundary scan is a static test technique, it does not directly aid timing performance testing for the delay faults. However, there are some scan test techniques which use internal scan in conjunction with the system clock to test for timing delays.

An example of boundary scan as an embedded MCM test methodology is the design verification and test of a gigabit memory unit (GMU) used in the solid-state recorder (SSR) discussed earlier in Section 9.3. It uses a boundary scan modification which enables a single scannable memory interface ASIC to test the interconnects for stuck-at-zero faults and shorts on control, address, and data signal paths to the memory devices [27]. In this design, the difficulty of probing fully populated memory bank substrates was the motivation behind the addition of a boundary scan through memory driver units (MDUs). These MDU devices are first placed on the routed substrate, and the boundary scan is used to test the MDU and substrate interconnects to detect the shorts to ground and between the adjacent interconnects. Then the 3-D memory stacks are mounted on the substrate and the MDU testing is repeated. Also, a modified March algorithm is performed on selected memory locations on each memory die in the stack. Typically, the boundary scan interconnect testing utilizes at least two scannable devices, one at each end of the interconnect. However, some design modifications to the typical boundary scan circuit were made to test for shorts on the substrate interconnects without requiring an-

other device on the scannable interconnect. The readback modifications saved the considerable time required to isolate each of the detected interconnect faults.

In addition to the DFT techniques, the known-good-die (KGD) concept is being actively pursued by MCM manufacturers. The Advanced Research Projects Agency (ARPA) has sponsored the KGD assurance technology guidelines study to identify the materials, processes, equipment, and information needed to ensure that the unpackaged die would have the same quality level and reliability as a packaged die. Currently, the KGD test flow of MCM manufacturers and IC chip suppliers ranges from a simple dc probing to complete ac/dc three-temperature testing and burn-in per MIL-STD-883 requirements. There is a need to establish temporary packaging of the IC chips in order to perform a screening prior to use in the MCM or chip-on-board applications. Cost-effective techniques and high-yield methods are being developed to aid the bare die preparation for test, analysis, and burn-in [28].

A bare chip test methodology has been developed based upon GE-HDI technology that uses a proprietary chip coating which, although not hermetic, protects the chip from testing, handling, probing, or process-induced damage. The overcoat polymer can be spray-coated onto bare chips which are placed edge to edge on a flat substrate, or after they have been placed into the milled substrate [29]. The overcoat can have vias drilled down to the chip pads. Metallization is done using a fully sputtered process and metal patterning with standard HDI photolithography. This process can be used to form temporary wire bonds on the overcoat polymer that are connected to the chip pads. The bare chips can then be placed into a standard chip carrier (e.g., PGA or LCC) and wire-bonded into place. These "temporarily packaged bare chips" can be tested and burned in with standard package fixtures and test equipment. After the completion of test and burn-in, passing chips are recovered for integration into the HDI modules by removing the wire bonds from temporary pads and by dissolving the die attach polymer in a solvent soak.

The HDI process allows implementation of a unique interconnect pattern on a multichip substrate to enable full testing of the complex microcircuits with a temporary interconnect overlay. In this approach, the chips are mounted on a sacrificial one or two layers of HDI substrate and interconnected to a unique net list (not the final circuit net list) that maximizes testability and fault isolation. This works well with the MCMs that use memory chips and have all pads available, since it allows running of algorithmic patterns such as Checkerboard, March (walking 1/0), and Galpat to detect for normal RAM failure modes.

Bare die testing is also accomplished by utilizing a modified wafer probe system and a high-speed digital tester for "at speed" performance characterization of the individual chips prior to assembly into MCMs [30]. An extension of the conventional wafer probe is the hot chuck probing technique, in which the probing is done at room or slightly elevated temperatures with each die on the wafer tested for continuity, functionality, and limited parametric performance [31]. The parametric tests include voltage and current levels at dc and sometimes ac at low frequencies such as 1 MHz. TI has entered into a partnership with a company for the development of a thin-film membrane probe technique for burn-in and testing of bare die IC chips. This technique is quite suitable for high-volume applications since it requires that the IC die be dropped into a specially designed support frame which is subsequently plugged into a test socket.

The MCMs used with bipolar chips utilize a multilayer ceramic (MLC) substrate [32]. For TTL logic devices, the ac defects and burn-in-related problems are not as significant as for CMOS devices. Therefore, for a high module level yield, 100% dc stuck-at fault coverage at the die level is usually the goal. The memory devices are not considered dc test problem. A level-sensitive scan design (LSSD) approach provides an effective means to test for transition and delay faults. Also, the circuit enhancements and design for quality (DFQ) can reduce the impact of circuit delay defects. For MCM reliability, a rigorous implementation of thermal management, power distribution, and power conditioning is necessary.

A well-designed screening and quality conformance inspection (QCI) program for the package level MCMs can provide a cost-effective means of controlling failures. For high-reliability space application memory MCMs, screening and QCI requirements are based upon military specifications for microcircuits (MIL-M-38510 and MIL-I-38535) and for hybrids (MIL-H-38534). These were reviewed in Chapter 6. An example of DFT methodology, chip level testing, packaged MCM screening, and QCI flow for memory VCOS (IBM Federal Systems) is discussed in the following section [7], [30].

9.4.1 VCOS DFT Methodology and Screening Flow (Example)

VCOS packaging technology for 2-D memory MCMs was discussed briefly in Section 9.2. This silicon-on-silicon technology may have 4–20 chips mounted on a single substrate, which implies that the integrated package may consist of 400,000 primitive logic blocks, 10,000 latches, and up to 256 signal I/Os. This presents a major testing challenge. A good chip level testing is considered critical to high MCM yield. Therefore, the wafer level test for die includes the following:

- High stuck-fault coverage (99%+) and the use of some nonstuck fault detection techniques for application of tests beyond those that target stuck-at faults, e.g., quiescent current (IDDQ) testing [37], [38].
- Use of level-sensitive scan design (LSSD) techniques with minimal test application time.
- AC performance assessment testing and ac fault coverage.

The VCOS package assembly begins with a 100% probe tested die. Following the KGD approach, the dies are optionally mounted on a sacrificial substrate to allow for burn-in, testing, and screening out of infant mortality failures. The die level electrical testing includes standard parametric tests (e.g., input/output voltage levels and leakage currents). The IDDQ tests are applied by scanning in an appropriate vector set to SRLs and primary inputs in order to detect

the faults missed by standard stuck-at fault testing. These are followed by standard LSSD testing for high stuck-at fault coverages and ac BIST patterns for performance assessment (rather than go/no-go testing). Figure 9-20 shows the memory VCOS die (chip) level burn-

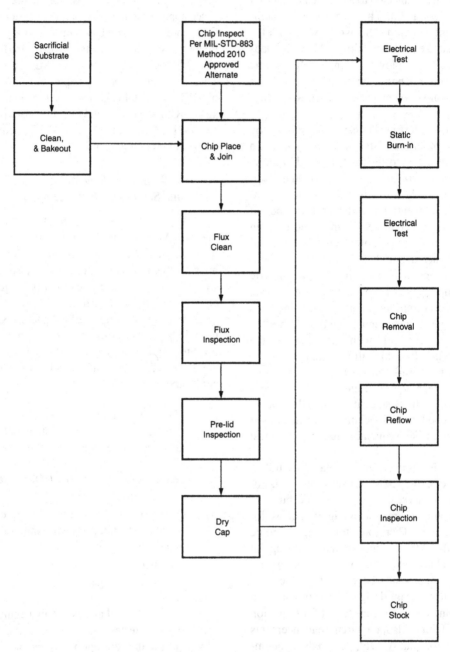

Figure 9-20. Memory VCOS die (chip) level burn-in and testing process flow. (From [7], with permission of IBM Federal Systems.)

in and testing process flow [7]. Figure 9-21 shows the memory VCOS module assembly level process flow [7].

A testing approach based on the IEEE 1149.1 boundary scan standard has conflicting requirements compared to the LSSD-based test-ing. IBM LORAL Federal Systems has used several MCM level test methodologies with various degrees of effectiveness, including the STUMPS, self-test using a multiple-input shift register (MISR) and a parallel shift sequence generator [33]. In this approach, a single chip on MCM is

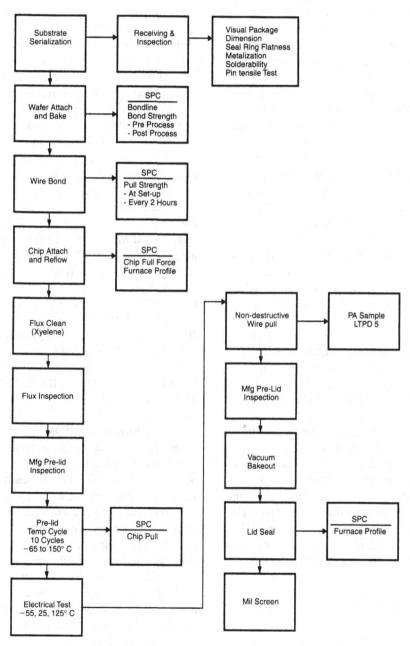

Figure 9-21. Memory VCOS module assembly level process flow. (From [7], with permission of IBM Federal Systems.)

devoted to supplying pseudorandom stimuli (vectors) to the remaining chips via the LSSD scan strings. The ac testing performed at the MCM level includes functional patterns, gross delay tests, and deterministically generated patterns. Although a variety of DFT methodologies existed, each of these approaches had some flaws. For memory VCOS testing, an effective strategy developed was based upon the LSSD, ac/dc BIST, boundary scan, and design extractability for diagnostic isolation. This was done by extending the concept of LSSD on a chip self-test (LOCST), and included a pseudorandom pattern generator and MISR for self-test [34]. The key to meeting test objectives was in proper partitioning and configuration of the SRLs in the following test modes: LSSD, self-test, and single string.

An MCM rework is often necessary to replace defective chips, and a key reliability concern is the removal and replacement of a chip from the module without affecting the reliability of the adjacent chip sites. In VCOS technology, a defective chip is removed by a hot gas reflow tool which heats the chip's back side until C4 solder reflows. A vacuum probe lifts the defective chip, and then the residual solder left on the silicon wafer is removed.

The VCOS memory modules for space applications are screened to QML Specification MIL-I-38535, Class V requirements that include the following tests per MIL-STD-883: internal visual, stabilization bake, temperature cycling, mechanical shock, particle impact noise detection (PIND) test, radiography, burn-in, electrical testing, and fine and gross leak testing. Figure 9-22 shows the memory VCOS module level screening flow per MIL-I-38535, Class V level [7]. Statistical process controls (SPCs) are incorporated in the process flow as indicated. The VCOS module level QCI is performed per MIL-I-38535, Groups A, B, C, D, and E requirements, which were discussed in Chapter 6.

9.5 MEMORY CARDS

Memory cards are lightweight plastic and metal cards containing the memory chips and associated circuitry for use in notebook PCs and mo-

bile communication applications. They are intended to serve as alternatives to the traditional hard disks and floppy drives, and offer significant advantages in size, weight, speed, and power consumption. The standardization at the hardware and data interchange level is provided by the Personal Computers Memory Card International Association (PCMCIA), and the Japan Electronic Industry Development Association (JEIDA) established memory card standards. Additional applications for the memory cards include printers, copiers, camcorders, medical instrumentation, and industrial controllers.

The first memory cards were introduced in the early 1980s as edgecard terminals that were susceptible to damage and shorting from frequent insertion and withdrawal [35]. The PCMCIA standard was later revised to specify pin and socket designs, and include input/output (I/O) card support. The storage capacity on these early cards was only 8–128 kb, but currently, 4–20 Mb range and even higher density memory cards are available from various manufacturers. Memory cards are commonly available in three standard formats: (1) 60-pin Joint Electron Devices Engineering Council (JEDEC) with a 16/18 b DRAM, (2) 68-pin PCMCIA and JEIDA cards, and (3) 88-pin JEIDA/JEDEC card with a 32/38 b DRAM. For the 68-pin cards which are available in three formats, type I contains the SRAM, PSRAM, MROM, OPTROM, and flash chips to perform memory functions.

The SRAM and ROM are currently the dominant IC card memory technologies. A ROM has the advantage of being inexpensive, but is not changeable, i.e., for each new software application update, ROM cards may need to be replaced. SRAM memory cards are reprogrammable, but batteries are required to maintain the data. In the last few years, flash chips have been introduced to the memory card market to address the need for a nonvolatile storage medium with low-power consumption. Flash memory density is about four times that for a SRAM, yielding lower cost per bit.

Memory card durability and reliability are important issues. Most of the major memory card manufacturers have reliability assur-

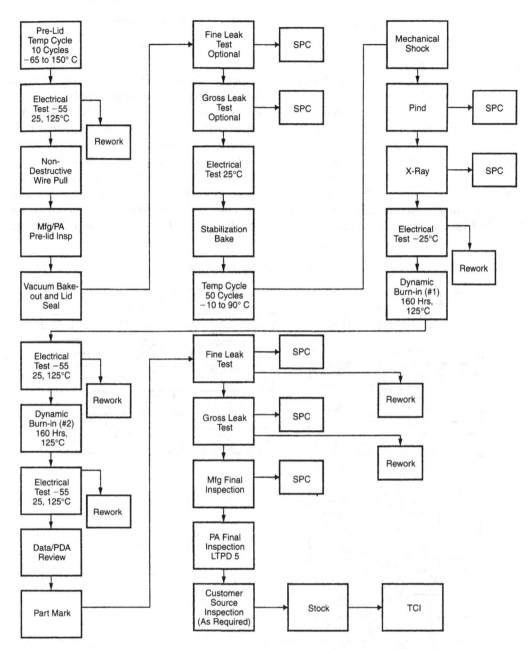

Figure 9-22. Memory VCOS module level screening flow per MIL-I-38535, Class V level. (From [7], with permission of IBM Federal Systems.)

ance programs for their products, and specify minimum insertion/withdrawal cycles (typically 10,000) and insertion withdrawal force (<2 kg). Some others feature nonrigid mounts which allow the cards to "float" inside the

package, thus eliminating stress on the solder joints during repeated insertion and withdrawal cycles. A few examples of SRAM and flash memory cards are discussed in the following sections.

9.5.1 CMOS SRAM Card (Example)

Hitachi developed a high-density CMOS SRAM card organized as 1 Mword $\times$ 8 b or 512 k word $\times$ 16 b [36]. It consists of eight pieces of 1 Mb SRAM devices sealed in TSOP packages and mounted on a memory card housed in a 68-pin, two-piece connector package. Figure 9-23 shows the pinout and pin description of this package [36]. Figure 9-24 shows a block diagram of this 8 Mb SRAM card [36]. It operates from a single +5 V supply, has a write protection switch, and conforms to JEIDA 4.1/PCMCIA interfacing standards.

9.5.2 Flash Memory Cards

The combination of nonvolatility and re-programmability makes flash memory cards an attractive alternative for low-power, portable, and high-performance system applications. An example is the Intel 4 Mbyte flash memory card which conforms to PCMCIA 1.0 International Standard, and provides standardization at the hardware and data interchange level [39]. It uses Intel's ETOX II flash memories, has 5 V read operation, and 12 V Erase/Write operation. The maximum specified time for a read operation is 200 ns, and 10 μs is the typical byte write

■ Pinout

Pin Name	Pin No.			Pin No.	Pin Name
GND	1	O	O	35	GND
D3	2	O	O	36	/CD1
D4	3	O	O	37	D11
D5	4	O	O	38	D12
D6	5	O	O	39	D13
D7	6	O	O	40	D14
/CE1	7	O	O	41	D15
A10	8	O	O	42	/CE2
/OE	9	O	O	43	NC
A11	10	O	O	44	NC
A9	11	O	O	45	NC
A8	12	O	O	46	A17
A13	13	O	O	47	A18
A14	14	O	O	48	A19
/WE	15	O	O	49	NC
NC	16	O	O	50	NC
Vcc	17	O	O	51	Vcc
NC	18	O	O	52	NC
A16	19	O	O	53	NC
A15	20	O	O	54	NC
A12	21	O	O	55	NC
A7	22	O	O	56	NC
A6	23	O	O	57	NC
A5	24	O	O	58	NC
A4	25	O	O	59	NC
A3	26	O	O	60	NC
A2	27	O	O	61	/REG
A1	28	O	O	62	BVD2
A0	29	O	O	63	BVD1
D0	30	O	O	64	D8
D1	31	O	O	65	D9
D2	32	O	O	66	D10
WP	33	O	O	67	/CD2
GND	34	O	O	68	GND

■ Pin Description

Pin Name	Function
/WE	Write Enable
/OE	Output Enable
A0 to A19	Address Input
D0 to D15	Data-input/output
/CE1, /CE2	Chip Enable
/CD1, /CD2	Card Detect
Vcc	+5V Supply
GND	Ground
WP	Write Protect
REG	Attribute memory select
BVD1, BVD2	Battery Voltage Detect
NC	No Connect

Figure 9-23. Pinout and pin description of Hitachi 8 Mb SRAM card (HB66A51216CA-25). (From [36], with permission of Hitachi Corporation.)

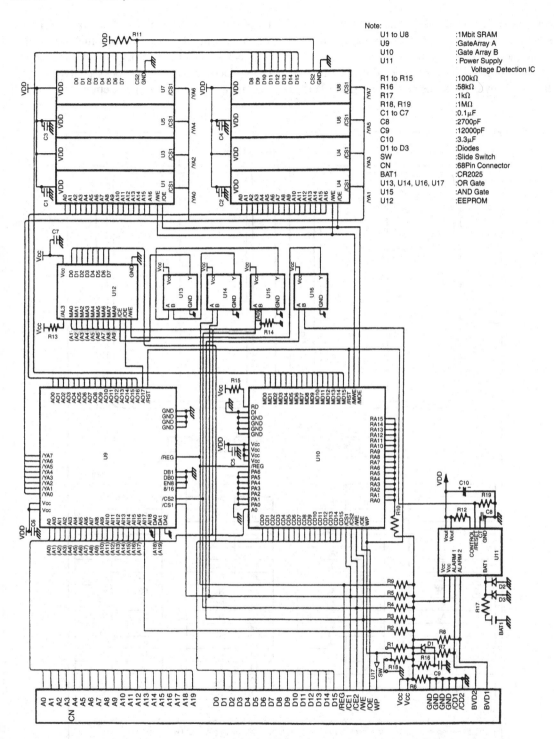

Figure 9-24. Block diagram of Hitachi 8 Mb SRAM card (HB66A51216CA-25). (From [36], with permission of Hitachi Corporation.)

(random) time to erased zones. Figure 9-25 shows the 4 Mbyte flash memory card dimensions and pin configurations [39].

Figure 9-26 shows a block diagram of the Intel 4 Mbyte flash memory card [39]. This flash memory card consists of an array of individual memory devices, each of which defines a physical zone. The memory erasure operation takes place as individual blocks equivalent in size to a 256 kbyte zone (memory devices).

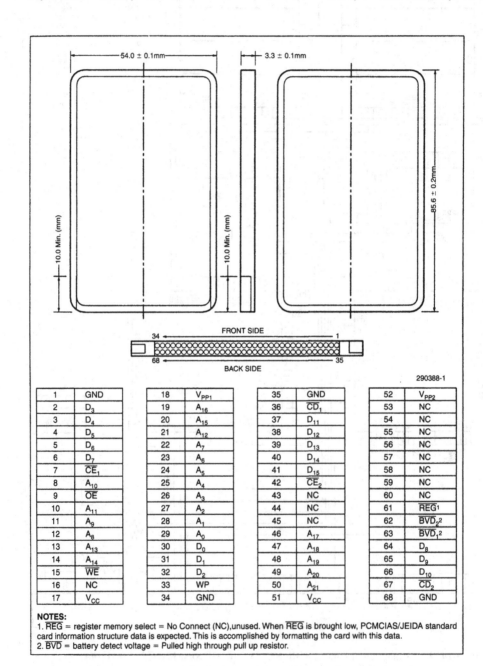

1	GND	18	V_{PP1}	35	GND	52	V_{PP2}
2	D_3	19	A_{16}	36	$\overline{CD}_1$	53	NC
3	D_4	20	A_{15}	37	D_{11}	54	NC
4	D_5	21	A_{12}	38	D_{12}	55	NC
5	D_6	22	A_7	39	D_{13}	56	NC
6	D_7	23	A_6	40	D_{14}	57	NC
7	$\overline{CE}_1$	24	A_5	41	D_{15}	58	NC
8	A_{10}	25	A_4	42	$\overline{CE}_2$	59	NC
9	$\overline{OE}$	26	A_3	43	NC	60	NC
10	A_{11}	27	A_2	44	NC	61	$\overline{REG}$1
11	A_9	28	A_1	45	NC	62	$\overline{BVD}_2$2
12	A_8	29	A_0	46	A_{17}	63	$\overline{BVD}_1$2
13	A_{13}	30	D_0	47	A_{18}	64	D_8
14	A_{14}	31	D_1	48	A_{19}	65	D_9
15	$\overline{WE}$	32	D_2	49	A_{20}	66	D_{10}
16	NC	33	WP	50	A_{21}	67	$\overline{CD}_2$
17	V_{CC}	34	GND	51	V_{CC}	68	GND

NOTES:
1. $\overline{REG}$ = register memory select = No Connect (NC),unused. When $\overline{REG}$ is brought low, PCMCIAS/JEIDA standard card information structure data is expected. This is accomplished by formatting the card with this data.
2. $\overline{BVD}$ = battery detect voltage = Pulled high through pull up resistor.

Figure 9-25. A 4 Mbyte flash memory card (Intel IMC004FLKA) dimensions and pin configurations. (From [39], with permission of Intel Corporation.)

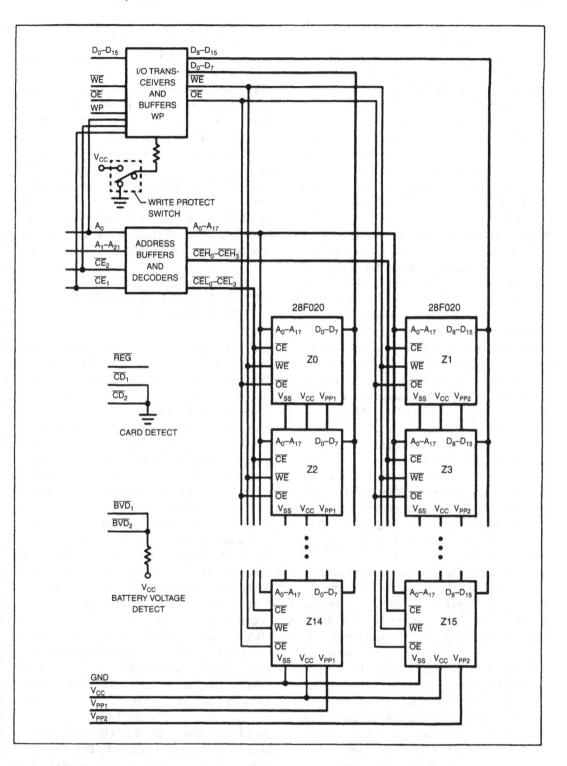

Figure 9-26. Intel 4 Mbyte flash memory card (IMC004FLKA) block diagram. (From [39], with permission of Intel Corporation.)

Multiple zones can be erased simultaneously if sufficient current is provided from both the V_{pp} and V_{cc} power supplies. These erased zones can be rewritten in bit- or byte-at-a-time and read randomly like a RAM. In the absence of high voltage on the $V_{pp1/2}$ pins, the flash memory card remains in the read-only mode.

For a read operation, two control functions must be logically active to obtain data at the outputs. The Card Enable ($\overline{CE}$) is the power control used for high and/or low zone(s) selection. The Output Enable ($\overline{OE}$) is the output control used to gate data from the output pins independently of the accessed zone selection. In the byte-wide mode, only one $\overline{CE}$ is required, whereas in the write-wide mode, both $\overline{CE}$ inputs have to be active low. With the Output Enable at logic high level (VIH), the memory card output is disabled and the output pins are placed in a high-impedance state. If only one $\overline{CE}$ input is at logic high level, the standby mode disables one-half of ×16 outputs read/write buffer.

The zone erasure and rewriting operations are performed via the command register on application of high voltage to $V_{pp1/2}$. The register contents for a given zone serve as input to that zone's internal finite state machine, which controls the erase and rewrite circuitry. The command register is a latch used to store the command, along with address and data information needed to execute the command. The command register is written by bringing the Write Enable ($\overline{WE}$) to an active-low level while $\overline{CE}$ inputs are also low.

The optimal use of flash memory cards is as disk emulators in mass storage applications. A company, SunDisk, is already producing 20 Mbyte flash cards that emulate disk drives under the PCMCIA/ATA (AT attachment) standard connection, and therefore can take advantage of existing disk controller software. Flash memory cards have lower power consumption requirements versus hard disk in all operational modes: ready, read, write, and standby. Also, reading data from a magnetic disk is very slow compared to a flash-memory-based solid-state disk (SSD).

Intel is also offering a 20 Mbyte flash memory card. Mitsubishi has developed an IC

packaging technology called the tape carrier package (TCP), which is half the thickness of a standard 1 mm thin small outline package (TSOP) currently being used for flash memory chips [40]. Two boards with TCP on both sides would double the density of a single-side TSOP board. SSD cards with densities of 40 Mbyte and higher are being developed.

Reliability assurance testing for these memory cards should include an extended life test (ELT); environmental testing such as temperature cycling and moisture resistance (85°C/85% RH); and mechanical stresses such as vibration, drop test, pressure/crush test, socketing (number of insertions) tests, and switch tests [41].

9.6 HIGH-DENSITY MEMORY PACKAGING FUTURE DIRECTIONS

ARPA has sponsored a project called Ultra-Dense, in which the goal is to achieve high performance through high packaging density for high-throughput computing requirements [42]. In this parallel digital signal processing system design, silicon-on-silicon MCM technology was used for first-level packaging of ICs. Subsequently, conductive adhesives are used for 3-D implementation of the system that included a digital signal processing (DSP) chip, an ASIC, and 8 of 256K SRAMs.

Memory MCM and 3-D (memory cubing) techniques are being used to increase memory density for high-performance computer and telecommunication systems, and in the military/space environment. The future trend is toward further memory densification using the largest memory chips available (e.g., 4 Mb SRAMs and 16 Mb DRAMs). Irvine Sensors Corporation (ISC) has signed an agreement to integrate their 3-D memory cubing process with IBM's silicon-on-silicon technology. A feasibility study was performed by using modified ISC cubing process to fabricate DRAM cubes that used 18–20 IBM 1 Mb DRAM chips. This cube interconnected to the substrate by IBM flip-chip technology was packaged in a metallized ceramic/polyimide pin grid array with 160 I/O

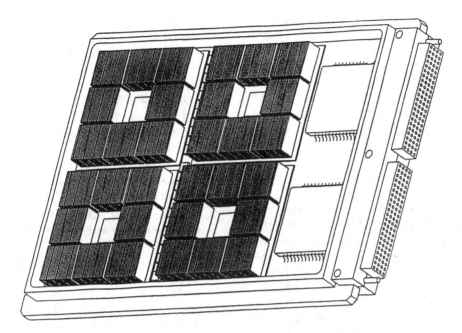

Figure 9-27. Schematic of high-density hermetic configuration (HDHC) 41 Gb memory board. (From [44], with permission of IBM Federal Systems.)

pins. Some design modifications that were made showed that the memory cube could replace conventionally packaged memories and result in considerable board space reduction [43].

Figure 9-27 illustrates a potential application in the fabrication of a high-density hermetic configuration (HDHC) board with 16 Mb chip cubes on both sides for a total memory capacity of 41 Gb [44]. The advanced solid-state recorder (SSR) planned for the earth observing system (EOS) mission has over 160 Gb storage capacity. The use of 64 Mb DRAM chips in the near future for the same application could easily quadruple that storage capacity.

REFERENCES

[1] CALCE Electronic Packaging Research Centre, Report on "Technical assessment for advanced interconnection qualification," Apr. 22, 1992.

[2] C. W. Ho *et al.,* "A low cost multichip module-D technology," in *Proc. ICEMM'93,* pp. 483–488; also, product information and photographs supplied by Honeywell, Inc.

[3] C. Meisser, "Modern bonding process for large-scale integrated circuits, memory modules and multichip modules in plastic packages," in *Proc. IEEE CHMT'91 IEMT Symp.*

[4] *NEC Memory Products Data Book,* 1991, and data sheets supplied by NEC Inc.

[5] *Hitachi IC Memory Data Book,* 1990.

[6] MIL-STD-1750A, "Military standard for avionics data bus."

[7] G. Rose, N. Virmani *et al.,* NASA Projects Parts Office (NPPO) Report, "VHSIC chip on silicon (VCOS) technology, development and qualification report," May 6, 1993.

[8] F. D. Austin *et al.,* "VCOS (VHSIC chips-on-silicon): Packaging, performance and applications," courtesy of IBM Corp./Federal Systems Division, Manassas, VA.

[9] IBM Fact Sheets, "64K × 16 radiation hardened SRAM memory modules," courtesy of IBM Corp./Federal Systems Division, Manassas, VA.

[10] R. T. Crowley *et al.*, "3-D multichip packaging for memory modules," in *Proc. Int. Conf. MCM'94*, pp. 474–479.

[11] J. Forthum and C. Belady, "3-D memory for improved system performance," in *Proc. 1992 Int. Electron. Packaging Conf.*, pp. 667–677.

[12] R. Scone and M. Suer, "The development of 3D memory for high and low-end systems," in *Proc. SEMI Technol. Symp.'92*, Japan, Dec. 1992.

[13] Irvine Sensors Corporation (ISC), product information and photographs supplied.

[14] Irvine Sensors Corporation, *Data Sheet for 4 Mbit SRAM Short Stack™ (512K × 8 SRAM)*.

[15] R. R. Some, "3D stack on active substrate, An enabling technology for high performance computer architectures," in *Semiconductor Device Technol., Vol. 2, Electro/92 Technol. Conf.*, May 1992.

[16] C. Val, "The 3D interconnection applications for mass memories and microprocessors," Tech. Paper supplied by Thomson CSF.

[17] C. Val and T. Lemoine, "3-D interconnection for ultra-dense multichip modules," *IEEE Trans. Components, Hybrids, Manufacturing Technol.*, vol. 13, pp. 814–821, Dec. 1990.

[18] D. Escobar *et al.*, "Memory densification by application of Z-plane lithography (MDAZL)," presented at the 1992 Government Microcircuits Appl. Conf. (GOMAC).

[19] J. Lykes, "Two- and three-dimensional high performance patterned overlay multi-chip module technology," Tech. Paper supplied by Phillips Laboratory (USAF).

[20] D. Frew, "High density memory packaging technology high speed imaging applications," Tech. Paper supplied by Texas Instruments (TI) Inc.

[21] R. Bruns *et al.*, "Utilizing three dimensional memory packaging and silicon-on-silicon technology for next generation recording devices," Tech. Paper supplied by TI Inc.

[22] Technical literature and photographs provided courtesy of TI Inc.

[23] D. B. Tuckerman *et al.*, "Laminated memory: A new 3-dimensional packaging technology for MCMs," in *Proc. 44th Electron. Components and Technol. Conf. (ECTC)*, 1994, pp. 58–63.

[24] R. J. Wagner *et al.*, "Design-for-test techniques utilized in an avionics computer MCM," in *Proc. IEEE ITC'93*, pp. 373-382.

[25] D. Chu *et al.*, "Multichip module enabler for high reliability applications," in *Proc. ICEMM*, 1992, pp. 102–105.

[26] J. S. Haystead, "In search of a military MCM test strategy," *Military Aerosp. Electron.*, Sept. 20, 1993.

[27] J. M. Aubert, "Boundary scan modification to enhance multichip module testing," Tech. Paper supplied by TI Inc.

[28] G. A. Forman *et al.*, "Die for MCMs: IC preparation for testing, analysis and assembly," in *Proc. ICEMM*, 1992, pp. 32–35.

[29] R. A. Fillion *et al.*, "Bare chip test techniques for multichip modules," in *Proc. ICEMM*, 1991, pp. 554–558.

[30] D. C. Keezer, "Bare die testing and MCM probing techniques," in *Proc. ICEMM*, 1991, p. 2023.

[31] S. Martin, "A practical approach to producing known-good-die," in *Proc. ICEMM*, 1993, pp. 139–151.

[32] C. E. Radke *et al.*, "Known good die and its evolution—Bipolar and CMOS," in *Proc. ICEMM*, 1993, pp. 152–159.

[33] P. H. Bardell *et al.*, "Self-testing of multiple chip logic modules," in *Proc. IEEE ITC 1982*, pp. 200–204.

[34] J. J. LeBlanc, "LOCST: A built-in self test technique," *IEEE Design and Test of Comput.*, pp. 45–52, Nov. 1984.

[35] J. Somerville and T. Grossi, "Memory cards mobilize portable computers," *Interconnection Technol.*, pp. 30–32, Aug. 1993.

[36] *Hitachi Data Sheets for 8 Mbit SRAM Card (HB66A51216CA-25)*, rev. 0, July 10, 1992.

[37] E. B. Eichelberger *et al.*, "Random-pattern coverage enhancement and diagnosis for LSSD self-test," *IBM J. Res. Develop.*, vol. 27, pp. 165–178, 1983.

[38] P. C. Maxwell *et al.*, "The effectiveness of IDDQ, functional and scan tests: How many fault coverages do we need?," in *Proc. IEEE ITC 1992*, pp. 168–177.

[39] *Intel Data Sheets for 4-Mbyte Flash Memory Card iMC004FLKA*, Oct. 1992.

[40] R. D. Hoffman, "Flash cards are ready—Almost," *EBN,* pp. 38, 42, Sept. 6, 1993.

[41] Intel Corp., Reliability Rep. RR-70, "Flash memory card quality/reliability data summary," Aug. 1992.

[42] M. Y. Lao *et al.,* "A versatile, IC process compatible MCM-D for high performance and low cost application," in *Proc. ICEMM,* 1993, pp. 107–112.

[43] C. L. Bertin *et al.,* "Evaluation of a 3-D memory cube system," presented at the 43rd Electron. Components and Technol. Conf., 1993.

[44] IBM VLSI Technology Development, Memory Product Development Roadmap Presentation.

Index

Address transition detection (ATD), 17, 62, 98, 353
  precharge circuit for ATD design, **356**
Advanced memory technologies, 387–411
  analog memories, 398–401, **399**
  experimental memory devices, 407–410
    Coulomb blockade effect, **409**
    GaAs n-p-n- thyristor/JFET memory cell,
      408–409
    neuron-MOS multiple-valued (MV) mem-
      ory technology, 409–410
    quantum-mechanical switch memories,
      407–408
    single-electron memory, 409
  ferroelectric random access memories
    (FRAMs), 389–397
    basic theory, 389–390
    ferroelectric nonvolatile RAM (FN-
      VRAM), 391–**392**
    FM 1208S FRAM memory, **394**
    FRAM cell and memory operation,
      390–392, **391**
    FRAM radiation effects, 395–397
    FRAM and "readout" capacitors, **401**
    FRAM reliability issues, 393–395
    FRAMs vs. EEPROMs, 397
    FRAM technology developments, 393
    GaAs FRAMs, 387, 397
  Introduction, 387–388
  magnetoresistive random access memories
    (MRAMs), 401–407
Advanced Micro Devices
  AM99C10A CMOS CAM, **39**–40
  4 Mb CMOS EPROM block diagram, **103–104**
Advanced Research Projects Agency (ARPA), 435

Advanced Spaceborne Computer Module (ASCM),
  416, 417, 421
Algorithmic test sequence (ATS), 143–**144, 207**
Alternate twin word activation (ATWA), 34
AMD, 5 V-only flash memory technology, **131**–132
Analog memories, 398–401, **399**
AT&T, operational life test (OLT) sequence,
  297–299, **298**
Atmel
  CMOS programmable and erasable read-only
    memory (PEROM), 127–128
  256 kb CMOS EEPROM, 117–**118**

Bathtub curve (typical), **250**
Berkeley Reliability Tool (BERT), 297
Bipolar SRAM technologies, 10, 17–24. *See also*
  static random access memories
    BiCMOS technology, 20–24, **21, 23**
    direct-coupled transistor logic (DCTL) technol-
      ogy, **18**–19
    emitter-coupled logic (ECL) technology, **19**–20
Breakdown-of-insulator-for-conduction (BIC) cell,
  91
British Telecom, HRD4 failure rate formulas, **293**
Buffer
  first-in first-out (FIFO), 35–**36**
  last-in first-out (LIFO), 35
Burn-in process, 269–270, 308–309. *See also* memory reliability
    dynamic burn-in, 310–311
    static burn-in, 310

Capacitor. *See also* trench capacitor
  ferroelectric, 389

Capacitor, (*continued*)
  modulated stacked capacitor (MOST), 57
  rough vs. smooth, 56–**57**
  stacked capacitor cells (STCs), 55–58, **56–57,**
    264, 265
Channel hot electron injection (CHE) programming,
  123, 128, **281**. *See also* flash memories
Chip-on-board (COB) technology, 416
Chip-on-lead (COL) technology, 417
CMOS DRAM, 45–50. *See also* dynamic random
  access memory (DRAM)
  bulk CMOS
    latchup considerations, 350–351
    radiation hardening, 358–**359**
  latchup tendencies, 252
  1 Mb DRAM example, 47–50
  radiation hardening example, 363–366,
    361–365
CMOS RAM. *See also* random access memories
  irradiated under worst case biasing, **353**
Cosmic ray. *See also* ionizing radiation; radiation
  hardening
  effect on semiconductor memory, 320–322, 324
  soft error creation by, 62, 335–336
Cosmic Ray Effects on MicroElectronics (CREME),
  324, 336
CROWN cell, 64 Mb DRAM, 66–**67**
"Cycle time", parameters of for memory, 15
Cypress Semiconductor, four-transistor memory cell,
  99–**100**

DC erase fallout, relation to ONO layer thickness,
  273
Defense Advanced Research Projects Agency
  (DARPA), 431
Dense-Pac Microsystems, stacked ceramic leadless
  chip carrier (LCC) package, 425
Direct-coupled transistor logic (DCTL) technology,
  18–19
Divided-shared bit line (DSB), 66
Dual-in-line packages (DIPs), 412
Dynamic random access memory (DRAM), 40–80
  advanced DRAM designs and architectures,
    62–69
    buried-capacitor (BC) cell, 69
    NAND-structured cell, **119**–120
    NAND-structured cell vs. conventional
      cell, **67**
    16 Mb DRAM example, 63–64
    ULSI DRAM developments, 64–69, 267
  application-specific DRAMs, 69–76
    cache DRAM (CDRAM), 69, 74, **75**
    enchanced DRAM (EDRAM), 69, 72

extended data-out DRAM (EDOD
  RAM), 72
glossary and summary of characteristics,
  75–76
high-speed DRAM (HSDRAM), 71–75
pseudostatic DRAM (PSRAM), 69
Rambus DRAM (RDRAM), 69, 72, 74
silicon file, 69–**70**
synchronous DRAM (SDRAM), 69, 72
synchronous graphics RAM (SGRAM),
  72–74, **73**
3-D RAM, 69, 75
video DRAM (VRAM), 69, 70–**71**
virtual-static DRAM (VSRAM), 69
window RAM (WRAM), 71
BiCMOS DRAM technology, 20–24, 58–60,
  59, 265–**266**
capacitor-over-bitline (COB), 55
CMOS DRAM, 45–50
  bulk CMOS latchup considerations,
    350–351
  compared to BiCMOS DRAM, 58–59
  1 Mb DRAM example, 47–50
compared to SRAM, 10
DRAM cell theory and advanced cell structures,
  50–58
  planar DRAM cell, 50–**51,** 55
  stacked capacitor cells (STX), 55–58,
    56–57, 264, 265
  trench capacitor cells, 52–55, **53, 56,**
    262–264, **263,** 265
DRAM technology development, 40–45
  one-transistor (1-T) DRAM cell, **41**
  three-transistor (3-T) DRAM cell, 40–**41**
embedded DRAM fault modeling and testing,
  156–157
failure rate calculations for, 294–296, **295**
megabit DRAM testing, 178–180, 237, 267
single-in-line memory modules (SIMMs), 416
soft-error failures in DRAMs, 60–62, 264–267
upset to, 333–335

Earth
  magnetosphere radiation, **323**
  South Atlantic Anomaly (SAA), 324
Electromigration-related failure, 255–256. *See also*
  memory reliability
Electronically erasable PROMs (EEPROMs),
  104–122
  architectures, 116–120, 412
  charge loss mechanisms, 272–273
  compared to FRAMs, 397
  FETMOS EEPROM evaluation, 278–**280**

functional model, **181**
NIT-SAMOS EPROM cell, **274**
nonvolatile SRAM (shadow RAM), 120–122
structure, **278**
technologies, 105
    floating-gate tunneling oxide (FLOTOX)
        technology, 104, 110–115, **112,
        275–277,** 279
    metal-nitride-oxide-silicon (MNOS) mem-
        ories, 105–109
    silicon-oxide-nitride-oxide semiconductor
        (SONOS) memories, 109–110
    textured-polysilicon technology, 115–116,
        275
Elevated quiescent supply currents (IDDQ) fault
  modeling and testing, 185–189. *See also* memory
  fault modeling and testing
    IDDQ measurement techniques, **188**
Emitter-coupled logic (ECL) technology, **19**–20
Error-correcting codes (ECC). *See also* memory design for testability
    augmented product code (APC), 237–**238**
    Bose-Chaudhury-Hocquenghem (BCH) code,
      237
    DEC code, 237
    DEC-TED codes, 233
    with differential cascade voltage switch
      (DCVS), 239
    Golay code, 237
    linear sum codes (LSCs), 233–235, **234**
    parity check scheme, 231–232, 235–237, **245**
    Reed-Solomon code, 237
    SBC-DBD codes, 233
    SEC-DEC codes, 242
    SEC-DED codes, 232
    SEC-DED-SBD codes, 232–233
    use
      on chips, 239–240
      for memory testing, 230–231
      for soft-error recovery, 62
Error-detection and correction (EDAC), 120, 430
European Synchrotron Radiation Facility (ESRF), 374
Experimental memory devices. *See* advanced mem-
  ory technologies

"Fast-page mode", compared to "nibble mode",
  48–49
Ferroelectric memories, 268, 283–287. *See also* advanced memory technologies
    aging vs. temperature effects, **285**
    fatigue characterization, 286
    ferroelectric nonvolatile RAM (FNVRAM),
      391–**392**

Ferroelectric random access memories (FRAMs),
  389–397
    basic theory, 389–390
    FM 1208S FRAM memory, **394**
    FRAM cell and memory operation, 390–392,
      391
    FRAM radiation effects, 395–397
    FRAM and "readout" capacitors, **401**
    FRAM reliability issues, 393–395
    FRAMs vs EEPROMs, 397
    FRAM technology developments, 393
    GaAs FRAMs, 387, 397
Field-funneling effect, 332–**333**
Fin structures, schematic, **57**
Flash memories (EPROM/EEPROM), 122–135
    advanced flash memory architectures,
      128–135
      "blocking" of chip, 128
      divided bitline NOR (DINAR) erasure,
        135
      8 Mb flash EPROM, **133**
      4 Mb flash EEPROM, **130**
      4 Mb flash technology, **131**–132
      4 Mb NAND EEPROM, 130–**131**
      NAND EEPROM-based, 124–126, **125**
      NOR EEPROM-based, 124–**125**
      1 Mb flash memory schematic and algo-
        rithm, 128–**129**
      second generation flash memories,
        132–133
      self-aligned shallow trench isolation (SA-
        STI), 135
      sidewall select-gate on source side
        (SISOS) cell, **129**
      16 Mb CMOS flash memory, **134**
      substrate hot electron (SHE) erasure, **135**
      256K CMOS block diagram, **126**
    channel hot electron injection (CHE) program-
      ming, 123, 128, **281**
    compared to FRAMs, 397
    compared to UVEPROM, 280–281
    ETOX flash technology, 123–**124,** 126–127,
      283
    flash memory cards, 442–446
    flash memory cells and technology develop-
      ments, 123–128
    4 Mb flash memory, 129–**130**
    4 Mb flash memory card, **444–445**
    hot hole injection issues, 282
    program disturb via substrate injection of ther-
      mal electrons (PDSITE), 282–283
    read disturb failure rate evaluation, 282
    reliability issues, 280–283

Floating-gate tunneling oxide (FLOTOX) technology, 104, 110–115, 279
    contrasted to textured-polysilicon EEPROM, 115–116, 276–277
    energy-band diagrams, **275**
    FLOTOX transistor, **112–113**
    stacked storage capacitor on FLOTOX (SCF), 121–**123**
Folded-bit-line adaptive sidewall-isolated cell (FAXIC), 53
Folded-bit-line architecture, 63
Folded capacitor cell structure (FCC), 52–53
Fowler-Nordheim tunneling effect
    in EEPROM, 105–105
    in flash memory, 135
Fujitsu
    breakdown-of-insulator-for-conduction (BIC) cell, 91
    4 Mb BiCMOS SRAM, 21–24, **22**
    1 Mb CMOS UV EPROM block diagram, **97**
    16 Mb CMOS SRAM, 32, **33**

GaAs SRAM. *See also* static random access memories
    architecture, 34–**35**
    fault modeling and testing, 155–156
Galileo Mission, SEP test flow diagram, **376**
Gated-resistor, use with hardened CMOS, **362**–363
General Electric (GE), 3-D memory MCMs, 431–432

Hamming Code, 120, 430
Harris Semiconductors
    low-temperature cofired ceramic (LTCC) tub, 425
    self-aligned poly-gate junction-isolated process, 91
    16K CMOS PROM, **92**–93
Harry Diamond Laboratory, radiation effects study, 395–396
High-density memory packaging technologies, 412–449
    high-density memory packaging future directions, 446–447
    introduction, 412–**417**
        chip carrier types, 413
        dual-in-line package (DIP), 412
        insertion mount and surface mount technology packages, **415**
        multichip modules (MCMs), 413, 416
        outlines and pin configurations for memory packages, **414**
        pin grid array (PGA), 412

        single-in-line package (SIP), 412
        small-outline package (SOP), 413
        surface mount technology (SMT), 412–413
        thin small-outline package (TSOP), 413
        zigzag-in-line package (ZIP), 412
    memory MCM testing and reliability issues, 435–440
        known-good-die (KGD), 435–436
        known-good-substrate (KGS), 435–436
    memory cards, 440–446
        CMOS SRAM card (example), 442
        flash memory cards, 442–446
    memory hybrids and MCMs (2-D), 417–424
        memory MCMs (Honeywell ASCM), 421
        memory modules (commercial), 417–421
        VLSI chip-on-silicon (VCOS) technology, 421–424, 437–440, **438, 439**
    memory stacks and MCMs (3-D), 424–435
        die stacked on edge approach, **426**–427
        die stacked in layers approach, 427
        laminated memory, **434**
        layered die packaging, 425–426
        solid-state recorder (SSR), **433–434**
        stacked module 3-D packaging approach, 427
        3-D memory cube technology (Thomson CSF), 430–431
        3-D memory MCMs (GE-HDI/TI), 431–432
        3-D memory stacks (Irvine Sensors Corp.), 427–430
        3-D memory stacks (nCHIPS), 432–435
Highly accelerated stress test (HAST), 309. *See also* memory reliability
Hitachi
    chip-on-lead (COL) technology, 417
    ECL 4 Mb BiCMOS DRAM, 60
    failure modes data compilation, **250–251**
    4 Mb CMOS ROM, 87
    4 Mb CMOS SRAM, 30–31
    1 Mb flash memory, 128
    16 Mb CMOS SRAM, 34
    16 Mb DRAM, 62–63
    36 Mb DRAM module outline and dimensions, **420–421**
Honeywell
    Advanced Spaceborne Computer Module (ASCM), 416
    memory for MCM-C for ASCM, 421–**423**
    MRAM magnetoresistive bit structure, **402**
    16K × HC7167, **403, 404**
    hot carrier, 254

IBM
    lead-on-chip (LOC) technology, 417
    16 Mb DRAM LUNA-C DDR
        block diagram, 63–**64**
        cross section, **65**
IBM Federal Systems
    MCM level test methodologies, 439
    high-density hermetic configuration (HDHC)
        41 Gb memory board, **447**
    VLSI chip-on-silicon (VCOS) technology, 421
IDT
    71256 (32K × 8b) functional block diagram,
        38–**39**
    8K × 8 dual-port SRAM, 36–**37**
Intel
    8 Mb flash EPROM, 132–133
    ETOX flash technology, 123–**124,** 126–127,
        283
    4 Mb FLASH MEMORY CARD, **444–445**
    64K array flash memory cell, **281**
    256K CMOS block diagram, **126**
Interlaced vertical parity (IVP)
    soft-error detection using, 235–237
    use with multiple access with re-compaction
        (MARC), **235**
Ionizing radiation. *See also* radiation hardening
    bulk displacement effects, 252
    charge generation and initial recombination,
        326
    effects
        on bipolar devices, 352
        on MOS transistor, **254**
        on semiconductor memory, 320–322
    soft error creation by, 62
    total dose effects, 325–330, 357, 367, 369, **372**
        dose rate transient effects, 376–377
        drain current vs. gate voltage for n-MOS
            device, **328**
        e-h pairs creation, 325–326
        flowchart for analysis of, **356**
        NASA/JPL radiation monitor chips,
            380–381
        n-channel MOSFET schematic cross-
            section, 326–**327**
        on PZT capacitor, 395–396
        SRAM circuit illustration of, **354**
    total ionizing dose (TID) monitoring, 302
Irvine Sensors Corp.
    "memory cubing", 416
    3-D memory stacks, 427–430, **428**
        4 Mb SRAM Short Stack (example),
            429–430
Isolation vertical cell (IVEC), 53

Latchup
    of 1 Mb and 256kb SRAM, **334**
    in bipolar CMOS circuit, 252
    bulk CMOS latchup considerations, 350–351
    of semiconductor memory, 322
    single-event latchup (SEL), 331–333
Lead-on-chip (LOC) technology, 417
Lead zirconate titanate (PZT) film, 283, 287,
    389–**390**
    total dose radiation-induced degradation, 395
Level-sensitive scan design (LSSD), 197–199, **198.**
    See also memory design for testability
        "shift register latch" (SRL), 197–199, **198**
Logic partitioning, 196
Low-temperature cofired ceramic (LTCC) tub, 425

Magnetoresistive random access memories
    (MRAMs), 401–407. *See also* random access
    memories technologies
    Honeywell MRAM magnetoresistive bit struc-
        ture, **402**
    MHRAM cell structure, **406**
March test algorithms, **149**–150, 156–**157**
Masking, for CMOS DRAM, 45–46
Memory cards, 440–446
"Memory cubing", 416
Memory design for testability and fault tolerance,
    195–248
    advanced BIST and built-in self-repair architec-
        tures, 216–228
        BIST and built-in self-repair (BISR),
            222–228
        BIST scheme using microprogram ROM,
            220–222
        column address-maskable test (CMT),
            219–220
        multibit and line mode tests, 216–219, **217**
    boundary scan testing (BST), **202**–203
    DFT and BIST for ROMs, 228–230
        built-in self-diagnostic (BISD) ROM, **229**
        exhaustive enhanced output data modifica-
            tion (EEODM), 228–229
    embedded memory DFT and BIST techniques,
        211–216
        BIST architecture using serial shift tech-
            nique, **215**
        circular self-test path (CSTP) technique,
            212–**213**
        flag scan register (FLSR) schematic, **212**
    general design for testability techniques,
        195–203
        ad hoc design techniques, 196–197
        structured design techniques, 197–203

Memory design for testability and fault tolerance,
  (*continued*)
    level-sensitive scan design (LSSD), 197–199
    logic partitioning, 196
    memory error-detection and correction tech-
      niques, 230–241
        augmented product code (APC), 237
        Bose-Chaudhury-Hocquenghem (BCH)
          code, 237
        DEC code, 237
        DEC-TED codes, 233
        error-correcting codes (ECC) use,
          230–231
        Golay code, 237
        linear sum codes (LSCs), 233–235, **234**
        minimum distance requirements,
          231–232
        parity check scheme, 231–232, 235–237,
          245
        Reed-Solomon code, 237
        SBC-DBD codes, 233
        SEC-DEC codes, 242
        SEC-DED codes, 232
        SEC-DED-SBD codes, 232–233
        coupling faults, 147–151
        embedded DRAM fault modeling and test-
          ing, 156–157
        GaAs SRAM fault modeling and testing,
          155–156
        miscellaneous faults, 155
        pattern-sensitive faults, 151–155
        stuck-at-fault model, **142**–145
    RAM pseudoRAM testing (PRT), 176–178
Memory performance parameters, 15
Memory reliability. *See* semiconductor memory reli-
  ability
Memory technologies. *See* advanced memory tech-
  nologies
Metallization corrosion-related failures, 256–257.
  See also memory reliability
Metal-nitride-oxide-silicon (MNOS) memories,
  105–109
Microchip Technology, 256 kb EEPROM, 120
Military application. *See also* radiation hardening
    memory packages, 412
    MOS technology use, 325
    PROM use, 91
    radiation hardening, 367–369
    reliability testing, 269, 310–313
    wire bond pull test, 257–258
Mitsubishi
    4 Mb CMOS SRAM, 31–32
    16 Mb DRAM with embedded ECC, **240**

MOS SRAM
    architectures, 14–15
    cell and peripheral circuit operation, 15–17
    for military and space applications, 325–330
        self-checking on-line testable static RAM,
          241
    memory fault-tolerance designs, 241–246
    RAM built-in self-test (BIST), 203–211
        BIST for pattern-sensitive faults, 209–210
        BIST using 13N March algorithm,
          207–209, **208**
        BIST using algorithmic test sequence,
          205–207, **206**
        BIST using built-in logic block observa-
          tion (BILBO), 210–**211**
    random access scan, 200–202, **201**
    scan path DFT, **199–200**
Memory fault modeling and testing, 140–194
    application-specific memory testing, 189–192
        double-buffered memory (DBM) testing,
          191–192
        general testing requirements, 189–191
    IDDQ fault modeling and testing, 185–189
    introduction, 140–141
        categories of failure, 140
        fault models discussion, 140–141
    megabit DRAM testing, 178–180, 237, 267
    nonvolatile memory modeling and testing, DC
      electrical measurements, 181–182
    nonvolatile memory modeling and testing,
      180–185
        AC (dynamic) and functional measure-
          ments, 182–185
        64K EEPROM, 183–185
        256K UVEPROM, 182–183
    physical fault modeling compared, to logic fault
      modeling, 140–141
    RAM electrical testing, DC and AC parametric
      testing, 158
    RAM electrical testing, 158–176
        functional testing and algorithms, 158–174
    RAM fault modeling, 142–157, **151**
        bridging faults, **145**–147
Multichip modules (MCMs), 413, 416
    memory MCM testing and reliability issues,
      435–440
        VCOS DFT methodology and screening
          flow (example), 437–440, **438, 439**
    and memory hybrids, 417–424
        memory MCMs (Honeywell ASCM), 421
        memory modules (commercial), 417–421
        VLSI chip-on-silicon (VCOS) technology,
          421–424

memory stacks and MCMs (3-D), 424–435
    3-D memory cube technology (Thomson CSF), 430–431
    3-D memory MCMs (GE-HDI/TI), 431–432
    3-D memory stacks (Irvine Sensors Corp.), 427–430
    3-D memory stacks (nCHIPS), 432–435
Multimedia applications, synchronous DRAM use in, 72–73
Multiple access with read-compaction (MARC), **235**

NASA/JPL
    MOSIS test chip diagnostic, **303**
    radiation monitor chips, **380–381**
Naval Research Labs, Cosmic Ray Effects on Micro-Electronics (CREME), 324, 336
NCHIP, Inc., 3-D memory stacks, 432–435
NEC
    MC-424000LAB72 DRAM module, **418**
    MC-424000LAB72F DRAM module, **419**
    "engraved storage electrode" process, 57
    1 Mb DRAM, **47–49**
    16 Mb CMOS flash memory, **134**
    16 Mb ROM, 87
    3.3 V 16 Mb CMOS SRAM, 32
Neural network, analog memory use in, 400–401
Noise, in RAM, 43–44
Nonvolatile memories (NVMs), 81–139
    EEPROM technologies
        dual-mode sensing scheme (DMS), 120
        floating-gate tunneling oxide (FLOTOX) technology, 110–115, **112–113, 275–277,** 279
        high-performance MNOS (Hi-MNOS), 106–**108**
        metal-insulator-semiconductor-insulator-semiconductor (MISIS), 111–112
        metal-nitride-oxide-silicon (MNOS) technology, 104–105, **105–109**
        NAND-structured cell, **119**–120
        silicon-oxide-nitride-oxide-semiconductor (SONOS), 104–105
        silicon-oxide-nitride-oxide semiconductor (SONOS) memories, 109–110
        textured-polysilicon technology, **115**–116, **275**
    electronically erasable PROMs (EEPROMs), 104–122
        EEPROM architectures, 116–120
        EEPROM technologies, 105–116
        electronically alterable ROM (EAROM), 105–**107**

EPROM compared to EEPROM, 83
Fowler-Nordheim tunneling effect, 104–105
use with ROMs, 120
erasable (UV)-programmable read-only memories (EPROMs), 93–104
    advanced EPROM architecture, 98–102
    cross-point SPEAR cell, **101**
    dielectric failure concerns, 252–253, 284, 352
    EPROM technological evolution, **96**
    EPROM technology developments, 96–97
    erase process, 95–96
    floating-gate EPROM cell, 93–96
    large-tilt-angle implanted p-pocket (LAP) cell, **101**–102
    one-time programmable (OTP) EPROMs, 103, 274
    oxide-nitride-oxide (ONO) structure, 102, 270–275, **271, 273**
    programming, 94–95
    radiation effects, 345–347
    source-side injection EPROM (SIEPROM), 100
    stacked-gate EPROM transistor, **93**–94
    use of locally oxidized silicon (LOCOS), 100–101
flash memories (EPROM/EEPROM), 122–135.
    See also flash memories
    advanced flash memory architectures, 128–135
    flash memory cells and technology developments, 123–128
    reliability issues, 280–283
HFIELDS 2-D simulation, 114
introduction, 81–83
    masked ROMs, 81
    programmable ROMs (PROMs), 81–82
modeling and testing, DC electrical measurements, 181–182
modeling and testing, 180–185
    AC and functional measurements, 182–185
programmable read-only memories, 87–93
    programmable low-impedance circuit element (PLICE), 91
programmable read-only memories (PROMs)
    bipolar PROMS, 87–91, **88–89**
    CMOS PROMs, 91–93, **92**
    128K bipolar PROM, 90–**91**
radiation characteristics, 344–347

Nonvolatile memories (NVMs), (continued)
    read-only memories (ROMS), 83–87
        high-density ROMs, 87
        NAND ROMs, **86**
        programmable and erasable read-only
            memory (PEROM), 127–128
        ROM cell structures, 85–**86**
        technology development and cell program-
            ming, 83–85, **84**
        use with EEPROM, 120
    reliability issues, 268–287
        BURN-IN PROCESS, 269–270, 308–309
        electrically erasable programmable read-
            only memories (EEPROMs), 275–280
        EPROM data retention and charge loss,
            270–275
        ferroelectric memories, 283–287
        flash memories, 280–283
        lead zirconate titanate (PZT) film, 283,
            287, 389–**390**
        programmable read-only memory (PROM)
            fusible links, 268–270, 312
      SNOS memory transistor showing memory
        stack, **344**
      SNOS technology, **121, 344**–347
      threshold voltage programming, 85
NSC, DM77/87S195 TTL PROM, **90**

P-channel MOS (PMOS), 327
    radiation-induced threshold voltage shifts, **349**
Personal Computers Memory Card International As-
    sociation (PCMCIA), 440
Pin grid array (PGA), 412
Plastic package failure, 256–248, 273–274. *See also*
    memory reliability
    highly accelerated stress test (HAST), 309
Programmable read-only memory (PROM), reliabil-
    ity issues, 268–270

Quality conformance inspection (QCI). *See also*
    semiconductor memory reliability
    for reliability testing, 249, 310–313
Quantum-mechanical switch memories, 407–408
    resonant-tunneling diodes (RTDs), **407**
    resonant-tunneling hot-electron transistors
      (RHETs), **407**

Radiation dosimetry, 377–378
Radiation hardening techniques, 347–367. *See also*
    ionizing radiation; semiconductor
    memory radiation effects
    charged particle motion in earth's magneto-
      sphere, **323**

    for PROM, 91–93
    radiation-hardened memories characteristics,
      363–366
    radiation-hardening design issues, 352–363
      one column of six-transistor cell memory,
        357
      single-event upset (SEU) hardening,
        358–363
      total dose radiation hardness, 321, 353–358
    radiation-hardening process issues, 347–352
      bipolar process radiation characteristics,
        352
      bulk CMOS latchup considerations,
        350–351
      CMOS SOS/SOI processes, 351–352
      field oxide hardening, 349
      gate electrode effects, 348–349
      gate oxide (dielectric) effects, 347–348
      postgate-electrode deposition processing,
        349
      substrate effects, 347
    radiation types overview, **321**
    for SONOS device, 109–110
Radiation hardness assurance, 367–39
Radiation testing
    dose rate transient effects, 376–377
    irradiation sources, 370–372, **371**
    neutron irradiation test, 377
    for PZT capacitor, 395–396
    radiation data test results, **396**
    single-event phenomenon (SEP) testing,
      372–376
    total dose testing, 321, 357, 367, 3790–**372**
      dose rate transient effects, 376–377
      drain current vs. gate voltage for n-MOS
        device, **328**
      flowchart for analysis of, **356**
      NASA/JPL radiation monitor chips,
        380–381
      SRAM circuit illustration, **354**
    wafer-level radiation testing, 378–381, **379,**
      437–438
      radiation test structure, 379–381, **380**
RAM, dual-port RAM, 36–38
Ramtron Corp.
    enhanced DRAM (EDRAM), 72
    FM 1208S FRAM memory, **394**
    FRAM development, 389–390
Random access memories (RAMs)
    CMOS RAM irradiated under worst case bias-
      ing, **353**
    ferroelectric random access memories
      (FRAMs), 389–397

Random access memories (RAMs) electrical testing.
  See also random access memory fault modeling
    DC and AC parametric testing, 158
      BUTTERFLY, 174
      CHECKERBOARD, 163
      CHEKCOL, 170–171
      DIAPAT, 168–169
      DUAL WAKCOL, 169–170
      GALCOL, 167–168
      GALDIA, 166–167
      GALPAT, 165–166
      HAMPAT, 171–172
      MARCH C, 164–165
      MASEST, 163–164
      MOVI, 172–173
      WAKCOL, 169
      WAKPAT, 165
      ZERO-ONE, 163
Random access memories (RAMs) electrical testing,
  158–176
    functional test pattern selection, 174–176
    RAM functional block diagram, **141**
    RAM pseudoRAM testing (PRT), 176–178, **177**
Random access memories (RAMs) fault modeling.
  141, 142–157. *See also* random access memory
  electrical testing
    bridging faults, **145**–147
    coupling faults (CFs), 147–151
      detection of, 148
      March test algorithms, **149**–150, 156–**157**
    fault models
      algorithmic test sequence (ATS), 143–**144,**
        207
      memory scan (MSCAN) test, 143
      multiple stuck line fault (MSF), 143
      sensitized path technique, **142**
      single-stuck fault (SSF) detection,
        142–143
      "soft fault" compared to single-stuck fault,
        143
      stuck-at fault (SA) model, 142–145
      transition fault (TF), 144
    miscellaneous faults, 155
    pattern-sensitive faults, 151–155
      five-cell neighborhood pattern, **153–154,**
        210
      five-cell tiling neighborhood marking, **154**
      neighborhood pattern sensitive faults
        (NPSFs), 152
      neighborhoods of base cell, **151, 210**
    RAM fault-tolerance methods
      binary-tree dynamic RAM (TRAM),
        244–245

deterministic model scrubbing, 243–344
      memory reconfiguration by graceful degra-
        dation scheme, 243
      probabilistic model scrubbing, 243
      standby reconfiguration method, 242–243
      wafer scale integration (WSI), 245
Random access memories (RAMs) technologies,
  10–80
    dynamic random access memory (DRAM),
      40–80
      advanced DRAM designs and architec-
        tures, 62–69
      BiCMOS DRAM technology, 20–24, **21,**
        23, 58–60, 265–**266**
      CMOS DRAMs, 45–50
      DRAM technology development, 40–45
      soft-error failures in DRAMs, 60–62,
        264–267
    ferroelectric memories (FRAM), 268,
      283–287
      aging vs. temperature effects, **285**
      fatigue characterization, 286
    introduction, 10–12
      volatile memory discussion, 10, 12
    magnetoresistive random access memories
      (MRAMs), 401–407
    refresh modes, 49–50
    static random access memories (SRAMs),
      12–40
      advanced SRAM architectures and tech-
        nologies, 28–35
      application-specific SRAM, 35–40
      bipolar SRAM technologies, 17–24
      floating-gate avalanche-injection MOS
        (FAMOS), 100–101, 110–112
      MOS SRAM architecture, 14–15, 325
      MOS SRAM cell and peripheral circuit op-
        eration, 15–17
      NMOS and PMOS radiation damage,
        327–**328**
      silicon-on-insulator technology, 24–28
      SRAM (NMOS and CMOS) cell struc-
        tures, 12–14
"Read-access time", parameters of for memory, 15
Redundancy scheme, use with error-correction cir-
  cuitry, 45
Reliability. *See* semiconductor memory reliability

Sandia National Laboratories, PZT study, 395
SEEQ Technology, *Q*-cell floating-gate memory
  transistor, 116–**117**
Semiconductor memory radiation effects, 320–386
  introduction, 320–322

Semiconductor memory radiation effects, (*continued*)
   radiation dosimetry, 377–378
   radiation effects, 322–347
      nonvolatile memory radiation characteris-
        tics, 344–347
      single-event phenomenon (SEP), 330–344
      space radiation environments, 322–325
      total dose effects, 321, 325–330, 395
   radiation hardening techniques, 347–367
      radiation-hardening process issues,
        347–352
   radiation hardness assurance, 367–369
   radiation testing, 369–381
      dose rate transient effects, 376–377
      single-event phenomenon (SEP) testing,
        372–376
      total dose testing, 321, **354**, 357, 367,
        370–**372**
      wafer level radiation testing and test struc-
        tures, 378–381, **379**, 437–438
Semiconductor memory reliability, 249–319
   British Telecom HRD4 failure rate formulas,
     293
   design for reliability, 296–300
      AT&T operational life test (OLT) se-
        quence, 297–299, **298**
      Berkeley Reliability Tool (BERT), 297
      one column of six-transistor cell memory,
       357
      Simulation Program with Integrated Cir-
       cuits Emphasis (SPICE), 297
   failure definition, 287
   failure rate equation summary, **292**
   general reliability issues, 249–252
      accelerated aging process, 257, 297–298
      assembly-and-packaging-related failures,
        257–258
      bathtub curve (typical), **250**
      conductor and metallization failures,
        255–256
      dielectric-related failures, 252–253
      "85/85" process, 257
      electrical overstress (EOS) effects, 253
      electromigration-related failure, 255–256
      electrostatic discharge (EDS) effects, 253
      failure causes and mechanisms, **250–251**
      hot carrier definition, 254
      metallization corrosion-related failures,
        256–257
      quality conformance inspection (QCI), 249
      semiconductor bulk failures, 252
      semiconductor-dielectric interface failures,
        253–255

      time-dependent dielectric breakdown
        (TDDB), 252, 284, 352
   nonvolatile memory reliability, 268–287
      burn-in process, 269–270, 308–309
      electrically erasable programmable read-
        only memories (EEPROMs), 275–280
      EPROM data retention and charge loss,
        270–275
      ferroelectric memories, 268, 283–287
      flash memories, 280–283
      lead zirconate titanate (PZT) film, 283,
        287, 389–**390**
      for military and space applications, 269
      michrome fusible link reliability, 268–270,
        312
      programmable read-only memory (PROM)
        fusible links, 268–270
   one column of six-transistor cell memory, **357**
   RAM failure modes and mechanisms, 258–268
      DRAM capacitor reliability, 262–264
      DRAM data-retention properties, **267**–268
      DRAM soft-error failures, 264–267
      144K SRAM NFET and PFET cross-
        sections, **259**
      RAM gate oxide reliability, 258–259
      RAM hot-carrier degradation, 260–262
   reliability definition, 287–292
      binomial distribution, 289
      exponential distribution, 291
      gamma distribution, 291
      lognormal distribution, 292
      normal (Gaussian) distribution, 289–291
      Poisson distribution, 289
      Weibull distribution, 291–292
   reliability modeling and failure rate prediction,
     287–296
      reliability definitions, 287–292
   reliability screening and qualification, 304–313
      highly accelerated stress test (HAST), 309
      n-channel MOS EPROM failure mecha-
        nisms, **305**
      quality conformance inspection (QCI),
        310–313
      reliability testing, 304–309
   reliability test structures, 300–304
      process monitor (PM) chips, 304
      16 Mb DRAM test structure, **301**–302
      statistical process control (SPC), 300–301
Sense amplifier
   distributed sense and restore amplifier, 46
   1 Mb CMOS EPROM circuit structure, **99**
   pseudodifferential sense amplifier, **98**
   16 Mb DRAM test structure, **301**

64 kb DRAM operating sequence, **44**–45
stabilized feedback current sense amplifier, 33–34
    use with dummy cell, 42–**43**
Sensing
    dual-mode sensing scheme (DMS), 120
    signal enhancement requirements, 51–52
Silicon-on-insulator (SOI) technology, 24–**28**
    SIMOX technology, 24–26
Silicon-on-sapphire technology (SOS), 24, 26–**27**
Silicon-oxide-nitride-oxide semiconductor (SONOS) memory, 109–110
SIMOX technology, 24–26. *See also* silicon-on-insulator technology
Simtek, 256 kb NVSRAM, **122**
Simulation Program with Integrated Circuits Emphasis (SPICE), 297
Single-bit-line cross-point cell activation (SCPA), use with 16 Mb SRAM, 32
Single-event latchup (SEL). *See also* latchup
    causes, 331–333
Single-event phenomenon (SEP), 321–322, 330–344
    radiation hardening testing, 372–376, **375**
    SEP in-orbit flight data, 337–344
        combined release and radiation effects satellite (CRRES), 322, **342**–344
        NASA Tracking and Data Relay Satellite Systems (TDRSS), 322, 338, **339**
        National Space Development Agency (NASDA), 338–339
        UoSAT satellites, 322, 339–342, **340, 341**
    SEP test flow diagram for NASA Galileo mission, **376**
Single-event upset (SEU), 252, 322, 330–333. *See also* upset
    critical charge plot, **332**
    field-funneling effect, 332–**333**
    modeling and error rate prediction, 335–337
    of 1 Mb and 256kb SRAM, **334**
    SEU radiation-hardening and testing, 358–363, 369, 374–**375**
Single-in-line memory modules (SIMMs), 416
SNOS technology, **121, 344**–347. *See also* nonvolatile memories (NVMs)
Soft-error failure (SEF)
    in DRAMs, 60–62, 264–267
        creation of by alpha-particle hit, 60–**61**, 63
Soft-error rate (SER)
    for BiCMOS process, 265–**266**
    detection of with interlaced vertical parity (IVP), 235–237
    measurement of, 61–62

Solar flare. *See also* ionizing radiation; radiation hardening
    effect on semiconductor memory, 342, 343
Sony, 9 ns 16 Mb CMOS SRAM, 33–34
Space application. *See also* radiation hardening
    memory packages, 412
    MOS technology use, 325
    PROM use, 91
    radiation hardening, 322, 334, 357–358, 363
    RADPAK packaging, 368–369
    reliability testing, 269, 310–313
    SRAM use, 334
    VCOS memory modules use, 440
    wire bond pull test, 257–258
Space environment
    EPROM affected by, 96
    radiation effects in, 322–325, 367–370, **368**
    soft error creation in, 62
Speech recording, with analog memory, 400
SRAM. *See* static random access memories
Stacked capacitor cells (STC), 55–58, **56–57,** 264, 265
    3-D stacked capacitor cell, 57–**58**
Stacked ceramic leadless chip carrier (LCC) package, 425
Static random access memories (SRAMs), 12–40
    "Address Transition Detection" (ATD) function, 17
    advanced SRAM architectures and technologies, 28–35
        4 Mb CMOS SRAM cross-section and block diagram, **31**
        Gallium Arsenide (GaAs) SRAM, 34–**35**
        1-4 Mb SRAM designs, 28–32
        16-64 Mb SRAM development, 32–34
        stacked-CMOS SRAM cell, 29–**30**
    application-specific SRAM, 35–40
        content-addressable memories (CAMs), 38–40
        dual-port RAM, 36–38, **37**
        four-port SRAM schematic, 37–**38**
        nonvolatile SRAM, 38
        serially accessed memory (line buffers), 35–36
    bipolar SRAM technologies, 17–24
        BiCMOS technology, 20–24, **21, 23**
        direct-coupled transistor logic (DCTL), **18**–19
        emitter-coupled logic (ECL), **19**–20
    clocked, compared to not clocked, 17
    compared to DRAM, 10
    compared to pseudostatic DRAM, 69

Static random access memories (SRAMs),
    (*continued*)
    GaAs SRAM fault modeling and testing,
        155–156
    MOS SRAM architectures, 14–15, 325
    MOS SRAM cell and peripheral circuit opera-
        tion, 15–17
        CMOS RAM cell (six-transistor), 15–**16**
        SRAM circuit elements (various), **17**
    NMOS and PMOS radiation damage,
        327–**328**
    nonvolatile SRAM (NVRAM), 120–122, **121**
    self-checking on-line testable SRAM, **241**
    SEU and latchup test results, **334**
    silicon-on-insulator (SOI) technology, 24–28
        radiation-hardening process issues,
        351–352
    Simtek 256 kb NVSRAM, **122**
    SNOS 8 kb static NVRAM, **121**
    SNOS memory transistor showing memory
        stack, **344**
    SRAM (NMOS and CMOS) cell structures,
        12–14
        CMOS SRAM cell configuration, **13**
        MOS SRAM cell schematic, 12–**13**
        p-channel MOS (PMOS), 327, **349**
    upset to, 333–335
Strategic Defensive Initiative Organization (SDIO),
    431
Surface mount technology (SMT), 412–413
"Surrounding gate transistor" (SGT), 34
Synchrotron, 374

Texas Instruments, 4 Mb flash memory, 129–**130**
Textured-polysilicon technology, 115–116, **275**
Thomson CSF, 3-D memory cube technology,
    430–431
Threshold voltage monitoring program (TMP), 99

Time-dependent dielectric breakdown (TDDB), 252,
    284, 352. *See also* memory reliability
Toshiba
    4 Mb NAND EEPROM, 130–**131**
    16 ns 1 Mb high-speed CMOS EPROM, 98–99
    "surrounding gate transistor" (SGT), 34
    256 kb flash EEPROM, **127**
Trench capacitor cells, 52–55, **53**, 262–264, **263**,
    265. *See also* capacitor
    compared to planar and stacked cells, 55–**56**
    dielectrically encapsulated trench (DIET), 53
    epitaxy-over-trench cell (EOT), **54**–55
    reliability issues, 263
    straight-line trench isolation (SLIT), 55
    substrate-plate-trench capacitor cell (SPT),
        53–**54**
    trench transistor cell with self-aligned contacts
        (TSAC), 55
    trench transistor cell (TTC), **54**

Upset. *See also* single-event upset
    creation of by alpha-particle hit, 60–**61**
    DRAM and SRAM upsets, 333–335
    single-event upset (SEU), 252, 330–333

Van de Graff accelerator, 374
Very high-speed integrated circuit (VHSIC), 416
VLSI chip-on-silicon (VCOS) technology, 421–**424**,
    423, 437–440, **438**, **439**

Wafer level radiation testing, and test structures,
    378–381, **379**, 437–438

X-ray
    moderate/hard X-ray simulator, 377
    soft X-ray (SXR) simulator, 376–377

Zigzag-in-line packages (ZIPs), 412